Helicobacter pylori

Edited by

Lyudmila Boyanova

Department of Medical Microbiology
Medical University of Sofia
Sofia
Bulgaria

Authors: Lyudmila Boyanova, Ivan Mitov and Borislav Vladimirov

Caister Academic Press

Caister Academic Press
Norfolk, UK

www.caister.com

British Library Cataloguing-in-Publication Data
A catalogue record for this book is available from the British Library

ISBN: 978-1-904455-84-4

Cover image adapted from Figure 2.1

Printed and bound in Great Britain

Contents

Contributors

Ivan Mitov
Head of the Department of Medical Microbiology
Medical University of Sofia
Sofia
Bulgaria

igmitov@gmail.com

Lyudmila Boyanova
Department of Medical Microbiology
Medical University of Sofia
Sofia
Bulgaria

l.boyanova@hotmail.com

Borislav Vladimirov
Clinic of Gastroenterology
University Hospital Queen Joanna-ISUL
Medical University of Sofia
Sofia
Bulgaria

borislavvladimirov@yahoo.com

Preface

Twenty-eight years ago, the revolutionary discovery of *Helicobacter pylori* by Marshall and Warren started a new era in the understanding and management of most gastroduodenal diseases and also influenced the entire field of medical science. Importantly, the discovery revealed associations between chronic bacterial infections and malignancy as well as between malignancy and antimicrobial therapy. Presently, the *Helicobacter* genus includes over 35 species, some of which may be associated with chronic gastric, hepatobiliary and intestinal diseases in humans. *H. pylori,* the most clinically important species in the genus, has co-evolved with humans since the Palaeolithic era. The chronic and often lifelong infection, although decreasing in most developed and even in some developing countries, still affects more than half of the global human population. In addition, host immune response cannot clear the infection. However, it is now well known that *H. pylori* eradication is curative for chronic gastritis, peptic ulcers and more than half of cases of mucosa-associated lymphoid tissue (MALT) lymphoma. The bacteria are also involved in pathogenesis of gastric cancer, the second leading cause of cancer death worldwide, and several extragastric diseases. *H. pylori* displays an enormous genetic diversity and some strains harbour numerous virulence factors. The clinical outcome of *H. pylori* infection is a result of the combined activity of *H. pylori* virulence factors and depends on the interactions between the bacteria and both host and environmental factors. The infection is diagnosed by various invasive and non-invasive tests with increasing use of molecular and proteomic and immunoproteomics methods, and can be cured by combined regimens of antibiotics and other agents. Non-antibiotic agents can contribute to the management of the infection. Regrettably, no vaccines are available yet for prophylaxis of the infection and the rising antibiotic resistance in *H. pylori* compromises treatment success. Many still unanswered questions about *H. pylori* pathogenesis, epidemiology, prophylaxis and treatment remain to be clarified, upgraded or optimized. As over 30,000 publications, including 18,900 articles since 2000, have been published on *H. pylori* and its associated diseases so far, it is difficult for most readers to follow them all. For this reason, the present book highlights many new data (mostly those published over the last decade) on the important or unanswered questions and interesting topics to motivate the microbiologists and research scientists but also the students and clinicians to start working or to continue working on *H. pylori* infection.

Lyudmila Boyanova

Current Books of Interest

Microbial Biofilms: Current Research and Applications	2012
Bacterial Glycomics: Current Research, Technology and Applications	2012
Non-coding RNAs and Epigenetic Regulation of Gene Expression	2012
Brucella: Molecular Microbiology and Genomics	2012
Molecular Virology and Control of Flaviviruses	2012
Bacterial Pathogenesis: Molecular and Cellular Mechanisms	2012
Bunyaviridae: Molecular and Cellular Biology	2011
Emerging Trends in Antibacterial Discovery: Answering the Call to Arms	2011
Epigenetics: A Reference Manual	2011
Metagenomics: Current Innovations and Future Trends	2011
Nitrogen Cycling in Bacteria: Molecular Analysis	2011
The Biology of Paramyxoviruses	2011
Helicobacter pylori	2011
Microbial Bioremediation of Non-metals: Current Research	2011
Lactic Acid Bacteria and Bifidobacteria: Current Progress in Advanced Research	2011
Viruses and Interferon: Current Research	2011
Essentials of Veterinary Parasitology	2011
Hepatitis C: Antiviral Drug Discovery and Development	2011
Streptomyces: Molecular Biology and Biotechnology	2011
Alphaherpesviruses: Molecular Virology	2011
Recent Advances in Plant Virology	2011
Vaccine Design: Innovative Approaches and Novel Strategies	2011
Salmonella: From Genome to Function	2011
PCR Troubleshooting and Optimization: The Essential Guide	2011
Insect Virology	2010
Environmental Microbiology: Current Technology and Water Applications	2011
Sensory Mechanisms in Bacteria: Molecular Aspects of Signal Recognition	2010
Bifidobacteria: Genomics and Molecular Aspects	2010
Molecular Phylogeny of Microorganisms	2010
Nanotechnology in Water Treatment Applications	2010
Iron Uptake and Homeostasis in Microorganisms	2010
Caliciviruses: Molecular and Cellular Virology	2010
Epstein-Barr Virus: Latency and Transformation	2010
Anaerobic Parasitic Protozoa: Genomics and Molecular Biology	2010
Lentiviruses and Macrophages: Molecular and Cellular Interactions	2010
Microbial Population Genetics	2010
Borrelia: Molecular and Cellular Biology	2010
Influenza: Molecular Virology	2010
RNA Interference and Viruses: Current Innovations and Future Trends	2010
Retroviruses: Molecular Biology, Genomics and Pathogenesis	2010
Metagenomics: Theory, Methods and Applications	2010

Historical Data

1

Lyudmila Boyanova

Abstract

The discovery of *Helicobacter pylori* by Marshall and Warren in 1982 was the start of a real revolution in gastroenterology with a strong impact on the whole field of medicine. In the 1870s, despite some reports of spiral bacteria in gastric tissues, stress and diet were thought to be the only causes of peptic ulcers. In 1979, Warren evaluated *Campylobacter*-like organisms (CLOs) in inflamed gastric tissues and in 1982 his co-worker, Marshall, isolated the bacteria. However, the discovery was initially met with much disagreement. For this reason, in 1985, Marshall performed self-inoculation by CLOs and proved their ability to cause gastritis. Originally called *Campylobacter pyloridis* and then corrected to *Campylobacter pylori*, the bacteria were renamed again owing to taxonomic data as *Helicobacter pylori* in a new genus, *Helicobacter*. In 2005, Warren and Marshall were jointly awarded the Nobel Prize in Physiology or Medicine. Since the year of *H. pylori* discovery, various invasive and non-invasive diagnostic tests, susceptibility testing methods and treatment regimens for the infection have been developed and later improved. Participation of *H. pylori* in the pathogenesis of gastric cancer and mucosa-associated lymphoid tissue (MALT) lymphoma has been proven. Moreover, *H. pylori* has been associated with several extragastric diseases. The enormous genetic diversity of the bacteria and their numerous virulence factors have been revealed and genomes of many strains have been sequenced. Presently, chronic gastritis and peptic ulcers are treated as bacterial infections by antibiotics combined with acid inhibitors. Improvement in eradication has been achieved by triple, quadruple, rescue, sequential or other regimens. The legacy that Marshall and Warren's discovery has given to human medicine has been the consequent detection of the link between chronic bacterial infections and malignancy and, fascinatingly, the option to heal a tumour (as MALT lymphoma) by treating the associated microorganisms. However, at present, the treatment of *H. pylori* infection is not easy and still no vaccines are available for prophylaxis. Although the global rate of *H. pylori* infection is gradually decreasing, the antibiotic resistance of the bacteria is increasing sharply in many countries. Regrettably, despite the fact that the infection is often asymptomatic, *H. pylori* still infects half of the global human population and every fifth infected person can develop severe disease. Therefore, it is important to stress that in the field of pathogenesis, epidemiology, diagnosis, prophylaxis and treatment of the infection, there are still approaches to be optimized and many questions to be answered.

Introduction

In the fourth century BC, Hippocrates described the clinic signs of some gastric diseases and first used the word 'cancer', but it was not until 1821 that Nepveu stated that some chronic gastritis may develop to gastric cancer (Kidd and Modlin, 1998). Empiric treatment of peptic ulcers with bismuth was tried by Kussmaul in 1868.

In 1874, Bottcher and Letulle observed bacteria in gastric ulcer specimens and believed that ulcers might be caused by the microorganisms

(Kidd and Modlin, 1998). Since 1874, spiral gastric bacteria have been found in both animals and humans (Rappin, 1881; Bizzozero, 1893; Krienitz, 1906; Kidd and Modlin, 1998; Konturek, 2003). In 1889, Jaworski described and named the spiral bacteria from human gastric washing specimens as *Vibrio rugula,* and also suggested their relationship with severe gastric diseases (Konturek, 2003). During that time it was thought that the stomach was sterile owing to its acidity. Nevertheless, in 1951–1958, peptic ulcers were treated with penicillin by Allende or with 'Elgaco', containing streptomycin, by Lykoudis, although both attempts were rejected by the medical society (Rigas *et al.*, 1999; Unge, 2002). Only in 1978, the American Gastroenterology Association stated that antiacids are ineffective to cure ulcers (Peterson *et al.*, 2002).

From 1979 to 2005

In 1979, Robert Warren, a pathologist from Perth, Australia, observed spiral-shaped *Campylobacter*-like organisms (CLOs) in human gastric biopsies by Gram and Warthin–Starry stains and detected gastric inflammation only in the CLO-positive cases (Chisholm and Owen, 2006). Barry Marshall, then a clinical fellow, tried to culture the bacteria. In 1982, after the Easter holidays, small colonies of the CLOs became visible after a longer (5-day) than usual incubation. In 1984, the presence of CLOs in gastric antrum was also reported by McNulty and Watson (1984), and Eldridge *et al.* (1984) detected antibodies against the bacteria. In 1985, distinctive structural and physiological characteristics of the spiral gastric bacteria were described in detail by Megraud *et al.* (1985) and Goodwin *et al.* (1985). In the 1980s, however, there was a belief that stress and dietary factors were the only causes of peptic ulcers (Pajares and Gisbert, 2006) and hence first reactions to the discovery were negative. Marshall's data met disapproval at a local meeting and his submitted abstract to an Australian journal was rejected (Marshall, 2002; Pajares and Gisbert, 2006). Nevertheless, Marshall continued his work, combining bismuth with metronidazole and amoxicillin or tetracycline to treat the CLO-positive patients. Finally, in 1983, the work on the 'unidentified curved bacilli' was reported in the *Lancet* (Marshall, 1983; Warren, 1983). The publication has since been cited in thousands of research articles .

In 1985, Marshall reported his self-inoculation to fulfil Koch's postulates. After endoscopy to prove a lack of gastritis, he ingested a suspension of two culture plates of the bacteria (Marshall, 2008). Soon, Marshall developed symptomatic gastritis, which he successfully cured by metronidazole and bismuth salts (Marshall *et al.*, 1985). Second self-infection was made by Morris but its eradication required 5 years (Morris and Nicholson, 1987). In 1987, the European *Helicobacter pylori* Study Group was founded to support international *H. pylori* research.

The observed gastric bacteria were named *Campylobacter pyloridis* and later the name was corrected to *Campylobacter pylori.* However, owing to differences between this species and *Campylobacter* spp. in ultrastructure, fatty acids, DNA percentage of guanine and cytosine content and other data, in 1989 *C. pylori* was re-renamed as *Helicobacter pylori* in a new genus, *Helicobacter* (Goodwin *et al.*, 1985; Goodwin *et al.*, 1989).

H. pylori virulence factors became an important subject of research. In 1988, more than half of strains were found to produce a vacuolating toxin (VacA) *in vitro* (Leunk *et al.*, 1988; Cover *et al.*, 1991). In 1991–1996, major *H. pylori* virulence factors, i.e. the cytotoxin-associated gene A (*cagA*) and *cag* pathogenicity island (*cag*PAI), were evaluated (Crabtree *et al.*, 1991a; Covacci *et al.*, 1993; Censini *et al.*, 1996). A key advance in the research was the first complete sequence of the genome of *H. pylori* strain 26 695 (Tomb *et al.*, 1997).

H. pylori treatment was gradually improved. Owing to poor results of double therapy (proton pump inhibitor, PPI + clarithromycin or amoxicillin), triple regimens were developed (Axon, 1989); however, *H. pylori* resistance to antibiotics also was reported (McNulty *et al.*, 1985). Now, triple regimens (PPI + clarithromycin + amoxicillin or metronidazole) are used as first-line therapy for *H. pylori* eradication (Malfertheiner *et al.*, 2007).

In 1991, studies established a link between the infection and gastric cancer and gastric lymphoma (Parsonnet *et al.*, 1991; Wotherspoon *et al.*, 1991).

In 1994, *H. pylori* was recognized as a cause of peptic ulcers by the US National Institutes of Health and as a first (strongest) group carcinogen by the World Health Organization (Anonymous, 1994).

Since *H. pylori* discovery, invasive [culture, histology, rapid urease test (RUT) and polymerase chain reaction (PCR)] methods and non-invasive [serology, urea breath test (UBT), stool antigen tests (SATs) and PCR] methods have been developed and then improved (Table 1.1). Guidelines for management of the infection were published in 1997 and in 2002, the updated Maastricht 2–2000 Consensus Report suggested a 'test and treat' strategy by non-invasive tests for dyspeptic adults aged <45 years without alarm symptoms (Malfertheiner *et al.*, 2002).

In 2005, the Warren and Marshall's discovery was rewarded by the Nobel Prize in Physiology or Medicine (Chisholm and Owen, 2006). This was the deserved triumph of the non-traditional thinking, hard work, determination, courage and dedication of the discoverers.

The work of Marshall and Warren has presented to the world a true discovery rather than rediscovery because it has proven unconditionally the previous, although accurate and courageous (during those years) suggestions. The gift that this discovery has given to the field of human medicine has been the subsequent detection of the link between chronic bacterial infections and malignancy and, importantly, the option to treat a tumour (the example is mucosa-associated lymphoid tissue, MALT, lymphoma) by eradicating the associated microorganisms.

Since 2005

The clinical importance of *H. pylori* infection has resulted in a vast interest and >30300 publications (of them, >2800 published since 2009) according to PubMed so far. It is already well known that *H. pylori* is the causative agent of chronic gastritis, peptic ulcers, gastric cancer and MALT lymphoma (Pajares and Gisbert, 2006) and is linked also to extragastric diseases, e.g. unexplained iron deficiency anaemia and idiopathic thrombocytopenic purpura (Malfertheiner *et al.*, 2007). *H. pylori* association with other extragastric disorders, e.g. ischaemic heart disease, atherosclerosis, primary Raynaud phenomena, rosacea, etc. is still under evaluation (de Korwin, 2008). Although the infection has been suggested to partially protect patients from oesophageal cancer and asthma, *H. pylori* should be considered as an important human pathogen (Graham *et al.*, 2007; McNeil, 2008).

Lately, antibiotic resistance in *H. pylori* has led to treatment failure in 20–30% of first-line regimens (Jafri *et al.*, 2008). In these cases, quadruple or 'rescue' regimens with amoxicillin and rifabutin or levofloxacin, or furazolidone have been suggested (Malfertheiner *et al.*, 2007). It has been stressed that rescue regimens should be based on the strain susceptibility pattern. Currently, new regimens, e.g. sequential therapy, four-drug, three-antibiotic 'concomitant' therapy or bismuth-containing quadruple therapy are used to improve eradication success (Graham *et al.*, 2008). Many non-antibiotic agents are evaluated as treatment enhancers (Lesbros-Pantoflickova *et al.*, 2007). Diagnostic and treatment guidelines have been upgraded owing to the rising *H. pylori* resistance to antibiotics (Graham and Shiotani, 2008). However, *H. pylori* eradication needs to be further improved. Since 2007, a positive RUT alone has been stated to be sufficient to start a treatment, SATs with monoclonal antibodies have been recommended and the use of the serology has been re-accepted for specific patient groups, e.g. for those with bleeding ulcer, atrophic gastritis and MALT lymphoma (Malfertheiner *et al.*, 2007). In recent years, new diagnostic tests have been developed and some of them are already commercially available. New SATs, immunoproteomics methods, biprobe real-time PCR tests, including also a non-invasive method for detection of clarithromycin resistance in *H. pylori* (Mégraud and Corti, 2009) have been some of the numerous advances in the detection of the bacteria, their virulence or resistance markers. In addition, OLGA staging has been suggested recently for histological estimation of the gastric cancer risk (see Chapter 3).

In 2005–2010, new *H. pylori* virulence factors, among them duodenal ulcer-promoting gene (*dupA*), tumour necrosis factor-α inducing protein (*Tipa*) gene and intermediate region of

Table 1.1 Examples of some advances in the field of *H. pylori* infection since 1983

Year	Advances	Reference
1982	Discovery and culture of *Campylobacter*-like bacteria, currently *H. pylori*	Marshall (1983), Warren (1983)
1985	Marshall's self-inoculation; work on structural and physiological characteristics of *H. pylori;* detection of *H. pylori* resistance to antibiotics	Marshall *et al.* (1985), Megraud *et al.* (1985), Goodwin *et al.* (1985), McNulty *et al.* (1985)
1986–1987	Use of serology and urea breath test (^{13}C UBT) for diagnosis	Booth *et al.* (1986), Graham *et al.* (1987)
1988	Evaluation of vacuolating toxin (VacA)	Leunk *et al.* (1988), Cover *et al.* (1991)
1989	Taxonomic change of *Campylobacter pyloridis/pylori* to *H. pylori;* use of triple therapy to treat the infection; assessment of anti-*H. pylori* activity of non-antibiotic agents (lactobacilli)	Axon (1989), Bhatia *et al.* (1989), Goodwin *et al.* (1989)
1990	Use of molecular techniques for diagnosis of the infection and work on vaccines	Hoshina *et al.* (1990), Pallen and Clayton (1990)
1991	*H. pylori* association with gastric tumours; study on cytokines in the immune response against the infection	Crabtree *et al.* (1991b), Parsonnet *et al.* (1991), Wotherspoon *et al.* (1991)
1991–1993	Evaluation of cytotoxin-associated gene A (*cagA)*	Crabtree *et al.* (1991a), Covacci *et al.* (1993)
1994	*H. pylori* inclusion in the group 1 carcinogens	Anonymous (1994)
1996–1997	Evaluation of *cag*PAI and molecular methods for clarithromycin resistance; first *H. pylori* genome sequencing; publication of first Maastricht Consensus Report	Censini *et al.* (1996), Malfertheiner *et al.* (1997), Szczebara *et al.* (1997), Tomb *et al.* (1997)
1998	Development and introduction of stool antigen test	Makristathis *et al.* (1998), Trevisani *et al.* (1998)
2000	Use of new regimens (sequential therapy) for therapy; involvement of proteomics	Nilsson *et al.* (2000), Zullo *et al.* (2000)
2002–2007	Transcriptomic evaluation of *H. pylori* gene expression	Graham *et al.* (2002), Scott *et al.* (2007)
2003	Use of diagnostic serum biomarkers and stool test for clarithromycin resistance	Di Mario *et al.* (2003), Fontana *et al.* (2003)
2005	Nobel Prize in Physiology or Medicine awarded to Warren and Marshall	Chisholm and Owen (2006)
2005–2007	Maastricht III Consensus Report to treat *H. pylori* infection	Malfertheiner *et al.* (2007)
2008	Advances in vaccine preparation; first direct evidence for the role of CagA as oncoprotein	Malfertheiner *et al.* (2008), Ohnishi *et al.* (2008)
2005–2010	Reports on new virulence factors, e.g. *dupA*, *Tipα* and *vacA* i alleles in *H. pylori*	Lu *et al.* (2005), Suganuma *et al.* (2005), Chung *et al.* (2010)
2010	Advance in studies on *H. pylori* protein interactions (complexome) and primary transcriptome and suggestion of the impact of A2143G mutation on the eradication	Bernarde *et al.* (2010), De Francesco *et al.* (2010), Sharma *et al.* (2010)

vacuolating cytotoxin gene (*vacA*), were detected (Lu *et al.*, 2005; Suganuma *et al.*, 2005; Chung *et al.*, 2010). Seven complete *H. pylori* genomes have been sequenced so far (see Chapter 4). Modern comparative genomics, transcriptomics (assessment of the entire set of RNA transcripts produced by the genome at a given time) and proteomics (evaluation of the complete set of the proteins expressed by the genome) have been developed to predict clinical outcomes of the infection and to develop vaccines or new anti-*H. pylori* therapeutic agents (Scott *et al.*, 2007; Wu *et al.*, 2008; Dong *et al.*, 2009). Transcriptomic data, obtained by high-throughput screening techniques with microarrays, have shown in depth the 'strategy' of *H. pylori* to adapt to the host. Proteomics, based on two-dimensional polyacrylamide gel electrophoresis (2-DE), has revealed the continuous change of the *H. pylori* proteome in response to the changes in the genome and in the gastric mucosa as well as the protein functions and inter-protein interactions. Comparative genomics has provided much information about the enormous genetic diversity of *H. pylori* and the species has been used to assess the historical human migrations, including the first migration out of Africa 58,000 years ago (Desikan, 2010).

Recently, a success in research on anti-*H. pylori* vaccine has been reached. An example is a vaccine, containing VacA, CagA and neutrophil-activating protein (Hp-NAP), which has shown safety, immunogenicity and T-cell memory induction in healthy subjects (Malfertheiner *et al.*, 2008). Mutants of *Escherichia coli* heat-labile enterotoxin, CpG oligodeoxynucleotides, chitosan and other polymers, liposomes and bacterial ghosts (empty cell envelopes) have been evaluated as possible adjuvant or delivery systems for the vaccines (Sijun and Yong, 2009).

Thanks to the revolutionary *H. pylori* discovery, the gastritis, peptic ulcers and many cases of MALT lymphoma are treated as bacterial infectious diseases (Malfertheiner *et al.*, 2007). Prevalence of the infection and the associated diseases has been found to decline in Western and, lately, in some developing countries (Luminari *et al.*, 2010). Regrettably, despite the fact that the infection is often asymptomatic, *H. pylori* still infects half of the global human population (Pajares and Gisbert, 2006) and every fifth infected person can develop severe disease. Although *H. pylori* eradication could not nullify the risk of gastric cancer, *H. pylori* screening and treatment in high-risk populations have been suggested (Trautmann *et al.*, 2006).

Questions to be further elucidated

Obviously, since the epoch-making *H. pylori* discovery, enormous success in understanding and treating the infection has been made but more studies are still needed on highly important topics, including:

- function of individual *cag* PAI and some other virulence genes (Bourzac *et al.*, 2006);
- infection transmission, especially the role of oral *H. pylori* and viable but non-culturable coccoid forms and biofilms (see Chapter 7);
- combination of *H. pylori* factors leading to development of MALT lymphoma;
- non-invasive tests for *H. pylori* resistance to different antibiotics; improvement of diagnostic tests in extragastric samples and specific patient groups, e.g. youngest children, elderly and treated patients, and those with atrophic gastritis or bleeding ulcer;
- choice of most reliable biomarkers for patients at risk for severe diseases;
- level of *H. pylori* impact on many suggested extragastric diseases (Graham *et al.*, 2007);
- management of patients with uninvestigated or non-ulcer dyspepsia;
- strategy, methods and appropriate populations for *H. pylori* screening and eradication;
- control of the rising *H. pylori* resistance to antibiotics and improvement of eradication success (Graham and Shiotani, 2008); and
- optimal use of non-antibiotic agents for both prophylaxis and treatment of the infection as well as development of effective vaccines and vaccination strategy (Sijun and Yong, 2009).

Further studies are strongly required to answer these questions and to suggest the most effective prophylaxis and therapy of *H. pylori* infection as a chronic and still very common infection, carrying lifelong risks of peptic ulcers and gastric malignancy in a part of the infected population.

References

Anonymous. (1994). Schistosomes, liver flukes and *Helicobacter pylori*. IARC Working Group on the Evaluation of Carcinogenic Risks to Humans. IARC Monogr. Eval. Carcinog. Risks Hum. *61*, 1–241.

Axon, A.T. (1989). *Campylobacter pylori*-therapy review. Scand. J. Gastroenterol. Suppl. *160*, 35–38.

Bernarde, C., Lehours, P., Lasserre, J.P., Castroviejo, M., Bonneu, M., Megraud, F., and Menard, A. (2010). Complexomics study of two *Helicobacter pylori* strains of two pathological origins: potential targets for vaccine development and new insight in bacteria metabolism. Mol. Cell. Proteomics. 9, 2796–2826.

Bhatia, S.J., Kochar, N., Abraham, P., Nair, N.G., and Mehta, A.P. (1989). *Lactobacillus acidophilus* inhibits growth of *Campylobacter pylori in vitro*. J. Clin. Microbiol. *27*, 2328–2330.

Bizzozero, G. (1893). Sulle ghiandole tubulari del tubo gastroenterico e sui rapporti del loro coll'epitelio de rivestimento della mucosa. Atti. R. Accad. Sci. Torino. *28*, 233–251.

Booth, L., Holdstock, G., MacBride, H., Hawtin, P., Gibson, J.R., Ireland, A., Bamforth, J., DuBoulay, C.E., Lloyd, R.S., and Pearson, A.D. (1986). Clinical importance of *Campylobacter pyloridis* and associated serum IgG and IgA antibody responses in patients undergoing upper gastrointestinal endoscopy. J. Clin. Pathol. *39*, 215–219.

Bourzac, K.M., Satkamp, L.A., and Guillemin, K. (2006). The *Helicobacter pylori cag* pathogenicity island protein *CagN* is a bacterial membrane-associated protein that is processed at its C terminus. Infect. Immun. *74*, 2537–2543.

Censini, S., Lange, C., Xiang, Z., Crabtree, J.E., Ghiara, P., Borodovsky, M., Rappuoli, R., and Covacci, A. (1996). *cag*, a pathogenicity island of *Helicobacter pylori*, encodes type I-specific and disease-associated virulence factors. Proc. Natl. Acad. Sci. U.S.A. *93*, 14648–14653.

Chisholm, S.A, and Owen, R.J. (2006). From Nobel to no cure: a case for monitoring antibiotic resistance in the gastric pathogen *Helicobacter pylori*. Expert Rev. Anti Infect. Ther. *4*, 349–351.

Chung, C., Olivares, A., Torres, E., Yilmaz, O., Cohen, H., and Perez-Perez, G. (2010). Diversity of VacA intermediate region among *Helicobacter pylori* strains from several regions of the world. J. Clin. Microbiol. *48*, 690–696.

Covacci, A., Censini, S., Bugnoli, M., Petracca, R., Burroni, D., Macchia, G., Massone, A., Papini, E., Xiang, Z., Figura, N., *et al.* (1993). Molecular characterization of the 128-kDa immunodominant antigen of *Helicobacter pylori* associated with cytotoxicity and duodenal ulcer. Proc. Natl. Acad. Sci. U.S.A. *90*, 5791–5795.

Cover, T.L., Puryear, W., Perez-Perez, G.I., and Blaser, M.J. (1991). Effect of urease on HeLa cell vacuolation induced by *Helicobacter pylori* cytotoxin. Infect. Immun. *59*, 1264–1270.

Crabtree, J.E., Taylor, J.D., Wyatt, J.I., Heatley, R.V., Shallcross, T.M., Tompkins, D.S., and Rathbone, B.J. (1991a). Mucosal IgA recognition of *Helicobacter pylori* 120 kDa protein, peptic ulceration, and gastric pathology. Lancet 338, 332–335.

Crabtree, J.E., Shallcross, T.M., Heatley, R.V., and Wyatt, J.I. (1991b). Mucosal tumour necrosis factor alpha and interleukin-6 in patients with *Helicobacter pylori* associated gastritis. Gut 32, 1473–1477.

De Francesco, V., Zullo, A., Ierardi, E., Giorgio, F., Perna, F., Hassan, C., Morini, S., Panella, C., and Vaira, D. (2010). Phenotypic and genotypic *Helicobacter pylori* clarithromycin resistance and therapeutic outcome: benefits and limits. J. Antimicrob. Chemother. *65*, 327–332.

Desikan, P. (2010). The incredible journey of mankind: *Helicobacter pylori* as the narrator. Indian J. Med. Microbiol. *28*, 193–195.

Di Mario, F., Moussa, A.M., Caruana, P., Merli, R., Cavallaro, L.G., Cavestro, G.M., Dal Bò, N., Iori, V., Pilotto, A., Leandro, G., *et al.* (2003). 'Serological biopsy' in first-degree relatives of patients with gastric cancer affected by *Helicobacter pylori* infection. Scand. J. Gastroenterol. *38*, 1223–1227.

Dong, Q.J., Wang, Q., Xin, Y.N., Li, N., and Xuan, S.Y. (2009). Comparative genomics of *Helicobacter pylori*. World J. Gastroenterol. *15*, 3984–3991.

Eldridge, J., Lessells, A.M., and Jones, D.M. (1984). Antibody to spiral organisms on gastric mucosa. Lancet *1*, 1237.

Fontana, C., Favaro, M., Pietroiusti, A., Pistoia, E.S., Galante, A., and Favalli, C. (2003). Detection of clarithromycin-resistant *Helicobacter pylori* in stool samples. J. Clin. Microbiol. *41*, 3636–3640.

Goodwin, C.S., McCulloch, R.K., Armstrong, J.A., and Wee, S.H. (1985). Unusual cellular fatty acids and distinctive ultrastructure in a new spiral bacterium (*Campylobacter pyloridis*) from the human gastric mucosa. J. Med. Microbiol. *19*, 257–267.

Goodwin, C.S., Armstrong, J.A., Chilvers, T., Peters, M., Colins, M.D., Sly, L., McConnell, W., and Harper, W.E.S. (1989). Transfer of *Campylobacter pylori* and *Campylobacter mustelae* to *Helicobacter* gen. nov. as *Helicobacter pylori* comb. nov. and *Helicobacter mustelae* comb. nov., respectively. Int. J. Syst. Bacteriol. *39*, 397–405.

Graham, D.Y., Klein, P.D., Evans, D.J.,Jr., Evans, D.G., Alpert, L.C., Opekun, A.R., and Boutton, T.W. (1987). *Campylobacter pylori* detected noninvasively by the 13C-urea breath test. Lancet *1*, 1174–1177.

Graham, J.E., Peek, R.M., Jr., Krishna, U., and Cover, T.L. (2002). Global analysis of *Helicobacter pylori* gene expression in human gastric mucosa. Gastroenterology *123*, 1637–1648.

Graham, D.Y., Yamaoka, Y., and Malaty, H.M. (2007). Contemplating the future without *Helicobacter pylori* and the dire consequences hypothesis. Helicobacter *12 (Suppl 2)*, 64–68.

Graham, D.Y., and Shiotani, A. (2008). New concepts of resistance in the treatment of *Helicobacter pylori* infections. Nat. Clin. Pract. Gastroenterol. Hepatol. *5*, 321–331.

Graham, D.Y., Lu, H., and Yamaoka, Y. (2008). Therapy for *Helicobacter pylori* infection can be improved: sequential therapy and beyond. Drugs *68*, 725–736.

Hoshina, S., Kahn, S.M., Jiang, W., Green, P.H., Neu, H.C., Chin, N., Morotomi, M., LoGerfo, P., and Weinstein, I.B. (1990). Direct detection and amplification of *Helicobacter pylori* ribosomal 16S gene segments from gastric endoscopic biopsies. Diagn. Microbiol. Infect. Dis. *13*, 473–479.

Jafri, N.S., Hornung, C.A., and Howden, C.W. (2008). Meta-analysis: sequential therapy appears superior to standard therapy for *Helicobacter pylori* infection in patients naive to treatment. Ann. Intern. Med. *148*, 923–931.

Kidd, M., and Modlin, I.M. (1998). A century of *Helicobacter pylori*: paradigms lost-paradigms regained. Digestion *59*, 1–15.

Konturek, J.W. (2003). Discovery by Jaworski of *Helicobacter pylori* and its pathogenetic role in peptic ulcer, gastritis and gastric cancer. J. Physiol. Pharmacol. *54 (Suppl 3)*, 23–41.

de Korwin, J.D. (2008). Does *Helicobacter pylori* infection play a role in extragastric diseases? Presse Med. *37*, 525–534.

Krienitz, W. (1906). Ueber das auftreten von mageninhalt bei carcinoma ventriculi. Dtsch. Med. Wochenschr. *22*, 872.

Lesbros-Pantoflickova, D., Corthésy-Theulaz, I., and Blum, A.L. (2007). *Helicobacter pylori* and probiotics. J. Nutr. *137* (Suppl. 2), 812S–8128S.

Leunk, R.D., Johnson, P.T., David, B.C., Kraft, W.G., and Morgan, D.R. (1988). Cytotoxic activity in broth-culture filtrates of *Campylobacter pylori*. J. Med. Microbiol. *26*, 93–99.

Lu, H., Hsu, P.I., Graham, D.Y., and Yamaoka, Y. (2005). Duodenal ulcer promoting gene of *Helicobacter pylori*. Gastroenterology *128*, 833–848.

Luminari, S., Cesaretti, M., Marcheselli, L., Rashid, I., Madrigali, S., Maiorana, A., and Federico, M. (2010). Decreasing incidence of gastric MALT lymphomas in the era of anti*Helicobacter pylori* interventions: results from a population-based study on extranodal marginal zone lymphomas. Ann. Oncol. *21*, 855–859.

Makristathis, A., Pasching, E., Schütze, K., Wimmer, M., Rotter, M.L., and Hirschl, A.M. (1998). Detection of *Helicobacter pylori* in stool specimens by PCR and antigen enzyme immunoassay. J. Clin. Microbiol. *36*, 2772–2774.

Malfertheiner, P., Mégraud, F., O'Morain, C., Bell, D., Bianchi Porro, G., Deltenre, M., Forman, D., Gasbarrini, G., Jaup, B., Misiewicz, J.J., *et al.* (1997). Current European concepts in the management of *Helicobacter pylori* infection – the Maastricht Consensus Report. The European *Helicobacter pylori* Study Group (EHpSG). Eur. J. Gastroenterol. Hepatol. *9*, 1–2.

Malfertheiner, P., Mégraud, F., O'Morain, C., Hungin, A.P., Jones, R., Axon, A., Graham, D.Y., Tytgat, G., and European *Helicobacter pylori* Study Group (EHpSG). (2002). Current concepts in the management of *Helicobacter pylori* infection – the Maastricht 2-2000 Consensus Report. Aliment. Pharmacol. Ther. *16*, 167–180.

Malfertheiner, P., Megraud, F., O'Morain, C., Bazzoli, F., El-Omar, E., Graham, D., Hunt, R., Rokkas, T., Vakil, N., Kuipers, E.J., and The European Helicobacter Study Group (EHSG). (2007). Current concepts in the management of *Helicobacter pylori* infection: the Maastricht III Consensus Report. Gut *56*, 772–781.

Malfertheiner, P., Schultze, V., Rosenkranz, B., Kaufmann, S.H., Ulrichs, T., Novicki, D., Norelli, F., Contorni, M., Peppoloni, S., Berti, D., *et al.* (2008). Safety and immunogenicity of an intramuscular *Helicobacter pylori* vaccine in noninfected volunteers: a Phase I study. Gastroenterology *135*, 787–895.

Marshall, B. (1983). Unidentified curved bacilli on gastric epithelium in active chronic gastritis. Lancet *1*, 1273–1274.

Marshall, B.J., Armstrong, J.A., McGechie, D.B., and Glancy, R.J. (1985). Attempt to fulfill Koch's postulates for pyloric campylobacter. Med. J. Aus. *142*, 436–439.

Marshall, B. (2002). The discovery that *Helicobacter pylori*, a sprial bacterium, caused peptic ulcer disease. In *Helicobacter* pioneers: First hand accounts from the scientists who discovered helicobacters, 1892–1982, Marshall, B., ed. (Blackwell Science Asia: Victoria, Australia), pp. 165–202.

Marshall, B. (2008). *Helicobacter pylori* – a Nobel pursuit? Can. J. Gastroenterol. *22*, 895–896.

McNeil, C. (2008). *Helicobacter pylori*: good side complicates efforts to combat bad side. J. Natl. Cancer Inst. *100*, 1748–1750.

McNulty, C.A., and Watson, D.M. (1984). Spiral bacteria of the gastric antrum. Lancet *1*, 1068–1069.

McNulty, C.A., Dent, J., and Wise, R. (1985). Susceptibility of clinical isolates of *Campylobacter pyloridis* to 11 antimicrobial agents. Antimicrob. Agents Chemother. *28*, 837–838.

Megraud, F., Bonnet, F., Garnier, M., and Lamouliatte, H. (1985). Characterization of '*Campylobacter pyloridis*' by culture, enzymatic profile, and protein content. J. Clin. Microbiol. *22*, 1007–1010.

Mégraud, F., and Corti, R. (2009). Bacterial resistance of *Helicobacter pylori* in the world in 2009. Acta Gastroenterol. Latinoam. *39*, 282–290.

Morris, A., and Nicholson, G. (1987). Ingestion of *Campylobacter pyloridis* causes gastritis and raised fasting gastric pH. Am. J. Gastroenterol. *82*, 192–199.

Nilsson, C.L., Larsson, T., Gustafsson, E., Karlsson, K.A., and Davidsson, P. (2000). Identification of protein vaccine candidates from *Helicobacter pylori* using a preparative two-dimensional electrophoretic procedure and mass spectrometry. Anal. Chem. *72*, 2148–2153.

Ohnishi, N., Yuasa, H., Tanaka, S., Sawa, H., Miura, M., Matsui, A., Higashi, H., Musashi, M., Iwabuchi, K., Suzuki, M., *et al.* (2008). Transgenic expression of *Helicobacter pylori* CagA induces gastrointestinal and hematopoietic neoplasms in mouse. Proc. Natl. Acad. Sci. U.S.A. *105*, 1003–1008.

Pajares, J.M., and Gisbert, J.P. (2006). *Helicobacter pylori*: its discovery and relevance for medicine. Rev. Esp. Enferm. Dig. *98*, 770–785.

Pallen, M.J., and Clayton, C.L. (1990). Vaccination against *Helicobacter pylori* urease. Lancet *336*, 186–187.

Parsonnet, J., Friedman, G.D., Vandersteen, D.P., Chang, Y., Vogelman, J.H., Orentreich, N., and Sibley, R.K. (1991). *H. pylori* infection and the risk of gastric carcinoma. N. Engl. J. Med. *325*, 1131.

Peterson, W.L., Harford, W., and Marshall, B.J. (2002). The Dallas experience with acute *Helicobacter pylori* infection. In *Helicobacter* pioneers: First hand accounts from the scientists who discovered helicobacters, 1892–1982, Marshall, B., ed. (Blackwell Science Asia: Victoria, Australia), pp. 143–150.

Rappin, G. (1881). Contribution à l'étude des bactéries de la bouche a l'état normal et dans la fièvre typhoïde. PhD thesis. Collège de France, Nantes, France.

Rigas, B., Feretis, C., and Papavassiliou, E.D. (1999). John Lykoudis: an unappreciated discoverer of the cause and treatment of peptic ulcer disease. Lancet *354*, 1634–1635.

Scott, D.R., Marcus, E.A., Wen, Y., Oh, J., and Sachs, G. (2007). Gene expression *in vivo* shows that *Helicobacter pylori* colonizes an acidic niche on the gastric surface. Proc. Natl. Acad. Sci. U.S.A. *104*, 7235–7240.

Sharma, C.M., Hoffmann, S., Darfeuille, F., Reignier, J., Findeiss, S., Sittka, A., Chabas, S., Reiche, K., Hackermüller, J., Reinhardt, R., *et al.* (2010). The primary transcriptome of the major human pathogen *Helicobacter pylori*. Nature *464*, 250–255.

Sijun, H., and Yong, X. (2009). *Helicobacter pylori* vaccine: mucosal adjuvant and delivery systems. Indian J. Med. Res. *130*, 115–124.

Suganuma, M., Kurusu, M., Suzuki, K., Nishizono, A., Murakami, K., Fujioka, T., and Fujiki, H. (2005). New tumor necrosis factor-alpha-inducing protein released from *Helicobacter pylori* for gastric cancer progression. J. Cancer Res. Clin. Oncol. *131*, 305–313.

Szczebara, F., Dhaenens, L., Vincent, P., and Husson, M.O. (1997). Evaluation of rapid molecular methods for detection of clarithromycin resistance in *Helicobacter pylori*. Eur. J. Clin. Microbiol. Infect. Dis. *16*, 162–164.

Tomb, J.F., White, O., Kerlavage, A.R., Clayton, R.A., Sutton, G.G., Fleischmann, R.D., Ketchum, K.A., Klenk, H.P., Gill, S., Dougherty, B.A., *et al.* (1997). The complete genome sequence of the gastric pathogen *Helicobacter pylori*. Nature *388*, 539–547.

Trautmann, K., Stolte, M., and Miehlke, S. (2006). Eradication of *H pylori* for the prevention of gastric cancer. World J. Gastroenterol. *12*, 5101–5107.

Trevisani, L., Sartori, S., Galvani, F., Rossi, M.R., Ruina, M., and Caselli, M. (1998). Detection of *Helicobacter pylori* in faeces with a new enzyme immunoassay method: preliminary results. Scand. J. Gastroenterol. *33*, 893–894.

Unge, P. (2002). *Helicobacter pylori* treatment in the past and in the 21st century. In *Helicobacter* pioneers: First hand accounts from the scientists who discovered helicobacters, 1892–1982, Marshall, B., ed. (Blackwell Science Asia: Victoria, Australia), pp. 203–213.

Warren, J.R. (1983). Unidentified curved bacilli on gastric epithelium in active chronic gastritis. Lancet *1*, 1273.

Wotherspoon, A.C., Ortiz-Hidalgo, C., Falzon, M.R., and Isaacson, P.G. (1991). *Helicobacter pylori*-associated gastritis and primary B-cell gastric lymphoma. Lancet *338*, 1175–1176.

Wu, M.S., Chow, L.P., Lin, J.T., and Chiou, S.H. (2008). Proteomic identification of biomarkers related to *Helicobacter pylori*-associated gastroduodenal disease: challenges and opportunities. J. Gastroenterol. Hepatol. *23*, 1657–1661.

Zullo, A., Rinaldi, V., Winn, S., Meddi, P., Lionetti, R., Hassan, C., Ripani, C., Tomaselli, G., and Attili, A.F. (2000). A new highly effective short-term therapy schedule for *Helicobacter pylori* eradication. Aliment. Pharmacol. Ther. *14*, 715–718.

Genus *Helicobacter*

2

Lyudmila Boyanova

Abstract

The *Helicobacter* genus belongs to the class *Epsilonproteobacteria*, order *Campylobacterales*, family *Helicobacteraceae* and already involves over 35 species. Recently, new gastric (*Helicobacter suis* and *Helicobacter baculiformis*) and enterohepatic (*Helicobacter equorum*) species have been reported. *Helicobacter pylori* is of primary importance in medicine; however, non-*pylori Helicobacter* species (NPHS), which naturally inhabit mammals (except humans) and birds, have been detected in human clinical specimens. NPHS encompass two (gastric and enterohepatic) groups, showing different organ specificity. Importantly, some species such as *Helicobacter hepaticus*, *Helicobacter mustelae* and, probably, *Helicobacter bilis* exhibit carcinogenic potential in animals. NPHS harbour many virulence genes and may cause diseases not only in animals but also in humans. Gastric NPHS such as *H. suis* (most often), *Helicobacter felis*, *Helicobacter bizzozeronii* and *Helicobacter salomonis* have been associated with chronic gastritis and peptic ulcers in humans and, importantly, with higher risk for mucosa-associated lymphoid tissue (MALT) lymphoma compared with *H. pylori*. Enterohepatic species, e.g. *Helicobacter hepaticus*, *Helicobacter bilis* and *Helicobacter ganmani*, have been detected by polymerase chain reaction (PCR) in, but still not isolated from, specimens of patients with hepatobiliary diseases. Moreover, NPHS may be associated with Crohn's disease, inflammatory bowel disease and ulcerative colitis. The significance of avian helicobacters (*Helicobacter pullorum*, *Helicobacter anseris* and *Helicobacter brantae*) has also been evaluated extensively. NPHS such as *Helicobacter cinaedi* and *Helicobacter canis* can cause severe infections, mostly in immunocompromised patients with animal exposure. Briefly, the role of NPHS in veterinary and human medicine is increasingly recognized. However, despite the growing interest in the possible association between NPHS and the chronic hepatobiliary or intestinal diseases in humans, more studies are still required to prove the suggested association. Several other topics such as the isolation of still uncultured species, antibiotic resistance and treatment regimens for NPHS infections and, last but not least, NPHS pathogenesis and possible carcinogenesis in humans should be additionally evaluated.

Introduction

Helicobacter genus is an important member of the class *Epsilonproteobacteria*, order *Campylobacterales*, family *Helicobacteraceae* (Garrity *et al.*, 2005). The family includes the genera *Helicobacter* (most important for human medicine), *Wolinella* and free-living *Sulfuricurvum*, *Sulfurimonas*, *Sulfurovum* and *Thiovulum*. The major medically significant species, *Helicobacter pylori*, is described in other chapters of this book.

NPHS are divided into two (gastric and enterohepatic) groups, showing different organ specificity (Table 2.1). NPHS inhabit the gastrointestinal tract of mammals and birds, and often are specific for a host or limited number of hosts, e.g. *Helicobacter cetorum* for some marine mammals, *H. suis* for pigs, *H. mustelae* for ferrets, *H. equorum* for horses and *H. anseris* and *H. brantae*

Table 2.1 Distribution and association of some NPHS with diseases according to Harper *et al.* (2003), Moura *et al.* (2003), Ceelen *et al.* (2005, 2006), Garrity *et al.* (2005), Fox *et al.* (2006), Kusters *et al.* (2006), Leemann *et al.* (2006), Whary *et al.* (2006), Nebbia *et al.* (2007), Pasquali *et al.* (2007), Taylor *et al.* (2007), Veijola *et al.* (2007), Baele *et al.* (2008b), Holst *et al.* (2008), Moyaert *et al.* (2008, 2009a,b), Pellicano *et al.* (2008), Rossi *et al.* (2008), Haesebrouck *et al.* (2009); Loman *et al.* (2009), Okoli *et al.* (2009, 2010), Shen *et al.* (2009); Wadström *et al.* (2009), Casswall *et al.* (2010) and Nakonieczna *et al.* (2010)

Species	Source (prevalence,%)	Some associated diseases in animals	Some reported diseases in humans[1]
Gastric spp.			
H. pylori	Humans (<20–80) and primates[2]	CG in non-human primates, CG in cats (single study)	CG, PU, GC, MALT lymphoma
'Ca. H. heilmannii'	Dogs (20–100), cats (20–100), wild felids, humans[2], monkey[2] etc.	CG, dyspepsia, MALT lymphoma in mice	CG, gastric ulcer, MALT lymphoma
H. bizzozeronii	Dogs (70), cats (35), humans[2]	CG in cats and dogs	As above
H. felis	Cats (63), dogs (47), rabbits (<10), humans[2], mice and cheetahs	CG in cats and dogs; PU, GC and MALT lymphoma in mice	As above
H. salomonis	Dogs (9), cats (2), rabbits (≤4), humans[2]	CG in cats and dogs	As above
H. suis	Pigs (60–80), humans[2], monkeys	CG and PU in pigs	As above
'Ca. H. bovis'	Cattle	NA	CG
H. mustelae	Ferrets (0–100)	CG, PU, GC, MALT lymphoma in ferrets	NA
H. acinonychis	Cheetahs (low), tigers, other big cats	CG, PU in big cats	NA
H. nemestrinae	Pigtailed macaques	NA	NA
H. baculiformis	Cats	NA	NA
H. cynogastricus	Dogs	NA	NA
H. cetorum	Whales, dolphins (50), probably South American fur seals	CG and oesophageal ulcers in whales and dolphins	NA (found in asymptomatic subjects)
Enterohepatic spp.			
H. aurati	Syrian hamsters (>50)	CG, GC in hamsters	NA
H. cinaedi	Dogs (8), cats (2), foxes, Syrian hamsters, rodents (2), macaques, birds, humans[2]	Diarrhoea in a dog and non-human primates and HBD (hepatitis in monkeys, gall stones in mice)	Diarrhoea, proctocolitis, meningitis, bacteraemia, septic arthritis and cellulitis, mainly in immunocompromised patients
H. fennelliae	Humans[2], dogs[2], macaques[2]	NA	As above, diarrhoea in children
H. hepaticus	Mice (7–59) and other rodents	Typhlocolitis, IBD, colon cancer, HBD (hepatitis, gall stones, hepatocellular cancer) in mice	HBD (cholelithiasis, cholecystitis)
H. bilis	Mice (2–17) and other rodents, dogs (19), cats[2] (6)	IBD, chronic HBD (hepatitis) in mice	HBD (bile duct malignancy)
H. canadensis	Birds (geese), rodents, swine, laboratory rabbits and humans[2]	NA	Diarrhoea, Crohn's disease and bacteraemia
H. canis	Dogs (11), cats (14), humans[2]	Diarrhoea, hepatitis in dogs	HBD (liver diseases), diarrhoea, bacteraemia, cellulitis

Table 2.1 continued

Species	Source (prevalence,%)	Some associated diseases in animals	Some reported diseases in humans[1]
H. pullorum	Poultry (32–78), humans[2]	IBD, diarrhoea, HBD (hepatitis) in poultry	Diarrhoea, IBD, Crohn's disease and HBD (liver disease)
H. anseris	Canada geese (21)	NA	NA
H. brantae	Canada geese (7)	NA	NA
H. ganmani	Mice (27–37)	Isolated from mouse intestines and liver	HBD (chronic liver diseases)
H. rodentium	Mice (6–38)	IBD and diarrhoea in susceptible mice, HBD (gall stones) in mice	HBD (chronic liver diseases)
H. mastomyrinus	Rodents (6)	HBD (liver diseases) and intestinal[3] inflammation	NA
H. cholecystus	Syrian hamsters	HBD and pancreatic diseases	NA
H. typhlonius	Mice (2–26)	IBD, typhlocolitis and tumours in mice	NA
H. trogontum	Rats, pigs, sheep	HBD (liver diseases) and typhlocolitis in mice	NA
H. marmotae	Woodchucks	HBD (liver diseases)	NA
H. pametensis	Birds and mammals	NA	NA
H. equorum	Horses (68 in foals, <8 in horses)	NA	NA

[1]Infections in humans have been uncommon.

[2]Relatively rare.

[3]Possible association.

CG, chronic gastritis; GC, gastric cancer; HBD, hepatobiliary diseases; IBD, inflammatory bowel disease; PU, peptic ulcer; NA, not available.

for Canadian geese (Haesebrouck *et al.*, 2009). The specificity highlights the co-evolution and perfect co-adaptation of the bacteria and the hosts. *Helicobacter* strain has been reported to cause cellulitis and septicaemia even in a pancake tortoise (*Malacochersus tornieri*; Stacy and Wellehan, 2010). *Helicobacter* genus already encompasses over 35 species (Moyaert *et al.*, 2008; Hayashi *et al.*, 2009) and is expected to expand further. Since 2006, many new species, for example, *H. suis, Helicobacter baculiformis, H. equorum, H. brantae* and '*Helicobacter macacae*', have been identified or suggested (Fox *et al.*, 2006, 2007; Hayashi *et al.*, 2009).

Morphology and physiology

Like the major human gastric pathogen, *H. pylori* (Fig. 2.1), most NPHS are Gram-negative spiral and microaerophilic bacteria, undergoing transition to coccoids with ageing (Moyaert *et al.*, 2007). Taxonomic changes evolve rapidly in the field of NPHS. For instance, the fusiform bacteria, showing corkscrew-like movement, periplasmic fibrils and bipolar tufts of sheathed flagella, initially classified as '*Flexispira rappini*', now include many species, among them are *H. bilis, Helicobacter muridarum, Helicobacter trogontum, H. baculiformis* and *H. aurati* (Hänninen *et al.*, 2005; Baele *et al.*, 2008a).

NPHS are motile by means of their flagella (1–>20 per cell, Table 2.2). Many NPHS harbour sheathed flagella (Garrity *et al.*, 2005), which show protective functions, for example against host antibodies (Skene *et al.*, 2007) and probably against antimicrobial peptides and gastric acidity. Multilayer periplasmic fibrils, located under the outer membrane, have been found in some gastric and enterohepatic species. The periplasmic

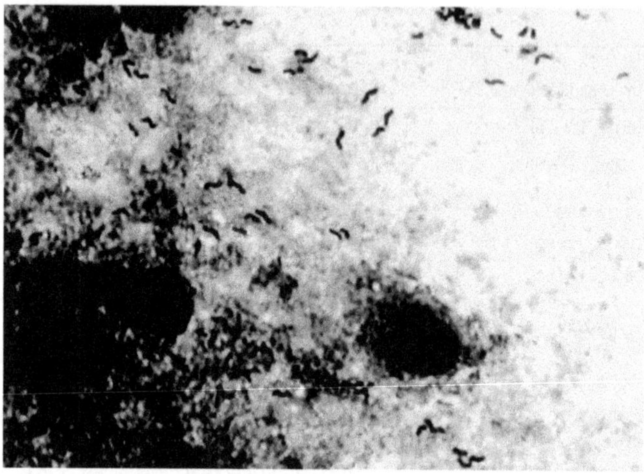

Figure 2.1 *H. pylori* in gastric mucosa of a symptomatic patient (×1000). See numerous spiral-shaped bacteria.

fibrils have most probably contractile properties (O'Rourke and Bode, 2001). Some NPHS are curved rods or short spirals, while other species, e.g. *H. felis, H. bizzozeronii, H. suis* and *H. cynogastricus*, are cork-screw-like (tightly coiled) bacteria (Baele *et al.*, 2004; Haesebrouck *et al.*, 2009; Fig. 2.2). Small bacteria (1.2–2.5 μm in length), e.g. *Helicobacter acinonychis, H. mustelae, 'Candidatus (Ca.) H. bovis'* and *H. ganmani*, or large (≥10 μm) bacteria such as *H. baculiformis* and *H. cynogastricus*, can be observed (Haesebrouck *et al.*, 2009). Most NPHS, especially the gastric species, exhibit urease activity (Garrity *et al.*, 2005). In addition, the majority of the species are oxidase and catalase positive, and microaerophilic bacteria except the anaerobic species *H. ganmani* (Table 2.2). NPHS perform a non-fermentative respiratory metabolism (Kusters *et al.*, 2006).

The genomes of NPHS have been extensively evaluated and those of some species have been sequenced. DNA guanine and cytosine (G + C) content of enterohepatic NPHS sequences (≤ 40 mol%) has been found to be lower than that (> 46 mol%) of gastric NPHS except *H. mustelae* (42 mol%) and than that of *H. pylori* (34 mol%; Hannula and Hänninen, 2007). Interestingly, the genome (size 1,553,928 bp, G + C content 38.2%) of *H. acinonychis*, a gastric species causing chronic gastritis and ulcers in cheetahs, lions, and tigers, is strikingly similar to that of *H. pylori* (1,596,366–1,667,867 bp and 39%, respectively). On the basis of total genomes and similarity between amino acids of *H. acinonychis* and *H. pylori*, it is possible to suggest that a likely host jump from prehistoric humans to big cats happened >100,000 years ago (Dong *et al.*, 2009). As in *H. pylori*, genomic islands have been found in *H. acinonychis* (*HacGI*) and *H. hepaticus* (*HHGI1*; Dong *et al.*, 2009). *H. acinonychis* strains also have exhibited a degenerate vacuolating cytotoxin (*vacA*) gene (Dailidiene *et al.*, 2004). However, the fascinating question, 'Who ate whom?', which Eppinger *et al.* (2006) have posed to explain the appearance of *H. acinonychis*, should be additionally evaluated.

Complete genomes of *H. hepaticus*, an enterohepatic species and hepatocarcinogen for mice, and *H. mustelae*, the gastric pathogen of ferrets, have been sequenced (Suerbaum *et al.*, 2003; O'Toole *et al.*, 2010). *H. hepaticus* has shown some genome resemblance to both closely related and distant species, e.g. *H. pylori* (by flagellar biosynthesis genes), *Campylobacter jejuni* (by *cdtABC* genes for the cytolethal distending toxin) and *V. cholerae* (by some genes of *HHGI1*; Suerbaum *et al.*, 2003). *H. mustelae*, although lacking orthologues of *H. pylori* cytotoxin-associated gene A (CagA) protein and VacA, has been found to possess homologues of *H. pylori* neutrophil-activating protein (NapA) and *Campylobacter jejuni* invasion antigen (CiaB) and major antigen (Peb4a). In addition, in a gerbil model of infection, *H. pylori* and *H. mustelae* have been observed to share

Table 2.2 Phenotypic characteristics of some *Helicobacter* species according to Ceelen *et al*. (2005), Van den Bulck *et al*. (2005b), Kusters *et al*. (2006), Moyaert *et al*. (2007), Veijola *et al*. (2007), Baele *et al*. (2008b), Holst *et al*. (2008) and Haesebrouck *et al*. (2009).

	Gastric species							Enterohepatic species							
Test	**1**	**2**	**3**	**4**	**5**	**6**	**7**	**8**	**9**	**10**	**11**	**12**	**13**	**14**	**15**
Urease	+	+	(+)	(+)	+	+	+	–	–	–	–	+	+	–	–
Catalase production	+	+	+	+	+	NA	NA	+	+	(+)	(+)	+	+	d	–
Nitrate reduction	–	–	+	+	+	NA	NA	V	+	–	+	+	+	+	–
Alkaline phosphate hydrolysis	+	+	V	V	V	NA	NA	–	–	d	d	–	–	–	+
γ-Glutamyl transpeptidase	+	+	+	+	+	NA	NA	–	NA	–	–	–	+	NA	+
Indoxyl acetate hydrolysis	d	–	d	d	d	NA	NA	+	–	+	d/–	+	–	–	+
Growth at 42°C	d	–	V	V	–	NA	NA	+	+	d	V	–	+	–	+
Growth on 1% glycine	–	–	–	d	–	NA	NA	+	–	–	–	+	+	–	–
Nalidixic acid (30 μg)	R	NA	R	R	R	NA	NA	R	R	S	S[R]	R	R	S	S
Cephalotin (30 μg)	S	NA	S	S	S	NA	NA	R	S	S	I	R	R	R	I
Length (μm)	2.5–5	2.3–6.7	5–7.5	5–10	5–7	5–10	1.5–2.5	NA	3–4	1.5–5	1.5–5	1.5–5	4–5	2.5	3–4
Periplasmic fibril	–	–	+	–	–	–	–	–	–	–	–	–	+	–	–
No. flagella per cell	4–8	4–10	14–20	10–20	10–23	10–20	≥4	1–2	1	2	1–2	2	3–14	2	2
Sheathed flagella	+	+	+	+	+	+	NA	–	–	+	+	+	+	–	+
Distribution of flagella	B	B	B	B	B	B	NA	B	M	B	B	B	B	B	B

+, 100% strains positive; –, 100% strains negative; (+), 80–94% strains positive; V, 42–66% strains positive; d, 7–33% strains positive; S, sensitive; R, resistant; I, intermediate; NA, not available; B, bipolar; M, monopolar.

Species: 1, *H. pylori;* 2, *H. suis;* 3. *H. felis*; 4, *H. bizzozeronii*; 5, *H. salomonis*; 6, '*Ca. H. heilmannii*'; 7, '*Ca. H. bovis*'; 8, *H. canadensis*; 9, *H. pullorum*; 10, *H. fennelliae*; 11, *H. cinaedi*; 12, *H. hepaticus*; 13, *H. bilis*; 14, *H. ganmani;* 15, *H. canis.*

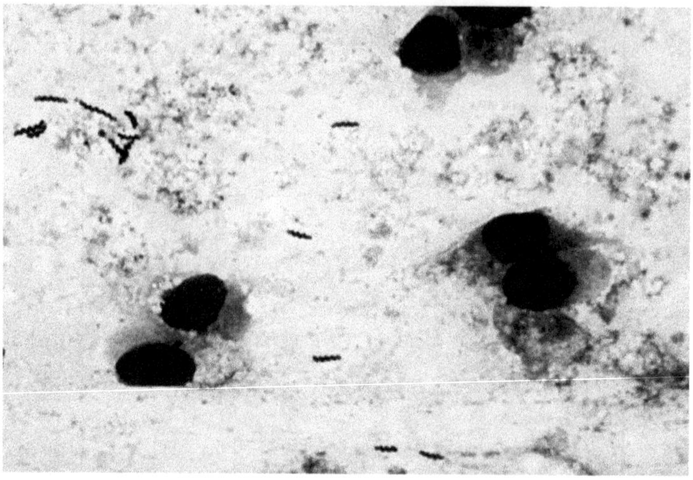

Figure 2.2 Gastric non-*H. pylori* helicobacters in the gastric mucosa of a symptomatic patient (×1000). See the bacteria with multiple spirals.

≥24 orthologues for colonization (O'Toole *et al.*, 2010). Furthermore, the *H. mustelae* genome also exhibits distinctive features, e.g. for immune evasion, it generates blood group A/B on its lipopolysaccharide, using a glycosyl transferase. Recently, the whole-genome sequencing of a *H. canadensis* strain has shown the presence of probable virulence factors, e.g. an immunoglobulin A protease and two homologues of *H. pylori* VacA (Loman *et al.*, 2009). Similarly to *C. jejuni*, *H. canadensis* has been found to use a capsular polysaccharide export system and to harbour genes for *N*-linked glycosylation pathway (Loman *et al.*, 2009). All these data reveal both genomic and functional similarities within NPHS or between NPHS and either *H. pylori* or *Campylobacter* spp. and, at the same time, many differences, especially in the virulence factors, owing to the close adaptation of every NPHS to its 'preferred' host.

Virulence factors of non-*pylori Helicobacter* species and immunity

Like *H. pylori*, the gastric NPHS use urease to maintain their periplasmic and cytoplasmic pH. They possess urease of two subunits, UreA and UreB, and urease encoding genes. As in *H. pylori*, NPHS proton-gated urea channel UreI is activated in low pH and accelerates urea entry into the bacterial cells (Haesebrouck *et al.*, 2009). However, in the gastric species *H. felis*, an independent urease system, UreA2B2, has also been detected. The urease activity has been blocked by *ureB* gene inactivation but has been reduced only slightly by *ureB2* inactivation (Pot *et al.*, 2007). The role of the second system remains to be clarified.

Urease is a virulence factor of the gastric NPHS, as for *H. pylori*. *H. mustelae* urease-negative mutant has lost the property of the wild-type strain to colonize the gastric tissues of ferrets (Haesebrouck *et al.*, 2009). In addition, specific surface rings, found in *H. mustelae*, have been associated with gastric inflammation (Haesebrouck *et al.*, 2009). Conversely, the urease of *H. hepaticus* has been found to be acid independent (Belzer *et al.*, 2005) as the enterohepatic species 'do not need' protection from gastric acid. However, although not needed for colonization, *H. hepaticus* urease has been linked to hepatic inflammation in mice (Ge *et al.*, 2008b).

NPHS catalase provides the necessary oxidative stress resistance to the bacteria. Only wild *H. hepaticus* strains but not catalase (*katA*) mutants can survive at 6% O_2 (Hong *et al.*, 2007). Importantly, a structural similarity (molecular mimicry) between *H. hepaticus* catalase and those of mammals could trigger autoimmune reactions. Antibodies to *H. hepaticus* catalase have been shown to bind murine hepatocyte catalase (Alyamani *et al.*, 2007). Another enzyme for *H. hepaticus* survival at 6% O_2 has been a quinone reductase (MdaB), encoded by *mdaB* gene (Hong *et al.*, 2008). In addition, the γ-glutamyltranspeptidase, which contributes to gastric cell apoptosis and suppression of T-cell proliferation in *H. pylori* (see Chapter 4), has been also used by some gastric NPHS (Haesebrouck *et al.*, 2009). Like *H. pylori, H. hepaticus* utilizes a hydrogenase, a virulence factor and constitutive nickel-dependent enzyme for oxidation of the molecular hydrogen. In the study of Mehta *et al.* (2005), over 5 months after inoculation, A/J mice infected with hydrogenase mutants exhibited lack of liver inflammation, unlike those infected by strains with functional enzyme. The authors suggested a role for proteins or enzymes coupled with hydrogen oxidation. Therefore, in *H. hepaticus,* the hydrogenase seems to contribute to long-term liver damage in mice.

Another property of many enterohepatic NPHS is the bile resistance, which has been found *in vitro* in *H. rappini, H. bilis, H. canis, H. cholecystus* and *H. pullorum,* and may protect the bacteria from the inhibitory activity of the bile in the intestinal and biliary tract (Mishra *et al.*, 2010). HefABC RND efflux system seems to be important for the bile resistance in *H. hepaticus*. Importantly, the *hefA* gene has been shown to contribute to *H. hepaticus* resistance not only to bile acids, but also to chemical agents and antibiotics (Belzer *et al.*, 2009; see below). However, the bile (0.1%) has been found to affect *H. hepaticus* cell envelope synthesis, motility and overall virulence (Okoli *et al.*, 2010).

Motility is needed for gastric colonization of NPHS. Flagellar filament has two flagellin subunits, FlaA (predominant) and FlaB (Haesebrouck *et al.*, 2009). *H. hepaticus flaA/B* mutants rapidly lose their property to colonize the gut mucosa of mice (Sterzenbach *et al.*, 2008; Haesebrouck *et al.*, 2009). While a gastric glycoprotein (mucin MUC1) reduces *H. pylori* attachment to gastric cells, *H. felis* colonization in mice can be changed by different mucins, e.g. MUC4 and MUC5 (Every *et al.*, 2008; Okoli *et al.*, 2009). In mice, soluble substances of *H. hepaticus* have been reported to weaken the innate immune response of intestinal epithelial cells to the bacterial lipopolysaccharide (LPS) by Toll-like receptor 4 (TLR4) and to bacterial flagellin-mediated activation by TLR5. *H. hepaticus* LPS has also been shown to decrease endotoxin tolerance to *Escherichia coli* LPS in the murine cells (Sterzenbach *et al.*, 2007). Therefore, the immune suppression, which *H. hepaticus* promotes, can disturb the overall balance of the intestinal flora.

The carcinogenic potential of NPHS has been extensively evaluated. An increase in hepatic steatosis by *H. hepaticus* infection has been reported (Okoli *et al.*, 2009). Furthermore, the infection has increased the liver cancer risk in mice in contact with other carcinogens (Fox *et al.*, 2010). Chronic *H. hepaticus* infection in mice also has promoted colon tumours due to increased bile production or altered bile absorption in the colon (Nagamine *et al.*, 2008).

Importantly, many NPHS can raise the release of pro-inflammatory cytokines not only in tissues/cells of the animal hosts but also in cell lines of human origin. The chronic inflammation can promote development of malignancy. Like *cag* pathogenicity island (*cag*PAI) in *H. pylori, H. hepaticus HHGI1* pathogenicity island is a major virulence factor. In mice, *HHGI1* has been shown to trigger the nuclear factor-κB (NF-κB) signalling pathways and the ensuing increase in secretion of pro-inflammatory cytokines such as interferon (IFN)-γ, tumour necrosis factor (TNF)-α and interleukin (IL) 17a. These cytokines are linked to the innate immunity and Th1 immune bias that in its turn, promotes inflammation in the host tissues (Ge *et al.*, 2008a; Fox *et al.*, 2010). As in *H. pylori* cag PAI, *HHGI1* genes have been highly variable in *H. hepaticus* strains (Ge *et al.*, 2008a). On HuCCT-1 human bile duct cancer cell line, *H. bilis* has stimulated the release of NF-κB, E2 transcription factor, vascular endothelial growth factor and cyclic AMP response with resultant increase in angiogenesis (Takayama *et al.*, 2010). *H. felis* and *H. bizzozeronii* infection has been associated with apoptosis of gastric epithelial cells, resulting in intestinal metaplasia, in gerbils (De Bock *et al.*, 2006). *H. felis* and *H. suis* have been shown to induce apoptosis also in both murine gastric epithelial cell line GSM06 and human AGS cell line (Haesebrouck *et al.*, 2009). Furthermore, *H. pullorum* has been reported to induce NF-κB-associated stimulation of IL-8 release, on both human AGS gastric cancer cell line and intestinal (CaCo-2 and HT-29) cell lines (Varon *et al.*, 2009). In animals, enterohepatic NPHS infections usually induce chronic and manifest inflammation and T helper 1 (Th1)-directed immune response, leading to hepatitis, gallstones and cancer (Harbour and Sutton, 2008; Shen *et al.*, 2009). Activity of enterohepatic (*H. hepaticus, H. bilis* and *H. pullorum*) and gastric (*H. felis* and *H. suis*) NPHS on cell lines from both animal and human origins is suggestive of probable involvement of the enterohepatic species in human diseases, as is the case with gastric NPHS.

Cytolethal distending toxin (CDT) is a key virulence factor in enterohepatic NPHS. Strains of *H. hepaticus, H. bilis, H. pullorum, H. canis* and *Helicobacter cinaedi* can produce a three-subunit CDT, similarly to species from close and distant genera such as *Campylobacter, Escherichia, Shigella, Haemophilus* and *Actinobacillus* spp. (Pratt *et al.*, 2006; Pellicano *et al.*, 2008; Shen *et al.*, 2009). The mechanism of action of *H. hepaticus* CDT on intestinal epithelial cells has been described in detail (Liyanage *et al.*, 2010). CDT has a *cdtB* gene-encoded DNase, its active CdtB subunit, which produces DNA double-strand breaks and permanent G2/M-phase cell cycle arrest in the host cells. Liyanage *et al.* (2010) have shown the toxin to cause cell cycle arrest and death of INT407 cells by mitochondrial apoptotic pathway with release of cytochrome c and activation of caspase 3/7 and caspase 9. In addition to the CdtB, the toxin has also two subunits (encoded by *cdtA* and *cdtC* genes) for binding to host cell targets (Shen *et al.*, 2009). CDT of *H. pullorum* also has been reported to cause hepatic cell death in mice (Okoli *et al.*, 2009). Importantly, CDT can affect many types of human cells. *H. hepaticus* CDT has been shown to induce apoptosis in human mononuclear and T-cell lines, to interact with antigen-presenting cells and to promote the persistence of the infection (Pratt *et al.*, 2006). CDT of *H. cinaedi* and other enterohepatic species have been found also to stimulate the release of pro-inflammatory cytokines, IFN-γ and TNF-α, and inducible nitric oxide synthase (iNOS) in the intestinal tract. They can promote the development of typhlocolitis in IL-10-deficient mice (Shen *et al.*, 2009). Briefly, like the virulence factors of *H. pylori*, those of the NPHS directly damage the host cells or modulate the immune response to increase the inflammation or to promote the persistence of the infection.

Interactions between *Helicobacter* spp. can occur and, importantly, can modify the course of *H. pylori* infection. Recombinations, like those observed between *H. pylori* and *H. acinonychis*, as well as the inter-species transfer of antibiotic resistance markers may change the bacterial virulence and susceptibility pattern of the species (Pot *et al.*, 2001; Dailidiene *et al.*, 2004; see also Chapter 9). Furthermore, in coinfections, *H. bilis* has been found to diminish *H. pylori*-induced pro-inflammatory effects in the gastric mucosa of C57BL/6 mice. This has been related to *H. bilis*-recognizing intestinal regulatory T cells (T_{reg}), which, if transported to the stomach, can suppress the Th1-biased immune response of following *H. pylori* infection (Lemke *et al.*, 2009). This explains, at least partly, why some African countries with common coinfections of *H. pylori* and NPHS have reported low gastric cancer incidence (Lemke *et al.*, 2009).

Microbiology

NPHS can be diagnosed by histology (haematoxylin–eosin and Whartin–Starry silver stain), culture, serological (immunoblot) and molecular methods. Isolation of the NPHS is difficult and may require 3–7 days in microaerophilic conditions. Growth of many NPHS can appear as a fine spreading film, which can initially be unrecognized, or as pinpoint colonies (Moyaert *et al.*, 2007; Baele *et al.*, 2008a,b). Some NPHS are slow-growing, so their isolation can be hampered by the coexisting bacteria in the specimen. Therefore, selective media and filter techniques are useful for NPHS isolation.

Like *H. pylori*, most NPHS are microaerophilic and grow best in humid atmosphere with 5% (3–7%) O_2 and 10% CO_2 at 37°C (Testerman *et al*, 2006). Conversely, Shiohara *et al.* (2009) have obtained large *H. felis* colonies by 10% CO_2 and higher (12%) O_2 concentration. As for *H. pylori*, use of fresh media is necessary for NPHS isolation. The media contain rich agar bases, blood or serum, growth supplement as Vitox (Oxoid, Basingstoke, UK), and often selective Skirrow's supplement

(10 μg/ml vancomycin, 5 μg/ml trimethoprim, and 2500 U/l polymyxin B) or its modifications. Serum has been required by *H. felis* (Testerman *et al*, 2006). Some NPHS have been successfully grown also in chemically defined (F-12) medium with or without serum (Testerman *et al*, 2006). *H. suis* has been cultured by moist plates, containing Brucella agar with 10% sheep blood or 20% fetal calf serum (Baele *et al.*, 2008b).

Selective media are useful to isolate NPHS from specimens with abundant normal flora, e.g. faeces. Appropriate selective medium for *H. felis, H. bizzozeronii* and *H. salomonis* has been brain heart infusion agar (BHIA) with 10% horse blood, Skirrow's and Vitox supplements and amphotericin B 5 μg/ml (Baele *et al.*, 2004). Recommended *H. suis* medium has been that, containing Brucella agar, 20% fetal bovine serum, amphotericin B 5 μg/ml, Skirrow's and Vitox supplements, 0.1% activated charcoal and HCl to achieve pH 5 (Baele *et al.*, 2008b; Kivistö *et al.*, 2010). The lower than usual pH has been used to reduce the contaminating bacteria in the specimen and the charcoal has been involved to neutralize the toxic fatty acids in the medium.

However, modified *H. pylori* medium (LabM, Lancashire, UK) with 20% fetal bovine serum, Skirrow's and Vitox supplements has provided better growth of *H. bizzozeronii* than *H. suis* medium, possibly owing to the inhibitory effect of the pH 5 (Kivistö *et al.*, 2010). Taylor *et al.* (2007) have isolated enterohepatic NPHS from mice by in-house-prepared medium with amphotericin B 50 μg/ml, vancomycin 100 μg/ml, polymyxin B 3.3 μg/ml, bacitracin 200 μg/ml and nalidixic acid 10.7 μg/ml. *H. baculiformis* has been isolated on fresh Columbia agar plates with 7% lysed horse blood, 7% lysed sheep blood and lower (5 μg/ml vancomycin, 3 μg/ml trimethoprim, 2500 U/l polymyxin B and 5 μg/ml amphotericin B) drug concentration than that of Skirrow's supplement (Baele *et al.*, 2008a). *H. pullorum* isolation may require enrichment (in Brucella broth with cefoperazone, amphotericin B and vancomycin) in microaerophilic atmosphere at 37°C before subculture on Campylosel agar (BioMérieux) for 48–72 h (González *et al.*, 2008). Importantly, the selective medium for *H. pullorum* should not contain polymyxin as the bacteria are susceptible to the agent (Garrity *et al.*, 2005). Briefly, many formulas for selective media for NPHS isolation have been suggested but variations have been found according to the species to be cultured and by the same media in different studies. Use of combination of several media and development of new selective media can resolve this problem.

The filter technique is appropriate to detect most NPHS, especially those susceptible to many antibiotics. By the technique, the specimen is inoculated onto a 0.45-μm pore-size filter, placed on a suitable agar medium. The filter is removed after 1-h microaerophilic incubation and the filtrate is inoculated onto the agar. Enterohepatic NPHS have been isolated by filters on blood trypticase soy agar (Taylor *et al.*, 2007). *H. pullorum* has been cultured by the filter technique and BHIA with 10% horse blood, amphotericin B 20 μg/ml and Vitox (Ceelen *et al.*, 2006). The authors have recommended use of only fresh specimens from caecal tissue instead of caecal contents because of the strong adherence of *H. pullorum* to the mucosa as well as combination of culture with a suitable molecular method.

Some phenotypic tests of NPHS are presented in Table 2.2. Many gastric NPHS are urease positive and can cause false-positive rapid urease test (RUT) results for *H. pylori* detection. RUT of NPHS also can be weakly positive or negative (Sykora *et al.*, 2003; Boyanova, 2007; Boyanova *et al.*, 2007). However, the morphological and phenotypic identification of NPHS is unreliable. Polymerase chain reaction (PCR)-based and immunoblot techniques have been developed for accurate diagnosis of the bacteria (Okoli *et al.*, 2009; Wadström *et al.*, 2009). In colonic tissue specimens of mice, agreement between results of immunoblot and PCR-denaturing gradient gel electrophoresis has been > 90% for *H. bilis* and *H. hepaticus,* and 82% for *H. ganmani* in the study of Wadström *et al.* (2009). However, the advantage of the PCR-based method has been additional detection of *H. typhlonius, H.* sp. *flexispira* and *H. rodentium.*

16S rRNA sequence analysis is a common method for bacterial species identification. However, 16S rRNA sequences of *H. felis, H. bizzozeronii* and *H. salomonis* have shown very high (>99%) similarity and therefore, the method is unreliable for their identification (Baele *et al.*, 2004). Similar results have been found in enteric NPHS (Rossi *et al.*, 2008). Dewhirst *et al.* (2005) have explained some discrepant data by a horizontal transfer of 16S rRNA gene fragments and formation of mosaic structures. The authors have performed 23S rRNA gene-derived phylogeny. Baele *et al.* (2004) have obtained a good discrimination of gastric NPHS by multiplex PCR test based on tRNA intergenic spacers and urease gene, combined with capillary electrophoresis. To identify gastric NPHS, sequencing of *ureB*, *hsp60* and *gyrB* genes and protein profiling have been used as well (Haesebrouck *et al.*, 2009). As *gyrB* gene evolves more quickly than rRNA genes and lacks intervening sequences, partial *gyrB* gene sequencing has been appropriate for NPHS phylogenetic analysis (Hannula and Hänninen, 2007). To identify *H. bilis*, *H. canis* and *H. cinaedi* in dogs and cats, Rossi *et al.* (2008) have used PCR-restriction fragment length polymorphism method (RFLP) analysis of *gyrB*.

Real-time PCR (RT-PCR) has been one of the best methods to detect NPHS genes encoding virulence or resistance factors. The advantages of RT-PCR over standard PCR include better detection level, higher speed of testing, better specificity, independence on gel electrophoresis (González *et al.*, 2008), lower risk for contamination and, most importantly, quantitative analysis of the results. RT-PCR has been performed for *H. pullorum* in chicken specimens with detection level of one colony-forming unit (CFU)/g (González *et al.*, 2008).

Recently, Alispahic *et al.* (2010) have identified *Helicobacter*, *Arcobacter* and *Campylobacter* species by matrix-associated laser desorption/ionization-time of flight MALDI-TOF mass spectrometry fingerprinting; however, the identification has been dependent on the age, storage conditions and media used. The authors have recommended the use of specimens stored at room temperature or at 4°C for <10 days as well as Campylosel agar rather than modified charcoal cefoperazone deoxycholate agar.

However, serological methods are not recommended for NPHS. Serological cross-reactivity between NPHS (*H. suis*, '*Ca. Helicobacter heilmannii*' and *H. bilis*) and *H. pylori* has been found (Al-Hamoudi *et al.*, 2006; Pisani *et al.*, 2008; Haesebrouck *et al.*, 2009).

Susceptibility testing of the NPHS can be done by molecular methods, E test and agar dilution method with prolonged (for 4–7 days) incubation (Baele *et al.*, 2008b). Van den Bulck *et al.* (2005a) have used high (3 McFarland standard) inocula and Mueller–Hinton II agar (Becton Dickinson, Cockeysville, MD, USA) with 10% horse blood for susceptibility testing of *H. felis*, *H. bizzozeronii*, and *H. salomonis*. The authors have obtained bacterial growth as single colonies, which are easier to be read. However, no standard methods for resistance breakpoints for NPHS have been recommended so far.

Baele *et al.* (2008b) reported susceptibility of *H. felis*, *H. bizzozeronii* and *H. salomonis* strains to ampicillin, clarithromycin, tetracycline and gentamicin along with metronidazole resistance (MIC 16 μg/ml) in 2 of 15 *H. felis* strains. Metronidazole resistance has been also reported in *H. bizzozeronii* strains from several patients (Kivistö *et al.*, 2010). Although gastric NPHS (*H. mustelae* and *H. acinonychis*) are often amoxicillin susceptible (MIC <0.25 μg/ml) like *H. pylori*, enterohepatic NPHSs, e.g. *H. rappini*, *H. bilis* and *H. hepaticus*, are often amoxicillin resistant (MICs ≥8 μg/ml; Belzer *et al.*, 2009). In Nepal, most (>91%) *H. hepaticus* strains are susceptible to nitrofurantoin, ciprofloxacin, cephalexin and certriaxone, whereas only 57% of the strains were amoxicillin susceptible in the study of Pradhan (2009). *H. hepaticus hefA* gene, encoding a TolC component of a supposed efflux system, has been linked to resistance to both amoxicillin and non-antibiotic agents (Belzer *et al.*, 2009). Pasquali *et al.* (2007) have reported ciprofloxacin resistance in *H. pullorum* to result from Thr84Ile replacement of GyrA, DNA gyrase, highlighting the risks for transfer of the resistance to other species.

Gastric *Helicobacter* species

Gastric NPHS possibly encompass over six species; however, if no species is identified, they should be reported as 'gastric non-*H. pylori* helicobacters' rather than '*H. heilmannii* group'

(Haesebrouck *et al.*, 2009). Gastric NPHS mainly include *H. heilmannii* type I, i.e. *H. suis* isolated from swine, and type II, including *H. bizzozeronii, H. felis, H. salomonis, H. cynogastricus, H. baculiformis* and '*Ca. Helicobacter heilmannii*', detected in cats and dogs (Baele *et al.*, 2008b; Kivistö *et al.*, 2010). '*Ca. Helicobacter heilmannii*' involves so-far uncultivatable bacteria in stomachs of cats, dogs and humans (Haesebrouck *et al.*, 2009). '*Ca. Helicobacter bovis*', although frequent in the abomasums of cattle, has been rare in human stomach (Haesebrouck *et al.*, 2009).

In natural hosts, the gastric NPHS can cause gastritis, peptic ulcers or tumours. Gastric NPHS infections in dogs and cats have usually been caused by *H. felis* and *H. bizzozeronii* (most often) as well as *H. salomonis* and '*Ca. Helicobacter heilmannii*', while those in pigs have been most often related to *H. suis* (Van den Bulck *et al.*, 2005b; Haesebrouck *et al.*, 2009). *H. cynogastricus* has been found in dogs and *H. baculiformis* has been identified in cats (Haesebrouck *et al.*, 2009).

Gastric NPHS may cause diseases not only in animals but also in humans. Gastric NPHS have been found, although seldom (in 0.1–6%, often in 1–2.3%), in gastric specimens of patients with chronic gastritis, peptic ulcer and MALT lymphoma (Ierardi *et al.*, 2001; Okiyama *et al.*, 2005; Van den Bulck *et al.*, 2005b; Haesebrouck *et al.*, 2009). The prevalence of gastric NPHS in children has been 0.3–1.1% (Sýkora *et al.*, 2004; Boyanova *et al.*, 2007; Qualia *et al.*, 2007). In untreated Bulgarian patients, gastric NPHS have been present in 0.1% of adults and 0.3% of children (Boyanova, 2007; Boyanova *et al.*, 2007). Worryingly, although the gastric NPHS are uncommon in humans, they carry an overall risk for MALT lymphoma up to 10 times higher (1.5–10%) than that by *H. pylori* infection (<1%; Joens *et al.*, 2010).

Within the gastric NPHS, *H. suis* has been detected most often in humans, and hence pigs can be an important source of human infections (Baele *et al.*, 2009). In Belgium, Van den Bulck *et al.* (2005b) have found single *H. suis* infections to be most common (30.9%), followed by *H. salomonis* (11.4%), *H. felis* (8.9%), *H. bizzozeronii* (2.4%) and mixed infections (16.3%), in NPHS-positive patients. However, in Belgian NPHS-positive patients, De Groote *et al.* (2005) detected canine/feline NPHS more often (in 48.5%) than *H. suis* (13.9%) as well as *H. pylori* (12.9%), 'Ca. *H. bovis*' (0.9%) and mixed infections (10.9%). Interestingly, *Helicobacter cetorum*-like DNA has been found in 16% of specimens from healthy Venezuelan subjects (García-Amado *et al.*, 2007). In a Finnish patient with chronic gastritis, colonization by *H. bizzozeronii* lasting over 10 months was reported (Kivistö *et al.*, 2010). Therefore, like *H. pylori*, NPHS can establish long-term infections.

Coinfections of gastric NPHS and *H. pylori* have also been observed. Among NPHS-positive subjects, coinfections were found in one of two untreated Bulgarian adults and two of eight Italian patients (Ierardi *et al.*, 2001; Boyanova *et al.*, 2007). In South Africa, *H. pylori* has been PCR detected in 83.3% of volunteers but *H. felis* has been common, found in 25.6% alone and in 23.3% in coinfections with *H. pylori* (Fritz *et al.*, 2006). Surprisingly, the authors have not found any association between *H. felis* presence and gastric pathology.

In brief, gastric NPHS infections are associated almost always with presence of chronic active gastritis, and, in some patients, with peptic ulcer and MALT lymphoma, not forgetting the higher risk of tumour associated with NPHS infection compared with *H. pylori* infection.

Gastric NPHS infections may be transmitted to humans by direct contact or faecal contamination. Many but not all patients with gastric NPHS infections have reported a previous contact with cats, dogs or animal products (Al-Hamoudi *et al.*, 2006). *Helicobacter* spp. have been detected in oral secretions of almost all (93%) Iranian stray cats, although without any agreement between the presence of the oral and the gastric strains (Shojaee Tabrizi *et al.*, 2010). Transmission of NPHS from a dog to a child with gastritis and that of *H. felis* from a pet dog to an elderly woman with peptic ulcers have been reported (De Bock *et al.*, 2007; Duquenoy and Luyer, 2009). Interestingly, the presence of *H. felis, H. bizzozeronii* and '*Ca. H. heilmannii*' in the dog gastric mucosa has been related only to mild lesions. Water could contribute also to NPHS transmission. Importantly, in water without light exposure at 25°C, *H. felis* has shown short (6–10 h) survival, while other gastric and enterohepatic NPHS (*H. mustelae, H. muridarum, H. canadensis, H. pullorum,* and *H. canis*) have survived for >24 h (Azevedo *et al.*, 2008). For this reason, the role of the water, where the NPHS can be present as viable but non-culturable forms such as in the case of *H. pylori,* merits to be further evaluated.

Gastric NPHS infections can be treated like *H. pylori* infections. Both gastritis and MALT lymphomas have been either improved or cured by NPHS eradication (Haesebrouck *et al.*, 2009). A 7-day triple regimen with lansoprazole, clarithromycin and amoxicillin has eradicated '*Ca. H. heilmannii*' infection in mice and has cured gastric MALT lesions (Matsui *et al.*, 2008). For '*Ca. H. heilmannii*' infection eradication in humans, the following regimens have been used successfully: a 2-week course of lansoprazole, clarithromycin and metronidazole (Yoshimura *et al.*, 2002), a 4-week course of omeprazole, sucralfate, metronidazole and amoxicillin (Sykora *et al.*, 2003) and a 10-day course by omeprazole, amoxicillin and clarithromycin (Al-Hamoudi *et al.*, 2006). However, the infection by a metronidazole-resistant *H. bizzozeroni* strain in a Finnish patient failed to respond to a 7-day treatment with lansoprazole, tetracycline and metronidazole (Kivistö *et al.*, 2010). These data highlight the benefit of susceptibility testing of NPHS to suggest appropriate treatment regimen but also to observe the resistance evolution in the species and possible interspecies transfer of the resistance between NPHS and *H. pylori* in case of mixed infections. The inter-species transfer of resistance to metronidazole and clarithromycin between *H. acinonychis* and *H. pylori* has been proven (Pot *et al.*, 2001) and, given that mixed infections of NPHS and *H. pylori* are common in some developing countries, the impact of the inter-species transfer should be evaluated much more extensively.

Development of vaccines against NPHS has not been successful so far. Intranasal vaccine, containing bacterial cell lysate of *H. suis* or *H. bizzozeronii* and *H. cynogastricus,* has shown incomplete protection of mice (Flahou *et al.*, 2009). Further work on this topic should be undertaken.

Enterohepatic *Helicobacter* species

NPHS have been found in intestines and hepatobiliary tract of animals, e.g. mice (*H. hepaticus, H. ganmani* and *H. bilis*), poultry (*H. pullorum*), dogs (*H. canis*) and Syrian hamsters (*H. cholecystus*), and have been linked to different chronic diseases (Silva *et al.*, 2003). *H. equorum,* whose pathogenicity is still uncertain, is common (67.8%) in foals but rare (<7.9%) in adult horses (Moyaert *et al.*, 2009a,b). Conversely, *H. hepaticus* has been associated with both hepatobiliary (hepatitis, liver tumours, cholesterol gallstones) and intestinal [inflammatory bowel disease (IBD) and colon cancer] diseases in susceptible mice (Taylor *et al.*, 2007). *H. bilis* also has been linked to hepatitis, typhlocolitis, IBD and colon cancer in mice (Taylor *et al.*, 2007; Harbour and Sutton, 2008). Most (80%) of the susceptible mice infected with *H. bilis* have developed cholesterol gallstones (Maurer *et al.*, 2005). In immunocompromised mice, *Helicobacter typhlonius* has been related to typhlocolitis, tumours and IBD, and mixed infection with *H. rodentium* and *H. typhlonius* has caused most severe inflammation (Okoli *et al.*, 2010). *H. pullorum* has been associated with enteritis and hepatitis in chickens and hens. Therefore, NPHS are not only 'normal flora' but causative agents of diseases in their natural host.

NPHS prevalence should be considered. *H. hepaticus, H. bilis, Helicobacter mastomyrinus* and *H. rodentium* or mixed infections have been common (in 88%) in mice from commercial/academic institutions worldwide (Taylor *et al.*, 2007). In healthy and diarrhoeal dogs and cats in Italy, the enterohepatic NPHS isolates have been *H. bilis* (15.3%), *H. canis* (12%) and *H. cinaedi* (5.8%) (Rossi *et al.*, 2008). Importantly, one-third of live Belgian broiler chickens have been *H. pullorum* positive (Ceelen et *al.*, 2006). *H. anseris* and *H. brantae* have been isolated from faeces of Canada geese in 20.6% and 7.2%, respectively (Fox *et al.*, 2006). Obviously, the enterohepatic NPHS are prevalent in their animal hosts and may be transmitted to humans.

Enterohepatic NPHS could be transmitted to humans by faecal–oral route through contaminated water or food (Fox *et al.*, 2006). In Egypt, the contamination of hen eggs has been high, e.g. 20%

by *H. pullorum*, 3.3% by *H. cinaedi* and 1.6% by *H. felis* and, in addition, 23.3% by *H. pylori* (Hameed and Amin, 2010). In Spain, *H. pullorum*-like DNA has been found in 1 of 10 chicken burger samples and in 7 of 10 chicken carcasses (González *et al.*, 2008).

Enterohepatic NPHS infection may play an important role in diseases of the hepatobiliary tract in humans. *H. bilis, H. hepaticus, H. rappini, H. pullorum, H. muridarum, H. canis* and *H. ganmani,* in addition to *H. pylori,* have been found in patients with gallbladder diseases (Bohr *et al.*, 2007; Casswall *et al.*, 2010; Mishra *et al.*, 2010). In these patients, *Helicobacter* species have been more common than in healthy controls. In addition, Estonian adults with chronic liver diseases have exhibited higher antibody titre to *H. bilis/H. hepaticus* compared with controls (Vorobjova *et al.*, 2006). In Japan, patients with pancreaticobiliary maljunction have shown 2.4 times higher prevalence of *H. bilis* in bile specimens, compared with control patients (Kosaka *et al.*, 2010). In the USA, *H. ganmani* 16S rDNA has been present in the liver of 31% of children with chronic liver diseases versus none of the controls (Tolia *et al.*, 2004). *H. pylori* DNA also has been PCR detected in about one-fifth of gallbladder tissues of Turkish patients with cholelithiasis and cholecystitis, although with negative culture (Yucebilgili *et al.*, 2009). DNA of *H. hepaticus, H. muridarum, H. canis* and *H. pylori* has been found by 16S rDNA sequencing in 13% of liver tissues from Swedish children and adolescents with autoimmune hepatitis and ulcerative colitis (Casswall *et al.*, 2010). In addition, NPHS have been detected in cases with autoimmune hepatitis (*H. ganmani* and *H. bilis*) and in those with liver failure and cholestasis (*H. hepaticus* and *H. pullorum;* Casswall *et al.*, 2010). There is an increasing evidence about the probable role of the enterohepatic NPHS in human hepatobiliary diseases. Recently, the participation of *H. rodentium* in hepatobiliary diseases in humans has been also suggested. In Poland, of 97 liver biopsies from patients with chronic liver diseases, Nakonieczna *et al.* (2010) often detected DNA similar to that of *H. rodentium. H. pullorum* has been associated with diarrhoea, gastroenteritis and liver disease in humans (Ceelen *et al.*, 2006). Moreover, it should be further assessed whether the recently reported *H. anseris* and *H. brantae* can cause diseases in humans like *H. canadensis*, isolated from diarrheic and bacteraemic patients (Loman *et al.*, 2009). Conversely, no *H. equorum* DNA has been detected in human stool specimens (Moyaert *et al.*, 2009a), probably because contact of the humans with horses is rare compared with that with poultry products. However, the associations are not universal. For instance, in Belgium, *H. pullorum* DNA has been found in similar (about 4%) proportions in faeces from both patients with gastrointestinal diseases and healthy subjects (Ceelen *et al.*, 2005).

Cancer of the liver and biliary tract has shown low incidence in North America and Europe but high incidence in areas of Africa and Asia (Tischoff and Tannapfel, 2007). A recent immunoblot study has demonstrated three-times higher seroprevalence to *H. hepaticus* in Japanese patients with bile tract cancer (38.8%) compared with healthy controls (13.1%; Shimoyama *et al.*, 2010). Hamada *et al.* (2009) have detected often *H. hepaticus* in bile from Japanese patients with cholelithiasis (41%) and cholecystitis with gastric cancer (36%). Conversely, in Germany, only one of 99 cases has been PCR positive for NPHS (*H. ganmani;* Bohr *et al.*, 2007). The low NPHS prevalence could explain the low incidence of gallbladder cancer in Germany (Bohr *et al.*, 2007). In a review of 15 studies, Pandey and Shukla (2009) have reported 2.8 times higher prevalence of *Helicobacter* spp. in patients with hepatobiliary tract cancers (56%) than in controls (20%). Common *Helicobacter* spp. have been *H. bilis,* followed by *H. pylori* and *H. pullorum* (Pandey and Shukla, 2009). Mechanisms, by which *Helicobacter* spp. cause diseases of hepatobiliary tract in humans should be further clarified (see above).

Possible therapeutic option in human medicine has been suggested by a study on enterohepatic NPHS. *H. hepaticus* infected mice with androgen interruption (castrated male A/JCr mice or those receiving the competitive androgen receptor antagonist, flutamide) have shown milder hepatitis than those in the controls (Theve *et al.*, 2008). Whether a hormonal therapy can improve hepatitis B in young boys, should be evaluated in other studies.

In brief, many studies have associated the PCR-detected enterohepatic NPHS with liver diseases in humans. However, no NPHS has been isolated from human hepatobiliary specimens so far possibly because of their low bacterial number and inhibition by bile. However, NPHS isolation from liver of primates with hepatitis has been reported for '*Helicobacter macacae*' in a baboon and for *H. cinaedi* in a rhesus monkey (Fox *et al.*, 2001; García *et al.*, 2006). The causative relationship between the enterohepatic NPHS and the chronic hepatobiliary diseases in humans is a clinically important topic for future studies.

NPHS also could be associated with chronic intestinal diseases in humans. Enterohepatic and gastric NPHS have been detected by PCR sequencing in most (59%) stool specimens from Australian children with Crohn's disease compared with only 9% of healthy controls (Man *et al.*, 2008). Crohn's disease has been associated with *H. pullorum* and *H. canadensis* in French patients (Laharie *et al.*, 2009). In Germany, enterohepatic NPHS, including *H. fennelliae* and *H. pullorum*, have been found in 12% of patients with Crohn's disease and in 17% of those with ulcerative colitis compared with only 4% of controls (Bohr *et al.*, 2004). In addition to *Mycobacterium avium* subspecies *paratuberculosis* and the adherent-invasive *Escherichia coli*, NPHS have also been suggested as possible IBD pathogens (Hansen *et al.*, 2010). As in the animals, the innate immunity and pro-inflammatory cytokines could participate in IBD pathogenesis in NPHS-infected humans.

Immunocompromised patients are prone to develop NPHS infections. *H. cinaedi*, the most frequently reported enterohepatic NPHS in humans, and *H. canis* can cause severe infections in immunocompromised patients. *H. cinaedi*, found in the intestines of mammals and birds, has been linked to gastroenteritis, proctitis, proctocolitis, cellulitis and bacteraemia in humans (often in immunosuppressed subjects) and to septicaemia and meningitis in a neonate (Holst *et al.*, 2008; Shen *et al.*, 2009). In Japan, the prevalence of *H. cinaedi* in blood cultures has been 0.22% of the positive samples and most positive patients have been immunocompromised or surgically operated (Matsumoto *et al.*, 2007). *H. canis* bacteraemia has been reported in a patient with gastric lymphoma and contact with dogs (Alon *et al.*, 2010). Bacteraemia and cellulitis by *H. canis* have been described in an immunocompetent patient as well (Leemann *et al.*, 2006). Briefly, although causing relatively rare infections in humans, NPHS should be considered as important opportunistic pathogens, mainly for immunocompromised subjects in contacts with animals.

Enterohepatic NPHS can be difficult to be treated. In humans, *H. cinaedi* bacteraemia can require a prolonged therapy because of common recurrences. Uçkay *et al.* (2006) have reported a bacteraemia in a patient caused by *H. cinaedi* strain resistant to erythromycin, trimethoprim–sulfamethoxazole and ciprofloxacin. In this patient, first treatment, consisting of 2 weeks ceftriaxone + gentamicin combination and following 5-week oral combination by clarithromycin + levofloxacin, has failed but after a recurrence, the second therapy, consisting of ceftriaxone + doxycycline for 2 weeks, followed by oral amoxicillin + metronidazole + doxycycline for 2 months and doxycycline for another month, has been curative.

Treatment of the infections in animals can fail as well. In a review, Kerton and Warden (2006) have reported amoxicillin-based triple therapy to be successful in 50–100% of mice depending on the route of administration. To eradicate *H. bilis* and *H. rodentium* in rats, triple amoxicillin-based regimen has been unsuccessful, while quadruple regimen with omeprazole, amoxicillin, metronidazole and clarithromycin has been successful possibly because of the presence of two antibiotics in addition to the amoxicillin (Whary and Fox, 2006; Kostomitsopoulos *et al.*, 2007). Resistance in enterohepatic NPHS, especially that to amoxicillin, should be taken into account to choose treatment regimens. It should be also considered that the immunocompromised state of the affected patients could hamper the cure, which may require aggressive and prolonged therapy.

In brief, the number of NPHS constantly increases. Many species possess numerous virulence factors and some have carcinogenic properties, and may cause severe diseases not only in animals but also in humans. In addition to *H. pylori*, gastric NPHS have been associated with chronic gastritis, peptic ulcers and, importantly, MALT lymphoma in humans, whereas enterohepatic NPHS, summarizing the growing evidence from many studies, may contribute to the development of chronic hepatobiliary diseases, including tumours, and chronic enteric diseases. Further studies on isolation of still uncultured species and on the prevalence, pathogenesis and significance of NPHS, especially their carcinogenic properties in humans, as well as on treatment or prophylaxis of the associated infections in both animals and humans would be of high medical interest.

References

Al-Hamoudi, W.K., Alpert, L., and Szilagyi, A. (2006). Acute symptomatic gastritis due to *Helicobacter heilmannii*. Helicobacter *11*, 446–450.

Alispahic, M., Hummel, K., Jandreski-Cvetkovic, D., Nöbauer, K., Razzazi-Fazeli, E., Hess, M., and Hess, C. (2010). Species-specific identification and differentiation of *Arcobacter, Helicobacter* and *Campylobacter* by full-spectral matrix-associated laser desorption/ionization time of flight mass spectrometry analysis. J. Med. Microbiol. *59*, 295–301.

Alon, D., Paitan, Y., Ben-Nissan, Y., and Chowers, M. (2010). Persistent *Helicobacter canis* bacteremia in a patient with gastric lymphoma. Infection *38*, 62–64.

Alyamani, E.J., Brandt, P., Pena, J.A., Major, A.M., Fox, J.G., Suerbaum, S., and Versalovic, J. (2007). *Helicobacter hepaticus* catalase shares surface-predicted epitopes with mammalian catalases. Microbiology *153*, 1006–1016.

Azevedo, N.F., Almeida, C., Fernandes, I., Cerqueira, L., Dias, S., Keevil, C.W., and Vieira, M.J. (2008). Survival of gastric and enterohepatic *Helicobacter* spp. in water: implications for transmission. Appl. Environ. Microbiol. *74*, 1805–1811.

Baele, M., Van den Bulck, K., Decostere, A., Vandamme, P., Hänninen, M.L., Ducatelle, R., and Haesebrouck, F. (2004). Multiplex PCR assay for differentiation of *Helicobacter felis, H. bizzozeronii*, and *H. salomonis*. J. Clin. Microbiol. *42*, 1115–1122.

Baele, M., Decostere, A., Vandamme, P., Van den Bulck, K., Gruntar, I., Mehle, J., Mast, J., Ducatelle, R., and Haesebrouck, F. (2008a). *Helicobacter baculiformis* sp. nov., isolated from feline stomach mucosa. Int. J. Syst. Evol. Microbiol. *58*, 357–364.

Baele, M., Decostere, A., Vandamme, P., Ceelen, L., Hellemans, A., Mast, J., Chiers, K., Ducatelle, R., Haesebrouck, F., Baele, M., *et al.* (2008b). Isolation and characterization of *Helicobacter suis* sp. nov. from pig stomachs. Int. J. Syst. Evol. Microbiol. *58*, 1350–1358.

Baele, M., Pasmans, F., Flahou, B., Chiers, K., Ducatelle, R., and Haesebrouck, F. (2009). Non-*Helicobacter pylori* helicobacters detected in the stomach of humans comprise several naturally occurring *Helicobacter* species in animals. FEMS Immunol. Med. Microbiol. *55*, 306–313.

Belzer, C., Stoof, J., Beckwith, C.S., Kuipers, E.J., Kusters, J.G., and van Vliet, A.H. (2005). Differential regulation of urease activity in *Helicobacter hepaticus* and *Helicobacter pylori*. Microbiology *151*, 3989–3995.

Belzer, C., Stoof, J., Breijer, S., Kusters, J.G., Kuipers, E.J., and van Vliet, A.H. (2009). The *Helicobacter hepaticus hefA* gene is involved in resistance to amoxicillin. Helicobacter *14*, 72–79.

Bohr, U.R., Glasbrenner, B., Primus, A., Zagoura, A., Wex, T., and Malfertheiner, P. (2004). Identification of enterohepatic *Helicobacter* species in patients suffering from inflammatory bowel disease. J. Clin. Microbiol. *42*, 2766–2768.

Bohr, U.R., Kuester, D., Meyer, F., Wex, T., Stillert, M., Csepregi, A., Lippert, H., Roessner, A., and Malfertheiner, P. (2007). Low prevalence of *Helicobacteraceae* in gall-stone disease and gall-bladder carcinoma in the German population. Clin. Microbiol. Infect. *13*, 525–531.

Boyanova, L., Lazarova, E., Jelev, C., Gergova, G., and Mitov, I. (2007). *Helicobacter pylori* and *Helicobacter heilmannii* in untreated Bulgarian children over a period of 10 years. J. Med. Microbiol. *56*, 1081–1085.

Boyanova, L. (2007). Detection of *Helicobacter pylori* infection in symptomatic Bulgarian adults. Clin. Microbiol. Infect. *13*, 908–914.

Casswall, T.H., Németh, A., Nilsson, I., Wadström, T., and Nilsson, H.O. (2010). *Helicobacter* species DNA in liver and gastric tissues in children and adolescents with chronic liver disease. Scand. J. Gastroenterol. *45*, 160–167.

Ceelen, L., Decostere, A., Verschraegen, G., Ducatelle, R., and Haesebrouck, F. (2005). Prevalence of *Helicobacter pullorum* among patients with gastrointestinal disease and clinically healthy persons. J. Clin. Microbiol. *43*, 2984–2986.

Ceelen, L.M., Decostere, A., Van den Bulck, K., On, S.L., Baele, M., Ducatelle, R., and Haesebrouck, F. (2006). *Helicobacter pullorum* in chickens, Belgium. Emerg. Infect. Dis. *12*, 263–267.

Dailidiene, D., Dailide, G., Ogura, K., Zhang, M., Mukhopadhyay, A.K., Eaton, K.A., Cattoli, G., Kusters, J.G., and Berg, D.E. (2004). *Helicobacter acinonychis*: genetic and rodent infection studies of a *Helicobacter pylori*-like gastric pathogen of cheetahs and other big cats. J. Bacteriol. *186*, 356–365.

De Bock, M., Decostere, A., Hellemans, A., Haesebrouck, F., and Ducatelle, R. (2006). *Helicobacter felis* and *Helicobacter bizzozeronii* induce gastric parietal cell loss in Mongolian gerbils. Microbes Infect. *8*, 503–510.

De Bock, M., Van den Bulck, K., Hellemans, A., Daminet, S., Coche, J.C., Debongnie, J.C., Decostere, A., Haesebrouck, F., and Ducatelle, R. (2007). Peptic ulcer disease associated with *Helicobacter felis* in a dog owner. Eur. J. Gastroenterol. Hepatol. *19*, 79–82.

De Groote, D., Van Doorn, L.J., Van den Bulck, K., Vandamme, P., Vieth, M., Stolte, M., Debongnie, J.C., Burette, A., Haesebrouck, F., and Ducatelle, R. (2005). Detection of non-pylori *Helicobacter* species in '*Helicobacter heilmannii*'-infected humans. Helicobacter *10*, 398–406.

Dewhirst, F.E., Shen, Z., Scimeca, M.S., Stokes, L.N., Boumenna, T., Chen, T., Paster, B.J., and Fox, J.G. (2005). Discordant 16S and 23S rRNA gene phylogenies for the genus *Helicobacter*: implications for phylogenetic inference and systematics. J. Bacteriol. *187*, 6106–6118.

Dong, Q.J., Wang, Q., Xin, Y.N., Li, N., and Xuan, S.Y. (2009). Comparative genomics of *Helicobacter pylori*. World J. Gastroenterol. *15*, 3984–3991.

Duquenoy, A., and Le Luyer, B. (2009). Gastritis caused by *Helicobacter heilmannii* probably transmitted from dog to child. Arch. Pediatr. *16*, 426–429.

Eppinger, M., Baar, C., Linz, B., Raddatz, G., Lanz, C., Keller, H., Morelli, G., Gressmann, H., Achtman, M., and Schuster, S.C. (2006). Who ate whom? Adaptive *Helicobacter* genomic changes that accompanied a host jump from early humans to large felines. PLoS Genet. *2*, e120.

Every, A.L., Chionh, Y.T., Skene, C.D., McGuckin, M.A., and Sutton, P. (2008). Muc1 limits *Helicobacter felis* binding to gastric epithelial cells but does not limit colonization and gastric pathology following infection. Helicobacter *13*, 489–493.

Flahou, B., Hellemans, A., Meyns, T., Duchateau, L., Chiers, K., Baele, M., Pasmans, F., Haesebrouck, F., and Ducatelle, R. (2009). Protective immunization with homologous and heterologous antigens against *Helicobacter suis* challenge in a mouse model. Vaccine *27*, 1416–1421.

Fox, J.G., Handt, L., Sheppard, B.J., Xu, S., Dewhirst, F.E., Motzel, S., and Klein, H. (2001). Isolation of *Helicobacter cinaedi* from the colon, liver, and mesenteric lymph node of a rhesus monkey with chronic colitis and hepatitis. J. Clin. Microbiol. *39*, 1580–1585.

Fox, J.G., Taylor, N.S., Howe, S., Tidd, M., Xu, S., Paster, B.J., and Dewhirst, F.E. (2006). *Helicobacter anseris* sp. nov. and *Helicobacter brantae* sp. nov., isolated from feces of resident Canada geese in the greater Boston area. Appl. Environ. Microbiol. 72, 4633–4637.

Fox, J.G., Boutin, S.R., Handt, L.K., Taylor, N.S., Xu, S., Rickman, B., Marini, R.P., and Dewhirst, F.E., Paster, B.J., Motzel, S., *et al.* (2007). Isolation and characterization of a novel helicobacter species, '*Helicobacter macacae*,' from rhesus monkeys with and without chronic idiopathic colitis. J. Clin. Microbiol. *45*, 4061–4063.

Fox, J.G., Feng, Y., Theve, E.J., Raczynski, A.R., Fiala, J.L., Doernte, A.L., Williams, M., McFaline, J.L., Essigmann, J.M., Schauer, D.B., *et al.* (2010). Gut microbes define liver cancer risk in mice exposed to chemical and viral transgenic hepatocarcinogens. Gut 59, 88–97.

Fritz, E.L., Slavik, T., Delport, W., Olivier, B., and van der Merwe, S.W. (2006). Incidence of *Helicobacter felis* and the effect of coinfection with *Helicobacter pylori* on the gastric mucosa in the African population. J. Clin. Microbiol. *44*, 1692–1696.

García, A., Xu, S., Dewhirst, F.E., Nambiar, P.R., and Fox, J.G. (2006). Enterohepatic *Helicobacter* species isolated from the ileum, liver and colon of a baboon with pancreatic islet amyloidosis. J. Med. Microbiol. *55*, 1591–1595.

García-Amado, M.A., Al-Soud, W.A., Borges-Landaéz, P., Contreras, M., Cedeño, S., Baéz-Ramírez, E., Domínguez-Bello, M.G., Wadström, T., and Gueneau, P. (2007). Non-*pylori Helicobacteraceae* in the upper digestive tract of asymptomatic Venezuelan subjects: detection of *Helicobacter cetorum*-like and Candidatus *Wolinella africanus*-like DNA. Helicobacter *12*, 553–558.

Garrity, G. M., Bell, J.A., and Lilburn, T. (2005). Family II. *Helicobacteraceae* fam. nov. In Bergey's Manual of Systematic Bacteriology, 2nd ed., vol. 2 (The *Proteobacteria*), part C (The *Alpha-*, *Beta-*, *Delta-*, and *Epsilonproteobacteria*), Brenner, D.J., Krieg, N.R., Staley J.T., and Garrity, G.M., eds. (Springer: New York), p. 1168.

Ge, Z., Sterzenbach, T., Whary, M.T., Rickman, B.H., Rogers, A.B., Shen, Z., Taylor, N.S., Schauer, D.B., Josenhans, C., Suerbaum, S., *et al.* (2008a). *Helicobacter hepaticus* HHGI1 is a pathogenicity island associated with typhlocolitis in B6.129-IL10 tm1Cgn mice. Microbes Infect. *10*, 726–733.

Ge, Z., Lee, A., Whary, M.T., Rogers, A.B., Maurer, K.J., Taylor, N.S., Schauer, D.B., and Fox, J.G. (2008b). *Helicobacter hepaticus* urease is not required for intestinal colonization but promotes hepatic inflammation in male A/JCr mice. Microb. Pathog. *45*, 18–24.

González, A., Piqueres, P., Moreno, Y., Cañigral, I., Owen, R.J., Hernández, J., and Ferrús, M.A. (2008). A novel real-time PCR assay for the detection of *Helicobacter pullorum*-like organisms in chicken products. Int. Microbiol. *11*, 203–208.

Haesebrouck, F., Pasmans, F., Flahou, B., Chiers, K., Baele, M., Meyns, T., Decostere, A., and Ducatelle, R. (2009). Gastric helicobacters in domestic animals and nonhuman primates and their significance for human health. Clin. Microbiol. Rev. 22, 202–223.

Hameed, K.G.A., and Amin, W.F. (2010). Using of PCR assay for identification of *Helicobacter* species in hens' eggs. Vet. World 3, 404–408.

Hamada, T., Yokota, K., Ayada, K., Hirai, K., Kamada, T., Haruma, K., Chayama, K., and Oguma, K. (2009). Detection of *Helicobacter hepaticus* in human bile samples of patients with biliary disease. Helicobacter *14*, 545–551.

Hänninen, M.L., Kärenlampi, R.I., Koort, J.M., Mikkonen, T., and Björkroth, K.J. (2005). Extension of the species *Helicobacter bilis* to include the reference strains of *Helicobacter* sp. flexispira taxa 2, 3 and 8 and Finnish canine and feline flexispira strains. Int. J. Syst. Evol. Microbiol. 55, 891–898.

Hannula, M., and Hänninen, M.L. (2007). Phylogenetic analysis of *Helicobacter* species based on partial *gyrB* gene sequences. Int. J. Syst. Evol. Microbiol. *57*, 444–449.

Hansen, R., Thomson, J.M., El-Omar, E.M., and Hold, G.L. (2010). The role of infection in the aetiology of inflammatory bowel disease. J. Gastroenterol. *45*, 266–276.

Harbour, S., and Sutton, P. (2008). Immunogenicity and pathogenicity of *Helicobacter* infections of veterinary animals. Vet. Immunol. Immunopathol. *122*, 191–203.

Harper, C.G., Whary, M.T., Feng, Y., Rhinehart, H.L., Wells, R.S., Xu, S., Taylor, N.S., and Fox, J.G. (2003). Comparison of diagnostic techniques for *Helicobacter cetorum* infection in wild Atlantic bottlenose dolphins (*Tursiops truncatus*). J. Clin. Microbiol. *41*, 2842–2848.

Hayashi, S., Shimomura, H., and Hirai, Y. (2009). Latest advances in non-*pylori Helicobacter* species. Nippon Rinsho *67*, 2271–2278.

Holst, H., Andresen, K., Blom, J., Højlyng, N., Kemp, M., Krogfelt, K.A., and Christensen, J.J. (2008). A case of *Helicobacter cinaedi* bacteraemia in a previously healthy person with cellulitis. Open Microbiol. J. *2*, 29–31.

Hong, Y., Wang, G., and Maier, R.J. (2007). A *Helicobacter hepaticus* catalase mutant is hypersensitive to oxidative stress and suffers increased DNA damage. J. Med. Microbiol. *56*, 557–562.

Hong, Y., Wang, G., and Maier, R.J. (2008). The NADPH quinone reductase MdaB confers oxidative stress resistance to *Helicobacter hepaticus*. Microb. Pathog. *44*, 169–174.

Ierardi, E., Monno, R.A., Gentile, A., Francavilla, R., Burattini, O., Marangi, S., Pollice, L., and Francavilla, A. (2001). *Helicobacter heilmannii* gastritis: a histological and immunohistochemical trait. J. Clin. Pathol. *54*, 774–777.

Joens, L.A., Haeselbrouck, F., and Pasmans, F. (2010). *Campylobacter* and Helicobacter In Pathogenesis of Bacterial Infections in Animals, 4th ed, Gyles, C. L., Prescott, J.F., Songer, G., and Thoen, C.O., eds. (Wiley-Blackwell: Ames, Iowa, USA), pp. 485–502.

Kerton, A., and Warden, P. (2006). Review of successful treatment for *Helicobacter* species in laboratory mice. Lab. Anim. *40*, 115–122.

Kivistö, R., Linros, J., Rossi, M., Rautelin, H., and Hänninen, M.-L. (2010). Characterization of multiple *Helicobacter bizzozeronii* isolates from a Finnish patient with severe dyspeptic symptoms and chronic active gastritis. Helicobacter *15*, 58–66.

Kosaka, T., Tajima, Y., Kuroki, T., Mishima, T., Adachi, T., Tsuneoka, N., Fukuda, K., and Kanematsu, T. (2010). *Helicobacter bilis* colonization of the biliary system in patients with pancreaticobiliary maljunction. Br. J. Surg. *97*, 544–549.

Kostomitsopoulos, N., Donnelly, H., Kostavasili, I., Paronis, E., Alexakos, P., and Karayannacos, P. (2007). Eradication of *Helicobacter bilis* and *H. hepaticus* from infected mice by using a medicated diet. Lab. Anim. (NY). 2007, *36*, 37–40.

Kusters, J.G., van Vliet, A.H.M., and Kuipers, E.J. (2006). Pathogenesis of *Helicobacter pylori* infection. Clin. Microbiol. Rev. *19*, 449–490.

Laharie, D., Asencio, C., Asselineau, J., Bulois, P., Bourreille, A., Moreau, J, Bonjean, P., Lamarque, D., Pariente, A., Soulé, J.C., *et al.* (2009). Association between entero-hepatic *Helicobacter* species and Crohn's disease: a prospective cross-sectional study. Aliment. Pharmacol. Ther. *30*, 283–293.

Leemann, C., Gambillara, E., Prod'hom, G., Jaton, K., Panizzon, R., Bille, J., Francioli, P., Greub, G., Laffitte, E., and Tarr, P.E. (2006). First case of bacteremia and multifocal cellulitis due to *Helicobacter canis* in an immunocompetent patient. J. Clin. Microbiol. *44*, 4598–4600.

Lemke, L.B., Ge, Z., Whary, M.T., Feng, Y., Rogers, A.B., Muthupalani, S., and Fox, J.G. (2009). Concurrent *Helicobacter bilis* infection in C57BL/6 mice attenuates pro-inflammatory *H. pylori*-induced gastric pathology. Infect. Immun. *77*, 2147–2158.

Liyanage, N.P., Manthey, K.C., Dassanayake, R.P., Kuszynski, C.A., Oakley, G.G., and Duhamel, G.E. (2010). *Helicobacter hepaticus* cytolethal distending toxin causes cell death in intestinal epithelial cells via mitochondrial apoptotic pathway. Helicobacter *15*, 98–107.

Loman, N.J., Snyder, L.A., Linton, J.D., Langdon, R., Lawson, A.J., Weinstock, G.M., Wren, B.W., and Pallen MJ. (2009). Genome sequence of the emerging pathogen *Helicobacter canadensis*. J. Bacteriol. *191*, 5566–5567.

Man, S.M., Zhang, L., Day, A.S., Leach, S., and Mitchell, H. (2008). Detection of enterohepatic and gastric helicobacter species in fecal specimens of children with Crohn's disease. Helicobacter *13*, 234–238.

Matsui, H., Aikawa, C., Sekiya, Y., Takahashi, S., Murayama, S.Y., and Nakamura, M. (2008). Evaluation of antibiotic therapy for eradication of 'Candidatus *Helicobacter heilmannii*'. Antimicrob. Agents Chemother. *52*, 2988–2989.

Matsumoto, T., Goto, M., Murakami, H., Tanaka, T., Nishiyama, H., Ono, E., Okada, C., Sawabe, E., Yagoshi, M., Yoneyama, A., *et al.* (2007). Multicenter study to evaluate bloodstream infection by *Helicobacter cinaedi* in Japan. J. Clin. Microbiol. *45*, 2853–2857.

Maurer, K.J., Ihrig, M.M., Rogers, A.B., Ng, V., Bouchard, G., Leonard, M.R., Carey, M.C., and Fox, J.G. (2005). Identification of cholelithogenic enterohepatic helicobacter species and their role in murine cholesterol gallstone formation. Gastroenterology *128*, 1023–1033.

Mehta, N.S., Benoit, S., Mysore, J.V., Sousa, R.S., and Maier, R.J. (2005). *Helicobacter hepaticus* hydrogenase mutants are deficient in hydrogen-supported amino acid uptake and in causing liver lesions in A/J mice. Infect. Immun. *73*, 5311–5318.

Mishra, R.R., Tewari, M., and Shukla, H.S. (2010). *Helicobacter* species and pathogenesis of gallbladder cancer. Hepatobiliary Pancreat. Dis. Int. *9*, 129–134.

Moura, S.B., Queiroz, D.M., Rocha, G.A., Comunian, L.B., and Cara, D.C. (2003). Hepatic changes in mice chronically infected with *Helicobacter trogontum*. Braz. J. Med. Biol. Res. *36*, 1209–1213.

Moyaert, H., Decostere, A., Vandamme, P., Debruyne, L., Mast, J., Baele, M., Ceelen, L., Ducatelle, R., and Haesebrouck, F. (2007). *Helicobacter equorum* sp. nov., a urease-negative *Helicobacter* species isolated from horse faeces. Int. J. Syst. Evol. Microbiol. *57*, 213–218.

Moyaert, H., Pasmans, F., Ducatelle, R., Haesebrouck, F., and Baele, M. (2008). Evaluation of 16S rRNA gene-based PCR assays for genus-level identification of *Helicobacter* species. J. Clin. Microbiol. *46*, 1867–1869.

Moyaert, H., Pasmans, F., Decostere, A., Ducatelle, R., and Haesebrouck, F. (2009a). *Helicobacter equorum*: prevalence and significance for horses and humans. FEMS Immunol. Med. Microbiol. *57*, 14–16.

Moyaert, H., Haesebrouck, F., Dewulf, J., Ducatelle, R., and Pasmans, F. (2009b). *Helicobacter equorum* is highly prevalent in foals. Vet Microbiol. *133*, 190–192.

Nagamine, C.M., Sohn, J.J., Rickman, B.H., Rogers, A.B., Fox, J.G., and Schauer, D.B. (2008). *Helicobacter hepaticus* infection promotes colon tumorigenesis in the BALB/c-Rag2(–/–) Apc(Min/+) mouse. Infect. Immun. *76*, 2758–2766.

Nakonieczna, J., Stalke, P., Al-Soud, W.A., Wadström, T., and Bielawski, K.P. (2010). Detection of *Helicobacter rodentium*-like DNA in the liver tissue of patients with chronic liver diseases by polymerase chain reaction-denaturing gradient gel electrophoresis and DNA sequence analysis. Diagn. Microbiol. Infect. Dis. *68*, 201–207.

Nebbia, P., Tramuta, C., Ortoffi, M., Bert, E., Cerruti Sola, S., and Robino, P. (2007). Identification of enteric *Helicobacter* in avian species. Schweiz. Arch. Tierheilkd. *149*, 403–407.

Okiyama, Y., Matsuzawa, K., Hidaka, E., Sano, K., Akamatsu, T., and Ota, H. (2005). *Helicobacter heilmannii* infection: clinical, endoscopic and histopathological features in Japanese patients. Pathol. Int. *55*, 398–404.

Okoli, A.S., Menard, A., and Mendz, G.L. (2009). *Helicobacter* spp. other than *Helicobacter pylori*. Helicobacter *14 (Suppl. 1)*, 69–74.

Okoli, A.S., Wilkins, M.R., Raftery, M.J., and Mendz, G.L. (2010). Response of *Helicobacter hepaticus* to bovine bile. J. Proteome Res. *9*, 1374–1384.

O'Rourke, J., and Bode, G. (2001). Morphology and ultrastructure. In *Helicobacter pylori*: Physiology and Genetics, Mobley, H., and Mendz, G., eds. (ASM Press: Washington DC), pp. 53–67.

O'Toole, P.W., Snelling, W.J., Canchaya, C., Forde, B.M., Hardie, K.R., Josenhans, C., Graham, R.L.J., McMullan, G., Parkhill, J., and Bentley, S.D. (2010). Comparative genomics and proteomics of *Helicobacter mustelae*, an ulcerogenic and carcinogenic gastric pathogen. BMC Genomics *11*, 164.

Pandey, M., and Shukla, M. (2009). *Helicobacter* species are associated with possible increase in risk of hepatobiliary tract cancers. Surg. Oncol. *18*, 51–56.

Pasquali, F., Rossi, M., Manfreda, G., and Zanoni, R. (2007). Complete nucleotide sequence of the *gyrA* gene of *Helicobacter pullorum* and identification of a point mutation leading to ciprofloxacin resistance in poultry isolates. Int. J. Antimicrob. Agents. *30*, 222–228.

Pellicano, R., Ménard, A., Rizzetto, M., and Mégraud, F. (2008). *Helicobacter* species and liver diseases: association or causation? Lancet Infect. Dis. *8*, 254–260.

Pisani, P., Whary, M.T., Nilsson, I., Sriamporn, S., Wadström, T., Fox, J.G., Ljungh, A., and Forman, D. (2008). Cross-reactivity between immune responses to *Helicobacter bilis* and *Helicobacter pylori* in a population in Thailand at high risk of developing cholangiocarcinoma. Clin. Vaccine Immunol. *15*, 1363–1368.

Pot, R.G., Kusters, J.G., Smeets, L.C., Van Tongeren, W., Vandenbroucke-Grauls, C.M., and Bart, A. (2001). Interspecies transfer of antibiotic resistance between *Helicobacter pylori* and *Helicobacter acinonychis*. Antimicrob. Agents Chemother. *45*, 2975–2976.

Pot, R.G., Stoof, J., Nuijten, P.J., de Haan, L.A., Loeffen, P., Kuipers, E.J., van Vliet, A.H., and Kusters, J.G. (2007). UreA2B2: a second urease system in the gastric pathogen *Helicobacter felis*. FEMS Immunol. Med. Microbiol. *50*, 273–279.

Pradhan, S.B. (2009). Study of *Helicobacter hepaticus* in gallbladders with cholelithiasis and its sensitivity pattern. Kathmandu Univ. Med. J. (KUMJ) *7*, 125–128.

Pratt, J.S., Sachen, K.L., Wood, H.D., Eaton, K.A., and Young, V.B. (2006). Modulation of host immune responses by the cytolethal distending toxin of *Helicobacter hepaticus*. Infect. Immun.*74*, 4496–4504.

Qualia, C.M., Katzman, P.J., Brown, M.R., and Kooros, K. (2007). A report of two children with *Helicobacter heilmannii* gastritis and review of the literature. Pediatr. Dev. Pathol. *10*, 391–394.

Rossi, M., Hänninen, M.L., Revez, J., Hannula, M., and Zanoni, R.G. (2008). Occurrence and species level diagnostics of *Campylobacter* spp., enteric *Helicobacter* spp. and *Anaerobiospirillum* spp. in healthy and diarrheic dogs and cats. Vet. Microbiol. *129*, 304–314.

Shen, Z., Feng, Y., Rogers, A.B., Rickman, B., Whary, M.T., Xu, S., Clapp, K.M., Boutin, S.R., and Fox, J.G. (2009). Cytolethal distending toxin promotes *Helicobacter cinaedi*-associated typhlocolitis in interleukin-10-deficient mice. Infect. Immun. *77*, 2508–2516.

Shimoyama, T., Takahashi, R., Abe, D., Mizuki, I., Endo, T., and Fukuda, S. (2010). Serological analysis of *Helicobacter hepaticus* infection in patients with biliary and pancreatic diseases. J. Gastroenterol. Hepatol. *25 (Suppl 1)*, S86-S89.

Shiohara, M., Kawakubo, M., Matsumoto, T., Kumagai, T., Yamauchi, K., Oana, K., Ota, H., and Kawakami, Y. (2009). Laboratory appraisal of optimal gaseous conditions for growth of zoonotic *Helicobacter felis* ATCC 49179. Microbiol. Immunol. *53*, 251–258.

Shojaee Tabrizi, A., Jamshidi, S., Oghalaei, A., Zahraei Salehi, T., Bayati Eshkaftaki, A., and Mohammadi, M. (2010). Identification of *Helicobacter* spp. in oral secretions vs. gastric mucosa of stray cats. Vet. Microbiol. *140*, 142–146.

Silva, C.P., Pereira-Lima, J.C., Oliveira, A.G., Guerra, J.B., Marques, D.L., Sarmanho, L., Cabral, M.M., and Queiroz, D.M. (2003). Association of the presence of *Helicobacter* in gallbladder tissue with cholelithiasis and cholecystitis. J. Clin. Microbiol. *41*, 5615–5618.

Skene, C., Young, A., Every, A., and Sutton, P. (2007). *Helicobacter pylori* flagella: antigenic profile and protective immunity. FEMS Immunol. Med. Microbiol. *50*, 249–256.

Stacy, B.A., and Wellehan, J.F.Jr. (2010). Fatal septicemia caused by *Helicobacter* infection in a pancake tortoise (*Malacochersus tornieri*). J. Vet. Diagn. Invest. *22*, 660–662.

Sterzenbach, T., Lee, S.K., Brenneke, B., von Goetz, F., Schauer, D.B., Fox, J.G., Suerbaum, S., and Josenhans, C. (2007). Inhibitory effect of enterohepatic *Helicobacter hepaticus* on innate immune responses of mouse intestinal epithelial cells. Infect. Immun. *75*, 2717–2728.

Sterzenbach, T., Bartonickova, L., Behrens, W., Brenneke, B., Schulze, J., Kops, F., Chin, E.Y., Katzowitsch, E., Schauer, D.B., Fox, J.G., *et al.* (2008). Role of the *Helicobacter hepaticus* flagellar sigma factor *FliA* in gene regulation and murine colonization. J. Bacteriol.*190*, 6398–6408.

Suerbaum, S., Josenhans, C., Sterzenbach, T., Drescher, B., Brandt, P., Bell, M., Droge, M., Fartmann, B., Fischer, H.P., Ge, Z., *et al.* (2003). The complete genome sequence of the carcinogenic bacterium *Helicobacter hepaticus*. Proc. Natl. Acad. Sci. U.S.A. *100*, 7901–7906.

Sykora, J., Hejda, V., Varvarovska, J., Stozicky, F., Gottrand, F., and Siala, K. (2003). *Helicobacter heilmannii* related gastric ulcer in childhood. J. Pediatr. Gastroenterol. Nutr. *36*, 410–413.

Sýkora, J., Hejda, V., Varvarovská, J., Stozický, F., Siala, K., and Schwarz, J. (2004). *Helicobacter heilmannii* gastroduodenal disease and clinical aspects in children with dyspeptic symptoms. Acta Paediatr. *93*, 707–709.

Takayama, S., Takahashi, H., Matsuo, Y., Okada, Y., and Takeyama, H. (2010). Effect of *Helicobacter bilis* infection on human bile duct cancer cells. Dig. Dis. Sci. *55*, 1905–1910.

Taylor, N.S., Xu, S., Nambiar, P., Dewhirst, F.E., and Fox, J.G. (2007). Enterohepatic *Helicobacter* species are prevalent in mice from commercial and academic institutions in Asia, Europe, and North America. J. Clin. Microbiol. *45*, 2166–2172.

Testerman, T.L., Conn, P.B., Mobley, H.L., and McGee, D.J. (2006). Nutritional requirements and antibiotic resistance patterns of *Helicobacter* species in chemically defined media. J. Clin. Microbiol. *44*, 1650–1658.

Theve, E.J., Feng, Y., Taghizadeh, K., Cormier, K.S., Bell, D.R., Fox, J.G., and Rogers, A.B. (2008). Sex hormone influence on hepatitis in young male A/JCr mice infected with *Helicobacter hepaticus*. Infect. Immun. *76*, 4071–4078.

Tischoff, I., and Tannapfel, A. (2007). Hepatocellular carcinoma and cholangiocarcinoma – different prognosis, pathogenesis and therapy. Zentralbl. Chir. *132*, 300–305.

Tolia, V., Nilsson, H.O., Boyer, K., Wuerth, A., Al-Soud, W.A., Rabah, R., and Wadström, T. (2004). Detection of *Helicobacter ganmani*-like 16S rDNA in pediatric liver tissue. Helicobacter *9*, 460–468.

Uçkay, I., Garbino, J., Dietrich, P.Y., Ninet, B., Rohner, P., and Jacomo, V. (2006). Recurrent bacteremia with *Helicobacter cinaedi*: case report and review of the literature. BMC Infect. Dis. *6*, 86.

Van den Bulck, K., Decostere, A., Gruntar, I., Baele, M., Krt, B., Ducatelle, R., and Haesebrouck, F. (2005a). *In vitro* antimicrobial susceptibility testing of *Helicobacter felis, H. bizzozeronii,* and *H. salomonis*. Antimicrob. Agents Chemother. *49*, 2997–3000.

Van den Bulck, K., Decostere, A., Baele, M., Driessen, A., Debongnie, J.C., Burette, A., Stolte, M., Ducatelle, R., and Haesebrouck, F. (2005b). Identification of non-*Helicobacter pylori* spiral organisms in gastric samples from humans, dogs, and cats. J. Clin. Microbiol. *43*, 2256–2260.

Varon, C., Duriez, A., Lehours, P., Ménard, A., Layé, S., Zerbib, F., Mégraud, F., and Laharie, D. (2009). Study of *Helicobacter pullorum* pro-inflammatory properties on human epithelial cells *in vitro*. Gut *58*, 629–635.

Veijola, L., Nilsson, I., Halme, L., Al-Soud, W.A., Mäkinen, J., Ljungh, A., and Rautelin, H. (2007). Detection of *Helicobacter* species in chronic liver disease and chronic inflammatory bowel disease. Ann. Med. *39*, 554–560.

Vorobjova, T., Nilsson, I., Terjajev, S., Granholm, M., Lyyra, M., Porkka, T., Prükk, T., Salupere, R., Maaroos, H.I., Wadström, T., *et al.* (2006). Serum antibodies to enterohepatic *Helicobacter* spp. in patients with chronic liver diseases and in a population with high prevalence of *H. pylori* infection. Dig. Liver Dis. *38*, 171–176.

Wadström, T., Hau, J., Nilsson, I., and Ljungh, A. (2009). Immunoblot analysis as an alternative method to diagnose enterohepatic *Helicobacter* infections. Helicobacter *14*, 172–176.

Whary, M.T., and Fox, J.G. (2006). Detection, eradication, and research implications of *Helicobacter* infections in laboratory rodents. Lab. Anim. (NY). *35*, 30–36.

Whary, M.T., Danon, S.J., Feng, Y., Ge, Z., Sundina, N., Ng, V., Taylor, N.S., Rogers, A.B., and Fox, J.G. (2006). Rapid onset of ulcerative typhlocolitis in B6.129P2-IL10tm1Cgn (IL-10-/-) mice infected with *Helicobacter trogontum* is associated with decreased colonization by altered Schaedler's flora. Infect. Immun. *74*, 6615–6623.

Yoshimura, M., Isomoto, H., Shikuwa, S., Osabe, M., Matsunaga, K., Omagari, K., Mizuta, Y., Murase, K., Murata, I., and Kohno, S. (2002). A case of acute gastric lesions associated with *Helicobacter heilmannii* infection. Helicobacter *7*, 322–326.

Yucebilgili, K., Mehmetoğlu, T., Gucin, Z., and Salih, B.A. (2009). *Helicobacter pylori* DNA in gallbladder tissue of patients with cholelithiasis and cholecystitis. J. Infect. Dev. Ctries. *3*, 856–859.

Microbiology and Characteristics of *Helicobacter pylori*

3

Lyudmila Boyanova

Abstract

H. pylori is a Gram-negative spiral and microaerophilic bacterium that undergoes coccoid transformation under hostile conditions. The coccoids and bacterial biofilms may participate in the infection transmission. Specific *H. pylori* characteristics such as its enormous genomic diversity, helical morphology, acid acclimation, Krebs cycle and lipopolysaccharide indicate the extreme adaptation of the bacteria to the gastric mucosa. Modern methods such as comparative genomics, transcriptomics and proteomics provide a deep insight on *H. pylori* genomic diversity and its expression according to the severity of the disease. Methods to diagnose the infection are invasive, e.g. rapid urease test (RUT), histology, culture and molecular methods or less invasive such as the string test, or non-invasive such as serology, urea breath test (UBT), stool antigen test (SAT) and some molecular methods. Choice of the test depends on the test characteristics, e.g. type, preparation, accuracy, protocol, cut-off, availability and price, and on many patient characteristics which indicate the likelihood of the infection. Recently, non-invasive methods such as immunoproteomics, new SATs and improved UBT and SAT protocols have been reported. The significance of antiCagA antibodies for the risk of severe diseases has been highlighted. For the same purpose, a new test for serum pepsinogens and anti-*H. pylori* antibodies has been introduced. Recent molecular methods have been implemented to detect *H. pylori* and its resistance to both macrolides and quinolones or to spot non-invasively clarithromycin resistance. Advances have been made also in the invasive methods. *In vivo* histology has been suggested and, importantly, an operating link for histological estimation of gastric cancer risk [OLGA (Operative Link for Gastritis Assessment) staging] has been published. However, further studies are strongly required on important issues, among them are applications of current non-invasive tests or development of new non-invasive tests for detection of *H. pylori* resistance to antibiotics, improvement of test accuracy in specific patient groups, use of most reliable biomarkers for severe diseases and strategy for *H. pylori* screening and eradication.

Morphology and physiology of *H. pylori*

H. pylori are Gram-negative and spiral-shaped (2–4 × 0.5–1 μm), motile, microaerophilic and slow-growing bacteria (Kusters *et al.*, 2006; Fig. 3.1). Although young cells are spiral-shaped, they undergo a coccoid transition in hostile conditions such as prolonged incubation, cell ageing, temperature variations, subcultures, contact to oxygen (O_2) and antibiotic treatment (Fig. 3.2). In unfavourable conditions, *H. pylori* first changes from spiral to 'U'-shape form, the ends of which later fuse to form the coccoid (Zeng *et al.*, 2008). The coccoids cannot be subcultured. Cellini *et al.* (2008) have reported mostly spiral bacteria when *H. pylori* have been culturable but mainly clustered coccoids in a biofilm when the bacteria have been polymerase chain reaction (PCR) positive only (Cellini *et al.*, 2008). Indeed, the PCR detects DNA from live but unculturable and

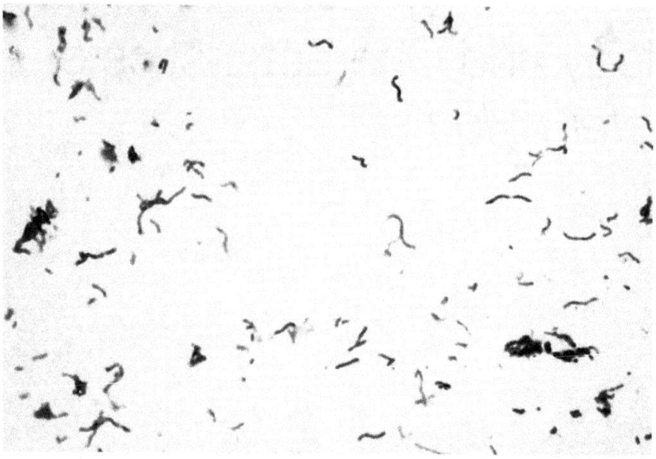

Figure 3.1 Pure culture of *H. pylori* with good spiral morphology of the cells (×1000).

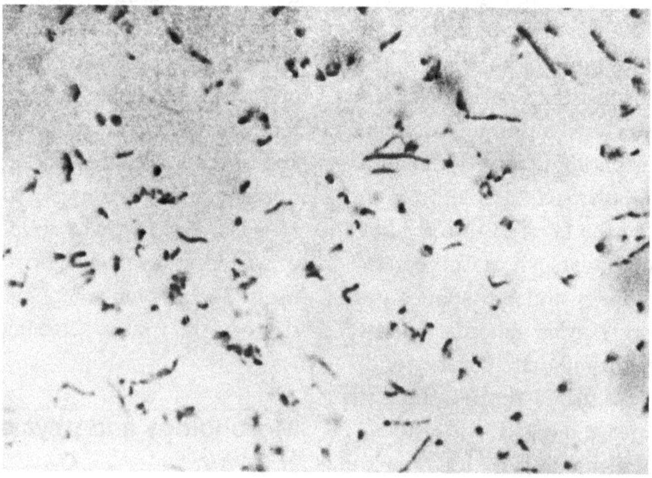

Figure 3.2 Pure culture of *H. pylori* showing U-like and coccoid cells (×1000).

dead bacteria. The coccoids and bacterial biofilms remain to be fully evaluated, as they may be associated with the infection transmission. Coccoids have been found to involve short-living (for ≤ 4 days) culturable bacteria or viable but non-culturable (VBNC) forms (Azevedo *et al.*, 2007). While the spiral bacteria have been viable, culturable and virulent, the coccoids have been often VBNC and virulent, although at less extent, and the degenerative coccoids have been pyknotic and dead (Zeng *et al.*, 2008). Although showing decreasing virulence, the coccoids have retained *ureA*, *ureB* genes and urease activity, flagella and adhering property, and, importantly, have been able to transform to spiral forms, to colonize mouse stomach and to induce development of gastritis (She *et al.*, 2003). The protective role of the coccoid transition has been suggested by Azevedo *et al.* (2007) as the transition has been found to occur only in mild but not in harsh hostile conditions.

Evaluating *H. pylori* peptidoglycan metabolism may help to search for new β-lactam antibiotics. Importantly, *H. pylori* penicillin-binding proteins (PBPs) are targets for β-lactams, e.g. amoxicillin, used often in anti-*H. pylori* eradication. *H. pylori* employs synthetases (PBP1, PBP2, and PBP3) and peptidoglycan hydrolases (Slt, MltD and AmiA, encoded by *amiA* gene; Chaput *et al.*, 2007). Slt is involved in *H. pylori* muropeptide release, while MltD contributes to the relocation of the peptidoglycan layers. The *slt* gene has been linked to nuclear factor B (NF-κB) activation and IL-8 synthesis in AGS cells that confirms the pro-inflammatory activity of the muropeptide (Boneca, 2005). Chaput *et al.* (2006) have observed the *amiA* gene to be responsible for accumulation of dipeptide (*N*-acetyl-D-glucosaminyl-β(1,4)-*N*-acetylmuramyl-L-Ala–D-Glu) motif and for the transition into coccoids. Coccoids have been able to escape detection by human intracellular Nod1 receptor in the gastric epithelial cells. They have not activated the associated NF-κB pathway and have not induced IL-8 release (Chaput *et al.*, 2006). This activity can contribute to *H. pylori* persistence and chronicity of the infection. In brief, *H. pylori amiA* gene is important for *H. pylori* coccoid transition and for the bacterial defence from the host innate immunity.

Spiral shape is a characteristic feature of relatively few bacterial genera, including *Helicobacter* and *Campylobacter*. *H. pylori* helical shape exhibits several specific features. *H. pylori* MreB protein (homologue of eukaryotic actin), which influences the helical morphology in other bacteria, performs other activities in *H. pylori*, affecting its urease activity and chromosome segregation and promoting a delay in cell division (Waidner *et al.*, 2009). Four genes, three LytM peptidoglycan peptidase homologues (*csd1*, *csd2*, *csd3*) and a *ccmA* homologue, have been shown to act cooperatively to generate, maintain or modify *H. pylori* spiral shape (Sycuro *et al.*, 2010). The gene products relax the peptide crosslinking of the peptidoglycan at some regions to provide curved and twisting *H. pylori* morphology. Sycuro *et al.* (2010) have observed that in mouse stomach, *csd1*, *csd2* and *ccmA* mutants have been outcompeted easily by wild-type strains. Apparently, *H. pylori* helical shape is essential for fitness of the strains and for a strong gastric colonization.

In addition, *H. pylori* cell envelope shows specific traits. Unlike most bacteria, *H. pylori* is able to extracts cholesterol from membranes of human epithelial cells, to modify the lipid by α-glucosylation with cholesterol-α-glucosyltransferase, encoded by *hp0421* gene, and to include the modified cholesterol into the bacterial outer membrane (Wunder *et al.*, 2006; Trainor *et al.*, 2011). *H. pylori* glucosylation of cholesterol counteracts both *H. pylori* phagocytosis and the T-cell activation and thus leads to immune evasion. In addition, the *H. pylori* cholesterol, together with a supposed bile salt efflux pump HefC, contributes to resistance to bile salts and ceragenins (positively charged aminosterol compounds, which mimic the activity of antimicrobial peptides; Trainor *et al.*, 2011).

Interestingly, *H. pylori* has a unique mechanism for lipopolysaccharide (LPS) biosynthesis. Unlike genes in other bacteria, *H. pylori* genes for LPS biosynthesis are present in different loci throughout the chromosome (Hug *et al.*, 2010). For the LPS biosynthesis, *H. pylori* uses a specific Wzk translocase (flippase), usually involved in protein *N*-glycosylation in other bacteria (Hug *et al.*, 2010). *H. pylori* LPS has 500-fold lower endotoxic activity than of *Escherichia coli* and provides only a negligible stimulation of macrophage release of pro-inflammatory cytokines, nitric oxide (NO) and prostaglandins (Moran, 2010). The poor activity of the lipid A is due to low phosphorylation and low acylation of its structures, interacting with immune receptors, compared with the endotoxin of enteric bacteria. By the resultant weak immune response, *H. pylori* can ensure its persistence in the host tissues (Moran, 2010). Moreover, Lewis antigens in the O chain of *H. pylori* LPS influence the infection. Structural similarity between *H. pylori* LPS antigens and Lewis blood group antigens on human gastric epithelial cells is a means to avoid the host immunity. Amount and position of Lewis antigens in *H. pylori* LPS show high variations among the strains, providing strain heterogeneity to adapt better to the host (Hug *et al.*, 2010). Lewis antigen expression is phase variable as fucosyltransferases show reversible 'on' and 'off' switches (Hug *et al.*, 2010). The important interaction of *H. pylori* LPS with DC-SIGN (dendritic cell-specific intercellular adhesion molecule-3-grabbing non-integrin) on dendritic cells is discussed in Chapter 4. Briefly, *H. pylori* LPS is able to modify the activity of the host immunity.

Hallmark of *H. pylori* genome is its enormous diversity owing to natural competence for transformation by exogenous DNA or by acquisition, deletion, rearrangement and point mutations in the DNA sequences. Strain variations also arise from phase-variations. *H. pylori* unique natural niche (the human stomach), its extensive DNA sequence diversity and the intrafamilial transmission of the infection, they altogether make the bacteria suitable to reveal the historical migrations of the humans. As a result, six major *H. pylori* populations (hpEurope, hpAsia2, hpNEAfrica, hpAfrica2, hpAfrica1 and hpEastAsia) have been described from a global collection of 769 isolates (Desikan, 2010).

A modern and powerful technique to evaluate and compare the genomes of different *H. pylori* strains is comparative genomics. The method, analysing open reading frames (ORFs) in the genome, determines the antigens that are associated with virulent strains only (Seib *et al.*, 2009). A disadvantage of the method is the lack of assessment of non-protein antigens and post-translational changes. As *H. pylori* genomic diversity, vacuolating cytotoxin (VacA), cytotoxin-associated gene A (CagA) protein and other factors are strongly linked to bacterial virulence; they are presented in Chapter 4.

Like other Gram-negative bacteria, *H. pylori* constantly produces and releases outer membrane vesicles. They carry membrane proteins, porins, adhesins and virulence factors, e.g. VacA and CagA on their surface (Mullaney *et al.*, 2009). The vesicles have been shown to stimulate the host immune response and, at high doses, to induce cell growth arrest and IL-8 release in the host cells (Ismail *et al.*, 2003). Moreover, uptake of the vesicles, especially of those from VacA-positive strains, has been observed in the gastric cells (Parker *et al.*, 2010). The possible participation of the *H. pylori* outer membrane vesicles in the formation of biofilms also has been suggested (Yonezawa *et al.*, 2009).

Although *H. pylori* has been thought to be non-invasive, it has been shown to induce autophagy in macrophages, dendritic and gastric epithelial cells and to replicate in large double-membrane vesicles (autophagosomes), which later, when fused with lysosomes, destroy the bacteria (Wang *et al.*, 2009). Importantly, the *H. pylori* presence in the autophagic vesicles of the host cells can reduce the antibiotic activity and hence treatment success of the infection, the activity of anti-*H. pylori* antibodies and also can inhibit the antigen presentation and the following T-cell immune response. Chu *et al.* (2010) have reported that intracellular killing of replicating bacteria has been achieved only by cell-penetrating antibiotics. Wang *et al.* (2009) have suggested *H. pylori* eradication to be active against the intracellular bacteria. All these data imply that a small population of *H. pylori* cells can be invasive and hence important for *H. pylori* pathogenesis, immune evasion and therapy of the infection, e.g. by clarithromycin.

H. pylori has 2–6 unipolar (lophotrichous), sheathed flagella (3 μm in length) with a bulb at the end, which protect the bacteria from the gastric acidity (Kusters *et al.*, 2006; Fig. 3.3). *In vitro*, *H. pylori* exhibits good motility only in the mid-exponential phase of growth, which is its growth phase in the gastric mucosa (Worku *et al.*, 1999). Unlike other bacteria, *H. pylori* lacks fimbrial adhesins. The flagella, encoded by *flaA* and *flaB* genes, contribute to the bacterial motility and penetration in gastric mucus layer (Algood and Cover, 2006). Unlike other bacteria, *H. pylori fliM*, *fliN*, *fliY*, and *fliG* genes encode four switch complex proteins i.e. FliM, FliN, FliY, and FliG, which all are required for the degree of flagellation of the cells. The switch proteins participate in switching between clockwise and counter-clockwise rotation of flagellar motor and hence the filament (Douillard *et al.*, 2010). The rotor switch component FliM participates in switching the directions of the flagellar rotation in response to binding of a phosphorylated chemosensory signalling protein, CheY (Lowenthal *et al.*, 2009).

H. pylori flagellar gene transcription is under the control of three RNA polymerase sigma factors (prokaryotic transcription initiation factors). The σ80 is responsible for transcription of early flagellar (class I) genes. The σ54 (RpoN) controls the transcription of middle flagellar (class II) genes needed for the basal body and flagellar hook formation. Late flagellar (class III) genes, needed the flagellar subunit formation as well as for rotation and chemotaxis, are under the control of σ^{28} (FliA) and the antisigma factor FlgM (de Reuse *et al.*, 2010). FhlA codes for a flagellar basal body protein and controls many genes such as FliA and RpoN regulons (de Reuse *et al.*, 2010). Hp0256, a FliJ homologue in *H. pylori*, is needed for full *H. pylori* motility, probably by interaction with

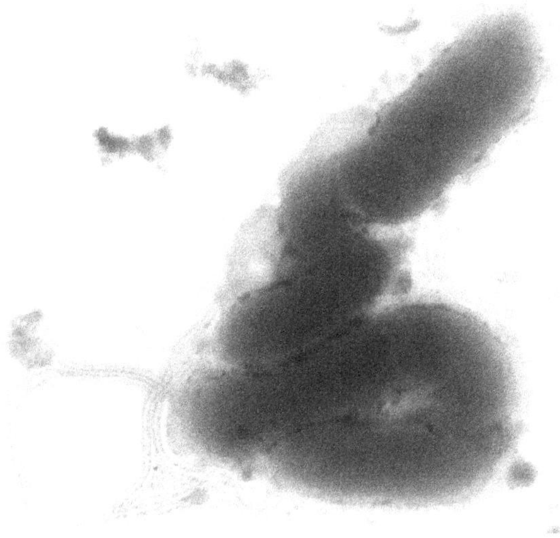

Figure 3.3 Electron microscopy picture of *H. pylori* (×10,000).

the switch components, and for the expression of adhesins (Douillard *et al.*, 2010).

The natural niche of *H. pylori* is extremely harsh by its high acidity. Both *H. pylori* arginase, hydrolysing L-arginine to L-ornithine and urea, and urease, hydrolysing urea to CO_2 and ammonium, participate in *H. pylori* nitrogen metabolism (Azizian *et al.*, 2010). *H. pylori* urease provides ammonium for nitrogen assimilation by glutamine synthetase. Importantly, the urease is crucial for the bacterial resistance to gastric acidity i.e. for *H. pylori* acid acclimation (*H. pylori* property to preserve periplasmic pH at 6.1 and cytoplasmic pH near neutrality despite the surrounding acidity; Scott *et al.*, 2007). *H. pylori* arginase is an additional source of intracellular urea for *H. pylori* and may reduce the gastric inflammation. Conversely, *H. pylori* urease damages host cells by the production of ammonia and influences the activity of the immune system (Stingl and De Reuse, 2005).

H. pylori urease is a large 1.1-MDa complex (a dodecameric nickel-containing enzyme) with 12 subunits of UreA and UreB. The enzyme is found in *H. pylori* cytoplasm and sometimes on its cell surface (owing to lysis of some bacteria). *H. pylori* urease exhibits large amount (10% of *H. pylori* protein content) and high affinity to the urea (Stingl and De Reuse, 2005). The enzyme has shown different kinetic properties as a result of two functional (multimeric and dissociated) forms, which may associate and dissociate often according to the urea concentrations in the gastric mucosa (Gang *et al.*, 2009). Control of the urease activity is very complex and performed by:

- transcription of acid acclimation *ureAB* genes by the low pH, presence of Ni^{2+} ions and nickel-responsive transcriptional regulator, NikR;
- nickel incorporation in urease to activate the urease apoenzyme by urease accessory proteins, UreEFGH, and two hydrogenase accessory proteins (HypA and HypB) that also participate in the nickel metabolism; and
- nickel availability in the cytoplasm, which is controlled by high-affinity nickel permease NixA, which expression is reduced at low pH, as well as by histidine-rich proteins (Hpn), which bind strongly to the nickel and are activated by NikR and also by pH-gated channel UreI, a 6-helice transmembrane protein in the inner membrane of the bacteria, which regulates the access of the urea to *H. pylori* cell according to the pH (Stingl and De Reuse, 2005; Sachs *et al.*, 2005; Scott *et al.*, 2007).

Mechanisms of *H. pylori* acid acclimation involve several steps (Scott *et al.*, 2007): urea and protons pass through *H. pylori* outer membrane porins and reach the periplasm. Acid activation of UreI leads to a 300-fold acceleration of urea entry into *H. pylori* cytoplasm. The ArsRS two-component system (ArsRS TCS) is also stimulated and activates the Ni^{2+} insertion genes, *ureE* and *ureG*, which, in turn, transform the urease apoenzyme into active urease. In addition to the acid resistance (urease and amidase), the ArsRS TCS regulates other proteins such as acetone carboxylase, involved in acetone metabolism, thioredoxin reductase, which confers resistance to oxidative stress, and Pfs, known as a quorum-sensing regulator (Loh *et al.*, 2010). Many homeostatic genes such as those for the ammonia producing enzymes (*amiE* and *amiF*), urea producing arginase (*rocF*), α-carbonic anhydrase and UreI are also activated by the ArsRS TCS. Following the urea hydrolysis, $2NH_3$ and H_2CO_3 are produced and then H_2CO_3 is converted into cytoplasmic CO_2 and H_2O by a cytoplasmic β-carbonic anhydrase. Both CO_2 and NH_3 penetrate *H. pylori* periplasm, where the CO_2 is converted by α-carbonic anhydrase into HCO_3^- and H^+. Bicarbonate preserves the periplasmic pH at 6.1, and the NH_3 neutralizes protons entering the periplasm, cytoplasm and possibly those present nearby the cell (Scott *et al.*, 2007). Role of *H. pylori* urease as a virulence factor is discussed in more detail in Chapter 4. *H. pylori* possesses also a hydrogenase, another nickel-containing enzyme, needed for hydrogen use as an energy source, which contributes to host colonization (Benoit and Maier, 2008).

Iron acquisition is usually associated with bacterial virulence. However, unlike other bacteria, *H. pylori* has exhibited preferential binding of iron-free forms of transferrin and lactoferrin, which implies a low ability to acquire iron from human serum (Senkovich *et al.*, 2010). This has been suggested to prevent extreme iron deficiency of the host and development of sepsis (Senkovich *et al.*, 2010). *H. pylori* has many proteins for iron transport and storage. Two of them are well evaluated, Pfr ferritin and Hp-NAP (neurtophil activating protein), which is a virulence factor (see Chapter 4). To degrade phosphonates for phosphorus metabolism, *H. pylori* uses specific pathways which are independent of phosphate starvation (Ford *et al.*, 2010). Interestingly, some *H. pylori* strains have also been shown to produce volatile sulfur compounds, which could promote development of halitosis (Lee *et al.*, 2006).

Bacterial quorum sensing with the participation of signalling molecule autoinducer-2 (AI-2) serves as 'bacterial Esperanto'. In many bacteria, LuxS acts as a quorum-sensing molecule synthase (Doherty *et al.*, 2010). In *H. pylori*, LuxS enzyme is involved in cysteine synthesis (by a reverse transsulfuration pathway), motility and fitness of the strains, which has been found in mice (Doherty *et al.*, 2010; Shen *et al.*, 2010). LuxS regulates *H. pylori* motility by flagellar gene expression and the flagellar synthesis by AI-2 production (Shen *et al.*, 2010).

As *H. pylori* is well adapted to the gastric mucosa, the bacteria lack some metabolic pathways such as those in the enteric bacteria. Within the sugars, *H. pylori* catabolizes only glucose and uses only O_2 as a terminal electron acceptor (Kusters *et al.*, 2006). *H. pylori* lacks some enzymes of Krebs cycle such as α-ketoglutarate dehydrogenase that produces succinyl CoA from α-ketoglutarate and succinyl CoA synthetase that produces succinate from succinyl CoA (Tsugawa *et al.*, 2008). In *H. pylori*, two different enzymes i.e. α-ketoglutarate oxidoreductase (KOR) and CoA transferase (CoAT), convert the α-ketoglutarate to succinyl CoA and succinate, and the KOR can also convert the α-ketoglutarate directly to succinate. In *H. pylori* VBNC forms, KOR activity has been found to increase, while that of CoAT has been reported to drop. Lack of KOR activity has resulted in killing of *H. pylori* owing to its inefficient transition into coccoids (Tsugawa *et al.*, 2008). KOR appears to be an essential enzyme for survival of the coccoids and its inactivation could be evaluated in attempt to obtain or enhance *H. pylori* eradication.

A modern method, transcriptomics, is used to evaluate the *H. pylori* adaptation to the constantly changing conditions in the gastric mucosa. Transcriptomics is based on evaluation, by means of DNA microarrays or cDNA sequencing, of the complete set of RNA transcripts, which *H. pylori* expresses under given conditions (Seib *et al.*, 2009). Transcriptomics shows well the mechanisms of *H. pylori* adaptation to favourable or unfavourable conditions.

To overcome oxidative DNA damage and immune clearance, *H. pylori* uses many enzymes (see Chapter 4). Transcriptomic study has shown that under high O_2 concentrations, 58 *H. pylori* genes have been activated and 40 have been suppressed (Kaakoush *et al.*, 2009). Oxygen stress has been found to induce changes in expression of the *ahpC* gene-encoded Ahp, which belongs to the peroxiredoxin family, *katA*-encoded catalase (KatA) and *sodB* gene product (SOD, superoxide dismutase, providing bacterial defence against oxidative damage; Zeng *et al.*, 2008). The role of these enzymes has been shown in the study of Wang *et al.* (2005), reporting high level of DNA fragmentation and high accumulation of free iron with toxic properties in mutant cells, lacking several of these enzymes. In another transcriptomic study of Kaakoush *et al.* (2009), upregulated nucleic acid metabolism has been observed at increased O_2 tension, and can be explained by *H. pylori* DNA injury via reactive oxygen species (ROS; Kaakoush *et al.*, 2009). An increase in the generation of precursor metabolites also has been found owing to the increasing energy needs of *H. pylori.* The reported activation of oxidative phosphorylation strongly indicates that *H. pylori* uses aerobic respiration (Kaakoush *et al.*, 2009).

In addition, *H. pylori* can express an autolysin, encoded by *lys* gene, with lytic activity against the cell walls of *H. pylori* and other bacteria (Marsich *et al.*, 2002). The enzyme has been suggested to participate in a so-called altruistic autolysis (although still dubious) of some bacteria. The autolysis has been suggested to release urease and other proteins, which may be adsorbed onto the outer membrane of neighbouring live bacteria to provide survival in hostile conditions and, in addition, to present antigens to the host immune system (Marsich *et al.*, 2002).

In brief, the specific *H. pylori* structural and metabolic features such as its enormous genomic diversity, helical morphology, acid acclimation, Krebs cycle, lipid A and lipopolysaccharide biosynthesis show its extremely high adaptation to its single natural niche, the gastric mucosa.

Growth requirements

H. pylori is a microaerophilic species, growing at 5–19% O_2 (optimum 5–6%), 5–10% CO_2 and high humidity at 37°C (range 34–40°C; Bury-Moné *et al.*, 2006; Kusters *et al.*, 2006). Fascinatingly, *H. pylori* can behave as a capnophile, growing in both microaerophilic and aerobic conditions at high bacterial density, although it is a strict microaerophile at low density (Bury-Moné *et al.*, 2006). This has been explained by Bury-Moné *et al.* (2006) by some CO_2-using enzymes, among them the carbonic anhydrases. However, this *H. pylori* property has not been shared by other gastric *Helicobacter* species.

Hydrogen is unnecessary for *H. pylori* growth but can enhance it. At 4% hydrogen, the growth appears after 48 h versus usually 60 h in the absence of the gas (Taneera *et al.*, 2002). *H. pylori* are neutralophilic bacteria, growing best at pH 6.0–7.0 *in vitro,* although as stated above, they survive in the presence of high gastric acidity in the stomach (Kusters *et al.*, 2006; Megraud and Lehours, 2007). Because of the usually poor *H. pylori* growth in liquid media, gastric biopsies for culture are plated most often on agar media. Nevertheless, Sainsus *et al.* (2008) have reported 87% sensitivity and 100% specificity of Ham's F12 medium with 5% horse serum and antibiotics for *H. pylori* isolation from biopsies.

Media for *H. pylori* isolation involve agar bases, e.g. Columbia, brain–heart infusion agar, Wilkins–Chalgren or Brucella agar bases with 7–10% blood (often sheep or human) or 7–10% fetal calf serum with or without growth supplements and antibiotics for the selective media. Chomvarin *et al.* (2006) have found Columbia agar base to be slightly superior to that of Brain heart infusion agar.

It is not to forget to use fresh and humid media. The media can be kept in the refrigerator at 4°C for up to one week if they are well enveloped to preserve their humidity.

To improve *H. pylori* growth, some authors have used growth supplements, e.g. 1% IsoVitaleX, 0.1% β-cyclodextrins, yeast extract (2–5 mg/ml), sodium pyruvate (0.5 mg/ml), porcine gastric mucin (2 mg/ml), activated charcoal (1–4 mg/ml) etc. (Taneera *et al.*, 2002; Degnan *et al.*, 2003; Douraghi *et al.*, 2010.) In liquid culture, β-cyclodextrin has delayed coccoid conversion up to 72 h of incubation and has enhanced the secretion of *H. pylori* proteins without affecting their expression (Douraghi *et al.*, 2010). However, the benefits of some supplements are doubtful. Adding ferric pyrohosphate (25 mg/ml) and

L-cystein (40 mg/ml) to blood agar has enhanced *H. pylori* growth but adding glucose has increased coccoid transition (Davood *et al.*, 2009). In our experience, the 1% IsoVitaleX (BBL Microbiology Systems, Cockeysville, MD, USA) appears to be been superior to the activated charcoal in agar media. Agar can be supplemented by triphenyltetrazolium chloride (40 µg/ml) to detect *H. pylori* colonies by their golden shine (Kusters *et al.*, 2006). This is useful, especially for early detection of the golden-red colonies and in mixed cultures. However, adding urea and phenol red to the media for the same purpose, can inhibit *H. pylori* growth by the rising pH of the media (Megraud and Lehours, 2007).

Best commercially available selective supplement for *H. pylori* is that of Dent (Oxoid, Basingstoke, UK), containing vancomycin 10 µg/ml, trimethoprim 5 µg/ml, cefsoludin 5 µg/ml and amphotericin B 5 µg/ml. In our experience, the Dent's supplement is superior to that of Skirrow for primary *H. pylori* isolation, possibly due to slightly inhibitory activity of some antibiotics such as polymyxin B in the Skirrow's supplement. Conversely, Degnan *et al.* (2003) found no inhibitory effect on *H. pylori* by the polymyxin B at 3500 U/l. Chemically defined media are inappropriate for routine work.

H. pylori colonies on blood agars are small (0.5–2 mm in size), smooth and translucent to pale greyish, sometimes showing a faint β-haemolysis with age (Fig. 3.4). The spiral shaped forms are typical for young strains in lag/exponential growth phases (usually up to 35–48 h after incubation), while in ageing cultures (>55 h after incubation), polymorphic, U-shaped forms and coccoids are often observed and finally, in the late decline phase (>60 h after incubation), all bacteria become coccoids (Worku *et al.*, 1999). However, in subcultures, *H. pylori* becomes laboratory-adapted and grows more quickly (in 2–3 days of incubation vs. often 3–7 days for the first incubation). Strains can survive 1 week on the medium at 4°C if kept in microaerophilic atmosphere. Long-term storage of strains (for decades) can be made at –70 to –80°C in 15–20% glycerol or 10% dimethyl sulfoxide-containing broth, or in liquid nitrogen (Megraud and Lehours, 2007). Freezing at –20°C and lyophilization of the strains often give poor results.

Diagnostic methods for *H. pylori* infection

Many tests for *H. pylori* detection are available. They are invasive [endoscopy based, e.g. rapid urease test (RUT), histology, culture and molecular methods] or less invasive (e.g. string test), or non-invasive such as serology, urea breath test (UBT), stool antigen test (SAT) and some molecular methods. All tests except serology reveal an active (present) infection. Importantly, each test carries some advantages and disadvantages. For instance, the low *H. pylori* density in bleeding

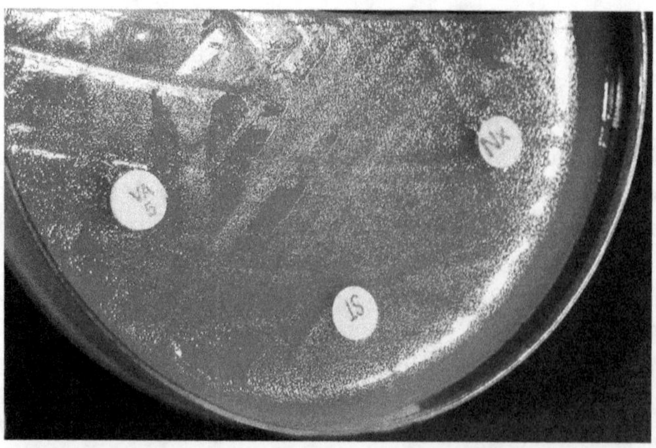

Figure 3.4 *H. pylori* growth on blood agar medium with typical resistance to vancomycin (5 µg/disc), nalidixic acid (30 µg/disc) and trimethoprim-sulfamethoxazole (25 µg/disc).

ulcer patients lessens the accuracy of the diagnostic tests except serology due to inhibition by the blood, antisecretory drugs used and gastric lavage (Lo *et al.*, 2005). In routine practice, commonly employed tests are histology, SAT, UBT and culture.

Accuracy of the diagnostic tests should be carefully evaluated. Likelihood ratio (LR, probability of having the disease/infection after a positive or negative test) is more suitable to show the accuracy of the test than only specificity and sensitivity (Lindsetmo *et al.*, 2008). High LR (>10) indicates that the test is appropriate to detect the disease/infection, while low LR (<0.1) indicate a negligible risk for the disease, however LRs of around 1 are uninformative. In a Spanish study, UBT has shown low (60%) specificity, while recalculated cut-off value has improved the positive LR (LR^+) to 8.6 and negative LR (LR^-) to 0.108 as well as the sensitivity and specificity to 90% (Calvet *et al.*, 2009).

Less invasive tests also have been developed. One of them uses an oro-gastric brush in a 5-mm-diameter plastic tube, which, after ingestion by the patient, brushes the gastric mucosa and is then removed (Graham *et al.*, 2005). By this technique, *H. pylori* has been isolated in all positive patients (Graham *et al.*, 2005). Another less invasive test is a string test, which, however, has exhibited suboptimal sensitivity (80%) compared with the culture (Velapatiño *et al.*, 2006).

In addition, Cho *et al.* (2008) have used a new invasive test for diagnosis of *H. pylori* infection, i.e. phenol red mucosal pH test. Phenol red solution, sprayed on the gastric mucosa, has produced different staining degrees, proportionate to those by UBT and histology (Cho *et al.*, 2008). Other new methods have been high-resolution magnifying endoscopy and confocal laser endomicroscopy, which can make feasible the virtual histology (Monteiro *et al.*, 2009).

Invasive methods

Biopsy specimens for the invasive tests

Gastric biopsy specimens are the most suitable specimens for invasive tests. Specimens from gastric juice, gastric mucus and vomitus have promoted low accuracy when compared with gastric biopsies (Boyanova *et al.*, 2007; Megraud and Lehours, 2007). In children, Boyanova *et al.* (2007) detected *H. pylori* in 63.9% of gastric biopsies compared with only 43.8% of gastric mucus specimens. For best performance of invasive tests, patients should not consume antibiotics ≥30 days and proton pump inhibitors (PPIs) ≥14 days before endoscopy as the antibiotics diminish the bacterial density and the PPIs inhibit the antral *H. pylori* by pH change (Megraud and Lehours, 2007). Automatic equipment for the endoscopy can eliminate the risks for bacterial or viral transmission (Megraud and Lehours, 2007).

Biopsy number is of primary importance for the diagnostics of *H. pylori* infection. Owing to patchy distributions of *H. pylori* in the gastric mucosa, multiple biopsy specimens per patient provide the most reliable diagnosis. Megraud and Lehours (2007) have recommended the use of ≥4 biopsy specimens (two from the antrum and two from the anterior and posterior corpus). Evaluation of only antral specimens may give false-negative results, for instance in case of intestinal metaplasia (Asaka *et al.*, 2010), as *H. pylori* cannot grow on intestinal-type epithelial cells.

Direct smear microscopy

H. pylori can be observed on preparations from gastric biopsy specimens by phase-contrast microscopy or by Gram staining with carbol fuchsin as a counterstain, or by simple carbol fuchsin staining. Owing to the patchy distribution of *H. pylori* in the gastric mucosa, numerous (if necessary, >10) microscopic fields should be observed. The method has shown a sensitivity of 74–80% but lower (65.7%) in treated patients (Boyanova, 2007). Both *H. pylori* and non-*pylori Helicobacter* spp. with large (6–10 μm) size and multiple spirals can be detected by the method (Boyanova *et al.*, 2007).

Rapid urease test (RUT)

H. pylori is a potent producer of urease and RUT is used as a specific test for the bacteria. Other urease-positive bacteria, which can be found in the stomach, produce less urease than *H. pylori* and, thus, RUT is highly specific after ≤2 h incubation. The ammonia produced by *H. pylori* owing to the hydrolysis of the urea, raise pH of RUTs and change quickly the colour of indicator. RUT detects the preformed *H. pylori* urease in the specimens; therefore, the test requires neither bacterial growth nor microaerophily. In untreated patients, usual sensitivity and specificity of RUT have been 83.4–94.5% and 95–100%, respectively, while in treated or bleeding ulcer patients, a low sensitivity (<70%) despite a good specificity (>90%) have been found (Lo *et al.*, 2005; Gisbert and Abraira, 2006; Asaka *et al.*, 2010). However, in bleeding ulcer patients, LR^+ has been very good (19.6) for culture, lower (9.5–9.6) for RUT and UBT and lowest (2.3–2.5) for SAT and serology. In these patients, LR^- has been poorest (0.31) for RUT and culture, poor (0.2–0.3) for SAT and serology and acceptable (0.11) for UBT (Gisbert and Abraira, 2006).

Older commercial RUTs such as CLO test (Delta-West Ltd, Bentley, Western Australia) are agar based. CLO test contains agar gel with urea, phenol red (pH indicator), buffer and a bacteriostatic agent. However, the agar-based RUTs have exhibited low (57–80%) sensitivity in untreated patients, falling additionally (to 50%) in treated patients after 1-h incubation (Megraud and Lehours, 2007; Chong *et al.*, 2008). In addition, the authors have detected late positive reactions (beyond 24h) by CLO test in 11% of the specimens. Importantly, only $\geq 10^4$–10^5 bacteria/ml can turn positive the CLO test (Megraud and Lehours, 2007). Nevertheless, in treaded patients, RUT sensitivity has been improved to 83.6% by two biopsies from antrum and corpus (Tepes, 2007).

Newer RUTs show better performance than that of CLO. PyloriTek (Serim Laboratories, Elkhart, IN, USA) is a strip RUT with a zone for urea hydrolysis and a zone for NH_3 capture and pH change. The test has shown high sensitivity and specificity (>95%) at 2h incubation compared with those of the CLO test (Megraud and Lehours, 2007). Importantly, a new 5-min RUT (UFT300, ABS Cernusco, sul Naviglio, Italy) has been reported to give a high (>94%) sensitivity even at 5 min and 100% specificity (Vaira *et al.*, 2010). HUITAI ultra-rapid biopsy RUT also has exhibited very high sensitivity and specificity (both >98%) and has turned the reaction positive usually at 1 min (Goh *et al.*, 2007). RUT can be prepared also in the laboratory, using 6–10% urea, reduced buffer amount and incubation at ≥37°C (Megraud and Lehours, 2007). However, in-house RUT sensitivity in untreated (56.3%) and especially treated adults (36.6%) has been low, using one-third of the biopsy specimen (Boyanova, 2007).

RUT has many advantages: it is simple, cheap and available, and does not require *H. pylori* growth. High (90.3%) sensitivity of UFT-300 at 1 min can render the test useful to establish *H. pylori* infection during endoscopic procedure (Vaira *et al.*, 2010; Zullo *et al.*, 2010). Therefore, for this purpose, the test merits to be further evaluated. Disadvantages of RUT are the false-negative results in patients with atrophic gastritis and intestinal metaplasia, in those on antibiotics and PPIs or in those after treatment. RUTs have been often negative in patients taking PPIs (in 82%) or even histamine-2 receptor blockers (60%; Yakoob *et al.*, 2008). CLO positivity has been very low (<50%) in patients with atrophic gastritis and intestinal metaplasia (Yoo *et al.*, 2007). Older age (>65 years) has decreased in-house RUT sensitivity in untreated and treated Bulgarian patients owing to the above-mentioned factors (Boyanova, 2007). Conversely, RUT can be false-positive in case of infection by non-*pylori* helicobacters. On the other hand, according to the Maastricht III consensus report, a positive RUT is a sufficient reason to start *H. pylori* therapy (Malfertheiner *et al.*, 2007) because of the high (often >95%) RUT specificity.

Histology

Histology is considered as a gold standard for *H. pylori* diagnosis. According to the updated Sydney classification, the histology reports acute or chronic inflammation, gastric atrophy, intestinal metaplasia and *H. pylori* density (Selgrad *et al.*, 2009). The updated Sidney system has recommended using two antral biopsies (from the greater and the lesser curvatures at 2–3 cm in front of the pylorus) and two specimens from the body (at the lesser curvature, at 4 cm above the angulus fold and at the greater curvature, at 8 cm to the cardia; Fischbach *et al.*, 2009). The sites are important, as the gastric antrum and the greater curvature usually exhibit high *H. pylori* density, while the lesser curvature and gastric body often show premalignant lesions (Fischbach *et al.*, 2009). As for the other invasive methods, the sensitivity of the histological methods increases with the number of biopsies used (Megraud and Lehours, 2007). Upper body greater curvature has been useful to detect *H. pylori* in gastric cancer because of less common atrophy and intestinal metaplasia compared with that at in other gastric regions (Kim *et al.*, 2009). There are also some specific recommendations. To preserve *H. pylori* morphology, the biopsies should be put instantly into 10% formaldehyde fixative, since Bouin fixative has been shown to alter the bacterial morphology (Megraud and Lehours, 2007). Tissues should be orientated before paraffin embedding to exhibit the surface gastric cells where *H. pylori* is easily detected (Megraud and Lehours, 2007).

Classical stains for *H. pylori* detection are numerous, for example, Genta (showing well both the bacteria and gastric lesions), Warthin–Starry, (exhibiting well *H. pylori*), modified Giemsa (accurate, simple and rapid technique), etc. but these stains are inferior to *H. pylori* immunostain (Kacar *et al.*, 2004; Selgrad *et al.*, 2009). In addition, haematoxylin and eosin stain has shown rather low accuracy owing to poor contrast between *H. pylori* and the adjacent tissue. Conversely, in optimal conditions, the sensitivity and specificity of Giemsa stain reach 87–96% (Asaka *et al.*, 2010; Table 3.1). Kacar *et al.* (2004) have reported highest sensitivity and specificity by *H. pylori* immunostain (100 and 100%), followed by Giemsa stain (97% and 90%) and haematoxylin and eosin stain (97% and 80%, respectively). Japanese guidelines for management of *H. pylori* infection have recommended using simultaneously special stains, e.g. Giemsa stain, to detect *H. pylori* and haematoxylin and eosin stain to detect gastric inflammation (Asaka *et al.*, 2010). Mixture of carbol fuschin and alcian blue staining has been reported lately as a simple, low-cost and helpful method to observe goblet cells in intestinal metaplasia (Yodavudh *et al.*, 2008). To reiterate, the low *H. pylori* density reduces the accuracy of the histological method. Lunet *et al.* (2009) found that modified Giemsa had a sensitivity of 96.2% in Portuguese compared with 66.3% in Mozambican patients as low *H. pylori* density was more frequent (20.6%) in the Mozambican than in the Portuguese patients (5%).

The histological method is reliable and accurate and can indicate the risk for gastric malignancy. An international group of pathologists has suggested a system for gastritis staging (from the lowest stage 0 to the highest stage IV) along a scale of increasing gastric cancer risk (Operative Link for Gastritis Assessment, OLGA staging), stages III and IV being associated with a high risk of gastric malignancy (Rugge *et al.*, 2008). In OLGA, gastritis stage is determined by both the extent of atrophy, scored histologically, and its topographical site. In each biopsy, atrophy (% of atrophic glands) is scored on a scale from score 0 (no atrophy) to score 3 (severe atrophy, ≥60%). OLGA recommends the use of five biopsy specimens: from the greater and lesser curvatures of the distal antrum (mucus-secreting area), from the lesser curvature at the incisura angularis (mucosecreting area), which shows well the early atrophy and metaplasia, and from the anterior and posterior walls of the proximal corpus (oxyntic area). In each compartment (mucosecreting and oxyntic), an overall atrophy score shows the level of atrophic changes (Rugge *et al.*, 2008).

However, the histology depends on the expertise of the pathologists and, like RUT, gives false-positive results in presence of non-*pylori* curved bacteria or false-negative results in patients with intestinal metaplasia (Yoo *et al.*, 2007), in treated patients or those, taking PPIs, owing to low *H. pylori* density or coccoids. Moreover, the debris may looks like *H. pylori* and may impede the diagnosis. Use of PPIs hampers more the accuracy of both RUT and histology than that of histamine-2 receptor blockers. In patients on PPIs, the histology results from antrum or corpus have been negative in 50–53% of cases, while in patients on histamine-2 receptor blockers, the results have been negative in 30% and 45%, respectively (Yakoob *et al.*, 2008). In bleeding ulcer patients, the sensitivity of the histological methods has been low (43–70%), and LR^+ (6.7) and LR^- (0.23) have been poor (Lo *et al.*, 2005; Gisbert and Abraira, 2006).

Table 3.1 Accuracy of diagnostic tests for *H. pylori* in dyspeptic patients according to some studies (studies with >30 cases were taken into account)

Test type	Patients	Sensitivity (%)	Specificity (%)	Reference
Rapid urease test (RUT)	Total children	75–100	84–100	Guarner *et al*. (2010)
	Total children (CLO)	83.4	99	Roma-Giannikou *et al*. (2010)
	Untreated adults (in-house RUT)	56.3	97.7	Boyanova (2007)
	Untreated adults (HUITAI RUT)	98.2	99	Goh *et al*. (2007)
	Treated adults (in-house RUT)	36.6	98.1	Boyanova (2007)
	Untreated adults (5-min RUT)	94.5	100	Vaira *et al*. (2010)
	Untreated adults (Jatrox Hp test)	94.7	100	Calvet *et al*. (2009)
	Untreated adults with BIU (in-house RUT)	66	95	Lo *et al*. (2005)
	Total with BIU	67	93	Gisbert and Abraira (2006)
Histology	Total children	66–100	94–100	Guarner *et al*. (2010)
	Total adults (Giemsa and modified Giemsa stains)	66.3–96.2	79–99	Lunet *et al*. (2009), Asaka *et al*. (2010)
	Untreated adults (Giemsa stain)	93.8	98.8	Calvet *et al*. (2009)
	Untreated adults with BIU (H&E)	43	95	Lo *et al*. (2005)
	Total with BIU (10 studies)	70	90	Gisbert and Abraira (2006)
Immunohistochemistry for East Asian CagA	Total patients	93.2	72.7	Kanada *et al*. (2008)
Fluorescence *in situ* hybridization (FISH)	Total children for Hp	84.1–94	100	Caristo *et al*. (2008), Guarner *et al*. (2010)
	Total adults for Hp	97	94	Morris *et al*. (2005)
	Total patients for ClaR	90–97	88–100	Morris *et al*. (2005), Vega *et al*. (2007)
Culture	Total children	55–96	100	Caristo *et al*. (2008), Guarner *et al*. (2010)
	Total children	84.6	100	Roma-Giannikou *et al*. (2010)
	Untreated adults	94	100	Boyanova (2007)
	Total adults	68–98	100	Asaka *et al*. (2010)
	Treated adults	96.3	100	Boyanova (2007)
	Untreated adults with BIU	37	100	Lo *et al*. (2005)
	Total with BIU (3 studies)	45	98	Gisbert and Abraira (2006)
Standard PCR, RFLP in gastric biopsy	Total patients	66–97	92–100	Burucoa *et al*. (2008), Sugimoto *et al*. (2009)
	Untreated adults with BIU	91	100	Lo *et al*. (2005)

RT-PCR in gastric biopsy	Total patients for Hp	98–100	92.5–100	Regnath *et al*. (2004), Schabereiter-Gurtner *et al*. (2004) Burucoa *et al*. (2008)
	Total patients for ClaR	82–96.2	97.6–100	Regnath *et al*. (2004) Schabereiter-Gurtner *et al*. (2004)
GenoType HelicoDR for ClaR or QnR	Total patients	94 (ClaR), 87 (QnR)	99 (ClaR), 98.5 (QnR)	Cambau *et al*. (2009)
Stool biprobe RT-PCR for Hp detection and ClaR	Untreated children for Hp	63	100	Lottspeich *et al*. (2007)
	Untreated children for Hp	85.9	98.3	Vécsei *et al*. (2010)
	Treated children for Hp	66.7	100	Vécsei *et al*. (2010)
	Total adults for Hp	98	98	Schabereiter-Gurtner *et al*. (2004)
	Total patients for ClaR	73–89.2	100	Schabereiter-Gurtner *et al*. (2004), Vécsei *et al*. (2010)
Serology	Total patients with BIU	88	69	Gisbert and Abraira (2006)
Serology for IgG in serum	Untreated children aged <5 years	35	88	Frenck *et al*. (2006)
	Untreated children aged ≥5 years	59	84	Frenck *et al*. (2006)
	Untreated children	88.7 (90.2[1])	93.4 (93.9[1])	Mégraud and EPTF-Hp[‡] (2005)
	Total children	50–100	70–98	Cherian *et al*. (2008), Guarner *et al*. (2010)
	Untreated adults	87.5	50	Yilmaz *et al*. (2006)
Serology for IgG in whole/capillary blood	Total children	76–100	79–100	Guarner *et al*. (2010)
	Untreated adults	90.6	85.1	Peng *et al*. (2009)
	Untreated adults with BIU	94	65	Lo *et al*. (2005)
CagA ELISA IgG	Untreated adults	54.2	75	Yilmaz *et al*. (2006)
East Asian CagA ELISA	Total patients	88	100	Yasuda *et al*. (2009)
Immunoblot	Total children	80–100	79–100	Leal *et al*. (2008) Guarner *et al*. (2010)
Urine EIA IgG	Total children	59–94	76–97	Mégraud and EPTF-Hp (2005), Leal *et al*. (2008), Guarner *et al*. (2010)
	Untreated children	63.2 (72.2[1])	97.3 (93.4[1])	Mégraud and EPTF-Hp (2005)
	Untreated adults	85.4–99	68–100	Megraud and Lehours (2007)
Rapid serological tests in urine	Untreated children	30.2	98.7	Mégraud and EPTF-Hp (2005)
	Total adults	79.5–100	83–97.4	Megraud and Lehours (2007), Nguyen *et al*. (2010)
Saliva EIA for IgG	Total children	64–80	86–99	Leal *et al*. (2008), Guarner *et al*. (2010)

Table 3.1 continued

Test type	Patients	Sensitivity (%)	Specificity (%)	Reference
Urea breath test	Untreated children	75–100	77.5–100	Frenck *et al*. (2006), Guarner *et al*. (2010)
	Untreated children	96.2	97.3	Mégraud and EPTF-Hp (2005)
	Untreated children aged <6 years	100	86	Frenck *et al*. (2006)
	Untreated children aged ≥6 years	97	95	Frenck *et al*. (2006);
	Treated children	94.1–100	92.3–100	Guarner *et al*. (2010)
	Untreated adults	100	85.1–95.7	Peng *et al*. (2009)
	Untreated adults	90.3	89.5	Calvet *et al*. (2009)
	Total patients	96.6–98	97–100	Gisbert and Pajares (2004), Ozdemir *et al*. (2008)
	Total adults with BIU	93–94	85–92	Lo *et al*. (2005), Gisbert and Abraira (2006)
Polyclonal antibody EIA-based stool antigen test (SAT)	Untreated children	67–100	61–100	Frenck *et al*. (2006), Guarner *et al*. (2010)
	Untreated children	72.9 (80.3[1])	97.3 (93.4[1])	Mégraud and EPTF-Hp (2005)
	Treated children	67–100	82–100	Guarner *et al*. (2010)
	Untreated adults	92.5–98.3	72.2–95	Tanaka and Takahashi (2004), Blanco *et al*. (2008)
	Total adults	63.6–85.7	67–92.6	Andrews *et al*. (2003), Erzin *et al*. (2004)
	Treated adults	87–90	97.5–97.7	Tanaka and Takahashi (2004), Deguchi *et al*. (2009)
Monoclonal antibody EIA-based SAT	Untreated children	93–98	88–100	Frenck *et al*. (2006), Guarner *et al*. (2010)
	Treated children	100	96.2–100	Guarner *et al*. (2010)
	Total children	100 (100[1])	76.2 (97.7[1])	Raguza *et al*. (2010)
	Untreated adults	78–98.2	76–100	Megraud and Lehours (2007) Calvet *et al*. (2009, 2010)
	Total adults	56–93	89.7–100	Andrews *et al*. (2003), Erzin *et al*. (2004), Trevisani *et al*. (2005)

Rapid monoclonal antibody-based SAT	Untreated or asymptomatic children	55–100	63.6–100	Cardenas *et al*. (2008), Guarner *et al*. (2010)
	Treated children	75–88.9	93.9–96	Guarner *et al*. (2010)
	Total children	74.6–87.5	63.6–97.8	Cherian *et al*. (2008), Kalach *et al*. (2009)
	Untreated adults	52.5–95.8	55.5–98	Blanco *et al*. (2008), Megraud and Lehours (2007) Calvet *et al*. (2010)
	Treated adults	91.6	98.4	Deguchi *et al*. (2009)
	Total adults	76.9–85	93–93.1	Trevisani *et al*. (2005), Krausse *et al*. (2008)

[1]With optimal cut-off.

BIU, bleeding ulcer; Hp, *H. pylori* detection; H&E, haemotoxylin and eosin; ClaR, clarithromycin resistance; QnR, quinolone resistance; EIA, enzyme immune assay; EPTF, European Paediatric Task Force on *H. pylori*; RT-PCR, real-time polymerase chain reaction; IgG, immunoglobulin G; CLO, *Campylobacter*-like organism; SAT, stool antigen test; RFLP, restriction fragment length polymorphism; ELISA, enzyme-linked immunosorbent assay.

For *H. pylori* detection, immunohistochemistry is superior to the classic strains and routine histochemistry because of the low number of false-positive results. By its specific anti-*H. pylori* antibodies, the immunostaining is able to distinguish both *H. pylori* and its coccoids from other bacteria (Asaka *et al.*, 2010). However, in specimens with low bacterial density and low inflammation, PCR is able to provide better *H. pylori* detection than the immunohistochemical method. By multiplex PCR with many primers for *H. pylori* genes, the bacteria have been detected in 52% of gastric biopsies from dyspeptic patients versus 20% by immunohistochemical staining or RUT in the study of Weiss *et al.* (2008). However, in Japan, the immunohistochemistry with anti-East Asian CagA antibodies has been useful to predict a risk for severe diseases (Kanada *et al.*, 2008).

During the last 10 years, fluorescence *in situ* hybridization (FISH) has been used to detect rapidly (within 3 h) both *H. pylori* and its clarithromycin resistance (Yilmaz and Demiray, 2007). FISH principle is based on hybridization and often utilizes fluorescence-labelled DNA probes, which bind to the homologous fragment in targeted DNA sequence (single gene or the whole chromosome) and then make it visible by fluorescence microscopy. The method has been highly sensitive (84.1–97%, often ≥94%) and specific (>94%) for *H. pylori* detection and also useful to detect coccoids (Morris *et al.*, 2005; Vega *et al.*, 2007; Caristo *et al.*, 2008; Guarner *et al.*, 2010). FISH probes are commercially available such as seaFAST (Combi-Kit; SeaPro Theranostics, Lelystad, the Netherland) and are used without expensive equipment except a fluorescence microscope (Yilmaz and Demiray, 2007). Advantages of FISH are numerous, e.g. its independence on DNA preparation, detection of both *H. pylori* and its antibiotic resistance, display of the location of the targeted sequences, specificity and rapidity, and low risk for inhibition, unlike PCR. However, if frozen biopsies are used, the standard protocol should be slightly modified (Vega *et al.*, 2007). A disadvantage of FISH is the possible difficulty in reading the results and the dependence, to some extent, on the observer (Morris *et al.*, 2005). In addition, the FISH method is more expensive than the classical histological methods.

Culture

Culture is the only direct method to identify *H. pylori* and serves as an alternative golden standard of the histological method (Kusters *et al.*, 2006). Specificity of the culture is absolute (100%) and in optimal conditions, the method reaches ≥90% sensitivity (Boyanova, 2007; Caristo *et al.*, 2008; Selgrad *et al.*, 2009). For this purpose, the patients should not consume antibiotics ≥4 weeks and PPIs ≥2 weeks before endoscopy. Use of two or more antral and two corpus biopsies is important for all patients but especially for those after treatment and those on PPIs. The specimens for culture should be collected prior to those for histology as fixatives may kill *H. pylori* (Megraud and Lehours, 2007).

H. pylori are fragile bacteria, dying quickly by drying, O_2 and room temperature. Biopsies can be transported or kept in a saline solution for several hours or, much better, in transport media, e.g. Portagerm pylori (bioMérieux, Marcy l'Etoile, France) or Stuart transport medium at 4°C for <24 h. In cool transport containers, biopsies can be transported by mail until 24 h. For storage of the biopsies for longer periods, they should be frozen at –70°C or in liquid nitrogen (Megraud and Lehours, 2007). L-Cysteine broth with 20% glycerol, a cryoprotective agent, also has been used for frozen biopsies (Davood *et al.*, 2009). However, freezing specimens from treated patients in Stuart transport medium at –20°C has been inappropriate (Boyanova, 2007). Dierikx *et al.* (2007) have developed 'serum- and animal tissue-free medium' for good *H. pylori* transport and growth.

Isolation of *H. pylori* from gastric biopsy specimens is not easy. In the laboratory, it is not to forget to homogenize the biopsy specimen because of the patchy distribution of the bacteria. This can be made by electrical/mechanical grinder or by sterile needles in several drops of sterile broth (Megraud and Lehours, 2007). Media for primary culture and subcultures of *H. pylori* are described above. Importantly, the specimens should be inoculated concurrently on selective and non-selective media as *H.*

pylori growth may be suppressed by antibiotics in the selective media or by overgrowing contaminant bacteria (oral flora or enteric bacteria if duodenal reflux is present) in the non-selective media.

Microaerophilic atmosphere for *H. pylori* incubation can be obtained by jar technique with commercial gas-generation envelopes such as Campy Pak (BBL), by chambers such as MACS 83 VA500 Microaerophilic Workstation (Don Whitley Scientific, Shipley, UK), by Anoxomat (MART, Nichtenvoorde, the Netherlands) or, for subcultures, in CO_2 incubators. Microaerophilic chambers maintain optimal microaerophilic atmosphere all the time but are rather expensive. Anoxomat (automated evacuation/replacement system) creates microaerophilic conditions more rapidly (in 1–3 min) than the envelopes (for some types, after several hours). Moreover, by envelopes, O_2 concentration continuously decreases (Megraud and Lehours, 2007). As *H. pylori* grows best in high (98%) humidity, moist sterile filter paper can be placed in the jar and, to ensure the circulation of the atmosphere, no more than 6–7 plates should be placed in a jar of 2.5 l. For primary *H. pylori* isolation, the inoculated plates should be read every day (Megraud and Lehours, 2007) or every other day from the third to the tenth or fourteenth day of microaerophilic incubation. For instance, in Bulgarian adult patients, 8- to 11-day incubation on selective and non-selective media has been needed for the growth of 6% and 9%, respectively, of strains from non-treated patients and for 12% of strains from treated patients on both media (Boyanova, 2007). Another important point for working with gas-generation envelopes is to keep the *H. pylori* cultures on the bench for <30 min as the bacteria may dye rapidly in air.

Gastric *H. pylori* identification is made by Gram staining, lack of aerobic growth and positive urease, oxidase and catalase tests (Boyanova, 2007). RUT from pure cultures turns positive in 1–2 min only. *H. pylori* density is important for both pathogenesis and treatment of the infection, although the density is not reported usually as a virulence factor. *H. pylori* density may be estimated by culture, real-time PCR (RT-PCR)-based technique and histology (Lai *et al.*, 2003; Boyanova, 2007). In Bulgarian non-treated adults, high *H. pylori* density has been more common (43.1%) in ulcer patients than in the rest (25.4%; Boyanova, 2007). *H. pylori* density is useful to be reported by culture or other methods as high histological density of *H. pylori* has been shown to reduce eradication success and ulcer healing in duodenal ulcer patients (Lai *et al.*, 2003).

The culture has many advantages: good sensitivity (under most favourable conditions) and absolute specificity; it serves as a base for assessing the virulence factors of *H. pylori,* for PCR typing of strains and for assessment of their resistance to a wide range of antibiotics. Disadvantages of the culture are the need for prolonged incubation (3–10 days), the dependence of the results on the specimen collection and transport and, as for most other methods, the reduced accuracy in some groups of patients, e.g. those with bleeding ulcers and intestinal metaplasia, and in treated patients and those on PPIs.

Polymerase chain reaction (PCR) and PCR-based methods for *H. pylori* detection

Most molecular methods have the advantage to be performed either on isolates or directly on gastric biopsies, to be highly accurate and rapid and to provide *H. pylori* identification or/and detection of the antibiotic resistance of the clinical strain. Isolates or biopsy specimens can be used for PCR to detect rapidly *H. pylori* species-specific genes such as 16S rRNA, *ureA*, *ureC* (glmM), 26K species-specific antigen gene and those, associated with the bacterial virulence such as *cagA* gene or *vacA* alleles. Molecular methods are especially useful for *H. pylori* detection in cases of contaminated specimens or transport delay, which have been found in 38% and 10%, respectively, of the specimens in the UK (Chisholm and Owen, 2008). Sensitivity (66–96%) of standard PCR in gastric biopsies has been comparable to that of the culture according to Sugimoto *et al.* (2009) but in the study of Chisholm and Owen, (2008), the use of multiplex PCR assay (HpHh) has improved *H. pylori* detection by 16.9% compared with the culture.

However, the extensive polymorphism of some *H. pylori* genes owing to its exceptional genomic diversity and the lack of virulence genes in some strains should be considered. Comparing primers for *H. pylori* detection in gastric and oral specimens by RT-PCR, Diouf *et al.* (2009) have reported low sensitivity by two *glmM* pairs used, poor specificity by *16S rRNA#295* pair and acceptable accuracy only by 16S rRNA#120 pair. By PCR for *H. pylori* in gastric biopsies, Sugimoto *et al.* (2009) have found false-positive results by all five primer pairs studied and good (>90%) sensitivity and specificity only by pairs Hp64-f/Hp64-r for *ureA* and Hp1/Hp2 for 16S rRNA. Problems with PCR also may arise owing to Taq polymerase inhibitors, especially if working with stool specimens, and contamination by exogenous DNA. Use of three completely separated rooms for PCR (for reagent preparation, for sample preparation and for the amplification) is recommended to avoid contamination (Megraud and Lehours, 2007). Many other factors are also important for the accuracy of the PCR. For instance, to improve PCR, DNA extraction is better to be performed by phenol technique or DNA extraction kits than by the easy and rapid boiling-water method (Megraud and Lehours, 2007). In our experience, by the last method, the DNA can undergo degradation in only several years.

Typing of strains is another advantage of the PCR-based methods. It can distinguish between reinfection and recrudescence, can detect the intrafamilial or extrafamilial distribution of the same strain and can reveal its antibiotic resistance. *H. pylori* strains can be typed by PCR-restriction fragment length polymorphism (PCR-RFLP) method or by randomly amplified polymorphic DNA (RAPD). In RAPD, DNA fragments are amplified with short randomly selected primers to generate the strain 'fingerprint' visualized by the following agarose gel electrophoresis. RFLP uses specific DNA digestion by restriction enzymes, which cut DNA at the place of specific short sequences, revealing a polymorphism of the length of restriction fragments. Point mutations also can change the length of the fragments. PCR-RFLP can be used to type *H. pylori* strains and to detect mutations associated with antibiotic resistance (Kumar *et al.*, 2010). By RFLP and RAPD, Matteo *et al.* (2007) have found 42 and 37 different profiles, respectively.

Real-time PCR (RT-PCR) is a powerful technique, which is used more and more often for rapid and highly accurate detection of *H. pylori* and its point mutations for antibiotic resistance. RT-PCR is based on fluorescent-labelled probes to detect and quantify one or several specific sequences in a DNA sample, tracing out the increase in amplified DNA after every amplification cycle. RT-PCR can be performed by an intercalating agent, e.g. SYBR green I dye or by fluorescence-resonance energy transfer (FRET) principle (see below). Non-specific fluorescent dyes, e.g. SYBR green I dye, bind to all double stranded DNA products (causing fluorescence), while sequence-specific DNA probes (labelled with a fluorescent reporter) fluoresce if hybridized with a complementary DNA. RT-PCR exhibits better detection level, better speed of testing than the standard PCR, and confirmation of the specificity by melting-point analysis. As all RT-PCR steps develop in a closed system, the risk for contamination is low. For *H. pylori* detection in gastric biopsies, RT-PCR has shown very high sensitivity (≥98%) and specificity (≥92%; Regnath *et al.*, 2004; Schabereiter-Gurtner *et al.*, 2004; Burucoa *et al.*, 2008). However, a problem can arise from difficulties in interpretation of the melting temperatures.

PCR-reverse hybridization-line probe assay (LiPA) has been used to detect *H. pylori* virulence or resistance-associated genes. PCR-LiPA uses biotin-labelled primers specific for the genes of interest and the biotin-labelled PCR product is used in the subsequent reverse hybridization PCR-LiPA assay (Ghose *et al.*, 2005). The method uses concurrent amplification of numerous genomic fragments. Positive hybridizations are visible as purple probe lines. PCR-LiPA has the advantage to detect the presence of mutants even at a very low level in the samples, mainly in cases of mixed infections. Another PCR-based reaction, nested PCR, using two pairs of PCR primers for a given locus, can be performed in gastric biopsies and paraffin-embedded tissues. The technique has the advantage to detect *H. pylori* in one-fifth of the histologically negative gastric biopsies (Zsikla *et al.*, 2006). However, use of nested PCR raises the possibility of contamination in the course of specimen extraction and manipulation (Megraud and Lehours, 2007). Recently, a new rapid (1-h) nanodiagnostic method has been introduced. The method uses isothermal helicase-dependent amplification of *ureC*, gold nanoparticle probes for hybridization and colorimetric detection of *H. pylori* DNA (Gill *et al.*, 2008). This method has shown high sensitivity (92.5%) and specificity (95.4%) compared with the culture (Gill *et al.*, 2008).

Detection of *H. pylori* virulence genes helps to assess the major risks for severe gastroduodenal diseases. For instance, *cagA* gene can be detected by standard or multiplex PCR, RT-PCR, RFLP, 3-mismatched PCR, PCR-LiPA, FISH etc. The *cagA* detection by standard PCR using one primer pair can be increased by 12% using second primer pair (Broutet *et al.*, 2001). Sequencing of *cagA* regions reveals numbers and types of EPIYA motifs (see Chapter 4). For the same purpose, Argent *et al.* (2005) have developed a PCR method with one forward and four reverse primers, and Monstein *et al.* (2010) have used a single primer pair by automated capillary electrophoresis and direct sequencing with universal sequencing primers. Loss of intact *cag*PAI has been detected by PCR for empty-site product of 550-bp size as well as by combinations within genes of the island (Matteo *et al.*, 2007; Miciuleviciene *et al.*, 2008). Some specific recommendations for detection of the virulence genes are discussed in Chapter 4.

Ryberg *et al.* (2008) have attempted to resolve an important problem, i.e. that of the insufficient bacterial or cell DNA yield in some human gastric biopsies. To detect *H. pylori* virulence genes and some host genes associated with gastric inflammation, the authors have developed multiple displacement amplification (MDA-DNA) technique, which performs amplification of the entire genomic DNA. MDA is based on isothermal, strand-displacing amplification. To perform the amplification, Ryberg *et al.* (2008) have used a random hexamer primer and a proofreading DNA polymerase, obtained from bacteriophage φ29. MDA-DNA has been found to be especially valuable to assess both *H. pylori* and human genes in the same clinical specimen at the same time.

Susceptibility testing methods

Phenotypic methods

Detection of *H. pylori* resistance to antibiotics is highly important owing to the continuous increase in the resistance rates and poor eradication success (Mégraud and Corti, 2009). In the guidelines for the management of *H. pylori* infection in Japan (a country with high *H. pylori* resistance to clarithromycin), Asaka *et al.* (2010) have recommended the wide use of *H. pylori* susceptibility testing. Methods to test *H. pylori* resistance are phenotypic, e.g. agar dilution method (ADM), epsilometer test (E test), breakpoint susceptibility testing (BST) and modified disc diffusion method (MDDM) or genotypic. Many genotypic methods are used, among them are PCR-RFLP, allele specific PCR (AS-PCR), RT-PCR, FISH, etc.

Phenotypic susceptibility testing methods for *H. pylori* are specific because of its atmosphere and growth needs. *H. pylori* heteroresistance (simultaneous presence of susceptible and resistant variants in a patient for the same antibiotic), observed for metronidazole, quinolones, amoxicillin and clarithromycin, hampers the detection of antibiotic resistance if only one biopsy is tested or if low inoculum density is used (see Chapter 9). Therefore, unlike those for most other bacteria, inocula with high density should be used to detect *H. pylori* resistance.

ADM is the test of choice for *H. pylori* susceptibility testing but other methods such as E test and MDDM also have been used (Megraud and Lehours, 2007; CLSI-NCCLS, 2010). For ADM, bacterial inoculum, corresponding to 2 McFarland standard or 10^7–10^8 colony forming units (CFU)/ml (according to CLSI) or 4 McFarland standard (0.5×10^9 to 1×10^9 CFU/ml according to the European *H. pylori* Study Group, EHSG), should be used. Recommended medium is Mueller–Hinton agar with 5% sheep blood aged ≥2 weeks (according to CLSI) or 10% horse blood (according to EHSG). Incubation is performed at 35–37°C in microaerophilic atmosphere for 3 days (Megraud and Lehours, 2007; CLSI-NCCLS, 2010). The drug concentration that completely inhibits the growth of *H. pylori* is the minimal inhibitory concentration (MIC) of the antibiotic.

E test (AB Biodisk, Solna, Sweden) is a simple

but relatively expensive method, showing directly MICs of an antibiotic by its gradient on plastic strips. The E test is an alternative of ADM except for metronidazole owing to possible false-positive results, overrating the presence of resistance. While the E test has shown excellent (>98% agreement within two $\log_2$ dilutions) agreement with ADM for amoxicillin and clarithromycin, metronidazole MICs by the E test have been higher against 5 of 10 *H. pylori* strains evaluated by Glupczynski *et al.* (2002). Therefore, E test results for metronidazole resistance should be proven by ADM. Conversely, tetracycline MICs have been higher by ADM than by E test in the study of Chisholm and Owen (2008). The authors have found mutations in 16S rRNA gene in strains with relatively low (0.75–2 μg/ml) tetracycline MICs. Reduced tetracycline susceptibility has been reported for these strains (Chisholm and Owen, 2008).

BST is an easy ADM version, using several concentrations per antibiotic, e.g. breakpoints for susceptible, intermediary (for clarithromycin) and resistant strains (Boyanova *et al.*, 2008). Unlike ADM, BST is suitable for susceptibility testing of few (1–6) strains, i.e. for routine use (Fig. 3.5). Boyanova *et al.* (2008) have assessed BST using pure *H. pylori* cultures with a good (93.3–100%) category agreement between the BST and E test or ADM results. As plates for ADM and BST are prepared in the laboratory, both tests should be controlled by non-selective plates to see the strain viability and by control strains with known susceptibility patterns.

Standard disc diffusion method is easy and cheap but unsuitable for slow growing bacteria in specific atmosphere. However, MDDM has been studied for *H. pylori*. *H. pylori* resistance to clarithromycin has been tested by erythromycin discs with 100% concordance between results for both drugs (Grignon *et al.*, 2002). Owing to different clustering of the MICs of susceptible and resistant strains, clarithromycin resistance can be detected by inhibitory zone diameters ≤17 mm for erythromycin and ≤20 mm for clarithromycin (Grignon *et al.*, 2002). Erythromycin discs (15 μg/disc) have given slightly better results than clarithromycin discs (15 μg). In brief, the *H. pylori* susceptibility to clarithromycin can be detected accurately by MDDM with erythromycin discs (Megraud and Lehours, 2007).

Although there is a lack of MDDM interpretive category standards for other drugs, some suggestions are given in Table 3.2. As for the other tests, the zone of the complete growth inhibition should be measured by MDDM. As for the metronidazole susceptibility testing of *H. pylori*, both E test and MDDM can give false results in a proportion (often about 10%) of the strains owing to the presence of mixed infections or owing to reasons associated with the methods used in different studies (McNulty *et al.*, 2002). Importantly,

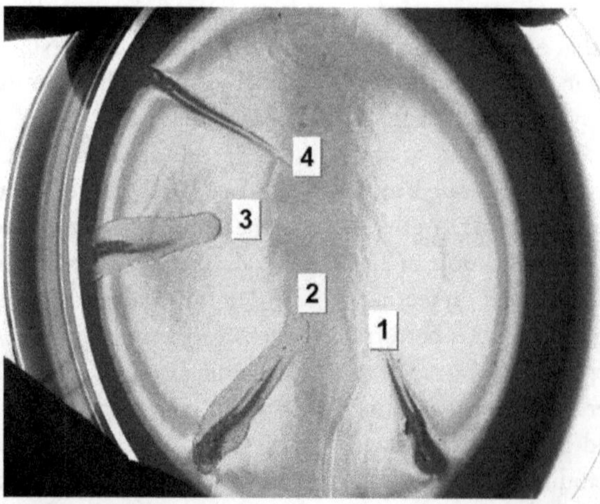

Figure 3.5 Breakpoint susceptibility testing of four *H. pylori* strains to metronidazole (8 μg/ml) on the same plate. See the good growth of strains 2 and 3 (resistant to the antibiotic concentration) and the lack of growth of strains 1 and 4 (susceptible to the drug).

Table 3.2 Breakpoints for *H. pylori* resistance to antibiotics used for agar dilution and modified disc-diffusion methods

				Category			
Agent	Agar	Method	Measure	S	I	R	Reference
ADM/E test							
Metronidazole	MHBA	ADM	μg/ml	≤8	NA	>8	Megraud and Lehours (2007)
Clarithromycin	MHBA	ADM	μg/ml	≤0.25	0.5	>0.5	Megraud and Lehours (2007)
Clarithromycin	MHBA	ADM	μg/ml	≤0.25	0.5	≥1	CLSI (2010)
Ciprofloxacin[1]	MHBA	ADM	μg/ml	≤1	NA	>1	Megraud and Lehours (2007)
Amoxicillin	MHBA	ADM	μg/ml	<0.25	0.25–0.5	>0.5	Megraud and Lehours (2007)
Tetracycline	MHBA	ADM	μg/ml	≤2	NA	>2	Megraud and Lehours (2007)
Tetracycline	CBA	E test	μg/ml	<4	NA	≥4	Chisholm and Owen (2008)
Rifabutin	MHBA	ADM	μg/ml	≤1	NA	>1	Megraud and Lehours (2007)
Rifampicin/ Rifabutin	Iso-sensitest	E test	μg/ml	≤1	NA	>4	Glocker *et al.* (2007)
MDDM (μg/disc)							
Metronidazole (5)	MHBA	MDDM	mm	>21	16–21	<16	McNulty *et al.* (2002)
Metronidazole (5)	CBA	MDDM	mm	≥10	NA	<10	McNulty *et al.* (2002)
Metronidazole (5)	CBA	MDDM	mm	≥20	NA	<20	Chisholm and Owen (2008)
Erythromycin[1] (15)	MHBA	MDDM	mm	>17	NA	≤17	Grignon *et al.* (2002)
Clarithromycin (15)	MHBA	MDDM	mm	>22	20–22	≤20	Grignon *et al.* (2002)
Clarithromycin (2)	MHBA	MDDM	mm	Zone	NA	No zone	McNulty *et al.* (2002)
Amoxicillin (2)	CBA	MDDM	mm	≥30	NA	<30	Chisholm and Owen (2008)
Amoxicillin (2)	MHBA	MDDM	mm	NA	NA	21	Sung *et al.* (2009)
Amoxicillin (10)	MHBA	MDDM	mm	NA	NA	32	Sung *et al.* (2009)
Tetracycline (10)	CBA	MDDM	mm	≥30	NA	<30	Chisholm and Owen (2008)
Ciprofloxacin[1] (1)	CBA	MDDM	mm	>20	>15–20	≤15	Chisholm and Owen (2009)
Rifampicin/rifabutin (5)	Iso-sensitest	MDDM	mm	≥21	NA	<21	Glocker *et al.* (2007)

[1]Erythromycin and ciprofloxacin, which are clinically ineffective, can be tested to detect the *H. pylori* susceptibility to clarithromycin and levofloxacin, respectively.

ADM, agar dilution method; MHBA, blood containing Mueller–Hinton agar; CBA, Columbia blood agar; MDDM, modified disc diffusion method; S, susceptible; I, intermediately susceptible; R, resistant.

for MDDM, only two discs should be placed on a plate (diameter, 90 mm) as *H. pylori* often displays enormous inhibitory zones. Types of media also can influence the results. According to McNulty *et al.* (2002), inhibitory zone diameter for *H. pylori* resistance by metronidazole discs (5 μg) is <16 mm on Mueller–Hinton blood agar but lower (<10 mm) on Columbia blood agar.

In brief, for all phenotypic susceptibility testing methods, is important not to forget to use young *H. pylori* cultures (aged ≤48 h) and high inocula (2–4 McFarland standard), and to read the results after 72 h microaerophilic incubation. Breakpoints (μg/ml) used for *H. pylori* resistance have been clarithromycin ≥1, metronidazole >8, amoxicillin >0.5, tetracycline >2, ciprofloxacin and levofloxacin >1, and rifabutin >1 (Megraud and Lehours, 2007; CLSI-NCCLS, 2010).

Molecular methods

Molecular methods, targeting *H. pylori* genomic mutations associated with antibiotic resistance, are accurate and rapid (requiring usually several hours), and can be performed on isolates, gastric biopsies or stool specimens (for some tests only). Clarithromycin resistance-associated mutations can be detected by many methods, e.g. PCR-sequencing, RFLP, FISH, RT-PCR, PCR-LiPA, AS-PCR, etc. (Owen, 2002; Megraud and Lehours, 2007.) PCR-RFLP amplifies the DNA region of the gene harbouring the mutations, then the synthesized fragments are digested by restriction endonucleases (BsaI enzyme for A2142G and BsbI for A2143G mutation), which cut at the specific mutation sites and thus, the size of the obtained fragments shows the presence/absence of the mutation (Owen, 2002). PCR with a 3′-mismatched specific primer has been used to detect A2142C mutation (Megraud and Lehours, 2007). However, RFLP is performed following the conventional PCR, which increases the time required by the method.

AS-PCR has been also used to detect the common mutations for clarithromycin resistance in *H. pylori* (Nishizawa *et al.*, 2007). AS-PCR is based on a preferential PCR amplification of one of the alleles of a given gene, e.g. the mutant allele, which confers clarithromycin resistance, and is able to distinguish single nucleotide polymorphism (SNP). For AS-PCR, two primers, chosen to mismatch or match the mutant allele at the 3′-end of the primer as well as a third primer, which acts in opposite direction, are selected. The method uses a single PCR reaction and requires about 3–4 hours. Furuta *et al.* (2007) have developed an AS-PCR method to distinguish clarithromycin resistance mutations A2142G, A2143G and the wild-type strain. The results have correlated well with the clarithromycin MICs. The advantage of the AS-PCR is the lack of need for restriction enzyme cleavage and RT-PCR equipment. However, the method often needs to be carefully optimized and can show background amplification due to non-specific amplification of DNA (Furuta *et al.*, 2007; Nishizawa *et al.*, 2007).

Biprobe RT-PCR with FRET probes is useful to detect *H. pylori* in gastric biopsies and the presence of the three common mutations for clarithromycin resistance (Oleastro *et al.*, 2003). FRET probes are two fluorescent probes in close (<10 nm) proximity: an acceptor probe 5′ labelled with fluorophore, e.g. LC-Red 640, hybridizing with the mutation site, and a donor probe 3′ labelled with fluorophore, e.g. fluorescein, hybridizing 3–5 bases upstream (Megraud and Lehours, 2007). Interaction of the two dyes occurs if both are bound to their targets. In this case, donor fluorophore excitation by a light source is transferred to the acceptor probe fluorophore, causing it to fluoresce. Melting curve analysis shows different melting curves for wild-type strains and clarithromycin-resistant mutants (Oleastro *et al.*, 2003). RT-PCR has been able to detect mutants among the strains at a level of only 10% (Oleastro *et al.*, 2003). Results of this method have shown 96.4% concordance with those of phenotypic susceptibility testing methods and have required only 2 h (Oleastro *et al.*, 2003). Therefore, the method is very reliable to detect the antibiotic resistance in *H. pylori*.

Scorpion probes combine a primer and a probe in a single molecule and can distinguish different alleles of a target in a single-step procedure (Burucoa *et al.*, 2008). Quadruplex RT-PCR with allele-specific Scorpion primers has been reported to detect both *H. pylori* and its three main point mutations for clarithromycin resistance in gastric biopsies (Burucoa *et al.*, 2008). The technique has shown impressive sensitivity (98.3%) and high specificity (92.5%) compared with the culture and perfect (99.8%) concordance with E test results (Burucoa *et al.*, 2008; Raymond *et al.*, 2010). Scorpion PCR has detected mixtures of susceptible and resistant clones, which RFLP has failed to detect (Burucoa *et al.*, 2008).

Another RT-PCR method for clarithromycin resistance in *H. pylori* is that with TaqMan probes. These are dual labelled hydrolysis probes, developed by Applied Biosystems (Foster City, CA, USA) to raise RT-PCR specificity. In Italy, TaqMan RT-PCR method has been used to detect the main clarithromycin resistance mutations in paraffin-embedded gastric biopsies (De Francesco *et al.*, 2010a).

However, few molecular methods for detection of *H. pylori* antibiotic resistance are commercially available and most of them are designed to detect the clarithromycin resistance only. DNA strip genotyping test (GenoType HelicoDR, Hain Lifescience GmbH, Nehren, Germany) has been recently developed to detect concurrently wild-type *H. pylori* strains or mutations to either clarithromycin or fluoroquinolones in both isolates and gastric tissues (Cambau *et al.*, 2009). The test comprises three steps: DNA amplification, hybridization and alkaline phosphatase reaction on a membrane strip. For GenoType HelicoDR, the probes are developed to hybridize with the sequences of *H. pylori* wild-type alleles or the mutated alleles. By this test, *H. pylori* suspensions with McFarland density of ≥1 (DNA concentration, ≥3.6 μg/ml) have provided a positive signal. The test has shown very high (94%) sensitivity and impressing specificity (99%) for the detection of clarithromycin resistance (Cambau *et al.*, 2009). GenoType HelicoDR appears to be very useful for wide laboratory use as the increasing *H. pylori* resistance to clarithromycin and levofloxacin is an important problem in many developed countries. Another test, an enzyme colorimetric-based DNA chip, also has been developed to detect clarithromycin resistance mutations in *H. pylori* with as high as 96.8% concordance with DNA sequencing (Xuan *et al.*, 2009).

The clarithromycin resistance can be detected also by FISH (see above). FISH can detect *H. pylori* by 16S rRNA probe, labelled with fluorochrome Cy3, fluorescing in red, and the clarithromycin-resistant mutants by 23S rRNA probe, labelled with fluorescein, fluorescing in green. The FISH has exhibited 90–97% sensitivity and 88–100% specificity for clarithromycin resistance (Morris *et al.*, 2005; Vega *et al.*, 2007), which render the method valuable for the histological laboratories.

Importantly, non-invasive methods for detection of clarithromycin resistance in *H. pylori* from stool specimens have been developed. Such tests are RFLP-nested PCR and the commercially available stool biprobe RT-PCR (*H. pylori* ClariRes assay; Ingenetix, Vienna, Austria; Rimbara *et al.*, 2005). Sensitivity of the latter method has been often 83.8–98% for the detection of *H. pylori* and 73–89.2% for that of its clarithromycin resistance (Schabereiter-Gurtner *et al.*, 2004; Vécsei *et al.*, 2010). Conversely, Lottspeich *et al.* (2007) have reported poor (63%) sensitivity by this method in German children. Post-treatment sensitivity of the stool biprobe RT-PCR in Austrian children also has been suboptimal (66.7%) in the study of Vécsei *et al.* (2010). Makristathis *et al.* (2007) have stressed that delayed transport, storage of samples at room temperature or repeated thawing of DNA extracts could reduce the sensitivity of the faecal molecular tests. Further evaluation or optimization of the promising ClariRes assay as a tool for non-invasive detection of *H. pylori* resistance to clarithromycin is strongly required. We can hope that non-invasive tests for the resistance to other antibiotics also will be designed.

Examples for the so-far commercially available tests for clarithromycin resistance in *H. pylori* are ClariRes assay based on FRET-melting curve analysis; seaFAST *H. pylori* based on FISH (see above), Seeplex ClaR-*H. pylori* PCR (Seegene Inc., Seoul, South Korea) based on dual-priming oligonucleotide-PCR and GenoType HelicoDR (see above; Mégraud and Corti, 2009).

Recently, an interesting and important link between the results of phenotypically and genotypically detected *H. pylori* resistance to macrolides has been reported. In Italy, De Francesco *et al.* (2010b) have PCR detected *H. pylori* clarithromycin resistance in 37.6% of strains versus 18.4% by E test. Importantly, when the genotypic resistance has been phenotypically absent, *H. pylori* eradication in the patients has been high (80%) but the resistance observed by both phenotypic and genotypic methods has been associated with reduced (45%) eradication rate and thus can predict treatment failure. The authors have stressed that the PCR methods not always predict the resistance leading to unsuccessful eradication (De Francesco *et al.*, 2010b). Therefore, the molecular and phenotypic susceptibility testing methods should be used together to avoid treatment failures by clarithromycin-based regimens (see more detail in Chapter 9).

To detect mutations in *gyrA* gene, resulting in quinolone resistance in *H. pylori*, many techniques such as PCR-sequencing, RFLP, AS-PCR and RT-PCR have been used (Glocker and Kist, 2004; Nishizawa *et al.*, 2007; Wang *et al.*, 2010). AS-PCR with seven forward and a reverse primers has distinguished wild *H. pylori* type and *gyrA* mutants at codon 87 Asn or 91 Asp (Nishizawa *et al.*, 2007). By electrophoresis, a specific band has been found in the susceptible strains and an additional band has indicated quinolone resistance for the five mutant *gyrA* types, however, without revealing the exact mutation.

In addition, RT-PCR with FRET-melting curve analysis has been useful to identify the main (aa87 and aa91) mutations associated with *H. pylori* resistance to fluoroquinolones at amino acid position 87 or 91 in a study of Glocker and Kist (2004). However, owing to high genetic heterogeneity of *H. pylori* strains from distant parts of the world, the authors have recommended the use of local DNA extracts in the different geographic areas (Glocker and Kist, 2004). Newer method to detect both quinolone and clarithromycin resistance in *H. pylori* pure cultures or gastric biopsies is GenoType HelicoDR (see above). It has shown good (87%) sensitivity and very high specificity (98.5%) for estimation of levofloxacin-resistant strains (Cambau *et al.*, 2009). Importantly, if isolates are used to detect the resistance-associated mutations to an antibiotic, it is important to use only colonies showing resistance to the agent. In most (80%) strains with heterogeneous resistance, quinolone mutations have been found by PCR-sequencing only with subcultures from the resistant single colonies inside the inhibitory zones of the E test (Bogaerts *et al.*, 2006).

H. pylori resistance to tetracycline has been evaluated by several methods, among them are RT-PCR, PCR-sequencing and PCR-RFLP followed by PCR-sequencing. However, in a Chilean study, the RFLP method alone has been unable to detect the tetracycline resistance and has required a following PCR-sequencing of the 535 bp PCR fragment (Toledo and López-Solís, 2010). In addition, Lawson *et al.* (2005) have developed a RT-PCR assay for tetracycline resistance-associated mutations but have not found any mutations in 16S rDNA in 8 of 18 *H. pylori* strains with reduced susceptibility to the agent. The authors have suggested the development and use of several probes to detect additional mutations. Nevertheless, in the study of Glocker *et al.* (2005), a RT-PCR assay for the tetracycline resistance has been able to detect as few as 10 bacteria/µl, including those with mixed mutations, and the method has performed well in 150 clinical specimens studied. Therefore, more studies for RT-PCR detection of tetracycline resistance in *H. pylori* are expected.

For *H. pylori* resistance to metronidazole, PCR-sequencing of *rdxA* gene or immunoblot with specific antiRdxA antibodies can be used, although this resistance is not due only to mutations in *rdxA* and *frxA* genes (Megraud and Lehours, 2007). In Taiwan, high MICs of metronidazole in resistant strains have been linked to mutations in *rdxA* rather than in *frxA* gene (Yang *et al.*, 2004). However, Chisholm and Owen (2003) have reported that the *rdxA* mutations do not always predict metronidazole resistance. Therefore, the detection of the *rdxA* mutations seems to be of limited diagnostic value.

As for the amoxicillin resistance, PCR-sequencing of the *pbp1* genes has been performed (Co and Schiller, 2006). Watanabe *et al.* (2005) have stressed the importance of the intermediate amoxicillin resistance in *H. pylori* (MICs 0.04–1 µg/ml), which rates have increased in Japan. Shen *et al.* (2008) have reported unstable amoxicillin resistance in *H. pylori* as a possible result of changes in the phenotype. In addition, testing of β-lactamase has been suggested to detect the possibly emerging β-lactamase producing strains (see Chapter 9). PCR-sequencing of *rpoB* fragments has been used for *H. pylori* resistance to rifampicin (Chisholm and Owen, 2009).

In brief, molecular methods have been developed and successfully used to detect clarithromycin and quinolone resistance in *H. pylori*; however, those for detection of resistance to other antibiotics have been fewer and some of them should be further optimized or modified.

Non-invasive methods

Non-invasive methods, especially UBT and SAT, are becoming more and more widely used to diagnose *H. pylori* infection as most of them are highly accurate and easily accepted by patients. However, Nurgalieva and Graham (2003) have stressed that in countries/populations with high prevalence of the infection, the non-invasive diagnostic tests tend to exhibit higher sensitivity and lower specificity than in other countries. In populations with low prevalence of the infection, the negative results have been considered accurate but probable false-positive tests should be taken into account (Nurgalieva *et al.*, 2006). Conversely, in populations of high prevalence of the infection, the positive results of the tests have been considered as reliable, unlike the negative results.

Urea breath test (UBT)

In *H. pylori* diagnostics, UBT serves as an alternative golden standard to the histological method (Kusters *et al.*, 2006). Highly sensitive diagnostic tests should be used in routine clinical practice, especially in countries with low infection prevalence and in children (Koletzko, 2005). Such a test is UBT. Within the non-invasive tests, it has shown often the highest sensitivity (usually >93%) with good specificity (85–95%), although the latter has been reduced in children aged ≤ 6 years (Table 3.1). Although no uniform protocol for UBT has been accepted so far, the test has been validated in paediatric populations (Koletzko, 2005; Gisbert and Pajares, 2004) and is a valuable non-invasive option for the diagnosis of *H. pylori* infection.

UBT is based on the potent *H. pylori* urease activity, which hydrolyses urea to ammonia and CO_2. Urea is labelled with ^{13}C (non-radioactive stable) or ^{14}C (radioactive) isotope. To perform UBT, firstly, the patients ingest a solution of labelled urea often with citric acid solution. If the patient is *H. pylori* positive, the urea is quickly hydrolysed by *H. pylori* urease in the stomach, and then the labelled CO_2 diffuses into the blood vessels and thereafter reaches the lungs (Fig. 3.6). This makes possible the assessment of the labelled CO_2 in the exhaled air of the patient (collected in plastic/glass tubes or breath bags) by suitable apparatus. If the patient is *H. pylori* negative, no labelled CO_2 is generated in the stomach and all labelled urea is removed by the urine and faeces (Megraud and Lehours, 2007).

The citric acid is often used in UBT protocols and is helpful to delay gastric emptying, probably due to its low (2.4) pH and to increase the urea hydrolysis, possibly through interplay with *H. pylori* proton-gated urea channel UreI, which sharply increases the urea flux across the inner membrane of *H. pylori* (Megraud and Lehours, 2007; Anania *et al.*, 2008). Citric acid solution usually provides better UBT results than orange juice (pH, 3.4), which, however, is more pleasant

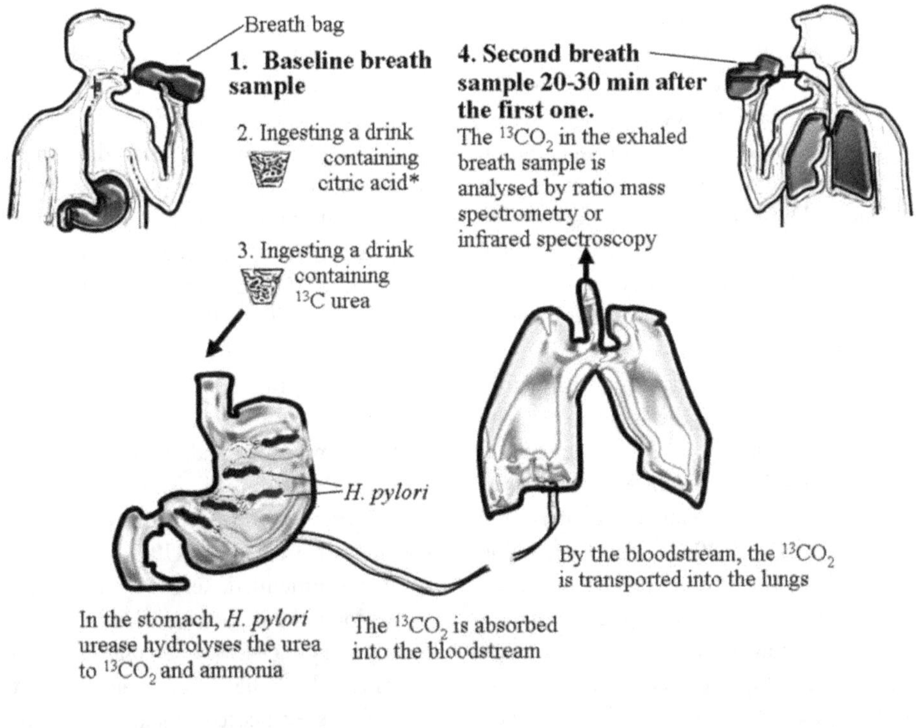

Figure 3.6 Mechanism of ^{13}C urea breath test (UBT). Citric acid has been omitted by several authors or has been taken together with the urea (Anania *et al.*, 2008).

for the children (Gisbert and Pajares, 2004). UBT sensitivity with orange juice has been lower by 12% compared with that of UBT with citric acid (Gisbert and Pajares, 2004). However, some studies have reported cost-effective protocol of ^{13}C UBT with good accuracy even without test meals before the test (Anania *et al.*, 2008). Campuzano-Maya (2007) has obtained similar accuracies of ^{13}C UBT by a 10-min, 50 mg ^{13}C UBT with no test meal (100%) and low cut-off value, and by a standard protocol with the test meal (98.6%).

Importantly, UBT should be done only after at ≥4 weeks without use of antibiotics use and 7–14 days without use of acid antisecretory agents as both factors can cause false-negative results owing to reduced *H. pylori* density. Patients are recommended to fast for >4–6 h, better overnight. Non-fasting patients often show higher UBT values compared with fasting subjects, perhaps due to ^{13}C-urea residues in the food, or lower values owing to poor contact of the urea with *H. pylori* (Gisbert and Pajares, 2004).

Dosage of urea is a key factor for UBT accuracy. The UBT may be carried out with relatively low (<100 mg) urea doses, e.g. 75 mg (widely used amount) or 50 mg (Gisbert and Pajares, 2004; Campuzano-Maya, 2007). In children, ^{13}C-urea has been administered in amounts of 45–100 mg or according to body weight (2 mg/kg but ≤ 100 mg; Anania *et al.*, 2008). As the endogenous CO_2 excretion in children is lower than in adults, a 50-mg dose of urea has been enough to achieve good ^{13}C UBT accuracy in children (Gisbert and Pajares, 2004). A new low-dose ^{14}C UBT test (Heliprobe, Stockholm, Sweden) has also shown very high sensitivity (96.6%) and specificity (100%; Ozdemir *et al.*, 2008).

For ^{13}C UBT, breath samples are taken before and often 20–30 min after the ingestion of the labelled urea. Th ratio $^{13}C/^{12}C$ is estimated in both samples and is shown as a difference between both values (delta over baseline, DOB) per 1000 (Megraud and Lehours, 2007). The basic value should be considered because the amount of the exhaled ^{13}C can vary, e.g. according to the patient diet. By the common protocol (with citric acid and 75 mg of urea), high accuracy (99% sensitivity and 98% specificity) has been obtained even when breath samples have been collected 10–15 min after the urea ingestion (Gisbert and Pajares, 2004). Nevertheless, in children, breath collection at 30 min is most suitable (Anania *et al.*, 2008).

^{14}C UBT reading is performed by scintillation counter with estimation of 2% of the administered dose (usually 1 μCi) as a cut-off value for positive tests (Megraud and Lehours, 2007). In the ^{13}C UBT, the exhaled breath samples are analysed by mass spectrometer (isotope ratio mass spectrometry, IRMS), infrared spectrometer (non-dispersive isotope-selective infrared spectrometry, NDIRS), laser-assisted ratio analyser (LARA) or a newly developed Fourier transform infrared spectroscope (FTIR; Gisbert and Pajares, 2004). NDIRS is less expensive than IRMS but sequentially processes fewer (<20) breath samples compared with IRMS (>200 samples) and needs a large air sample, which can pose problems in small children (Anania *et al.*, 2008). The LARA system has provided the quickest results compared with IRMS and NDIRS (Anania *et al.*, 2008). For ^{13}C UBT with DOB values >5‰, IRMS, NDIRS and FTIR techniques have been thought to be accurate (Motta *et al.*, 2009).

Urease-positive oropharyngeal bacteria may cause false-positive UBT results. Breath samples taken read 10–15 min after the urea administration have been found to be less specific than those read at 30 min (Gisbert and Pajares, 2004; Peng *et al.*, 2009). To avoid false-positivity, either patient mouth washing after drinking the ^{13}C urea solution or a ^{13}C urea capsule, releasing the urea straight in the stomach, have been suggested (Peng *et al.*, 2009). In Taiwan, the accuracy of capsule UBT (98%) has been found to be higher than that of the standard UBT (93%; Peng *et al.*, 2009). For ^{14}C UBT, the citric acid also can sensitize the test and the urea can be given also in a capsule.

^{13}C UBT DOB cut-off values are still a matter of controversy (Selgrad *et al.*, 2009). For INFAI tests, DOB cut-off of >4‰ has been suggested by the provider. In 2232 untreated and treated Canadian patients, ^{13}C UBT cut-off of 3‰ has been recommended, i.e. 3.1‰ for untreated and 2.9‰ for treated patients (Mauro *et al.*, 2006). In Italy, high (94.7%) sensitivity and (95.7%) specificity have been reported by 4.5‰ DOB cut-off (De Francesco *et al.*, 2010a). In Taiwan, Peng *et al.* (2009) have used 4.8‰ DOB cut-off value in adults, obtaining 100% sensitivity and 85.1% specificity of ^{13}C UBT. The variations may be explained by many reasons; among them are *H. pylori* densities in the stomach of the patients in different countries, DOB cut-off values, UBT protocols used and previous therapy for *H. pylori* eradication or other infections in some patients.

If lower doses of urea are used, lower cut-off value for DOB should be used. Many protocols have employed 75 mg of urea and cut-off points of 3.5–5‰ or 50 mg urea and cut-off points reduced to 2.5–3.5‰ (Gisbert and Pajares, 2004). Importantly, increasing the cut-off increases the specificity but can lessen the sensitivity of the test. Receiver operating curves (ROCs) have been used to establish the optimal cut-off points for the non-invasive tests.

A drawback of UBT is its diminished specificity in youngest children. For instance, in Egyptian children, ^{13}C UBT test, also analysed by ROC-established optimal cut-off point (6.2‰), has provided lower (86%) UBT specificity in children aged <6 years than in older children (95%) (Frenck *et al.*, 2006). Nevertheless, in Pakistani children, UBT has shown better parameters (79% sensitivity and 80% specificity) than SAT (77% sensitivity and 73% specificity; Hafeez *et al.*, 2007). Dondi *et al.* (2006) used a high (8‰) DOB cut-off in Italian children aged ≤5 years to increase the specificity of ^{13}C UBT to 98.1%. However, in South Korean children aged <6 years, false-positive ^{13}C UBT results were found in 12.6% of the cases by DOB and in 9% by the urea hydrolysis rate (UHR, μg/min; Yang *et al.*, 2008). UHR has been measured to adjust DOB values for the endogenous CO_2 release and thus to improve the ^{13}C UBT accuracy in patients. In US children aged 2–5 years, UHR for ^{13}C UBT estimate (positive at ≥10 μg/min) has given better performance (sensitivity and specificity of >95%) than DOB (Elitsur *et al.*, 2009).

While cut-off of about 4‰ has been used in many pre-treatment studies, lower post-treatment values (near 2–3‰) have been employed. ^{13}C UBT has been useful to detect the success of *H. pylori* eradication, usually, 4–6 weeks after the end of therapy (Selgrad *et al.*, 2009). Shirin *et al.* (2005) have shortened the period to 14 days by a high dose (4 g) of citric acid. By post-treatment cut-off points of 4% and 2.5%, the UBT sensitivity has increased from 80% to 94.3%, respectively, in combination with high (>95%) specificity (Gisbert and Pajares, 2004). Commonly suggested cut-off for post-treatment DOB has been 2–2.9‰ (Selgrad *et al.*, 2009).

Importantly, UBT detects the urease activity, which mirrors the overall *H. pylori* density in the stomach regardless of the bacterial susceptibility to the antibiotics. Although a correlation between DOB values of UBT and presence of antibiotic resistance in *H. pylori* has been reported (Kawai *et al.*, 2008), De Francesco *et al.* (2010a) have reconsidered the suggested association, obtaining similar DOB values in both susceptible and resistant strains for clarithromycin, metronidazole, and levofloxacin.

In bleeding ulcer patients, the UBT performance has been usually suboptimal (Lo *et al.*, 2005; Gisbert and Abraira, 2006). In these patients, if the first UBT is negative, a second UBT after the end of PPI use should be recommended. Only the second UBT has turned positive in 83% of cases with bleeding ulcers (Gisbert *et al.*, 2007).

Advantages of UBT are its non-invasiveness, high accuracy in both untreated and treated patients and applicability for control of eradication 1 month after treatment. For ^{13}C UBT, the specimens are transported easily and the results are independent on inter-observer interpretations (Cirak *et al.*, 2007). UBT also possesses several disadvantages; for example, ^{14}C-urea is radioactive, although with negligible radiation, so the test is not recommended for pregnant women and children. In addition, hospitals should be permitted to store and remove the ^{14}C-urea. In contrast, ^{13}C-urea is not radioactive but is more expensive than ^{14}C and requires costly machines for the sample analysis. UBT can give false-negative results owing to low bacterial density in bleeding ulcer patients, in those on PPIs and in those who had taken antibiotics before the test. Moreover, UBT cannot reveal the presence of ulcers or malignancy. False UBT positivity can be due to bacterial overgrowth, endogenous CO_2 production and other factors. False-positive UBT rates, associated with urease from oral or non-*pylori* gastric bacteria, have ranged from 3.9% to 14.9% (Osaki *et al.*, 2008; Peng *et al.*, 2009). UBT can give a false-positive result in children aged ≤6 years and in patients with atrophic gastritis, in the latter case, the hypochlorhydric stomach can be easily colonized by urease-positive enteric bacteria or staphylococci (Osaki *et al.*, 2008; Monteiro *et al.*, 2009). In case of positive UBT result close to the cut-off point, it is better to perform another test to detect the eradication or to repeat the UBT after follow-up (Asaka *et al.*, 2010).

Serological tests

H. pylori infection leads to seroconversion in almost all (98%) subjects, therefore, serological tests can detect the infection (Megraud and Lehours, 2007). There are different tests for anti-*H. pylori* antibody detection according to antigens (different strain sources, whole-cell antigens, purified mixtures etc.); test type [usually enzyme immune assay (EIA) such as enzyme-linked immunosorbant assay (ELISA) and immunoblot (Western blot)], preparation (commercially available or in-house tests), immunoglobulin to be detected (IgG and rarely IgA or IgM) and patient specimen (serum, saliva or urine; Leal *et al.*, 2008).

Serum-based ELISA is the most widely used serological method worldwide and many commercially available ELISAs are in use, most of them using patented combinations of *H. pylori* antigens. Most serological tests have been developed to detect IgG as tests for the early appearing IgM class of antibodies have shown utterly low (6.8%) sensitivity due to the chronic nature of the infection and those for IgA have given low sensitivity (42.6%) in children (Leal *et al.*, 2008; She *et al.*, 2009; Table 3.1). The serological tests are quantitative (for antibody titre) such as ELISA and qualitative (positive or negative) such as rapid immunochromatographic tests.

Importantly, ELISA for *H. pylori* IgG gives poor results in children aged <6 years owing to the relatively weak immune response in this age group (Leal *et al.*, 2008). Varying *H. pylori* antigens according to geographical areas and ethnicity, and poor handling of the serum (repeated freezing and thawing) additionally affect the sensitivity of the tests. In Egyptian children, the serology has shown poor sensitivity (50%), while the ^{13}C UBT and monoclonal antibody (MAb)-EIA-based SAT have been highly sensitive (≥93%) and specific (89% for UBT and 88% for the MAb-based SAT; Frenck *et al.*, 2006). In children aged <6 years, the sensitivity of the serological methods has been only 35% (Frenck *et al.*, 2006). Moreover, Cherian *et al.* (2008) have stressed the poor performance of the serology in African refugee children in Australia, failing to detect about one-third of the positive cases. It has been suggested to use lower cut-off in children than in adults (She *et al.*, 2009). Treatment of *H. pylori* infection in children, based only on the antibody detection, has been discouraged and needs to be confirmed by another test (Leal *et al.*, 2008).

Immunoblot uses specific *H. pylori* antigens separated by gel electrophoresis and transferred to filter-paper strip. For example, Western blot test (Euroimmun Medizinische Labordiagnostika, Lübeck, Germany) uses *H. pylori* extracts with molecular weights of 120 kDa (CagA), 95 kDa (VacA), 67 kDa (flagellar sheath protein), 66 kDa (UreB), 57 kDa (heat-shock protein homologue); 29 kDa (UreA), etc. (Yilmaz *et al.*, 2006.) Immunoblot and in-house ELISA with local strains usually provide better results than the commercial tests. According to a meta-analysis in children, the sensitivity and specificity have been 91.3% and 89% for immunoblot and 94% and 96.4% for in-house ELISAs with whole-cell antigen tests but only 79.2% and 92.4%, respectively, for commercial ELISA-IgG assays (Leal *et al.*, 2008). LR^+ of the in-house ELISA with whole-cell antigen tests has been higher (19.9) than that of the immunoblot (8.2), whereas LR– have been good (<0.06) for both tests (Leal *et al.*, 2008). Like the performance of the commercial ELISA tests, that of the immunoblot test can be affected by geographic variations of *H. pylori* antigens, spontaneous clearance of the infection in children, cross-reaction with other *Helicobacter* or *Campylobacter* spp. or different criteria for the interpretation of the results (see Chapter 7).

The presence of CagA antibodies usually indicates an elevated risk for peptic ulcers and gastric cancer (Camorlinga-Ponce *et al.*, 2008; Mitchell *et al.*, 2008), despite that not all subjects infected with *cagA*⁺ *H. pylori* strains develop serum CagA antibody. Host recognition of CagA expression by *H. pylori* is crucial for CagA seropositivity

(Shimoyama *et al.*, 2005). In Chilean patients, human IgG antibodies have recognized CagA in 59.4% and VacA in 68.1% of cases (Lienlaf *et al.*, 2010). In Japan, CagA seronegativity has been linked to low gastric infiltration and less virulent strains lacking intact *cag*PAI (Shimoyama *et al.*, 2005). Importantly, the CagA positivity can be detected better by PCR and immunoblot than by ELISA. In Japan, CagA seropositivity has been 79.2% by immunoblot versus only 58.5% by ELISA (Shimoyama *et al.*, 2005). Similarly, the sensitivity of *H. pylori* detection in Australian patients with non-cardia gastric cancer has increased by 15% using CagA immunoblot compared with ELISA (Mitchell *et al.*, 2008).

Association of antiCagA IgG with intestinal metaplasia has been high but has been marginal with duodenal ulcer and low with gastric cancer owing to reduced *H. pylori* density with the growing mucosal atrophy (Camorlinga-Ponce *et al.*, 2008). AntiCagA antibodies have been associated with intestinal metaplasia and young gastric cancer in Mexico and with corpus atrophy in Estonia (Camorlinga-Ponce *et al.*, 2008; Vorobjova *et al.*, 2008). In Japan, *H. pylori* IgG antibody-positive subjects have a 5-fold elevated risk of gastric cancer, while the presence of CagA antibodies doubles the risk and, importantly, even IgG negative/CagA antibody-positive subjects are at 3-fold increased risk of the cancer (Sasazuki *et al.*, 2006). This is due to the long-standing persistence of CagA antibodies in the human organism compared with the surface antibodies to *H. pylori*. AntiCagA antibodies have been detected by immunoblot in 87% of the patients even ≥10 years following successful *H. pylori* eradication (Veijola *et al.*, 2008). However, Yilmaz *et al.* (2006) have reported poor performance of CagA ELISA IgG (57.1% accuracy) in Turkey and have suggested combining several serological tests, e.g. ELISA IgG, CagA ELISA IgG and immunoblot, in countries with high *H. pylori* prevalence. The authors have observed suboptimal overall results with sensitivity, specificity, LR^+ and LR^- of 87.5%, 50%, 1.75 and 1.75 for ELISA IgG and 54.2%, 75%, 0.72 and 2.17, respectively, for CagA ELISA IgG (Yilmaz *et al.*, 2006).

It is known that combination of virulence factors such as intact *cag* PAI, East Asian-type 3′ *cagA* and *vacA* s1/m1/i1 in most *H. pylori* strains from East Asia accounts for the high gastric cancer incidence there (see Chapter 4). Recently, Yasuda *et al.* (2009) have developed ELISA with MAb against the highly virulent East Asian CagA. The test, evaluated on 32 strains, has shown good sensitivity (88%) and specificity (100%).

Rapid serological tests such as Quickvue (Quidel Corporation, San Diego, CA, USA), Assure *H. pylori* Rapid Test (Genelabs Diagnostics, Singapore) and Pyloriset Screen (PS, Orion Diagnostica, Espoo, Finland) can be performed easily in the office of the clinicians. However, the rapid tests have shown often suboptimal specificity of <80% (65–100%) and sensitivity of <90% (76–100%; Lo *et al.*, 2005; Peng *et al.*, 2009; Guarner *et al.*, 2010). Assure *H. pylori* Rapid Test accuracy has been reported by Peng *et al.* (2009) to be lower (88%) than that of ^{13}C UBT (93%) and capsule ^{13}C UBT (98%). Generally, the rapid serological tests do not perform as well as the EIA-based tests. For instance, Pyloriset EIA-G (Orion Diagnostics, Espoo, Finland) has been found to be one of the most accurate tests in Europe (Megraud and Lehours, 2007).

Tests for anti-*H. pylori* IgG in urine, e.g. standard ELISA (Urinelisa; Otsuka, Tokyo, Japan) and a rapid test (Rapirun; Otsuka Diagnostic, Tokyo, Japan), have performed well in Asia but with poor sensitivity (30.2% for Rapirun and 63.2% for Urinelisa) in European children (Mégraud and European Paediatric Task Force on *Helicobacter pylori*, 2005). Generally, these tests are not suitable for routine practice.

Briefly, advantages of classical serological tests are their low cost, speed of the rapid tests, good patient comfort and usefulness for epidemiological studies. Immunoblot and in-house ELISA with whole-cell antigen from local strains tests provide good accuracy and reliable results. Maastricht III Consensus report (Malfertheiner *et al.*, 2007) has re-recommended the use of *H. pylori* serology in cases where other tests can show false-negative results, e.g. in patients with bleeding ulcers, gastric atrophy, mucosa-associated lymphoid tissue (MALT) lymphoma, treated patients and in those taking PPIs. In bleeding ulcer patients, the invasive tests have shown low sensitivity (≤70% for RUT and histology, and 45% for culture) and

SAT has exhibited low (70%) specificity (Gisbert and Abraira, 2006). In these patients, however, Gisbert and Abraira (2006) have observed low (69%) specificity by serology but high (≥92%) sensitivity and specificity by UBT (Gisbert and Abraira, 2006). Disadvantage of serological tests is their relatively poor (80–84%) accuracy (Malfertheiner *et al.*, 2007). Before use, the cut-off of some tests should be validated locally because of the high antigenic variability of *H. pylori*. Caution is needed in interpretation of results in treated patients. Rapid serological tests are unsuitable to control the eradication success. Quantitative tests are appropriate for this purpose, although for many tests, many months after treatment as the antibody titre declines slowly. Moreover, the serological tests cannot indicate the presence of peptic ulcers and malignancy and cannot distinguish between present and past infections owing to immunological memory (Kusters *et al.*, 2006). In countries/populations with low *H. pylori* prevalence, the result of a single positive test should be proven by another test (Nurgalieva and Graham, 2003). Moreover, the serological tests can be false-negative in cases of fresh infection and in immunosuppressed patients or false-positive, mostly in developing countries, in patients infected with *Campylobacter* or non-*pylori* helicobacters owing to cross-reactions between fagellar and other antigens (see Chapter 2). Use of rapid urinary and salivary immune tests is not recommended (Megraud and Lehours, 2007), although some authors have found these tests acceptable for routine purposes. In Vietnam, Rapirun *H. pylori* antibody detection kit has shown accuracy of 84.5% (Nguyen *et al.*, 2010).

Modern methods such as immunoproteomics have been developed to predict the risk for severe diseases (Khoder *et al.*, 2009; Michel *et al.*, 2009). Immunoproteomics is a combination of proteomics and immunoblot. The proteomics is based on assessment of the entire set of proteins expressed by *H. pylori* under given conditions and their interactions in the proteome, using two-dimensional polyacrylamide gel electrophoresis (2D-PAGE), mass spectrometry and chromatographic technique to identify the proteins (Seib *et al.*, 2009). A disadvantage of the method is the failure to spot proteins with low quantities or solubility and those expressed only *in vivo*. By the immunoblot, immunoreactive antigens are estimated by separation of the bacterial proteins according to their electrical charge and size by 2D-PAGE, blotting onto membranes and testing with patient sera (Lahner *et al.*, 2010). Combining proteomics with immunoblot, the immunoproteomics is able to identify immunoreactive antigenic patterns, which are indicative for the severe or less severe outcomes of *H. pylori* infection or those for asymptomatic infections and hence can select within all antigens those appropriate for diagnostic use and vaccine preparation.

Michel *et al.* (2009) have used *H. pylori* multiplex serology based on recombinantly expressed and affinity-purified *H. pylori* proteins in combination with fluorescent bead technology (Luminex). In German patients, the authors have reported *H. pylori* serological reactivity to ≥4 proteins with sensitivity of 89% and specificity of 82%. Agreement with immunoblot has been good for CagA, moderate for UreA and weak for VacA (Michel *et al.*, 2009). In another German study, the antibodies to CagA and heat shock protein Hsp60 have been good predictors for gastric cancer (Gao *et al.*, 2009). Three *H. pylori* proteomic biomarkers (NapA, a RNA-binding protein, and a DNA-binding histone-like protein) have been shown to distinguish between duodenal ulcer and gastric cancer and therefore have been suggested to detect at-high risk patients for the cancer (Khoder *et al.*, 2009).

Combined evaluation of serum pepsinogens A and C, gastrin-17 and anti-*H. pylori* antibodies, used in a recently marketed serological kit, GastroPanel (Biohit Plc., Helsinki, Finland), can assess both *H. pylori* infection, corpus atrophy in adults or gastric inflammation in children. In Swedish adults, 96% agreement has been found between the results of GastroPanel and those of the histological methods for corpus atrophy (Storskrubb *et al.*, 2008). In Italian children, GastroPanel negative and positive predictive values have been 89.6% and 66.8%, respectively, so a negative test has been suggested to avoid endoscopy in children (Guariso *et al.*, 2009). In Italian adults, GastroPanel has shown 80% sensitivity and 96% specificity when compared with histology for corpus-prevalent chronic atrophic gastritis

(Lombardo *et al.*, 2010). These promising results should be confirmed in further studies worldwide.

Stool antigen test (SAT)

H. pylori stool antigen tests (SATs) are either polyclonal antibody (PAb) based or monoclonal antibody (MAb) based. They are either ELISA (EIA)-type or rapid (immunochromatographic) tests. For all SATs, fresh stool specimens should be used the same day or frozen at –20°C to –30°C, if testing is delayed. EIA-type SATs are read quantitatively (by spectrophotometry with ELISA reader), while reading of the rapid SATs is only visual. For best performance of SAT, patients are discouraged from taking antibiotics in the month before the test, PPIs in the 2 weeks before and histamine H_2 receptor antagonists at least 1 day before the test (Shah, 2007).

EIA-type MAb-SATs have shown higher sensitivity than PAb-SATs in both untreated (95 vs. 83%) and treated patients (91 vs. 76%, respectively) and good performance (LR^+ and LR^- of 24 and 0.07 before treatment, and 17 and 0.1, respectively, after treatment; Gisbert *et al.*, 2006; Cirak *et al.*, 2007). In bleeding ulcer patients, the sensitivity, specificity, LR^+ and LR^- of SAT have been inferior (87%, 70%, 2.3 and 0.2, respectively) to those in other patients (Gisbert and Abraira, 2006). PAb-SATs have shown variations (often from 67 to 98.3%) in sensitivity and cross-reactions with non-*pylori Helicobacter* species (Forné *et al.*, 2000). Nevertheless, recently, Mohammadian *et al.* (2008) have reported an easy, rapid, and economical production of highly specific PAb against *H. pylori* alkyl hydroperoxide reductase for EIA-type SAT.

Example of EIA-type PAb-SAT is HpSA test (Premier Platinum HpSA, Meridian Diagnostics, Cincinnati, OH, USA) and those of EIA-type MAb-SAT are HpStAR (Dako, Germany)/Femtolab (Connex, Martinsried, Germany) and Amplified IDEIA Hp StAR (DakoCytomation, Cambridge, UK). In Brazilian children aged below 7 years, Amplified IDEIA Hp StAR has shown 100% sensitivity but lower (76.2%) specificity (Raguza *et al.*, 2010). However, ROC curve can be used to improve SAT performance. After changing cut-off according to the ROC curve, the specificity of Amplified IDEIA Hp StAR in children has been increased by 21.5% without a decrease in sensitivity (Raguza *et al.*, 2010).

Recently, many rapid SATs have been developed. Examples are ImmunoCard Stat! HpSA test (Meridian Diagnostic, Cincinnati, Ohio), RAPID Hp Star test (DakoCytomation) and helicoCARE direct test (CARE diagnostica, Moellersdorf, Austria). Usually, for the rapid tests, the results are read in 5–15 min. However, the sensitivity (52–95%) and specificity (63–98%) of the rapid MAb-SAT have been highly variable (Table 3.1). In Turkish bleeding ulcer patients, the sensitivity of two rapid MAb-SATs was only 33–60% (Demiray *et al.*, 2006). For the rapid tests, the role of observer is very important. The EIA-type MAb-SAT, Amplified IDEIA Hp StAR, is more sensitive (90%) than the rapid ImmunoCard STAT! HpSA (69–74% according to the observer) and more specific (89%) than the immunochromatographic test RAPID Hp StAR (76–80% according to the observer; Calvet *et al.*, 2010). Good performance (sensitivity, specificity, LR+ and LR- of 90.3–93.5%, 89.4- 91.9%, 8.8–11.1% and 0.07–0.11%, according to the reader), comparable to that of IDEIA Hp StAR, has been reported in untreated patients by a modified rapid MAb-SAT for *H. pylori* antigen, RAPID Hp StAR (Prell *et al.*, 2009). However, post-treatment sensitivity of the rapid test has been inferior (71.4–85.7%) to that of IDEIA Hp StAR (100%). Importantly, the sensitivity of the test has been improved when all very weak lines have been deemed positive results (Prell *et al.*, 2009) as in our observations.

The sensitivity of rapid SAT test is reading time dependent. The sensitivity of Rapid Hp StAR has been found to be higher (76.9%) at 30 min than at 20 min (59.1%), whereas specificity has been reported to be higher (93.1%) at 20 min than at 60 min (79.5%; Krausse *et al.*, 2008). More studies are required to determine the best reading time of some rapid SATs. In treated patients, Blanco *et al.* (2008) found good agreement between UBT and HpStAR EIA (90%) or rapid MAb-SAT *H. pylori* Letitest (85%).

Recently, a commercially available rapid SAT (TestMate; Wakamoto Pharmaceutical, Tokyo, Japan), using MAb against *H. pylori* catalase, has been evaluated. With this test, high (>98%) specificity and excellent sensitivity (100% in North

American children and 91.6% in treated Japanese adults) have been reported (Cardenas *et al.*, 2008; Deguchi *et al.*, 2009). The test detects *H. pylori* antigen corresponding to 10^4–10^5 bacteria/ml and merits extensive evaluation. Probably, this test will be established as accurate option for routine diagnosis of *H. pylori* infection, given other studies are performed in more countries and more patients.

Several factors influence SAT accuracy. False-positive results can be found in patients with bleeding ulcers, while false-negative results can be found in those who had taken antibiotics and antisecretory drugs two weeks before the test or in those with corpus atrophy or constipation (Megraud and Lehours, 2007). As constipation may degrade *H. pylori* antigens, it should be recorded during patient examination and, if necessary, the test should be repeated later. A 24-hour delay in transport of the stool specimen also can lead to a false-negative SAT result. Whereas 4 weeks after eradication regimens have been enough to assess eradication failure by invasive tests and UBT, a longer (35–60-day) period has been suggested for SAT (Nurgalieva and Graham, 2003). The accuracy of SAT differs according to the geographic areas and patient groups. The accuracy of MAb-SAT was poor (61%) in aboriginal Australian children (Ritchie *et al.*, 2009). As for UBT, the performance of SAT can be low in young children. The sensitivity of a rapid MAb-SAT (Immunocard STAT!HpSA) was 100% in French children aged over 10 years but only 75% in those aged below 5 years (Kalach *et al.*, 2005).

The advantages of EIA-type MAb-SAT are its good accuracy, applicability for control of eradication and lower cost than UBT. Unlike UBT, SATs do not require any expensive apparatus. The disadvantages of SAT as a non-invasive test are the lack of information on possible gastric lesions and suboptimal (57% to <82%) specificity in treated patients 1 month after the therapy. SAT result is dependent on the storage of stool sample (at 2–8°C for <72 h and at −20°C to −80°C for longer storage). Sensitivity of SAT has decreased by 21% with storage of the samples at room temperature for 2–3 days (Malfertheiner *et al.*, 2007). Conversely, recalculating cut-off values by ROC curves has shown the sensitivity of SAT (PAb-based SAT) and urine EIA IgG ELISA to improve by 7.4–9% with slight decrease (by 3.9%) in specificity (Mégraud and European Paediatric Task Force on *Helicobacter pylori,* 2005). In any case, the EIA-type MAb-SATs are a valuable diagnostic option in many countries, especially in those lacking possibility to use UBT.

In conclusion, the choice of diagnostic test for *H. pylori* infection depends on numerous factors, e.g. patient characteristics such as age, disease, ethnicity, country and previous therapy, reflecting the likelihood of the infection, and test characteristics, e.g. type, preparation, accuracy, cut-off, availability and price. Low *H. pylori* density in the stomach of patients on PPIs, bismuth or antibiotics or in those with MALT lymphoma, bleeding ulcer and gastric atrophy often results in false-negative results of all tests except the serology (Nurgalieva and Graham, 2003). For instance, the sensitivity of CLO test (80.4%) and culture (56.2%) and specificity of serology (67.5%) have been low in cases with atrophic gastritis and intestinal metaplasia (Shin *et al.*, 2009). In such cases, at least two of these tests should be combined. Test sensitivity in bleeding ulcer patients has fallen from 78–94% to 21–66% for culture, from 89–90% to 21–43% for histology and from 82–94% to 29–37% for RUT, while PCR has been sensitive (often 79% to 91%) (Lo *et al.*, 2005; Peng *et al.*, 2009). In developed countries and children, it is important to confirm a positive non-invasive test by another test. For 'test and treat' strategy (*H. pylori* detection by non-invasive test before therapy), it has been recommended to use UBT or MAb-SAT for first diagnostics in dyspeptic adults aged <45 years with no alarming symptoms or before planned non-steroid anti-inflammatory drug therapy. The serology has been recommended in addition to the invasive tests for patients with probable low *H. pylori* density and for epidemiological trials. UBT or MAb-SATs have been stated to be appropriate to control the success of *H. pylori* eradication (Malfertheiner *et al.*, 2007; Selgrad *et al.*, 2009).

It should not be forgotten that the accuracy of non-invasive tests can be improved by recalculation of their cut-off values. However, suggesting an optimized cut-off value could be misleading if a small sample or inappropriate gold standard are used (Nurgalieva and Graham, 2003). As for the invasive tests, the histology is needed for patients aged >45 years and those with 'alarming symptoms' (pain, rectal bleeding, melaena, anaemia or weight loss) to not miss ulcers or tumours. Susceptibility testing of strains is recommended in countries/regions with high *H. pylori* resistance rates (Megraud and Lehours, 2007). In addition to the classical tests, PCR methods on biopsy or stool samples can detect *H. pylori* genes for antibiotic resistance.

Diagnostics in children is an important topic. Recommendations for *H. pylori* diagnosis in children have been published in the USA, Canada and Europe (Guarner *et al.*, 2010). Commercial non-invasive tests have been considered as accurate to detect child infections but the histology has been strongly recommended as the only method, which reveals both *H. pylori* and gastric lesions. In children, overall pre-treatment and post-treatment sensitivities of UBT (>75%) and MAb-SAT (>95%) have been reported by Guarner *et al.* (2010). In children with recurrent abdominal pain or iron deficiency anaemia, endoscopy is needed for the diagnosis. Within the non-invasive tests, UBT with DOB of 3.5% for children aged over 2 years, SAT with MAb and serology have been used, the latter test with highly controversial results in some populations and age groups (Frenck *et al.*, 2006; Cherian *et al.*, 2008).

In elderly patients, the UBT has exhibited the best accuracy (Salles and Mégraud, 2007). In this group, the serology can underrate the infection owing to low antibody titre caused by immunosuppression, antibiotic use for other infections or protein malnutrition. While SAT can be affected by chronic constipation and RUT, culture and histology can be suboptimal owing to PPI or antibiotic use and gastric atrophy, the histology is important to detect possible ulcers or malignancy in elderly patients (Salles and Mégraud, 2007). In addition, the serology is recommended for bleeding ulcer patients as RUT, culture, histology and UBT show suboptimal sensitivity and SAT exhibits relatively low specificity in this patient group (Megraud and Lehours, 2007). In MALT lymphoma patients, several tests, including the serology, should be used to provide optimal diagnosis (Asaka *et al.*, 2010) and treatment of the infection, which can cure at least half of the patients (see Chapter 10).

Newly developed (immunoproteomics and GastroPanel) or improved (by ROC curve) tests are useful for the accurate diagnosis of *H. pylori* infection and related predisposing factors. Nevertheless, further studies are needed on many topics. Among them are application of the current non-invasive tests or development of new non-invasive tests for *H. pylori* resistance to antibiotics; improvement of test performance in specific patient groups, e.g. young children, patients with atrophic gastritis, MALT lymphoma and treated patients; use of most reliable biomarkers for peptic ulcer and gastric cancer and a global strategy and target populations for *H. pylori* screening.

References

Algood, H.M., and Cover, T.L. (2006). *Helicobacter pylori* persistence: an overview of interactions between *H. pylori* and host immune defenses. Clin. Microbiol. Rev. *19*, 597–613.

Anania, C., Pacifico, L., Olivero, G., Osborn, J.F., Bonaiuto, E., and Chiesa, C. (2008). Breath tests in pediatrics. Clin. Chim. Acta *397*, 1–12.

Andrews, J., Marsden, B., Brown, D., Wong, V.S., Wood, E., and Kelsey, M. (2003). Comparison of three stool antigen tests for *Helicobacter pylori* detection. J. Clin. Pathol. *56*, 769–771.

Argent, R.H., Zhang, Y., and Atherton, J.C. (2005). Simple method for determination of the number of *Helicobacter pylori* CagA variable-region EPIYA tyrosine phosphorylation motifs by PCR. J. Clin. Microbiol. *43*, 791–795.

Asaka, M., Kato, M., Takahashi, S., Fukuda, Y., Sugiyama, T., Ota, H., Uemura, N., Murakami, K., Satoh, K., Sugano, K., and Japanese Society for *Helicobacter* Research. (2010). Guidelines for the management of *Helicobacter pylori* infection in Japan: 2009 revised edition. Helicobacter *15*, 1–20.

Azevedo, N.F., Almeida, C., Cerqueira, L., Dias, S., Keevil, C.W., and Vieira, M.J. (2007). Coccoid form of *Helicobacter pylori* as a morphological manifestation of cell adaptation to the environment. Appl. Environ. Microbiol. *73*, 3423–3427.

Azizian, H., Bahrami, H., Pasalar, P., and Amanlou, M. (2010). Molecular modeling of *Helicobacter pylori* arginase and the inhibitor coordination interactions. J. Mol. Graph. Model. *28*, 626–635.

Benoit, S.L., and Maier, R.J. (2008). Hydrogen and nickel metabolism in *Helicobacter* species. Ann. N. Y. Acad. Sci. *1125*, 242–251.

Blanco, S., Forné, M., Lacoma, A., Prat, C., Cuesta, M.A., Latorre, I., Viver, J.M., Fernández, G., Molinos, S., and Domínguez, J. (2008). Comparison of stool antigen immunoassay methods for detecting *Helicobacter pylori* infection before and after eradication treatment. Diagn. Microbiol. Infect. Dis. *61,* 150–155.

Bogaerts, P., Berhin, C., Nizet, H., and Glupczynski, Y. (2006). Prevalence and mechanisms of resistance to fluoroquinolones in *Helicobacter pylori* strains from patients living in Belgium. Helicobacter *11,* 441–445.

Boneca, I.G. (2005). The role of peptidoglycan in pathogenesis. Curr. Opin. Microbiol. *8,* 46–53.

Boyanova, L. (2007). Detection of *Helicobacter pylori* infection in symptomatic Bulgarian adults. Clin. Microbiol. Infect. *13,* 908–914.

Boyanova, L., Lazarova, E., Jelev, C., Gergova, G., and Mitov, I. (2007). *Helicobacter pylori* and *Helicobacter heilmannii* in untreated Bulgarian children over a period of 10 years. J. Med. Microbiol. *56,* 1081–1085.

Boyanova, L., Gergova, G., Nikolov, R., Davidkov, L., Kamburov, V., Jelev, C., and Mitov I. (2008). Prevalence and evolution of *Helicobacter pylori* resistance to 6 antibacterial agents over 12 years and correlation between susceptibility testing methods. Diagn. Microbiol. Infect. Dis. *60,* 409–415.

Broutet, N., Marais, A., Lamouliatte, H., de Mascarel, A., Samoyeau, R., Salamon, R., and Mégraud, F. (2001). *cagA* status and eradication treatment outcome of anti*Helicobacter pylori* triple therapies in patients with nonulcer dyspepsia J. Clin. Microbiol. *39,*1319–1322.

Burucoa, C., Garnier, M., Silvain, C., and Fauchère, J.L. (2008). Quadruplex real-time PCR assay using allele-specific scorpion primers for detection of mutations conferring clarithromycin resistance to *Helicobacter pylori*. J. Clin. Microbiol. *46,* 2320–2326.

Bury-Moné, S., Kaakoush, N.O., Asencio, C., Mégraud, F., Thibonnier, M., De Reuse, H., and Mendz, G.L. (2006). Is *Helicobacter pylori* a true microaerophile? Helicobacter *11,* 296–303.

Calvet, X, Sánchez-Delgado, J., Montserrat, A., Lario, S., Ramírez-Lázaro, M.J., Quesada, M., Casalots, A., Suárez, D., Campo, R., Brullet, E., *et al.* (2009). Accuracy of diagnostic tests for *Helicobacter pylori*: a reappraisal. Clin. Infect. Dis. *48,* 1385–1391.

Calvet, X., Lario, S., Ramírez-Lázaro, M.J., Montserrat, A., Quesada, M., Reeves, L., Masters, H., Suárez-Lamas, D., Gallach, M., Sánchez-Delgado, J., Martínez-Bauer, E., *et al.* (2010). Comparative accuracy of 3 monoclonal stool tests for diagnosis of *Helicobacter pylori* infection among patients with dyspepsia. Clin. Infect. Dis. *50,* 323–328.

Cambau, E., Allerheiligen, V., Coulon, C., Corbel, C., Lascols, C., Deforges, L., Soussy, C.J., Delchier, J.C., and Megraud, F. (2009). Evaluation of a new test, genotype HelicoDR, for molecular detection of antibiotic resistance in *Helicobacter pylori*. J. Clin. Microbiol. *47,* 3600–3607.

Camorlinga-Ponce, M., Flores-Luna, L., Lazcano-Ponce, E., Herrero, R., Bernal-Sahagún, F., Abdo-Francis, J.M., Aguirre-García, J., Muñoz, N., and Torres, J. (2008). Age and severity of mucosal lesions influence the performance of serologic markers in *Helicobacter pylori*-associated gastroduodenal pathologies. Cancer Epidemiol. Biomarkers Prev. *17,* 2498–2504.

Campuzano-Maya, G. (2007). An optimized ^{13}C-urea breath test for the diagnosis of *H pylori* infection. World J. Gastroenterol. *13,* 5454–5464.

Cardenas, V.M., Dominguez, D.C., Puentes, F.A., Aragaki, C.C., Goodman, K.J., Graham, D.Y., and Fukuda, Y. (2008). Evaluation of a novel stool native catalase antigen test for *Helicobacter pylori* infection in asymptomatic North American children. J. Pediatr Gastroenterol. Nutr. *46,* 399–402.

Caristo, E., Parola, A., Rapa, A., Vivenza, D., Raselli, B., Dondi, E., Boldorini, R., and Oderda, G. (2008). Clarithromycin resistance of *Helicobacter pylori* strains isolated from children gastric antrum and fundus as assessed by fluorescent *in situ* hybridization and culture on four-sector agar plates. Helicobacter *13,* 557–563.

Cellini, L., Grande, R., Di Campli, E., Traini, T., Giulio, M.D., Lannutti, S.N., and Lattanzio, R. (2008). Dynamic colonization of *Helicobacter pylori* in human gastric mucosa. Scand. J. Gastroenterol. *43,* 178–185.

Chaput, C., Ecobichon, C., Cayet, N., Girardin, S.E., Werts, C., Guadagnini, S., Prévost, M.C., Mengin-Lecreulx, D., Labigne, A., and Boneca, I.G. (2006). Role of AmiA in the morphological transition of *Helicobacter pylori* and in immune escape. PLoS Pathog. *2,* e97.

Chaput, C., Labigne, A., and Boneca, I.G. (2007). Characterization of *Helicobacter pylori* lytic transglycosylases Slt and MltD. J. Bacteriol. *189,* 422–429.

Cherian, S., Burgner, D.P., Carson, C.F., Sanfilippo, F.M., Cook, A.G., and Forbes, D.A. (2008). Diagnosis of *Helicobacter pylori* infection in a high-prevalence pediatric population: a comparison of 2 fecal antigen testing methods and serology. J. Pediatr. Gastroenterol. Nutr. *47,* 130–135.

Chisholm, S.A., and Owen, R.J. (2003). Mutations in *Helicobacter pylori rdxA* gene sequences may not contribute to metronidazole resistance. J. Antimicrob. Chemother. *51,* 995–999.

Chisholm, S.A., and Owen, R.J. (2008). Application of polymerase chain reaction-based assays for rapid identification and antibiotic resistance screening of *Helicobacter pylori* in gastric biopsies. Diagn. Microbiol. Infect. Dis. *61,* 67–71.

Chisholm, S.A., and Owen, R.J. (2009). Frequency and molecular characteristics of ciprofloxacin- and rifampicin-resistant *Helicobacter pylori* from gastric infections in the UK. J. Med. Microbiol. *58,* 1322–1328.

Cho, Y.S., Chae, H.S., Jang, S.N., Kim, J.S., Son, H.S., Kim, H.K., Kim, B.W., Han, S.W., Choi, K.Y., Lee, H.K., *et al.* (2008). Comparison of the ^{13}C-urea breath test and the endoscopic phenol red mucosal pH test in the quantification of *Helicobacter pylori* infection loading. Korean J. Intern. Med. 23, 134–139.

Chomvarin, C., Kulsantiwong, P., Chantarasuk, Y., Chantrakooptungool, S., and Kanjanahareutai, S. (2006). Comparison of media and antibiotic supplements for isolation of *Helicobacter pylori* from gastric biopsies. SouthEast Asian J. Trop. Med. Public Health. 37, 1163–1169.

Chong, V.H. (2008). Poor sensitivity of the signify *H. pylori* test. Curr. Microbiol. *57*, 519–520.

Chu, Y.T., Wang, Y.H., Wu, J.J., and Lei, H.Y. (2010). Invasion and multiplication of *Helicobacter pylori* in gastric epithelial cells and implications for antibiotic resistance. Infect. Immun. *78*, 4157–4165.

Cirak, M. Y., Akyön, Y., and Mégraud, F. (2007). Diagnosis of *Helicobacter pylori*. Helicobacter *12 (Suppl 1)*, 4–9.

CLSI-National Committee for Clinical Laboratory Standards. (2010). Performance standards for antimicrobial susceptibility testing. Twentieth informational supplement. Approved standard M100-S20. M7, NCCLS (CLSI :Wayne, PA), p. 107.

Co, E.M., and Schiller, N.L. (2006). Resistance mechanisms in an *in vitro*-selected amoxicillin-resistant strain of *Helicobacter pylori*. Antimicrob. Agents Chemother. *50*, 4174–4176.

Davood, E., Ashraf, M.M., Hatef, S.A., and Ahmad, Z.H. (2009). Optimization of *Helicobacter pylori* culture in order to prepare favorable antigens. J.Bacteriol. Res. *1*, 101–104.

De Francesco, V., Zullo, A., Perna, F., Giorgio, F., Hassan, C., Vannella, L., Cristofari, F., Panella, C., Vaira, D., and Ierardi, E. (2010a). *H. pylori* antibiotic resistance and ^{13}C urea breath test values. J. Med. Microbiol. *59*, 588–591.

De Francesco, V., Zullo, A., Ierardi, E., Giorgio, F., Perna, F., Hassan, C., Morini, S., Panella, C., and Vaira, D. (2010b). Phenotypic and genotypic *Helicobacter pylori* clarithromycin resistance and therapeutic outcome: benefits and limits. J. Antimicrob. Chemother. *65*, 327–332.

Degnan, A.J., Sonzogni, W.C., and Standridge, J.H. (2003). Development of a plating medium for selection of *Helicobacter pylori* from water samples. Appl. Environ. Microbiol. *69*, 2914–2918.

Deguchi, R., Matsushima, M., Suzuki, T., Mine, T., Fukuda, R., Nishina, M., Ozawa, H., and Takagi, A. (2009). Comparison of a monoclonal with a polyclonal antibody-based enzyme immunoassay stool test in diagnosing *Helicobacter pylori* infection after eradication therapy. J. Gastroenterol. *44*, 713–716.

Demiray, E., Yilmaz, Ö., Şarkiş, C., Soytürk, M., and Şimşek, I. (2006). Comparison of invasive methods and two different stool antigen tests for diagnosis of *H pylori* infection in patients with gastric bleeding. World J. Gastroenterol. *12*, 4206–4210.

Desikan, P. (2010). The incredible journey of mankind: *Helicobacter pylori* as the narrator. Indian J. Med. Microbiol. *28*, 193–195.

Dierikx, C.M., Martodihardjo, J., Kuipers, E.J., Hensgens, C.M., Kusters, J.G., Suzuki, H., de Groot, N., and van Vliet, A.H. (2007). Serum- and animal tissue-free medium for transport and growth of *Helicobacter pylori*. FEMS Immunol. Med. Microbiol. *50*, 239–243.

Diouf, A., Martinez-Gomis, J., Miquel, M., Quesada, M., Lario, S., and Sixou, M. (2009). Comparison of four different primer sets for detection of *Helicobacter pylori* in gastric biopsies and oral samples by using real-time PCR. Pathol. Biol. (Paris) *57*, 30–35.

Doherty, N.C., Shen, F., Halliday, N.M., Barrett, D.A., Hardie, K.R., Winzer, K., and Atherton, J.C. (2010). In *Helicobacter pylori*, LuxS is a key enzyme in cysteine provision through a reverse transsulfuration pathway. J. Bacteriol. *192*, 1184–1192.

Dondi, E., Rapa, A., Boldorini, R., Fonio, P., Zanetta, S., and Oderda, G. (2006). High accuracy of noninvasive tests to diagnose *Helicobacter pylori* infection in very young children. J. Pediatr. *149*, 817–821.

Douillard, F.P., Ryan, K.A., Lane, M.C., Caly, D.L., Moore, S.A., Penn, C.W., Hinds, J., and O'Toole, P.W. (2010). The *Hp0256* gene product is involved in motility and cell envelope architecture of *Helicobacter pylori*. BMC Microbiol. *10*, 106.

Douraghi, M., Saberi Kashani, S., Zeraati, H., Esmaili, M., Oghalaie, A., and Mohammadi, M. (2010). Comparative evaluation of three supplements for *Helicobacter pylori* growth in liquid culture. Curr. Microbiol. *60*, 254–262.

Elitsur, Y., Tolia, V., Gilger, M.A., Reeves-Garcia, J., Schmidt-Sommerfeld, E., Opekun, A.R., El-Zimaity, H., Graham, D.Y., and Enmei, K. (2009). Urea breath test in children: The United States prospective, multicenter study. Helicobacter *14*, 134–140.

Erzin, Y., Altun, S., Dobrucali, A., Aslan, M., Erdamar, S., Dirican, A., and Kocazeybek, B. (2004). Comparison of two different stool antigen tests for the primary diagnosis of *Helicobacter pylori* infection in Turkish patients with dyspepsia. Helicobacter *9*, 657–662.

Fischbach, W., Malfertheiner, P., Hoffmann, J.C., Bolten, W., Bornschein, J., Götze, O., Höhne, W., Kist, M., Koletzko, S., Labenz, J., *et al.* (2009). S3-guideline '*Helicobacter pylori* and gastroduodenal ulcer disease' of the German society for digestive and metabolic diseases (DGVS) in cooperation with the German society for hygiene and microbiology, society for pediatric gastroenterology and nutrition e. V., German society for rheumatology, AWMF-registration-no. 021/001. Z. Gastroenterol. 47, 1230–1263.

Ford, J.L., Kaakoush, N.O., and Mendz, G.L. (2010). Phosphonate metabolism in *Helicobacter pylori*. Antonie Van Leeuwenhoek 97, 51–60.

Forné, M., Domínguez, J., Fernández-Bañares, F., Lite, J., Esteve, M., Galí, N., Espinós, J.C., Quintana, S., and Viver J.M. (2000). Accuracy of an enzyme immunoassay for the detection of *Helicobacter pylori* in stool specimens in the diagnosis of infection and posttreatment check-up. Am. J. Gastroenterol. *95*, 2200–2225.

Frenck, R.W., Jr., Fathy, H.M., Sherif, M., Mohran, Z., El Mohammedy, H., Francis, W., Rockabrand, D., Mounir, B.I., Rozmajzl, P., and Frierson, H.F. (2006). Sensitivity and specificity of various tests for the diagnosis of *Helicobacter pylori* in Egyptian children. Pediatrics *118*, e1195-e1202.

Furuta, T., Soya, Y., Sugimoto, M., Shirai, N., Nakamura, A., Kodaira, C., Nishino, M., Okuda, M., Okimoto, T., Murakami, K., *et al.* (2007). Modified allele-specific primer-polymerase chain reaction method for analysis of susceptibility of *Helicobacter pylori* strains to clarithromycin. J. Gastroenterol. Hepatol. *22*, 1810–1815.

Gang, J.G., Yun, S.K., and Hwang, S.Y. (2009). *Helicobacter pylori* urease may exist in two forms: evidence from the kinetic studies. J. Microbiol. Biotechnol. *19*, 1565–1568.

Gao, L., Michel, A., Weck, M.N., Arndt, V., Pawlita, M., and Brenner, H. (2009). *Helicobacter pylori* infection and gastric cancer risk: evaluation of 15 *H. pylori* proteins determined by novel multiplex serology. Cancer Res. *69*, 6164–6170.

Ghose, C., Perez-Perez, G.I., van Doorn, L.J., Domínguez-Bello, M.G., and Blaser, M.J. (2005). High frequency of gastric colonization with multiple *Helicobacter pylori* strains in Venezuelan subjects. J. Clin. Microbiol. *43*, 2635–2641.

Gill, P., Alvandi, A.H., Abdul-Tehrani, H., and Sadeghizadeh, M. (2008). Colorimetric detection of *Helicobacter pylori* DNA using isothermal helicase-dependent amplification and gold nanoparticle probes. Diagn. Microbiol. Infect. Dis. *62*, 119–124.

Gisbert, J.P., and Pajares, J.M. (2004). Review article: C-urea breath test in the diagnosis of *Helicobacter pylori* infection: a critical review. Aliment. Pharmacol. Ther. *20*, 1001–1017.

Gisbert, J.P., and Abraira, V. (2006). Accuracy of *Helicobacter pylori* diagnostic tests in patients with bleeding peptic ulcer: a systematic review and meta-analysis. Am. J. Gastroenterol. *101*, 848–863.

Gisbert, J.P., de la Morena, F., and Abraira, V. (2006). Accuracy of monoclonal stool antigen test for the diagnosis of *H. pylori* infection: a systematic review and meta-analysis. Am. J. Gastroenterol. *101*, 1921–1930.

Gisbert, J.P., Esteban, C., Jimenez, I., and Moreno-Otero, R. (2007). 13C-urea breath test during hospitalization for the diagnosis of *Helicobacter pylori* infection in peptic ulcer bleeding. Helicobacter *12*, 231–237.

Glocker, E., and Kist, M. (2004). Rapid detection of point mutations in the *gyrA* gene of *Helicobacter pylori* conferring resistance to ciprofloxacin by a fluorescence resonance energy transfer-based real-time PCR approach. J. Clin. Microbiol. *42*, 2241–2246.

Glocker, E., Berning, M., Gerrits, M.M., Kusters, J.G., and Kist, M. (2005). Real-time PCR screening for 16S rRNA mutations associated with resistance to tetracycline in *Helicobacter pylori*. Antimicrob. Agents Chemother. *49*, 3166–3170.

Glocker, E., Bogdan, C., and Kist, M. (2007). Characterization of rifampicin-resistant clinical *Helicobacter pylori* isolates from Germany. J. Antimicrob. Chemother. *59*, 874 – 879.

Glupczynski, Y., Broutet, N., Cantagrel, A., Andersen, L.P., Alarcon, T., Lopez-Brea, M., and Mégraud, F. (2002). Comparison of the E test and agar dilution method for antimicrobial suceptibility testing of *Helicobacter pylori*. Eur. J. Clin. Microbiol. Infect. Dis. *21*, 549–552.

Goh, K.L., Cheah, P.L., Navaratnam, P., Chin, S.C., and Xiao, S.D. (2007). HUITAI rapid urease test: a new ultra-rapid biopsy urease test for the diagnosis of *Helicobacter pylori* infection. J. Dig. Dis. *8*, 139–142.

Graham, D.Y., Kudo, M., Reddy, R., and Opekun, A.R. (2005). Practical rapid, minimally invasive, reliable nonendoscopic method to obtain *Helicobacter pylori* for culture. Helicobacter *10*, 1–3.

Grignon, B., Tankovic, J., Megraud, F., Glupczynski, Y., Husson, M.O., Conroy, M.C., Emond, J.P., Loulergue, J., Raymond, J., and Fauchere, J.L. (2002). Validation of diffusion methods for macrolide susceptibility testing of *Helicobacter pylori*. Microb. Drug Resist. *8*, 61–66.

Guariso, G., Basso, D., Bortoluzzi, C.F., Meneghel, A., Schiavon, S., Fogar, P., Farina, M., Navaglia, F., Greco, E., Mescoli, C., *et al.* (2009). GastroPanel: evaluation of the usefulness in the diagnosis of gastro-duodenal mucosal alterations in children. Clin. Chim. Acta. *402*, 54–60.

Guarner, J., Kalach, N, Elitsur, Y., and Koletzko, S. (2010). *Helicobacter pylori* diagnostic tests in children: review of the literature from 1999 to 2009. Eur. J. Pediatr. *169*, 15–25.

Hafeez, A., Bilal, R., Haseeb, H.A., Khan, U.F., Latif, Z., and Hassan, M. (2007). Comparison of diagnostic accuracy of non-invasive tests for *Helicobacter pylori* infection in children. J. Coll. Physicians Surg. Pak. *17*, 261–264.

Hug, I., Couturier, M.R., Rooker, M.M., Taylor, D.E., Stein, M., and Feldman, M.F. (2010). *Helicobacter pylori* lipopolysaccharide is synthesized via a novel pathway with an evolutionary connection to protein *N*-glycosylation. PLoS Pathog. *6*, e1000819.

Ismail, S., Hampton, M.B., and Keenan, J.I. (2003). *Helicobacter pylori* outer membrane vesicles modulate proliferation and interleukin-8 production by gastric epithelial cells. Infect. Immun. *71*, 5670–5675.

Kaakoush, N.O., Baar, C., Mackichan, J., Schmidt, P., Fox, E.M., Schuster, S.C., and Mendz, G.L. (2009). Insights into the molecular basis of the microaerophily of three *Campylobacterales*: a comparative study. Antonie Van Leeuwenhoek *96*, 545–557.

Kacar, F., Çulhacı, N., Yükselen, V., Meteoğlu, I., Dikicioğlu, E., and Levi, E. (2004). Histologic demonstration of *Helicobacter pylori* in gastric biopsies: which is the best staining method? Int. J. Pathol. 3.

Kalach, N., Nguyen, V.B., Bergeret, M., Boutros, N., Dupont, C., and Raymond, J. (2005). Usefulness and influence of age of a novel rapid monoclonal enzyme immunoassay stool antigen for the diagnosis of *Helicobacter pylori* infection in children. Diagn. Microbiol. Infect. Dis. *52*, 157–160.

Kalach, N., Dehecq, E., Gosset, P., Decoster, A., Bergeret, M., Dupont, C., and Raymond, J. (2009). Usefulness and influence of age of a novel Rapid HpStAR stool antigen for the diagnosis of *Helicobacter pylori* infection in children. Diagn. Microbiol. Infect. Dis. *65*, 450–453.

Kanada, R., Uchida, T., Tsukamoto, Y., Nguyen, L.T., Hijiya, N., Matsuura, K., Kodama, M., Okimoto, T., Murakami, K., Fujioka, T., *et al.* (2008). Genotyping of the *cagA* gene of *Helicobacter pylori* on immunohistochemistry with East Asian CagA-specific antibody. Pathol. Int. *58*, 218–225.

Kawai, T., Kawakami, K., Kataoka, M., Itoi, T., Takei, K., Moriyasu, F., Takagi. Y., Aoki, T., Serizawa, H., Rimbard, E., *et al.* (2008). A study of the relationship between *Helicobacter pylori* microbial susceptibility, ^{13}C-urea breath test values. Hepatogastroenterology 55, 786–790.

Khoder, G., Yamaoka, Y., Fauchère, J.L., Burucoa, C., and Atanassov, C. (2009). Proteomic *Helicobacter pylori* biomarkers discriminating between duodenal ulcer and gastric cancer. J. Chromatogr. B. Analyt. Technol. Biomed. Life Sci. *877*, 1193–1199.

Kim, C.G., Choi, I.J., Lee, J.Y., Cho, S.J., Nam, B.H., Kook, M.C., Hong, E.K., and Kim, Y.W. (2009). Biopsy site for detecting *Helicobacter pylori* infection in patients with gastric cancer. J. Gastroenterol. Hepatol. *24*, 469–474.

Koletzko, S. (2005). Noninvasive diagnostic tests for *Helicobacter pylori* infection in children. Can. J. Gastroenterol. *19*, 433–439.

Krausse, R., Müller, G., and Doniec, M. (2008). Evaluation of a rapid new stool antigen test for diagnosis of *Helicobacter pylori* infection in adult patients. J. Clin. Microbiol. *46*, 2062–2065.

Kumar, S., Kumar, A., and Dixit, V.K. (2010). Diversity in the *cag* pathogenicity island of *Helicobacter pylori* isolates in populations from North and South India. J. Med. Microbiol. *59*, 32–40.

Kusters, J.G., van Vliet, A.H.M., and Kuipers, E.J. (2006). Pathogenesis of *Helicobacter pylori* infection. Clin. Microbiol. Rev. *19*, 449–490.

Lahner, E., Bernardini, G., Santucci, A., and Annibale, B. (2010). *Helicobacter pylori* immunoproteomics in gastric cancer and gastritis of the carcinoma phenotype. Expert Rev. Proteomics 7, 239–248.

Lai, Y.C., Wang, T.H., Huang, S.H., Yang, S.S., Wu, C.H., Chen, T.K., and Lee, C.L. (2003). Density of *Helicobacter pylori* may affect the efficacy of eradication therapy and ulcer healing in patients with active duodenal ulcers. World J. Gastroenterol. 9, 1537–1540.

Lawson, A.J., Elviss, N.C., and Owen, R.J. (2005). Real-time PCR detection and frequency of 16S rDNA mutations associated with resistance and reduced susceptibility to tetracycline in *Helicobacter pylori* from England and Wales. J. Antimicrob. Chemother. *56*, 282–286.

Leal, Y.A., Flores, L.L., García-Cortés, L.B., Cedillo-Rivera, R., and Torres, J. (2008). Antibody-based detection tests for the diagnosis of *Helicobacter pylori* infection in children: a meta-analysis. PLoS One 3, e3751.

Lee, H., Kho, H.S., Chung, J.W., Chung, S.C., and Kim, Y.K. (2006). Volatile sulfur compounds produced by *Helicobacter pylori*. J. Clin. Gastroenterol. *40*, 421–426.

Lienlaf, M., Morales, J.P., Díaz, M.I., Díaz, R., Bruce, E., Siegel, F., León, G., Harris, P.R., and Venegas, A. (2010). *Helicobacter pylori* HopE and HopV porins present scarce expression among clinical isolates. World J. Gastroenterol. *16*, 320–329.

Lindsetmo, R.-O., Johnsen, R., Eide, T.J., Gutteberg, T., Husum, H.H., and Revhaug, A. (2008). Accuracy of *Helicobacter pylori* serology in two peptic ulcer populations and in healthy controls. World J. Gastroenterol. *14*, 5039–5045.

Lo, C.C., Lai, K.H., Peng, N.J., Lo, G.H., Tseng, H.H., Lin, C.K., Shie, C.B., Wu, C.M., Chen, Y.S., Huang, W.K., *et al.* (2005). Polymerase chain reaction: a sensitive method for detecting *Helicobacter pylori* infection in bleeding peptic ulcers. World J. Gastroenterol. *11*, 3909–3914.

Loh, J.T., Gupta, S.S., Friedman, D.B., Krezel, A.M., and Cover, T.L. (2010). Analysis of protein expression regulated by the *Helicobacter pylori* ArsRS two-component signal transduction system. J. Bacteriol. *192*, 2034–2043.

Lombardo, L., Leto, R., Molinaro, G., Migliardi, M., Ravarino, N., Rocca, R., and Torchio, B. (2010). Prevalence of atrophic gastritis in dyspeptic patients in Piedmont. A survey using the GastroPanel test. Clin. Chem. Lab. Med. *48*, 1327–1332.

Lottspeich, C., Schwarzer, A., Panthel, K., Koletzko, S., and Rüssmann, H. (2007). Evaluation of the novel *Helicobacter pylori* ClariRes real-time PCR assay for detection and clarithromycin susceptibility testing of *H. pylori* in stool specimens from symptomatic children. J. Clin. Microbiol. *45*, 1718–1722.

Lowenthal, A.C., Hill, M., Sycuro, L.K., Mehmood, K., Salama, N.R., and Ottemann, K.M. (2009). Functional analysis of the *Helicobacter pylori* flagellar switch proteins. J. Bacteriol. *191*, 7147–7156.

Lunet, N., Peleteiro, B., Carrilho, C., Figueiredo, C., and Azevedo, A. (2009). Sensitivity is not an intrinsic property of a diagnostic test: empirical evidence from histological diagnosis of *Helicobacter pylori* infection. BMC Gastroenterol. *9*, 98.

Makristathis, A., Hirschl, A.M., Rüssmann, H., and Koletzko, S. (2007). Detection and clarithromycin susceptibility testing of *Helicobacter pylori* in stool specimens by real-time PCR: how to get accurate test results. J. Clin. Microbiol. *45*, 2756.

Malfertheiner, P., Megraud, F., O'Morain, C., Bazzoli, F., El-Omar, E., Graham, D., Hunt, R., Rokkas, T., Vakil, N., Kuipers, E.J., and the European Helicobacter Study Group (EHSG). (2007). Current concepts in the management of *Helicobacter pylori* infection: the Maastricht III Consensus Report. Gut *56*, 772–781.

Marsich, E., Zuccato, P., Rizzi, S., Vetere, A., Tonin, E., and Paoletti, S. (2002). *Helicobacter pylori* expresses an autolytic enzyme: gene identification, cloning, and theoretical protein structure. J. Bacteriol. *184*, 6270–6279.

Matteo, M.J., Granados, G., Pérez, C.V., Olmos, M., Sanchez, C., and Catalano, M. (2007). *Helicobacter pylori cag* pathogenicity island genotype diversity within the gastric niche of a single host. J. Med. Microbiol. *56*, 664–669.

Mauro, M., Radovic, V., Zhou, P., Wolfe, M., Kamath, M., Bercik, P., Croitoru, K., and Armstrong, D. (2006). ^{13}C urea breath test for *Helicobacter pylori*: determination of the optimal cut-off point in a Canadian community population. Can. J. Gastroenterol. *20*, 770–774.

McNulty, C., Owen, R., Tompkins, D., Hawtin, P., McColl, K., Price, A., Smith, G., Teare, L., and PHLS

Helicobacter Working Group. (2002). *Helicobacter pylori* susceptibility testing by disc diffusion. J. Antimicrob. Chemother. *49,* 601–609.

Mégraud, F., and European Paediatric Task Force on *Helicobacter pylori.* (2005). Comparison of non-invasive tests to detect *Helicobacter pylori* infection in children and adolescents: results of a multicenter European study. J. Pediatr. *146,* 198–203.

Megraud, F., and Lehours, P. (2007). *Helicobacter pylori* detection and antimicrobial susceptibility testing. Clin. Microbiol. Rev. *20,* 280–322.

Mégraud, F., and Corti, R. (2009). Bacterial resistance of *Helicobacter pylori* in the world in 2009. Acta Gastroenterol. Latinoam. *39,* 282–290.

Michel, A., Waterboer, T., Kist, M., and Pawlita, M. (2009). *Helicobacter pylori* multiplex serology. Helicobacter *14,* 525–535.

Miciuleviciene, J., Calkauskas, H., Jonaitis, L., Kiudelis, G., Tamosiūnas, V., Praskevicius, A., Kupcinskas, L., and Berg, D. (2008). *Helicobacter pylori* genotypes in Lithuanian patients with chronic gastritis and duodenal ulcer. Medicina (Kaunas) *44,* 449–454.

Mitchell, H., English, D.R., Elliott, F., Gengos, M., Barrett, J.H., Giles, G.G., and Forman, D. (2008). Immunoblotting using multiple antigens is essential to demonstrate the true risk of *Helicobacter pylori* infection for gastric cancer. Aliment. Pharmacol. Ther. *28,* 903–910.

Mohammadian, T., Doosti, M., Paknejad, M., Siavoshi, F., Massarrat, S., and Soukhtanloo, M. (2008). Production of polyclonal antibody against alkyl hydroperoxide reductase of *Helicobacter pylori* and its antigenicity. Hybridoma (Larchmt) *27,* 481–485.

Monstein, H.J., Karlsson, A., Ryberg, A., and Borch, K. (2010). Application of PCR amplicon sequencing using a single primer pair in PCR amplification to assess variations in *Helicobacter pylori* CagA EPIYA tyrosine phosphorylation motifs. BMC Res. Notes *3,* 35.

Monteiro, L., Oleastro, M., Lehours, P., and Mégraud, F. (2009). Diagnosis of *Helicobacter pylori* infection. Helicobacter *14* (Suppl. 1), 8–14.

Moran, A.P. (2010). The role of endotoxin in infection: *Helicobacter pylori* and *Campylobacter jejuni.* Subcell. Biochem. *53,* 209–240.

Morris, J.M., Reasonover, A.L., Bruce, M.G., Bruden, D.L., McMahon, B.J., Sacco, F.D., Berg, D.E., and Parkinson, A.J. (2005). Evaluation of seaFAST, a rapid fluorescent *in situ* hybridization test, for detection of *Helicobacter pylori* and resistance to clarithromycin in paraffin-embedded biopsy sections. J. Clin. Microbiol. *43,* 3494–3496.

Motta, O., De Caro, F., Quarto, F., and Proto, A. (2009). New FTIR methodology for the evaluation of $^{13}C/^{12}C$ isotope ratio in *Helicobacter pylori* infection diagnosis. J. Infect. *59,* 90–94.

Mullaney, E., Brown, P.A., Smith, S.M., Botting, C.H., Yamaoka, Y.Y., Terres, A.M., Kelleher, D.P., and Windle, H.J. (2009). Proteomic and functional characterization of the outer membrane vesicles from the gastric pathogen *Helicobacter pylori.* Prot. Clin. Appl. *3,* 785–796.

Nguyen, L.T., Uchida, T., Tsukamoto, Y., Trinh, T.D., Ta, L., Ho, D.Q., Matsuhisa, T., Uchida, M., Takayama, A., Hijiya, N., *et al.* (2010). Evaluation of rapid urine test for the detection of *Helicobacter pylori* infection in the Vietnamese population. Dig. Dis. Sci. *55,* 89–93.

Nishizawa, T., Suzuki, H., Umezawa, A., Muraoka, H., Iwasaki, E., Masaoka, T., Kobayashi, I., and Hibi T. (2007). Rapid detection of point mutations conferring resistance to fluoroquinolone in *gyrA* of *Helicobacter pylori* by allele-specific PCR. J. Clin. Microbiol. *45,* 303–305.

Nurgalieva, Z.Z., and Graham, D.Y. (2003). Pearls and pitfalls of assessing *Helicobacter pylori* status. Dig. Liver Dis. *35,* 375–377.

Nurgalieva, Z.Z., Opekun, A.R., and Graham, D.Y. (2006). Problem of distinguishing false-positive tests from acute or transient *Helicobacter pylori* infections. Helicobacter *11,* 69–74.

Oleastro, M., Ménard, A., Santos, A., Lamouliatte, H., Monteiro, L., Barthélémy, P., and Mégraud, F. (2003). Real-time PCR assay for rapid and accurate detection of point mutations conferring resistance to clarithromycin in *Helicobacter pylori.* J. Clin. Microbiol. *41,* 397–402.

Osaki, T., Mabe, K., Hanawa, T., and Kamiya, S. (2008). Urease-positive bacteria in the stomach induce a false-positive reaction in a urea breath test for diagnosis of *Helicobacter pylori* infection. J. Med. Microbiol. *57,* 814–819.

Owen, R.J. (2002). Molecular testing for antibiotic resistance in *Helicobacter pylori.* Gut 50, 285–289.

Ozdemir, E., Karabacak, N.I., Degertekin, B., Cirak, M., Dursun, A., Engin, D., Unal, S., and Unlü, M. (2008). Could the simplified ^{14}C urea breath test be a new standard in noninvasive diagnosis of *Helicobacter pylori* infection? Ann. Nucl. Med. *22,* 611–616.

Parker, H., Chitcholtan, K., Hampton, M.B., and Keenan, J.I. (2010). Uptake of *Helicobacter pylori* outer membrane vesicles by gastric epithelial cells. Infect. Immun. 78, 5054–5061.

Peng, N.J., Lai, K.H., Lo, G.H., and Hsu, P.I. (2009). Comparison of noninvasive diagnostic tests for *Helicobacter pylori* infection. Med. Princ. Pract. *18,* 57–61.

Prell, C., Osterrieder, S., Lottspeich, C., Schwarzer, A., Rüssmann, H., Ossiander, G., and Koletzko, S. (2009). Improved performance of a rapid office-based stool test for detection of *Helicobacter pylori* in children before and after therapy. J. Clin. Microbiol. *47,* 3980–3984.

Raguza, D., Machado, R.S., Ogata, S.K., Granato, C.F., Patrício, F.R., and Kawakami, E. (2010). Validation of a monoclonal stool antigen test for diagnosing *Helicobacter pylori* infection in young children. J. Pediatr. Gastroenterol. Nutr. *50,* 400–403.

Raymond, J., Lamarque, D., Kalach, N., Chaussade, S., and Burucoa, C. (2010). High level of antimicrobial resistance in French *Helicobacter pylori* isolates. Helicobacter *15,* 21–27.

Regnath, T., Enninger, A., and Schalasta, G. (2004). Real-time PCR assay for rapid detection of clarithromycin-resistant *Helicobacter pylori* in gastric biopsy specimens. Z. Gastroenterol. *42*, 1371–1375.

de Reuse, H., Heimesaat, M.M., and Bereswill, S. (2010). Helicobacteromics – genomics of a highly variable, well-adapted and persistent pathogen. In *Helicobacter pylori* in the 21st Century, Sutton, P., and Mitchell, H., eds. (Cab International: Wallingford, UK), pp. 269–291.

Rimbara, E., Noguchi, N., Yamaguchi, T., Narui, K., Kawai, T., and Sasatsu, M. (2005). Development of a highly sensitive method for detection of clarithromycin-resistant *Helicobacter pylori* from human feces. Curr. Microbiol. *51*, 1–5.

Ritchie, B., Brewster, D., Tran, C.D., McNeil, Y., Zacharakis, B., Davidson, G.P., and Butler, R.N. (2009). Lack of diagnostic accuracy of the monoclonal stool antigen test for detection of *Helicobacter pylori* infection in young Australian aboriginal children. Pediatr. Infect. Dis. J. *28*, 287–289.

Roma-Giannikou, E., Roubani, A., Sgouras, D.N., Panayiotou, J., Van-Vliet, C., Polyzos, A., Roka, K., and Daikos, G. (2010). Endoscopic tests for the diagnosis of *Helicobacter pylori* infection in children: Validation of rapid urease test. Helicobacter *15*, 227–232.

Rugge, M., Correa, P., Di Mario, F., El-Omar, E., Fiocca, R., Geboes, K., Genta, R.M., Graham, D.Y., Hattori, T., Malfertheiner, P., *et al.* (2008). OLGA staging for gastritis: a tutorial. Dig. Liver Dis. *40*, 650–658.

Ryberg, A., Borch, K., Sun, Y.Q., and Monstein, H.J. (2008). Concurrent genotyping of *Helicobacter pylori* virulence genes and human cytokine SNP sites using whole genome amplified DNA derived from minute amounts of gastric biopsy specimen DNA. BMC Microbiol. *8*, 175.

Sachs, G., Weeks, D.L., Wen, Y., Marcus, E.A., Scott, D.R., and Melchers, K. (2005). Acid acclimation by *Helicobacter pylori*. Physiology (Bethesda) *20*, 429–438.

Sainsus, N., Cattori, V., Lepadatu, C., and Hofmann-Lehmann, R. (2008). Liquid culture medium for the rapid cultivation of *Helicobacter pylori* from biopsy specimens. Eur. J. Clin. Microbiol. Infect. Dis. *27*, 1209–1217.

Salles, N., and Mégraud, F. (2007), Current management of *Helicobacter pylori* infections in the elderly. Expert Rev. Anti Infect. Ther. *5*, 845–856.

Sasazuki, S., Inoue, M., Iwasaki, M., Otani, T., Yamamoto, S., Ikeda, S., Hanaoka, T., Tsugane, S., and Japan Public Health Center Study Group. (2006). Effect of *Helicobacter pylori* infection combined with CagA and pepsinogen status on gastric cancer development among Japanese men and women: a nested case-control study. Cancer Epidemiol. Biomarkers Prev. *15*, 1341–1347.

Schabereiter-Gurtner, C., Hirschl, A.M., Dragosics, B., Hufnagl, P., Puz, S., Kovách, Z., Rotter, M., and Makristathis, A. (2004). Novel real-time PCR assay for detection of *Helicobacter pylori* infection and simultaneous clarithromycin susceptibility testing of stool and biopsy specimens. J. Clin. Microbiol. *42*, 4512–4518.

Scott, D.R., Marcus, E.A., Wen, Y., Oh, J., and Sachs, G. (2007). Gene expression *in vivo* shows that *Helicobacter pylori* colonizes an acidic niche on the gastric surface. Proc. Natl. Acad. Sci. U.S.A. *104*, 7235–7240.

Seib, K.L., Dougan, G., and Rappuoli, R. (2009). The key role of genomics in modern vaccine and drug design for emerging infectious diseases. PLoS Genet. *5*, e1000612.

Selgrad, M., Kandulski, A., and Malfertheiner, P. (2009). *Helicobacter pylori:* diagnosis and treatment. Curr. Opin. Gastroenterol. *25*, 549–556.

Senkovich, O., Ceaser, S., McGee, D.J., and Testerman, T.L. (2010). Unique host iron utilization mechanisms of *Helicobacter pylori* revealed with iron-deficient chemically defined media. Infect Immun. *78*, 1841–1849.

Shah, R. (2007). Dyspepsia and *Helicobacter pylori.* BMJ *334*, 41–43.

She, F.F., Lin, J.Y., Liu, J.Y., Huang, C., and Su, D.H. (2003). Virulence of water-induced coccoid *Helicobacter pylori* and its experimental infection in mice. World. J. Gastroenterol. *9*, 516–520.

She, R.C., Wilson, A.R., and Litwin, C.M. (2009). Evaluation of *Helicobacter pylori* immunoglobulin G (IgG), IgA, and IgM serologic testing compared with stool antigen testing. Clin. Vaccine Immunol. *16*, 1253–1255.

Shen, J., Deng, D.J., Ke, Y., and Zhang, J.Z. (2008). Detection of point mutation in an *in vitro*-selected amoxicillin-resistant strain of *Helicobacter pylori.* Zhonghua Liu Xing Bing Xue Za Zhi 29, 166–172.

Shen, F., Hobley, L., Doherty, N., Loh, J.T., Cover, T.L., Sockett, R.E., Hardie, K.R., and Atherton, J.C. (2010). In *Helicobacter pylori* auto-inducer-2, but not LuxS/MccAB catalysed reverse transsulphuration, regulates motility through modulation of flagellar gene transcription. BMC Microbiol. *10*, 210.

Shimoyama, T., Fukuda, S., Nakasato, F., Yoshimura, T., Mikami, T., and Munakata, A. (2005). Relation of CagA seropositivity to *cag*PAI phenotype and histological grade of gastritis in patients with *Helicobacter pylori* infection. World J. Gastroenterol. *11*, 3751–3755.

Shin, C.M., Kim, N., Lee, H.S., Lee, H.E., Lee, S.H., Park, Y.S., Hwang, J.H., Kim, J.W., Jeong, S.H., Lee, D.H., *et al.* (2009). Validation of diagnostic tests for *Helicobacter pylori* with regard to grade of atrophic gastritis and/or intestinal metaplasia. Helicobacter *14*, 512–519.

Shirin, H., Levine, A., Shevah, O., Shabat-Sehayek, V., Aeed, H., Wardi, J., Birkenfeld, S., Eliakim, R., and Avni, Y. (2005). Eradication of *Helicobacter pylori* can be accurately confirmed 14 days after termination of triple therapy using a high-dose citric acid-based ^{13}C urea breath test. Digestion *71*, 208–212.

Stingl, K., and De Reuse, H. (2005). Staying alive overdosed: How does *Helicobacter pylori* control urease activity? Int. J. Med. Microbiol. *295*, 307–315.

Storskrubb, T., Aro, P., Ronkainen, J., Sipponen, P., Nyhlin, H., Talley, N.J., Engstrand, L., Stolte, M., Vieth, M., Walker, M., *et al.* (2008). Serum biomarkers provide an accurate method for diagnosis of atrophic gastritis in a general population: The Kalixanda study. Scand. J. Gastroenterol. *43*, 1448–1455.

Sugimoto, M., Wu, J.Y., Abudayyeh, S., Hoffman, J., Brahem, H., Al-Khatib, K., Yamaoka, Y., and Graham, D.Y. (2009). Unreliability of results of PCR detection of *Helicobacter pylori* in clinical or environmental samples. J. Clin. Microbiol. *47*, 738–742.

Sung, H., Kang, J.-O., Lee, M.A., Lee, J., Lee, H.K., Lee, M.-K., Lim, J.-H., Kim, M.-N., and *Helicobacter* Study Group. (2009). Clarithromycin and amoxicillin susceptibility testing of *Helicobacter pylori* by disk diffusion method. Korean J. Clin. Microbiol. *12*, 30–36.

Sycuro, L.K., Pincus, Z., Gutierrez, K.D., Biboy, J., Stern, C.A., Vollmer, W., and Salama, N.R. (2010). Peptidoglycan crosslinking relaxation promotes *Helicobacter pylori*'s helical shape and stomach colonization. Cell *141*, 822–833.

Tanaka, A., and Takahashi, S. (2004). *Helicobacter pylori* stool antigen test. Nippon Rinsho *62*, 464–469.

Taneera, J., Moran, A.P., Hynes, S.O., Nilsson, H.O., Al-Soud, W., and Wadström, T. (2002). Influence of activated charcoal, porcine gastric mucin and beta-cyclodextrin on the morphology and growth of intestinal and gastric *Helicobacter* spp. Microbiology *148*, 677–684.

Tepes, B. (2007). Comparison of two invasive diagnostic tests for *Helicobacter pylori* after antimicrobial therapy. Scand. J. Gastroenterol. *42*, 330–332.

Toledo, H., and López-Solís, R. (2010). Tetracycline resistance in Chilean clinical isolates of *Helicobacter pylori*. J. Antimicrob. Chemother. *65*, 470–473.

Trainor, E.A., Horton, K.E., Savage, P.B., Testerman, T.L., and McGee, D.J. (2011). Role of the HefC efflux pump in *Helicobacter pylori* cholesterol-dependant resistance to ceragenins and bile salts. Infect. Immun. *79*, 88–97.

Trevisani, L., Sartori, S., Rossi, M.R., Ruina, M., Matarese, V., Gullini, S., and Abbasciano, V. (2005). Evaluation of a new rapid immunoassay for the detection of *Helicobacter pylori* in faeces: a prospective pilot study. Aliment. Pharmacol. Ther. *21*, 485–489.

Tsugawa, H., Suzuki, H., Nakagawa, I., Nishizawa, T., Saito, Y., Suematsu, M., and Hibi, T. (2008). Alpha-ketoglutarate oxidoreductase, an essential salvage enzyme of energy metabolism, in coccoid form of *Helicobacter pylori*. Biochem. Biophys. Res. Commun. *376*, 46–51.

Vaira, D., Gatta, L., Ricci, C., Perna, F., Saracino, I., Fiorini, G., Castelli, V., and Holton, J. (2010). A comparison amongst three rapid urease tests to diagnose *Helicobacter pylori* infection in 375 consecutive dyspeptic. Intern. Emerg. Med. *5*, 41–47.

Vécsei, A., Innerhofer, A., Binder, C., Gizci, H., Hammer, K., Bruckdorfer, A., Riedl, S., Gadner, H., Hirschl, A.M., and Makristathis, A. (2010). Stool polymerase chain reaction for *Helicobacter pylori* detection and clarithromycin susceptibility testing in children. Clin. Gastroenterol. Hepatol. *8*, 309–312.

Vega, A.E., Alarcón, T., Domingo, D., and López-Brea, M. (2007). Detection of clarithromycin-resistant *Helicobacter pylori* in frozen gastric biopsies from pediatric patients by a commercially available fluorescent *in situ* hybridization. Diagn. Microbiol. Infect. Dis. *59*, 421–423.

Veijola, L., Oksanen, A., Sipponen, P., and Rautelin, H. (2008). Evaluation of a commercial immunoblot, Helicoblot 2.1, for diagnosis of *Helicobacter pylori* infection. Clin. Vaccine Immunol. *15*, 1705–1710.

Velapatiño, B., Balqui, J., Gilman, R.H., Bussalleu, A., Quino, W., Finger, S.A., Santivañez, L., Herrera, P., Piscoya, A., Valdivia, J., *et al.* (2006).Validation of string test for diagnosis of *Helicobacter pylori* infections. J. Clin. Microbiol. *44*, 976–980.

Vorobjova, T., Maaroos, H.I., and Uibo, R. (2008). Immune response to *Helicobacter pylori* and its association with the dynamics of chronic gastritis in the antrum and corpus. APMIS *116*, 465–476.

Waidner, B., Specht, M., Dempwolff, F., Haeberer, K., Schaetzle, S., Speth, V., Kist, M., and Graumann, P.L. (2009). A novel system of cytoskeletal elements in the human pathogen *Helicobacter pylori*. PLoS Pathog. *5*, e1000669.

Wang, G., Conover, R.C., Olczak, A.A., Alamuri, P., Johnson, M.K., and Maier, R.J. (2005). Oxidative stress defense mechanisms to counter iron-promoted DNA damage in *Helicobacter pylori*. Free Radic. Res. *39*, 1183–1191.

Wang, Y.-H., Wu, J.-J., and Lei, H.-Y. (2009). When *Helicobacter pylori* invades and replicates in the cells. Autophagy *5*, 540–542.

Wang, L.-H., Cheng, H., Hu, F.-L., and Li, J. (2010). Distribution of *gyrA* mutations in fluoroquinolone-resistant *Helicobacter pylori* strains. World J. Gastroenterol. *16*, 2272–2277.

Watanabe, K., Tanaka, A., Imase, K., Tokunaga, K., Sugano, H., Kai, A., Ishida, H., Itoh, T., and Takahashi, S. (2005). Amoxicillin resistance in *Helicobacter pylori:* studies from Tokyo, Japan from 1985 to 2003. Helicobacter *10*, 4–11.

Weiss, J., Tsang, T.K., Meng, X., Zhang, H., Kilner, E., Wang, E., and Watkin, W. (2008). Detection of *Helicobacter pylori* gastritis by PCR: correlation with inflammation scores and immunohistochemical and CLOtest findings. Am. J. Clin. Pathol. *129*, 89–96.

Worku, M.L., Sidebotham, R.L., Walker, M.M., Keshavarz, T., and Karim, Q.N. (1999). The relationship between *Helicobacter pylori* motility, morphology and phase of growth: implications for gastric colonization and pathology. Microbiology *145*, 2803–2811.

Wunder, C., Churin, Y., Winau, F., Warnecke, D., Vieth, M., Lindner, B., Zähringer, U., Mollenkopf, H.J., Heinz, E., and Meyer, T.F. (2006). Cholesterol glucosylation promotes immune evasion by *Helicobacter pylori*. Nat. Med. *12*, 1030–1038.

Xuan, S.H., Zhou, Y.G., Shao, B., Cui, Y.L., Li, J., Yin, H.B., Song, X.P., Cong, H., Jing, F.X., Jin, Q.H., *et al.* (2009). Enzymic colorimetry-based DNA chip: a rapid and accurate assay for detecting mutations for clarithromycin resistance in the 23S rRNA gene of *Helicobacter pylori*. J. Med. Microbiol. *58*, 1443–1448.

Yakoob, J., Jafri, W., Abbas, Z., Abid, S., Islam, M., and Ahmed, Z. (2008). The diagnostic yield of various tests for *Helicobacter pylori* infection in patients on acid-reducing drugs. Dig. Dis. Sci. *53,* 95–100.

Yang, Y.J., Wu, J.J., Sheu, B.S., Kao, A.W., and Huang, A.H. (2004). The *rdxA* gene plays a more major role than *frxA* gene mutation in high-level metronidazole resistance of *Helicobacter pylori* in Taiwan. Helicobacter *9,* 400–407.

Yang, H.R., Ko, J.S., and Seo, J.K. (2008). Does the diagnostic accuracy of the ^{13}C-urea breath test vary with age even after the application of urea hydrolysis rate? Helicobacter *13,* 239–244.

Yasuda, A., Uchida, T., Nguyen, L.T., Kawazato, H., Tanigawa, M., Murakami, K., Kishida, T., Fujioka, T., and Moriyama, M. (2009). A novel diagnostic monoclonal antibody specific for *Helicobacter pylori* CagA of East Asian type. APMIS *117,* 893–899.

Yilmaz, O., Sen, N., Küpelioğlu, A.A., and Simşek, I. (2006). Detection of *H. pylori* infection by ELISA and Western blot techniques and evaluation of anti CagA seropositivity in adult Turkish dyspeptic patients. World J. Gastroenterol. *12,* 5375–5378.

Yilmaz, O., and Demiray, E. (2007). Clinical role and importance of fluorescence *in situ* hybridization method in diagnosis of *H pylori* infection and determination of clarithromycin resistance in *H pylori* eradication therapy. World J. Gastroenterol. *13,* 671–675.

Yodavudh, S., Tangjitgamol, S., and Puangsa-art, S. (2008). Mixture of carbol fuchsin and alcian blue staining of gastric tissue for the identification of *Helicobacter pylori* and goblet cell intestinal metaplasia. SouthEast Asian J. Trop. Med. Public Health. *39,* 659–666.

Yonezawa, H., Osaki, T., Kurata, S., Fukuda, M., Kawakami, H., Ochiai, K., Hanawa, T., and Kamiya, S. (2009). Outer membrane vesicles of *Helicobacter pylori* TK1402 are involved in biofilm formation. BMC Microbiol. *9,* 197.

Yoo, J.Y., Kim, N., Park, Y.S., Hwang, J.H., Kim, J.W., Jeong, S.H., Lee, H.S., Choe, C., Lee, D.H., Jung, H.C., *et al.* (2007). Detection rate of *Helicobacter pylori* against a background of atrophic gastritis and/or intestinal metaplasia. J. Clin. Gastroenterol. *41,* 751–755.

Zeng, H., Guo, G., Mao, X.H., Tong, W.D., and Zou, Q.M. (2008). Proteomic insights into *Helicobacter pylori* coccoid forms under oxidative stress. Curr. Microbiol. *57,* 281–286.

Zsikla, V., Hailemariam, S., Baumann, M., Mund, M.T., Schaub, N., Meier, R., and Cathomas, G. (2006). Increased rate of *Helicobacter pylori* infection detected by PCR in biopsies with chronic gastritis. Am. J. Surg. Pathol. *30,* 242–248.

Zullo, A., Hassan, C., Trapani, S., and Tammaro, G. (2010). Rapid urease test for *H. pylori* diagnosis: pros and cons. Intern. Emerg. Med. *5,* 257–258.

Helicobacter pylori Virulence Factors

4

Lyudmila Boyanova

Abstract

H. pylori infection outcomes strongly depend on the strain virulence. *H. pylori* displays enormous genetic diversity by frequent mutations, intra- or intergenomic recombinations and natural transformation, and additional phase variations by slipped-strand mispairing. *H. pylori* spiral shape, urease, motility, lipopolysaccharide (LPS) and outer membrane adhesins enable the establishment of the infection. *H. pylori* genetic diversity and induced immunomodulation contribute to the infection chronicity. Prevalence of virulence factors varies according to the patient disease, ethnicity, age and country. *H. pylori* VacA causes vacuolation, pore formation, disruption of endolysosomal activity, apoptosis in gastric cells and immunomodulation. CagA oncoprotein alters cell-signalling pathways and induces cellular morphological changes, chromosomal instability and proliferation as well as interleukin 8 (IL-8) release and proto-oncogene activation. CagA type D EPIYA and increased number of C repeats have been linked to high SHp-2 phosphatase activity and hence high risk for gastric cancer. In East Asia, nearly all strains have been highly virulent, carrying intact *cag*PAI, East Asian CagA and *vacA* s1/i1/m1 type, which can help to explain the high gastric cancer incidence there. Although the infection outcomes have shown strong association with *cagA*, *cag*PAI and *vacA* status of the strains, mainly in Western countries, the combined activity of all *H. pylori* virulence factors, including also *dupA, oipA, iceA, homB, babA, sabA, hopQ* and other genes/gene status appears to be crucial for the infection pathogenesis. Tipα protein, Hp-NAP, heat-shock proteins, LPS mimicry and interaction with toll-like receptors influence the infection course as well. By the complex and well-coordinated interplay of its virulence factors, *H. pylori* adapts to the changing environment and can either increase or suppress the gastric inflammation. Moreover, microevolution of the virulence genes emerges in the individual patient over years. Briefly, the direct effects of *H. pylori* virulence factors and the chronic gastric inflammation can lead to development of peptic ulcers or malignancy. Targeting the virulent strains in a country or region is important to explain better the clinical significance of some virulence factors and their interaction, to choose local diagnostic markers, to imply aggressive eradication strategies in the concerned patients and to provide new agents and improved regimens to control the infection.

Introduction

H. pylori infection virtually always leads to development of chronic gastritis but most (80%) infected individuals remain asymptomatic; therefore, some strains are much more virulent compared with the others. Infection outcome depends also on host factors, among them are gene polymorphisms of IL-1 and tumour necrosis factor-α (TNF-α), and on environmental or dietary (e.g. salt or meat intake) factors (Israel and Peek, 2010).

Genome and strain diversity

Knowledge on *H. pylori* has sharply increased in the genomic era. Many *H. pylori* genes strongly influence the onset of gastric inflammation and the ensuing diseases. *H. pylori* is exceptionally well adapted to the human gastric mucosa but harbours an open pan-genome (all genes/orthologous groups in a genome), showing that the bacteria, acquiring or diversifying genes, continue to change (Fischer *et al.*, 2010).

Genomes (size, 1.5–1.8 Mb) of various *H. pylori* strains have been sequenced, about 1100–1237 core genes (present in all/almost all strains) have been described and auxiliary genes have been found in 22–27% of the genome (Dong *et al.*, 2009; McClain *et al.*, 2009; Fischer *et al.*, 2010). McClain *et al.* (2009) have observed 33 genes for *H. pylori* viability in the core genome and >40 genes for colonization that highlight again the fascinating adaptation of the bacteria to the host. In seven *H. pylori* genomes evaluated, many (40.9%) genes have been absent from ≥1 genomes (Fischer *et al.*, 2010). In five genomes studied, amino acid sequences have been identical in only 11 of 1237 core genes (McClain *et al.*, 2009). Indeed, during the long-lasting interplay between the bacteria and the hosts, *H. pylori* genetic diversity changes constantly.

H. pylori shows enormous genetic diversity with high rate of spontaneous mutations, intragenomic recombinations, assisted by repetitive elements in the genome, or intergenomic recombinations by non-self DNA during natural transformation in mixed infections (Costa *et al.*, 2009; Dong *et al.*, 2009). Typical *H. pylori* mutation rate has been 1×10^{-6}, i.e. higher than that of the species in *Enterobacteriaceae* and similar to that of *Escherichia coli* mutator strains. Moreover, the recombination rate at any nucleotide has been shown to reach 6.9×10^{-5} (Kulick *et al.*, 2008; Dong *et al.*, 2009). Whole mismatch repair system, controlling confidentiality of replication, has not been found in *H. pylori*, which could help to explain the high mutation rates.

The genetic diversity in *H. pylori* is assisted by a family of hypervariable (RD) genes with extensive repetitive DNA in *H. pylori*. Single RD gene has been shown to encompass a conserved region, a variable mid-region with heptad repeats and a 5′ region with one of both observed alleles (Shak *et al.*, 2009). Patterns of RD gene repeats have been linked to both slipped-strand mispairing and domain duplications. High diversity among strains has been observed owing to the RD gene number, location, and arrangement (Shak *et al.*, 2009). Functional role of the cell-surface products of the RD genes should be further elucidated. Like in other bacteria, gene fusion (overlapping genes) also has been found in *H. pylori*. Thirty-two fusion genes have been either separated or fused in two genomes studied (Sakharkar *et al.*, 2006). This may provide a selective benefit, although still unspecified, to the strains.

H. pylori genome possesses more (30) restriction and modification (R-M) genes than other bacteria (4.3 R-M systems per genome; Vale *et al.*, 2009). R-M systems spot and cut the corresponding DNA sequence by a restriction endonuclease and methylate the sequence by a DNA methyltransferase to avoid its cleavage. Distribution of methyltransferases has shown the major roads of human migrations. Two methyltransferases (M. NaeI and M. HhaI) have been found everywhere, implying a coevolution of *H. pylori* and humans since the first human movement out of Africa (Vale *et al.*, 2009).

H. pylori has been suggested to lack RecBCD pathway of the enteric bacteria, used for homologous recombination and DNA double-strand break repair. However, *H. pylori* AddAB enzymes with double helicase and exonuclease activity, similar to RecBCD enzyme, have been essential for DNA repair and robust gastric colonization (Amundsen *et al.*, 2009). The AddAB, with the participation of RecA, initiates the repair of double-stranded DNA breaks by homologous recombination (Dorer *et al.*, 2010). *AddA (Hp1553)* gene has been found to confer resistance to the damage of DNA owing to oxidative stress. Briefly, the activities of both RecA and AddAB have been prerequisites for *H. pylori* colonization of the stomach (Dorer *et al.*, 2010)

It is known that Holliday junctions (between four DNA strands) are usually produced during the process of homologous recombination (Loughlin *et al.*, 2003). *H. pylori* Holliday junctions have been generated by RecG and RuvAB helicases and have been resolved by RuvC endonuclease at the site of the branch point. RuvC has been found to contribute to *H. pylori* survival in macrophages and to the persistence of the infection (Loughlin *et al.*, 2003). Although the gastric bacteria harbour RecG and RuvABC homologues, the *H. pylori* RecG restricts the recombinational repair (Kang and Blaser, 2008). Both the high levels of recombination and the antirepair property of RecG increase the genetic diversity of *H. pylori* strains. It is also interesting that an antimutator DNA glycosylase (MutY), encoded by *mutY* homologue Hp0142, protects *H. pylori* against specific cytosine to adenine transversions and enhances the bacterial fitness during the stationary phase of growth (Kulick *et al.*, 2008). In brief, the competition between repair and antirepair gene products in *H. pylori* generates its remarkable strain diversity and close adaptation to the individual host.

Homologous natural transformation is another important means for *H. pylori* genomic diversity, its adaptation and exchange of markers for antibiotic resistance. *H. pylori* natural transformation is specific as DNA is not imported by type IV pilus-based apparatus but is taken up by a transport system associated with a type IV secretory system (T4SS), the *comB* system (Zhong *et al.*, 2007). Three types of *H. pylori* T4SS have been described: the first one is the T4SS encoded by *cag* pathogenicity island (*cag* PAI), the second one is the *comB* system, which mediates the active import of free DNA, and the third system is *tfs3* cluster, which role should be better evaluated (Zhong *et al.*, 2007). As a barricade to horizontal transfer of non-self genes, *H. pylori* seems to use its R-M systems (Stingl *et al.*, 2010).

H. pylori takes up DNA usually at the cell poles by a two-step mechanism. The first step represents the increased DNA uptake in *H. pylori* and transport of dsDNA across the bacterial outer membrane by the VirB4-homologous ATPase ComB4 (ATPase). This ComB4 stimulates the uptake of foreign dsDNA relatively weakly but with poor specificity for DNA sequences (Stingl *et al.*, 2010). The second step involves the transport of DNA from the bacterial periplasm to the cytoplasm by ComEC channel (Stingl *et al.*, 2010). For the overall process, many proteins or protein complexes are essential such as ComB8, ComB9, and ComB10 for the natural competence (property to uptake free DNA from the environment) and ComB7 for the stability of ComB protein complex (Zhong *et al.*, 2007). In mixed infections by several *H. pylori* strains, the activated natural competence can lead to spread of antibiotic resistance genes and selection of variants with highest possible adaptation to the environment.

Dorer *et al.* (2010) have found an association between *H. pylori* natural competence and DNA damage. DNA damage may induce natural competence in bacteria such as *H. pylori* (Dorer *et al.*, 2010). However, after DNA damage, unlike other bacteria, *H. pylori* has been reported to activate both the *comB* system and a RecA-dependent expression of a protein resembling lysozyme. This protein has been able to damage adjacent cells to free DNA for import. Homologous recombination is important to create genetic diversity among strains and to promote resistance to quinolones, which damage the DNA (Dorer *et al.*, 2010).

Each *H. pylori*-positive subject may carry a different strain, although with slight dissimilarity within a family (Kusters *et al.*, 2006). Many strains possess different virulence factors, e.g. *cag* PAI, that increases the risk for peptic ulcers and gastric malignancy. Presence of hypervariable regions of *cag*PAI and other plasticity zones (PZs) suggests that *H. pylori* strains can acquire, lose or change genetic structures during the chronic infection, mostly inside the PZs (Romo-González *et al.*, 2009). Kraft *et al.* (2006) and de Reuse *et al.* (2010) have observed *H. pylori* adaptation to the host mainly owing to recombinations rather than loss or gain of genes. *H. pylori* also exhibits reversible phenotypic diversity of some virulence genes by small genetic variations (slipped-strand mispairing), e.g. in sialic acid-binding adhesin (*sabA*), *sabB*, outer-membrane protein *hopZ*, inflammatory outer membrane protein (*oipA*) and genes for lipopolysaccharide (LPS) biosynthetic pathway (Kusters *et al.*, 2006).

Microevolution of virulence genes in a strain can emerge in different gastric sites in a patient or his family members, infected by the same strain, over years (Shak *et al.*, 2009). Microevolution has been found within *cag*PAI genes, vacuolating cytotoxin (*vacA*) alleles, *oipA* and *dupA* genes. Alvi *et al.* (2007) have reported a microevolution in *cag*PAI genes (*cagA*, *cagE*, and *cag7*) in a duodenal ulcer patient over 10 years, although without relation to the pathology. Matteo *et al.* (2007) have detected coexisting strain subclones with different *cag*PAI genotypes in ¼ of Argentinean patients. Although most of the emerging variants are of little importance for the infection pathogenesis, acquisition of a *vacA* m1 fragment has been reported in a variant (Alvi *et al.*, 2007). In brief, the enormous genetic diversity of *H. pylori,* including the microevolution of its genes, provides its close adaptation to the human gastric mucosa and may influence the associated diseases (Costa *et al.*, 2009).

In the stomach, *H. pylori* can be found free in the mucus, attached to the cells or intracellularly. Few (1% of cells) intracellular bacteria may be important for the infection pathogenesis (Dubois and Boren, 2007). *H. pylori* infectious dose possibly ranges from 10^4 to 10^{10} bacteria (Graham *et al.*, 2004). Spiral shape and flagella provide *H. pylori* corkscrew movement through the gastric mucus and many adhesins allow its selective adherence to gastric cells. *H. pylori* produces a strong urease for protection from gastric acidity and may use numerous virulence factors to change the host environment (Fig. 4.1).

H. pylori virulence genes are strain-specific, phase-variable and genes with variable structures or genotypes (Yamaoka, 2008a; Oleastro *et al.*, 2009). Strain-specific genes, e.g. *cag*PAI and duodenal ulcer promoting (*dupA*) genes, are present in part of the strains. Phase-variable genes, e.g. *oipA* and *sabA,* are found in all strains, but, owing to phase variations, only some strains generate active gene products. Genes with variable structures are also always present in the strains but their activity and protein amount differ among the strains; examples are *vacA* allele mosaic and *cagA* 3′ region (Yamaoka, 2008a; Oleastro *et al.*, 2009). For *H. pylori,* the virulence factors may be beneficial to improve the nutrient delivery and conditions for survival (Atherton and Blaser, 2009). For the host, however, the activities of the virulence factors and the chronic gastric inflammation can determine the severe outcome of the infection, i.e. peptic ulcers, gastric cancer and mucosa-associated lymphoid tissue (MALT) lymphoma. Some

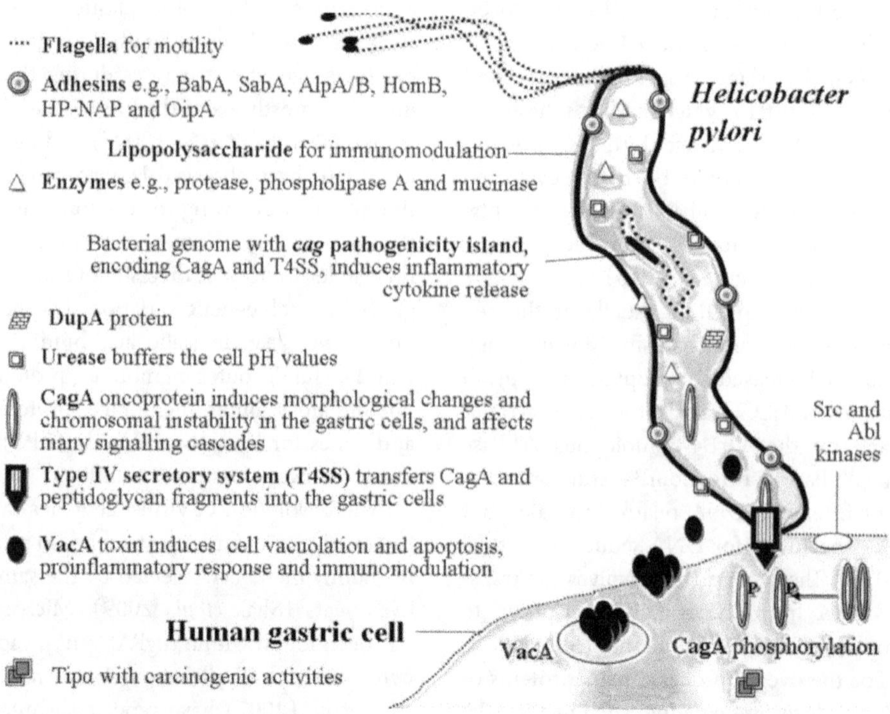

Figure 4.1 Virulence factors of *H. pylori.*

virulence factors can affect also the success of *H. pylori* eradication.

Urease

H. pylori survives in acid environment with gastric luminal pH 1–4.5 by acid acclimation (Wen *et al.*, 2009; see Chapter 3). Urea can pass through the bacterial outer membrane and can enter the cytoplasm (Proença-Modena *et al.*, 2009). *H. pylori* urease hydrolyses the urea to CO_2 and ammonium, which neutralize the acid. The urease, however, neutralizes the bacterial cytoplasmic and periplasmic pH rather than creates buffered 'clouds' around the bacteria. *H. pylori* urease and flagella are essential to colonize the stomach in animal models (Algood and Cover, 2006).

The urease consists of 12 UreA and 12 UreB subunits, encoded by the *ureAB* operon (Pflock *et al.*, 2006). UreA and UreB are immunogenic proteins and have been used to develop vaccines for *H. pylori* infection. To establish colonization, *H. pylori* needs many factors, among them are the urease activity, urea flux through a specific pH-gated urea channel (UreI channel), which is unique to gastric *Helicobacter* spp., and periplasmic α-carbonic anhydrase (Scott *et al.*, 2010). The anhydrases have been shown to contribute to *H. pylori* colonization and gastric inflammation in animal models (Bury-Moné *et al.*, 2008). Expression of UreI protein (encoded by *ureIEFGH* operon), which supplies urea for the cytoplasmic urease, is vital for *H. pylori* survival in low pH. UreI channel encompasses six transmembrane protein segments, located in the cytoplasmic membrane of *H. pylori* and bound by periplasmic and cytoplasmic loops (Sachs *et al.*, 2006). However, the channel exhibits a COOH terminus exposed to the periplasm. In unison with the UreI channel, UreEFGH accessory proteins participate in subunit assembly and nickel incorporation in the urease apoenzyme (Kusters *et al.*, 2006). Transcription of *ureAB* and *ureIEFGH* operons and accessory proteins such as UreD, necessary for Ni^{2+} inclusion into the active sites of the urease, is upregulated in low pH (Pflock *et al.*, 2006; Khan *et al.*, 2009). Interaction between accessory proteins HypA and UreE is essential for urease maturation (Benoit *et al.*, 2007). This multifaceted mechanism is interesting to be presented in more detail.

At low pH, the urease combines with the UreI channel transporter protein in *H. pylori* inner membrane, leading to activation of UreA and UreB apoenzyme complex due to Ni^{2+} insertion with the participation of UreF-UreH and UreE-UreG assembly proteins (Sachs *et al.*, 2006). It is intriguing that when the external pH decreases to <6.5, the UreI channel opens and increases 300 times the urea flux across the membrane compared with simple diffusion (Scott *et al.*, 2007; Atherton and Blaser, 2009). Thus, UreI channel renders high amount of urea accessible to the urease in bacterial cytoplasm. The ammonia, generated by the urea hydrolysis, maintain both periplasmic and cytoplasmic pH close to neutral and the HCO_3^+, produced in the cytoplasm and periplasm from the CO_2 with the assistance of the β-carbonic- and α-carbonic anhydrases contributes to the maintenance of intracellular pH at ≥6.1 (see Chapter 3). Finally, both *H. pylori* periplasm and cytoplasm are well buffered. When neutral pH is reached, UreI channel closes to avoid the extreme urease activity and overalkalinization of the bacteria (Scott *et al.*, 2007). The open probability of the UreI channel is negligible at neutral pH but high in low pH, e.g. up to 50% at pH 5.9 (Scott *et al.*, 2010).

H. pylori adaptation to the acid is also dependent on *rocF* (arginase), *nixA* (nickel importer) and *nikR* (nickel-responsive transcription factor) genes (Proença-Modena *et al.*, 2009). Nickel is a prerequisite for *H. pylori* urease activity. In *H. pylori* inner membrane, *nixA* gene encodes a permease with high affinity for nickel. This activity is noteworthy, as *nixA* mutation has reduced both urease activity and *H. pylori* colonization in mouse models. However, in the presence of excess nickel, which is inhibitory for *H. pylori*, NixA synthesis is repressed by NikR (Maier *et al.*, 2007). However, while NikR acts as a repressor for *nixA* gene, it is also an activator for the urease operon. Moreover, cadmium–zinc–nickel efflux complex (CznC) protein, responsible for nickel export, can reduce the urease activity. Metal-efflux CznC protein is also essential for host colonization in animal models (Duckworth *et al.*, 2009). Furthermore, Stingl *et al.* (2008) have observed a physical interaction between the urease and HypB, the accessory protein for Ni^{2+} incorporation into the hydrogenase, which has been shown to participate in urease activation. This implies a competition between the urease and the hydrogenase for the nickel incorporation. In brief, the activation of the urease of *H. pylori* in appropriate conditions is a very complex but well-coordinated mechanism.

Strikingly, after acid exposure, >100 *H. pylori* genes are upregulated and a smaller number are downregulated (Wen *et al.*, 2003; Pflock *et al.*, 2006). The regulation is linked, at least partly, to ArsRS two-component signal transduction system (ArsRS TCS). The two-component systems detect and respond to various environmental stimuli. *H. pylori* ArsRS TCS harbours a sensor kinase (ArsS, Hp0165 in *H. pylori* 26695) and OmpR-like response regulator (ArsR, Hp0166; Loh *et al.*, 2010). When the pH in the environment changes, ArsS protein is stimulated by phosphorylation and its phosphate is transferred to ArsR, which then binds to and modulates the expression of the numerous target genes (Loh *et al.*, 2010). For instance, the enzymes for ammonia production are the urease and amidases. Aliphatic amidase (AmiE, encoded by *amiE* gene) and formamidase (AmiF, encoded by *amiF* gene) hydrolyse and convert short-chain amides to ammonia and organic acids (Proença-Modena *et al.*, 2009). The amidases are important for *H. pylori* nitrogen metabolism. Expression of urease, AmiE, and AmiF as main ammonium producers is due to acid exposure and is controlled by the ArsRS TCS.

Interestingly, some acid-responsive genes are regulated by both ArsRS and other regulatory systems, e.g. the pH-responsive regulon of Hp0703-Hp0244 (FlgRS) two-component system, part of the regulatory system for flagellar gene expression. Fascinatingly, Hp0165 sensor kinase of the ArsRS TCS of *H. pylori* acts only after a moderate drop (mild acidity) in periplasmic pH. Conversely, the Hp0244 sensor kinase of the FlgRS promotes *H. pylori* survival at high acidity (Scott *et al.*, 2010). In other words, the ArsRS TCS acts on signalling pathways, which regulate expression of many acid acclimation genes at pH 4.5–5, while Hp0244 is unnecessary at pH 4.5 but indispensable at pH 2.5 after 30 min (Wen *et al.*, 2009). In addition to the flagellar genes, Hp0244 regulates 22 genes, e.g. *ureA*, *ureB*, *ureI* and *amiE*, at pH 4.5 without urea but many more (86) genes, including also *ureF*, *rocF* (arginase) and *ansB* (asparaginase), at pH 2.5 with 30 mM urea (Wen *et al.*, 2009). Moreover, the urease expression is modified by both ArsRS and the nickel regulator NikR, and AmiE expression is regulated by both ArsRS and the ferric uptake regulator (Fur; Pflock *et al.*, 2005, 2006; Loh *et al.*, 2010). Fur is important for the infection pathogenesis, as *fur* mutation has diminished the *H. pylori* survival at low pH (Lee *et al.*, 2010). Overlap in regulatory pathways for acidity, flagella, metal and heat-shock stress suggests the presence of a master regulator (de Reuse *et al.*, 2010).

In addition to the acid resistance, *H. pylori* urease exerts many other effects such as interaction with the host immune system. Both the induced ammonia and hypochlorous acid production form neutrophils are toxic for the gastric cells (Oyake *et al.*, 2006). *H. pylori* urease has been able to activate *in vitro* peripheral blood mononuclear leukocytes and mucosal macrophages, to induce the release of pro-inflammatory cytokines and to contribute to chemotaxis for human monocytes. Gobert *et al.* (2002) have reported that *H. pylori* urease stimulates the macrophage inducible NO synthase as an important factor of the innate immunity. Moreover, *H. pylori* UreB is able to activate the nuclear factor-κ (NF-κB) and to stimulate the release of IL-8. IL-8 is a neutrophil chemotactic agent and hence the major cytokine in *H. pylori* pathogenesis (Rizwan *et al.*, 2008). For diagnostic purpose, UreA, translation elongation factor EF-G (FusA) and catalase (KatA) have been used in combination to distinguish antibody response of duodenal ulcer patients from that of other patients (Lin *et al.*, 2007). In brief, the urease of *H. pylori* is crucial for the survival of the bacteria in the acid conditions of the stomach but also damages the gastric mucosa, increasing the inflammation.

Lipopolysaccharide

Bacterial lipopolysaccharide (LPS) contains lipid A, conserved saccharide core region and highly variable saccharide O-antigen chain. However,

H. pylori lipid A shows low endotoxic and inflammatory activities compared with those of most enteric bacteria (Hildebrandt and McGee, 2009). Special feature of *H. pylori* O-antigen chain is the display of fucosylated oligosaccharide Lewis antigens, which, by functional similarity to those on the surface of some human erythrocytes and gastric cells, help the bacteria to escape the immune recognition (Nilsson *et al.*, 2008). The similarity is only functional since the sequence resemblance between the bacterial and human α1,3/4 and α1,2-fucosyltransferases has been low (Moran, 2008).

O-antigen chain of *H. pylori* has a region with many internal Le^x units and terminal units Le^x or Le^y or, in some strains, extra units such as Le^a, Le^b, Le^c, sialyl -Le^x, H-1 antigens, etc. (Moran, 2008.) Most (>80%) *H. pylori* strains contain Le^x and/or Le^y epitopes (type 2 carbohydrates) in the LPS, unlike Le^a, Le^b, H-1 (type 1 carbohydrates), Le^c and sialyl Le^x (Moran, 2008; Lehours *et al.*, 2009). Strains with type 1 expression have been rare in Europe and North America, more common (for Le^b, up to 24%) in Latin America and widespread (≤35%) in Asia, where both Le^a and Le^b may be detected (Sheu and Wu, 2008). Presence of Le^x, which binds to its gastric receptor, galectin-3, is a prerequisite for colonization (Moran, 2008; Lindén *et al.*, 2010). Strong *H. pylori* colonization and neutrophil infiltration have been found in infections by strains with both 1 and 2 type epitopes compared with those with type 2 epitopes only (Sheu and Wu, 2008).

No less than five glycosyltransferase genes generate phase variants of the Lewis antigens, among them are *H. pylori futA* and *futB* genes, which encode α1,3-fucosyltransferases (Nilsson *et al.*, 2008). Number of repeated heptads in the variable heptad-repeat region of FutA and FutB determines the dimension of the O-antigen polymers, which then are fucosylated (Skoglund *et al.*, 2009). FutC transports an additional fucose to the Le^y or Le^b antigens (Lehours *et al.*, 2009). Differences in fucosylation of the Lewis antigens appear to be due to phase variation through slipped-strand mispairing, i.e. a reversible switching 'on' and 'off' of the LPS epitopes of a given strain (Skoglund *et al.*, 2009). The phase variation results from translational frameshifts in the glycosyltransferase genes during replication. Importantly, presence of *jhp0562* gene (for a glycosyltransferase) has been deemed a risk factor for peptic ulcer in French children (Oleastro *et al.*, 2010a). The clinical significance of *jhp0562* gene merits further clarification.

Gastric environment and the low pH induce adaptation of LPS in *H. pylori* strains during infection in a single host over time, although LPS diversity can be small. Strains have shown different glycosylation intensity and dimensions of fucosylated O-antigen chains within and between patients. This has been found also in mouse models (Skoglund *et al.*, 2009). The display of Lewis antigens on *H. pylori* surface as a functional mimicry can result in altered host recognition, gastric autoimmunity, partly leading to gastric atrophy and modulation of host immunity. This can promote strong bacterial adhesion to the gastric cells and hence high-level colonization and increased bacterial internalization in the cells (Moran, 2008). *H. pylori* immune recognition by surfactant protein D has been found to depend on amounts of LPS fucosylation (Moran, 2008).

Typical host response to *H. pylori* infection is the pro-inflammatory T helper (Th) type 1-biased response, causing chronic gastritis. However, Lewis antigen can modify the response during infection. The target of Le^x and Le^y are DC-SIGN (dendritic cell-specific intercellular adhesion molecule-3-grabbing non-integrin) on dendritic cells (Skoglund *et al.*, 2009). The interaction of *H. pylori* with the dendritic cells has been shown to stimulate gastric regulatory T-cell (T_{reg} cell) immune response, inhibiting Th17 cells, and the level of associated pro-inflammatory IL-17, IL-21, and IL-22 release (Fietta and Delsante, 2009; Kao *et al.*, 2010). The overall interaction can protect the bacteria from the immune clearance and can establish immune suppression and hence a long-lasting infection.

Mucosa-associated lymphoid tissue (MALT) lymphoma strains exhibit a specific Lewis antigen profile. The *cag*PAI negative strains with only Le^y have been strongly linked to MALT lymphoma compared with those with both Le^x and Le^y (Lehours *et al.*, 2009). The link between Le^y in *cag*PAI negative MALT strains with the malignancy merits further evaluation.

Toll-like receptors (TLRs) are transmembrane proteins of the innate immunity for detection of pathogen-associated molecular patterns (PAMPs) and response to the bacterial structures such as LPS (Mandell *et al.*, 2004). TLR-mediated signalling pathways activate NF-κB and promote subsequent pro-inflammatory cytokine release. *H. pylori* has been thought to escape both immune detection and elimination as TLR2 that typically recognizes lipoproteins is scarce in the stomach and LPS has weak agonist activity against TLR4, which usually recognizes Gram-negative LPS, owing to lipid A modifications (Mandell *et al.*, 2004). In a study on four gastric cell lines, *H. pylori* LPS, unlike that of *E. coli*, has induced neither IL-12 nor interferon-γ (IFN-γ) production by human mononuclear cells and thus, has reduced the antitumour properties of these cells (Chochi *et al.*, 2008). However, it has been shown that *H. pylori* induces LPS-TLR4 pathway and contributes to development of gastric malignancy by MEK1/2-ERK1/2 mitogen-activated protein kinase cascade in gastric cells (Yokota *et al.*, 2010). The cascade activation has been shown to depend on LPS epitopes, being stronger by low antigenic epitopes, often observed in gastric cancer patients, than by highly antigenic epitopes, often present in subjects with chronic gastritis (Yokota *et al.*, 2010). TLR9, which recognizes not the cell surface PAMPs but unmethylated CpG-DNA on endosomal compartment, seems to be also of importance. Single nucleotide polymorphism in TLR9 promoter has been found as a risk factor for pre-malignant changes in the gastric mucosa (Ng *et al.*, 2010). In addition, suppression of constitutive nitric oxide (NO) synthase activity and strong upregulation in the inducible NO synthase have been linked to *H. pylori* LPS. Amazingly, ghrelin, a hormone that stimulates appetite, has been found to be protective against gastric mucosal proapoptotic events (Slomiany and Slomiany, 2010).

In brief, *H. pylori* LPS is a virulence factor on host immunity and induces immune responses that may be necessary for long-term protection against *H. pylori* infection. However *H. pylori* LPS can induce either immunosuppressing or immunostimulating activities in the host and can determine the persistence of the infection as well as its severe clinical outcome.

cag Pathogenicity island (*cag*PAI)

In *H. pylori* genome, some regions, the so-called plasticity zones (PZs), possess lower (34–35%) G + C content, which looks like a 'foreign' portion of DNA compared with the chromosome (39%) as a whole. This can result from horizontal DNA transfer from distant bacterial species (Yamaoka, 2008a). These regions show high genetic variations in the isolates. Genes of some PZs have been reported to be located inside conjugative transposons or their residues (Kersulyte *et al.*, 2009). Conjugative transfer can spread the PZs of *H. pylori* genomes (Fischer *et al.*, 2010). XerCD-family tyrosine recombinase (ex-XerT) has been found to participate in the PZ excision and import of foreign DNA (Fischer *et al.*, 2010). The highly variable genes in the PZs strongly influence the pathogenesis of the infection and some of them are markers of its clinical outcomes (Yamaoka, 2008a, Lima *et al.*, 2010).

PZs in *H. pylori* pathogenicity island (*cag*PAI) show homology to T4SS structures in distant Gram-negative species such as *Agrobacterium*, *Legionella* and *Bordetella* (Backert *et al.*, 2010b). The *cag*PAI (37–40 kb) encodes 27–31 proteins (Olbermann *et al.*, 2010). The island can be divided into two regions, upstream (left) *cag*II region, harbouring for example *cagT, virB11* and *LEC* genes, and downstream (right) *cag*I region, containing *cagA, cagE, cagL, cagM* and many other genes. Three *cag*PAI features influence most strongly *H. pylori* pathogenesis: the *cagA* gene, the T4SS for transfer of CagA and peptidoglycan fragments into gastric cells and the stimulation of inflammatory cytokine release such as that of IL-8 from the cells (Nguyen *et al.*, 2008). Pro-inflammatory cytokine IL-8, named also neutrophil-activating factor, recruits and activates neutrophils and lymphocytes, leading to inflammation. It also increases gastrin release with resultant proliferation of epithelial cells. Importantly, IL-8 activates nuclear factor-kB (NF-kB) and COX-2 expression, and shows pro-angiogenic activity, leading to tumorigenesis (Robinson *et al.*, 2007; Table 4.1).

Most *cag*PAI genes are important for CagA translocation and more than half of them participate in the induction of IL-8 release (Nguyen *et al.*,

Table 4.1 Some known and putative virulence factors of *H. pylori* according to Ashour *et al.* (2001), Ando *et al.* (2002), Lehours *et al.* (2004, 2009), Lin *et al.* (2004, 2010a), Gong *et al.* (2005, 2010), Mattar *et al.* (2005), Olfat *et al.* (2005), Kusters *et al.* (2006), Wang *et al.* (2006, 2010), Zhao *et al.* (2007); Loh *et al.* (2008); Nilsson *et al.* (2008); Oleastro *et al.* (2008, 2009, 2010a); Rizwan *et al.* (2008), Suganuma *et al.* (2008); Tabassam *et al.* (2008), Yamaoka (2008a), Boyanova *et al.* (2009), Giannakis *et al.* (2009), Odenbreit *et al.* (2009), Proença-Modena *et al.* (2009);, Shak *et al.* (2009), Tanaka *et al.* (2009b), Lima *et al.* (2010), Sycuro *et al.* (2010) and Watanabe *et al.* (2010)

Virulence genes	Known or putative role of the gene product	Associated diseases
*cag*PAI (including *cagA*, *cagE*, *cagT* and *virB11*)	Expression of a type IV secretory system (*cag*-T4SS) and translocation of CagA and peptidoglycan fragments to the gastric cells with induction of morphological changes, release of pro-inflammatory cytokines, and activation of proto-oncogenes	PU and GC (*cagA*$^+$, *cagE*$^+$ and *cagT*$^+$ strains); GC in Western countries (high number of CagA EPIYA-C segments); GC (*virB11*)
vacA	Formation of membrane channels/pores in epithelial cell membranes, disruption of endo-lysosomal activity, development of pro-inflammatory response (IL-8 and PGE_2), disruption of cell signalling, induction of gastric cell apoptosis and modulation of immune cells. Type s1/i1/m1 is most active	PU, atrophic gastritis and GC (s1/m1 and s1/m2 types); PU in Western countries (*vacA* type); GC (*vacA* s1/i1/m1 and s1/i1/m2 strains)
babA	Adhesion (binds to fucosylated blood-group antigens Lewis b on gastric cells	DU and GC (*babA2* allele)
sabA	Adhesion (binds to sialyl Le^x and sialyl Le^a antigens)	GC and ML, negative association with DU
oipA	Adhesion, induction[1] of IL-8 production, cytoskeletal re-organization	PU and GC (OipA expression)
napA	Adhesion to mucin, release of oxygen radicals from neutrophils, antioxidant protection of *H. pylori*, induction of IL-12 release and Th1 type immune response	Injury to human gastric mucosa
iceA	Allele *iceA1* encodes CATG-specific restriction endonuclease	PU, GC and ML (*iceA1* allele[1])
jhp0947-jhp0950[2]	Unknown specificity	DU, GC and ML
jhp0940[2]	Induction of TNF-α and IL-8 secretion by human macrophages	Unknown
jhp0562	Encodes a glycosyltransferase for the synthesis of the lipopolysaccharide	PU in children
dupA[2]	Encodes a molecule, which is homologous to VirB4 ATPase	DU and reduced risk for GC[1]
homB	Adhesion, induction of IL-8 production	PU (two *homB* copies in the strain) and GC
hsp genes	Protection from stress factors. Hsp60 induces IL-8 and IL-6 release, angiogenesis and metastasis. HspA (GroES), a nickel chaperone, induces secretion of pro-inflammatory cytokines	GC (Hsp60 and antiGroES seropositivity)
hopQ	Antiadhesive properties, downregulation of *H. pylori*-induced gastric epithelial cell alterations	Gastric atrophy[1] and *cagA*$^+$ strains (type I *hopQ* allele)
hopZ	Adhesion, fitness of the strains	ML (*hopZ* 'off' status)
Tipα	Induction of expression of TNF-α and chemokine genes through NF-kB activation and tumour development	GC (higher amounts of TNF-α inducing protein than in CG)
ggt	Increase in IL-8 release, oxidative DNA damage and apoptosis of gastric cells	PU patients (high activity of the enzyme)
alpA and *alpB*	Bacterial adhesion, induction of the secretion of ILs, activation of NF-κB in gastric epithelial cells by East Asian strains	Unknown

Table 4.1 continued

Virulence genes	Known or putative role of the gene product	Associated diseases
pgdA	Peptidoglycan deacetylation; impeded host immune detection	Unknown
ureA and *ureB*	Maintain periplasmic and cytoplasmic pH (production of ammonia)	None
ureI	As above (transport of the urea into the bacterial cell)	None
nixA	As above (Ni^{2+} transport to the cytoplasm. Required for urease activity)	None
amiE/amiF	As above (produce nitrogen for anabolism)	None
rocF	As above (conversion of L-arginine to L-ornithine and urea)	None
nikR/arsRS	As above (regulation of acid resistance)	None
csd1–3 and *ccmA*	Required for helical shape of the bacteria	Unknown
flaA and *flaB*	Colonization of gastric mucosa (bacterial motility)	None
futA, *futB* and *futC*	Encode α1,3-fucosyltransferases. Role in Lewis antigen expression in lipopolysaccharide	Unknown
trx	Colonization of gastric mucosa	None
katA	Encodes catalase (antioxidant protection of *H. pylori)*	None
tlpB/cheY	Colonization of gastric mucosa (pH taxis)	None
RD gene family	Provides extensive variation between strains (persistence of infection)	Unknown

[1]Non-universal association. For instance, the presence of many virulence factors has not been linked to disease outcomes in East Asia.

[2]In the plasticity region.

T4SS, type IV secretory system; DU, duodenal ulcer; PU, peptic ulcer; GC, gastric cancer; CG, chronic gastritis; ML, mucosa-associated lymphoid tissue lymphoma; IL, interleukin; PGE_2, prostaglandin E_2; Hsp, heat-shock protein; Tipα, TNF-α-inducing protein; ggt-γ, glutamyltranspeptidase.

2008). The *cag*PAI⁺ strains also stimulate a DNA/RNA-editing enzyme in host cells, resulting in increase in mutations in *p53* tumour suppressor gene (Matsumoto *et al.*, 2007), which are, therefore, common in patients with gastric cancer. The *cag*PAI⁺/VacA producing strains also hamper the phagosome/lysosome fusion and hence the bacterial killing (Romero-Adrián *et al.*, 2010). Recent study has shown an ancient *cag*PAI acquisition by *H. pylori* before the first human migration out of Africa (Olbermann *et al.*, 2010). The *cag*PAI strains have been nearly universal (≥95%) in hpEastAsia and hpAfrica1, lower (58%) in hpEurope or absent in hpAfrica2 populations (Olbermann *et al.*, 2010). These data can help to explain the differences in the incidence of gastric cancer among the main geographic regions.

The *cag*-T4SS is a multi-subunit apparatus that looks like a molecular syringe (100–200 nm), crossing *H. pylori* inner and outer membranes and protruding from the cell surface by a pilus (filament; Kwok *et al.*, 2007). The *cag*-T4SS needle base is covered by CagT (VirB7 homologue) and CagW (VirB9) proteins (Backert *et al.*, 2010b). CagT, encoded by *cagT* gene in the *cag*II region, stabilizes the complex by a disulfide bridge with a VirB9 homologue (Proença Módena *et al.*, 2009). The *cag*-T4SS pilus carries CagC, CagX (VirB9) and CagT and is coated, more or less, by a large, highly antigen variable and acid-stable protein, CagY (VirB10), which protects the pilus (Delahay *et al.*, 2008). The pilus can be also ornamented by CagA, CagL and, most probably, CagI (Jiménez-Soto *et al.*, 2010). *H. pylori cagX, cagT*, and *cagY* mutants have lost ability to insert CagA into the host gastric cells and to colonize mice well. Changeable CagY (VirB10 homologue) size owing to deletions or duplications in repetitive motifs inside CagY repeat region contributes to immune evasion of the bacteria (Delahay *et al.*, 2008; Atherton and Blaser, 2009).

The most abundant protein encoded by *cag*PAI is the CagC protein, which acts as the pilus subunit (Kutter *et al.*, 2008). A newly evaluated protein of the *cag*-T4SS, Cag3 possibly serves as a bridge between *H. pylori* outer and inner membranes (Pinto-Santini and Salama, 2009). Cagα (an ATPase, VirB11 homologue) in complex with Hp1451 regulator protein acts as a gating molecule for CagA translocation to the inner membrane (Hilleringmann *et al.*, 2006; Torres and Backert, 2008). CagF is a CagA-binding secretion chaperone, which controls a step before CagA recognition and is a 'doorway' for CagA entry into the secretion channel (Pattis *et al.*, 2007). CagD, located both intra- and extracellularly, is vital for CagA translocation and stimulation of IL-8 release (Cendron *et al.*, 2009). VirB11 protein (ATPase) with six monomers also participates in the protein transport (Lima *et al.*, 2010). Recently, intensive interplay between coupling protein homologue Cagβ and Cag-specific translocation factor CagZ has been observed (Jurik *et al.*, 2010). Briefly, *cag*-T4SS components have specific 'tasks' and act in a perfect cooperation, although for many genes, the functions are not well clarified.

This T4SS is activated by contact with human gastric cells (Backert *et al.*, 2010b). After adherence, *cag*-T4SS injects the CagA oncoprotein into the gastric cell cytoplasm and strongly affects the cell functions (Torres and Backert, 2008). Specific *cag*-T4SS adhesin is CagL tip protein, which mimics the activities of fibronectin (Tegtmeyer *et al.*, 2010). Dong *et al.* (2009) have reported that using Arg-Gly Asp (RGD) motif, CagL attaches to, activates the receptor (integrin α5β1) on the gastric cells, and then injects CagA into them through the hollow membrane channel. CagL also increases the gastric inflammation as, all alone, it triggers cell spreading and activates several tyrosine kinases (Tegtmeyer *et al.*, 2010). Lately, Jiménez-Soto *et al.* (2010) have reported a strong RGD-independent binding of CagA, CagI and CagY to the host cell β1 integrin receptor and the role of its conformation changes. The authors have assessed the presence of CagA at the tip of the pilus and that of CagY at both bacterial cell surface and the pilus. Therefore, the *cag*-T4SS proteins can employ different successful means of binding to the host cell.

Presence of *cagA* gene does not indicate necessarily a presence of intact *cag*PAI. To 'work' well, *cag*PAI should be intact (*cag*PAI$^+$) but in many *H. pylori* strains, it may be absent or damaged and hence inactive (Kusters *et al.*, 2006). Presence of functional *cag*PAI has been associated with peptic ulcer and gastric cancer. Intact *cag*PAI has been found in about 50–60% of strains from Western countries but in >90% of those in Japan, a country with high gastric cancer incidence (Busler *et al.*, 2006). Conversely, in India and Turkey, countries with low gastric cancer incidence, few strains have been *cag*PAI$^+$ (Kauser *et al.*, 2004; Ali *et al.*, 2005). Kauser *et al.* (2004) have reported the *cag*PAI to be highly (57.1%) conserved in Japanese strains, least conserved in European (3–15.8%) and African strains and poorly conserved in Peruvian (18.6%) and Indian (12%) strains. In Sweden, the intact *cag*PAI has increased five times the risk of severe diseases (Nilsson *et al.*, 2003). In India, much more patients with gastric ulcer (97.8%) and gastric cancer (85.7%) have harboured *cag*PAI$^+$ strains than the non-ulcer (7.1%) and duodenal ulcer patients (6.9%; Ali *et al.*, 2005). Similarly, in Lithuania, *cag*PAI$^+$ strains have been found in 75.7% of duodenal ulcer patients versus only 45.5% of other patients (Miciuleviciene *et al.*, 2008). Therefore, the intact *cag*PAI is a crucial virulence factor of *H. pylori*.

The prevalence of *cag*PAI varies according to patient age and specimens. Only one-third of Argentinean patients have harboured *cag*PAI$^+$ strains in all gastric niches, other strains have shown empty-site or damaged *cag*PAI in several niches (Matteo *et al.*, 2007). Interestingly, in North America, a lower proportion of children (41%) than adults (62%) has harboured *cag*PAI$^+$ strains (Talarico *et al.*, 2009). The gain or loss of virulence factors with age requires further evaluation.

Cytotoxin-associated gene A (*cagA*)

The highly immunogenic 120–145-kDa *H. pylori* protein, CagA, is encoded by the *cagA* gene (Kusters *et al.*, 2006). When the bacterial CagA is injected into the host cells, it is localized to the inner surface of cell membrane, where it undergoes tyrosine phosphorylation. The phosphorylation is initially (during the first 2 h) mediated by SRC family tyrosine kinases (SFKs) and then continuously by c-ABL kinase. The activated kinases phosphorylate CagA at the 3′ region of *cagA* gene on specific tyrosine residues within repeating Glu-Pro-Ile-Tyr-Ala (EPIYA) motif, which is an important *H. pylori* virulence factor (see below).

After being phosphorylated, the CagA binds to and deregulates a host Src homology 2-containing protein-tyrosine phosphatase-2 (SHp-2) via the tyrosine-phosphorylated EPIYA motifs, leading to stimulation of other phosphorylases such as extracellular signal-regulated kinase (ERK) and change in many cell-signalling pathways (Atherton and Blaser, 2009). In the host, the SHp-2 tyrosine phosphatase, considered as an oncoprotein, participates in control of adhesion, cell growth and movement (Umeda *et al.*, 2009). CagA multimerization is needed for the CagA–SHp-2 interaction. CagA domain, named CagA multimerization motif, is important for the multimerization, which is a prerequisite for formation of the CagA–SHp-2 complex and deregulation of β-catenin signal (Nguyen *et al.*, 2008). Although exhibiting many activities, CagA has a short half-life of around 200 min in the gastric cells; moreover, its stability depends on its interaction with partitioning-defective 1 (PAR1; Ishikawa *et al.*, 2009). *H. pylori cagA*$^+$ strains have been found close to the cells or in intercellular spaces, whereas *cagA* negative strains have been found in mucous gel or on apical epithelial surface (Camorlinga-Ponce *et al.*, 2004).

CagA leads to alterations in host cellular cytoskeletal and tissue structures, cellular proliferation and, in addition, it contributes to delivery of peptidoglycan fragments, which induce pro-inflammatory cytokine release (Kersulyte *et al.*, 2009). It is known that intracellular cytoplasmic protein and pathogen-recognition receptor, named nucleotide-binding oligomerization domain protein I (Nod1), is involved in the innate immunity (Brandt *et al.*, 2005). *H. pylori* peptidoglycan, delivered to the gastric cell by *cag*-T4SS, is recognized by Nod1 receptors, leading to activation of nuclear factor (NF)-kB and following stimulation of IL-8 secretion. In *H. pylori* pathogenesis, the transcription factor NF-kB is essential. The activated NF-kB targets the genes for pro-inflammatory cytokines/chemokines responsible for the gastric inflammation and acts as a tumour promoter by antiapoptotic gene expression. CagA itself, injected into host epithelial cells, also can induce IL-8 induction, independently of phosphorylation (Brandt *et al.*, 2005). In the cells, CagA targets TNF receptor-associated factor 6 (TRAF6), transforming growth factor-β-activated kinase 1 (TAK1) or Met to stimulate NF-kB through TAK1 or Akt (Lamb and Chen, 2010).

Mainly in Western countries, CagA has increased the risk for peptic ulceration, and non-cardia gastric cancer. Briefly, CagA activities (Fig. 4.2) can be grouped into:

- membrane-associated activities, e.g. relocalization of two apical junction proteins (ZO1-zonula occludens-1 and JAM-junctional adhesion molecule), leading to altered apical–junctional complex and damaged cytoskeleton, cell proliferation and differentiation, and potential tumorigenesis. Ectopic assembly of tight junction components has been observed at the sites of *H. pylori* attachment (Backert *et al.*, 2010b). Mechanism depends on CagA inhibition of PAR1b/MARK2 kinase, which is key regulator of epithelial cell polarity (Hatakeyama, 2009). When CagA binds PAR1, it reduces its kinase activity and disrupts the cell polarity (Backert *et al.*, 2010b);
- phosphorylation-dependent activities such as:
  - actin-cytoskeletal rearrangements and cell scattering through suppression of focal adhesion kinase (FAK) activity by SHp-2, following its activation by the phosphorylated CagA, and stimulation of extracellular signal-regulated kinase;
  - suppression of Src kinases and related dephosphorylation of host cell proteins such as vinculin, ezrin and cortactin (Yamazaki *et al.*, 2005; Nguyen *et al.*, 2008;

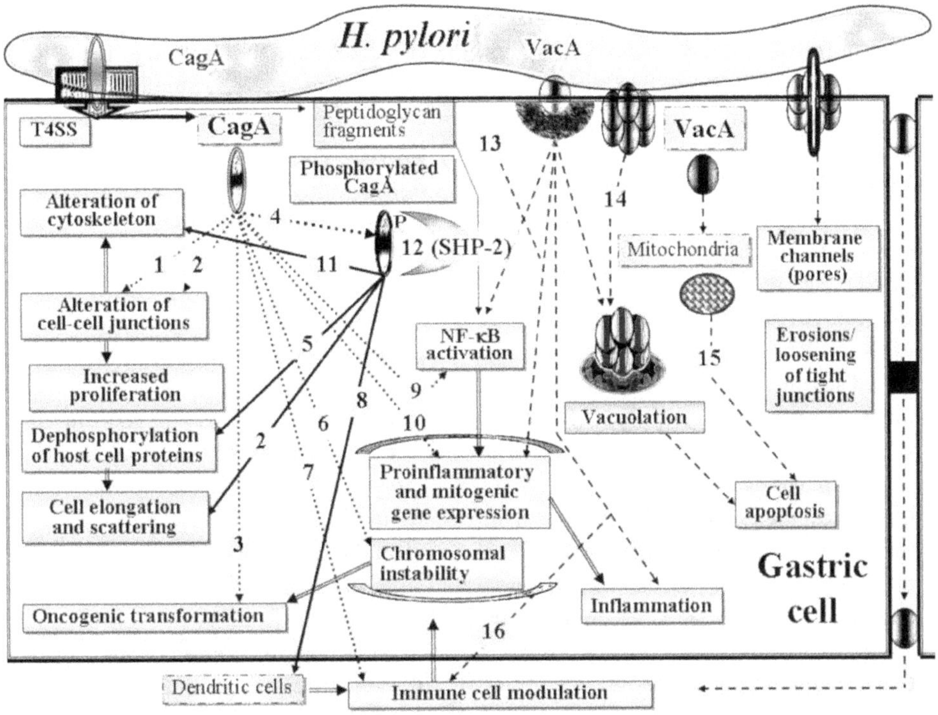

Figure 4.2 Effects of CagA and VacA. 1, PAR1b/MARK2 kinase; 2, activation of ERK; 3, E-cadherin/β-catenin; 4, SFKs and c-ABL kinase; 5, inhibition of Src family kinases; 6, inhibition of PAR1 family kinases; 7, suppression of p53 accumulation and inhibition of JAK-STAT signalling in IL-3-dependent B cells; 8, downregulation of TBK1 activity, reduced phosphorylation of IRF3 and nuclear translocation; 9, NF-kB/IL-8-activating pathway; 10, Ras/Raf/MEK/ERK/NF-kB signalling pathway; 11, inhibition of focal adhesion kinase activity by SHp-2; 12, host Src homology 2-containing protein-tyrosine phosphatase 2 (SHP-2); 13, p38 MAPK/ATF-2; 14, V-ATPase, dynamin, syntaxin 7, etc.; 15, cytochrome c release; 16, inhibition of IL-2 release.

Lu *et al.*, 2009). This leads to 'hummingbird phenotype' of the gastric cells (see below);

- induction of ERK signalling pathways, which can also promote increased cellular proliferation of the gastric epithelial cells (GECs). Activation of ERK can follow the interaction of phosphorylated CagA with SHp-2 protein (Backert *et al.*, 2010a);
- suppression of dendritic cell functions (found in mice) by downregulation of TBK1 enzymatic activity, reduced phosphorylation of IRF3 and following nuclear translocation. CagA protein, tyrosine-phosphorylated in dendritic cells, inhibits their maturation and the expression of pro-inflammatory cytokines, TNF-*α* and IL-12p40, and promotes expression of anti-inflammatory IL-10 cytokine. Suppression of the immune responses leads to persistence of the bacteria in the host (Tanaka *et al.*, 2010);
- direct translocation of CagA into B cells, stimulation of extracellular signal-regulated kinase, p38 mitogen-activated protein kinase and expressions of Bcl-2 and Bcl-XL, which prevent apoptosis and may lead to development of MALT lymphoma (Lin *et al.*, 2010b). It is known that some members of the Bcl-2 family are antiapoptotic and others are proapoptotic. However, there is still a controversy about the association between CagA and the MALT lymphoma (Lehours *et al.*, 2009; Thiberge *et al.*, 2010); and
- binding (with still unidentified role) to tyrosine phosphatase (Shp1),

phosphoinositide-2 (PI3)-kinase, adaptor protein (Grb2 and Grb7) and Ras GTPase activating protein Ras-GAP (Backert *et al.*, 2010a), which functions should be further evaluated.

- phosphorylation-independent activities such as:
  - generation of chromosomal instability in gastric cells (see below);
  - cell scattering and proliferation by activation of Ras/ERK signalling cascade through son of sevenless (SOS), leading to possible tumorigenesis (Nguyen *et al.*, 2008);
  - induction NF-κB and NF-κB/IL-8-activating pathway (Brandt *et al.*, 2005). CagA stimulates pro-inflammatory IL-8 release by the epithelial cells via Ras/Raf/MEK/ERK/NF-κB signalling pathway independent of SHp-2 (Brandt *et al.*, 2005; Nguyen *et al.*, 2008). However, CagA induction of IL-8 secretion is strain-dependent, i.e. only some strains can induce the secretion (Brandt *et al.*, 2005);
  - activation of oncogenic genes by accumulation of β-catenin in the cell cytoplasm and nucleus following interaction of CagA with the epithelial cell adhesion protein, E-cadherin (Nguyen *et al.*, 2008). It also has been shown that CagA WW domain, binding to gastric tumour suppressor RUNX3, blocks its activity (Tsang *et al.*, 2010); and
  - modulation of B-cell activities, e.g. avoidance of p53-dependent B-cell apoptosis by inhibition of p53 accumulation and suppression of the proliferation by inhibition of JAK-STAT signalling in IL-3-dependent B cells. This decreases immune response (Umehara *et al.*, 2003) and can promote development of MALT lymphoma.

In brief, the phosphorylated CagA modulates GTPase activation, inducing cellular actin-cytoskeletal rearrangements, and increased motility and elongation of the gastric cells. The non-phosphorylated CagA influences many signalling cascades, resulting in altered cell polarity, disrupted cell-to-cell junctions, cell proliferation, inflammation, and reduced apoptosis (Backert *et al.*, 2010a). The combined activities of CagA may lead to severe inflammation and damage of the gastric mucosa, activation of proto-oncogenes and risks for peptic ulcer or gastric cancer. Umeda *et al.* (2009) have revealed the dual role of CagA as oncoprotein. CagA promotes cell proliferation and, simultaneously, induces mitotic damage and chromosomal instability by disturbing the microtubule-based mitotic spindle owing to CagA-induced inhibition of PAR1 kinases. This may initiate tumorigenesis (Umeda *et al.*, 2009).

CagA also inhibits the clathrin-independent pinocytic endocytosis pathway and hence VacA entry and function in the host cells (Akada *et al.*, 2010). Inhibition of the endocytosis also may affect the cancer-cell signalling pathways.

The *cagA* gene has been present in most *H. pylori* strains from Japan versus only 50–70% of the strains in Western countries. Prevalence of *cagA*$^+$ strains in patients with peptic ulcers or gastric cancer has been usually >66.7% and up to 100% (Table 4.2). In East Asia (Japan, South Korea and China) and some other countries, e.g. Nigeria and South Africa, a high (often ≥90%) prevalence of *cagA*$^+$ strains has been found resulting usually in lack of significant association with the diseases (Kato *et al.*, 2000; Smith *et al.*, 2002; Ko *et al.*, 2008; Tanih *et al.*, 2010). For instance, 94% of the strains from South Korean children have been *cagA*$^+$ (Ko *et al.*, 2008). The high prevalence of *cagA*$^+$ *H. pylori* strains in East Asia correlates with the high gastric cancer incidence there.

In European strains, the frequency of *cagA*$^+$ has varied from 51.3% in France to 93.8% in Serbia-Montenegro, being usually high (>77%) in south-east European countries such as Bulgaria, Greece, Serbia, Montenegro and Turkey (Apostolopoulos *et al.*, 2002; Ribeiro *et al.*, 2003; Sokic-Milutinovic *et al.*, 2004; Caner *et al.*, 2007; Boyanova *et al.*, 2009). However, peptic ulcer, gastric cancer and, especially, MALT lymphoma can be observed in some patients with *cagA*-negative strains (Lu *et al.*, 2005a). Prevalence of *cagA*$^+$ strains may vary according to the patient ethnicity. All strains from black US children have been *cagA*$^+$ versus 56.5% of those in white children (Yamaoka *et al.*, 2010). Nevertheless, outcomes of *H. pylori*

Table 4.2 Prevalence of *cagA*⁺ *H. pylori* strains in some countries

Country	No. of strains	Overall *cagA* prevalence (%)	*cagA* prevalence in diseases PU	GC	Other	Association with diseases	Reference
Europe							
Bulgaria	108	81.5	92.3	NA	75.4	PU	Boyanova *et al.* (2009)
Germany	91	70.3	NA	79	69 (CG), 63 (ML)	None	Koehler *et al.* (2003)
Greece	62	77.4	NA	NA	NA	NA	Apostolopoulos *et al.* (2002)
Ireland	103	68	NA	NA	NA	NA	Taneike *et al.* (2009)
Italy	193	72	75	94.2	58.1	PU and GC	Basso *et al.* (2008)
North America							
Cuba	130	73.2	87 (DU), 57.6 (GU)	NA	70.6	None	Torres *et al.* (2009)
Mexico	143	NA	NA	NA	52	NA	Paniagua *et al.* (2009)
USA[1]	61	66	NA	NA	66	NA	Podzorski *et al.* (2003)
USA[1]	37	70.3	NA	NA	NA	NA	Yamaoka *et al.* (2010)
USA	NA	69.8	NA	NA	NA	Intestinal metaplasia	Schneller *et al.* (2006)
USA (African)	25	88	NA	NA	NA	NA	Chung *et al.* (2010)
South America							
Argentina	299	40.8	31.6	70	35.3–43.4	None	Vega *et al.* (2010)
Brazil	142	64.5 (children), 82.5 (adults)	78.3 (children), 87.5 (adults)	94.4 (adults)	56.4 (children), 55 (adults)	DU and GC	Ashour *et al.* (2001)
Brazil	94	73.4	86	NA	59.4–73.3	PU	Gatti *et al.* (2006)
Chile	66 (41 patients)	24.2	NA	NA	NA	NA	García *et al.* (2006)
Chile	53 (CG)	49	NA	NA	49	Severe CG	Araya *et al.* (2004)
Costa Rica	76	77.6	NA	NA	NA	High GC incidence regions	Con *et al.* (2009)
Venezuela	121	50–86	NA	NA	NA	NA	Ghose *et al.* (2005)
Uruguay	30	56.7	NA	NA	NA	NA	Chung *et al.* (2010)
Asia							
Bahrain	120	31.6	100	NA	9.9	PU	Bindayna and Al Mahmeed (2009)
China	115	73.6 (children), 82.3 (adults)	NA	NA	NA	NA	Li *et al.* (2009)
China	87	98	NA	NA	NA	NA	Wang *et al.* (2003)
China	30	86.7	NA	NA	NA	NA	Chung *et al.* (2010)
Japan[1,2]	40	92.5	95 (DU), 80 (GU)	NA	93.3	None	Kato *et al.* (2000)
Jordan	110	26.4	NA	NA	NA	NA	Nimri *et al.* (2006)
Pakistan	224	55.8	67.2	77	42	PU and GC	Yakoob *et al.* (2009)

Table 4.2 continued

Country	No. of strains	Overall *cagA* prevalence (%)	*cagA* prevalence in diseases			Association with diseases	Reference
			PU	GC	Other		
South Korea	136	94.1	94.5	NA	92.3	None	Park *et al.* (2001)
South Korea[1]	33	94	NA	NA	94	None	Ko *et al.* (2008)
Taiwan	234	80.8	82	76	88	None	Lin *et al.* (2004)
Turkey	91	73.6	85.7	84.8	50	DU	Erzin *et al.* (2006)
Turkey	46	93.5	93.8	NA	93.3	None	Caner *et al.* (2007)
Turkey	35	48.6	NA	NA	NA	NA	Chung *et al.* (2010)
Africa							
Nigeria	41	92.7	94.7 (DU)	NA	90.9	None	Smith *et al.* (2002)
Tunisia	281	61.6	67.9	NA	58.4	None	Ben Mansour *et al.* (2010)

[1]Children.

[2]CagA anitbody positive.

DU, duodenal ulcer; PU, peptic ulcer; GU, gastric ulcer; GC, gastric cancer; CG, chronic gastritis; ML, mucosa-associated lymphoid tissue lymphoma; NA, not available.

infection depend not only on CagA but also on the intact *cag*PAI and CagA EPIYA sequences.

Importantly, *cagA* gene has been associated with the success of anti-*H. pylori* therapy. Low virulent *cagA*-negative/*vacA* s2 type strains, which are prevalent in non-ulcer patients, carry a higher risk for eradication failure than the virulent *cagA*$^+$/*vacA* s1 strains, which are common in ulcer patients (see Chapter 9).

Dietary factors appear to influence *H. pylori* virulence. For instance, high salt intake can increase the bacterial virulence. Co-culture of GECs with *H. pylori* at high salt concentrations has shown increased tyrosine-phosphorylated CagA and higher IL-8 secretion by the cells compared with those at low salt concentrations (Loh *et al.*, 2007). Another study has shown strain-specific variations by salt shock and increased expression of *vacA* but not *cagA* transcript in one strain (Gancz *et al.*, 2008). This topic is not only interesting but also important for daily dietary practice and deserves additional evaluation.

Western and East Asian CagA

Importantly, cytotoxin-associated (*cagA*) gene has been highly divergent in East Asian strains compared with Western strains (McClain *et al.*, 2009). Variations in *cagA* activity depend on the number of repeating sequences in the 3′ region of the gene (Yamaoka *et al.*, 2008a). Phosphorylation of CagA occurs on the tyrosine residues within the EPIYA motifs. They are flanked by repetitive DNA sequences, which participate in recombinations.

CagA EPIYA involve EPIYA-A, B, C, and D motifs according to the amino acid sequences that surround the EPIYA sequence (Jones *et al.*, 2009). EPIYA-A and EPIYA-B motifs have been present in all strains worldwide. The EPIYA-C motif has been additionally present within Western strains (in Europe, North America, Australia, South Asia and Africa). EPIYA-D motif has been very common in East Asia (Japan, China and Korea; Argent *et al.*, 2008b; Backert *et al.*, 2010a). It is known that Asian countries have been considered as high risk for gastric cancer, e.g. Japan, Korea and China, intermediate risk, e.g. Vietnam, or low risk, e.g. Thailand and Indonesia (Uchida *et al.*, 2009). Therefore, two major types of CagA protein (East Asian and Western) are known. CagA proteins usually have EPIYA-A and EPIYA-B sites, followed by 1–3 (up to 5) EPIYA-C repeats in Western CagA or EPIYA-D motifs in East Asian CagA (Nguyen *et al.*, 2008; Shokrzadeh *et al.*,

2010). Among East Asian isolates, EPIYA-ABD has been associated most often with the gastric cancer (Jones *et al.*, 2009).

East Asian CagA exhibits a particular virulence. CagA type D EPIYA (East Asian CagA) and increased number of C repeats (in some Western strains) enhance the CagA-induced SHp-2 phosphatase activity and the resulting risk of gastric cancer. East Asian CagA with EPIYA-D motifs shows stronger tyrosine phosphorylation and higher affinity for SHp-2, which results in stronger damage and high IL-8 secretion in GECs, and higher inflammation and gastric atrophy than the Western CagA (Argent *et al.*, 2008b). East Asian CagA exhibits a six amino acid sequence overlying EPIYA-D (Y-A-T-I-D-F), matching to that of consensus ligand-binding motif for SH-2 domain of SHp-2, and hence strong transforming activities (Nguyen *et al.*, 2008; Tanaka *et al.*, 2009a). EPIYA-C differs from SHp-2-binding sequence only by an amino acid at $pY^{+}5$ position (Shokrzadeh *et al.*, 2010).

High prevalence of East Asian CagA within East Asian strains explains, at least partly, the high incidence of gastric cancer in Japan, Korea and China (Atherton and Blaser, 2009). Unlike East Asian strains, the majority of those from South Asia (India, Thailand and Pakistan) have harboured Western CagA in association with lower cancer incidence (Nguyen *et al.*, 2008). Although asymptomatic subjects in Japan and Thailand have shown similar rates of *H. pylori* infection by stool positivity, East Asian CagA rate has been only 7.2% in Thailand versus 63.1% in Japan (Hirai *et al.*, 2010). In India, only 34.8–46.5% of strains have harboured intact *cag*PAI and most strains have had Western-type CagA (Kumar *et al.*, 2010).

The 'hummingbird phenotype', i.e. the appearance of highly elongated cells with long protrusions, has been associated with gastric cancer. The increased motility of these cells has been suggested to participate in gastric cancer growth and invasion (Lu *et al.*, 2009). The induction of 'hummingbird phenotype' and CagA oncogenic properties have been associated with CagA type ABD >ABCCC >ABCC and >ABC (Mégraud, 2008). In Western countries, the amount of CagA tyrosine phosphorylation has been linked to increased number of CagA EPIYA-C motifs, which have stimulated the phosphorylation-dependent CagA activity and have been the main factor for intestinal metaplasia and gastric cancer (Basso *et al.*, 2008).

Interestingly, *H. pylori* strains with multiple numbers of CagA EPIYA-C segments have been less capable to survive in low pH. Strains from Italian patients with benign diseases have harboured higher numbers of EPIYA-C segments in gastric corpus than in the antrum (Basso *et al.*, 2008). In the same country, *cagA*$^{+}$ strains with ≥2 EPIYA-C segments have been associated with a strong risk for gastric cancer, >30-fold higher than the risk associated with strains with one segment (Basso *et al.*, 2008). In Iran, only 3.3% of strains have harboured four copies of the segments, which could explain the low incidence of gastric cancer in that country (Shokrzadeh *et al.*, 2010). However, it was found that *H. pylori* eradication in patients with precancerous lesions did not reduce the proportion of people developing cancer over 7.5 years; therefore, CagA is believed to have damaged the gastric cells at earlier stage of the infection (Wong *et al.*, 2004).

Number of EPIYA motifs may differ between children and adults. In 37 US children, 70% of strains were *cagA*$^{+}$, as in adults; however, all *cagA* genes from children harboured three EPIYA motifs, while 21% of those from the adults harboured more motifs (Yamaoka *et al.*, 2010). Further studies are required to assess whether the more acid-susceptible strains with many EPIYA-C motifs in adults imply an increased risk for gastric cancer or these strains simply predominate in case of gastric atrophy. Conversely, in China, almost all (97.4%) *cagA*$^{+}$ strains from children have exhibited the virulent EPIYA-ABD, while those from adults have shown ABD in 64.7% and AAD in 3.9% (Li *et al.*, 2009).

Cytotoxin-associated gene E (*cagE*) and other *cag*PAI genes

The *cagE* (*picB*, VirB3/4 homologue) gene, located upstream from *cagA* in the right half of *cag*PAI, encodes a protein, which mediates cytoskeletal rearrangements in host gastric cells, and according to some studies, activates NF-κB and participates in IL-8 release in the cells (Su *et al.*, 2003; Kauser *et al.*, 2005a). Moreover, CagE,

CagM, and CagL have been found to repress the transcription of α subunit of parietal cell H,K-ATPase and hence the proton pump expression in the gastric mucosa, leading to hypochlorhydria, which increases the gastric cancer risk (Saha *et al.*, 2010). As the presence of *H. pylori* biofilms on the gastric mucosa has been reported, it is also interesting that *cagE* gene has influenced the biofilm pattern *in vitro* as *cagE* mutants have shown increased biofilm growth of *H. pylori* in the presence of gastric mucin (Cole *et al.*, 2004). Therefore, the *cagE* mutant could provide higher resistance to clearance and phagocytosis in the bacteria in biofilms compared with the planktonic bacteria; however, these results need to be proven in more trials.

In some studies, not only *cagA* but also other *cag*PAI genes, e.g. *cagE, cagM, cagT, virB11, cagY/7* (VirB10 homologue) and *LEC* genes have been associated with severe diseases. In Brazil, *cagA* and *cagE* genes have been linked to higher frequency of peptic ulcers, and both *cagE* and *virB11* genes have been associated with gastric cancer, in which cases, *cagE*$^+$/*virB11*$^+$ has been the most frequent (50%) genotype (Proença Módena *et al.*, 2007; Lima *et al.*, 2010). In another Brazilian study, *virB11* gene has been found in almost all strains; however, *cagT* gene has been associated with peptic ulcers (Pacheco *et al.*, 2008). Again in Brazil, *cagA, cagM, cagT* and *LEC* have been linked to peptic ulcers but only *tnpA* and *LEC* genes have been associated with gastric cancer (Mattar *et al.*, 2010). In Turkey, *cagA, cagE* and *vacA* s1a genotypes have been associated with duodenal ulcer and gastric cancer, *cagE* and *vacA* s1a types have been independent predictors of duodenal ulcer, and *cagE* and blood group antigen-binding adhesin *(babA2)* types have been risk factors for gastric cancer (Erzin *et al.*, 2006). In Costa Rica, CagY/7 has been associated with gastric cancer (Occhialini *et al.*, 2001). It should be noted that most of the above-mentioned genes except *cagE* have not been widely evaluated, which can obstruct the understanding of their clinical significance.

Nevertheless, the association of *cag*PAI genes with the diseases is not universal and has been absent in East Asia and some other countries, e.g. Thailand and Malaysia (Tan *et al.*, 2005; Chomvarin *et al.*, 2008). As for many other virulence genes, in Japan, *H. pylori cagE*$^+$ and *cagA*$^+$ strains have been highly prevalent (>96%) despite the infection outcomes (Ikenoue *et al.*, 2001). In US children and in Colombia, frequencies of *cagA*$^+$ and *cagE*$^+$ strains have been similar, while in Turkish and Thai strains, *cagA*$^+$ genotype has been more common (Podzorski *et al.*, 2003; Quiroga *et al.*, 2005; Erzin *et al.*, 2006; Chomvarin *et al.*, 2008). From diagnostic viewpoint, the *cagE* gene can predict better the presence of intact *cag*PAI than the *cagA* gene as suggested in a Japanese study (Ikenoue *et al.*, 2001).

Vacuolating cytotoxin (VacA)

Unlike *cagA*, vacuolating cytotoxin (*vacA*) gene is present in all *H. pylori* strains; however, the biologically active form of VacA has usually been found in only 40–50% of them (Atherton and Blaser, 2009) owing to sequence variations within the *vacA* gene, causing 'mosaic design' of signal (s), mid (m) and intermediate (i) alleles.

VacA precursor is a 135–145-kDa protoxin, which, following a proteolysis during the process of secretion, produces 33-amino-acid signal sequence, mature 88–90-kDa secreted toxin, 12-kDa secreted peptide and carboxy-terminal domain associated with the bacterial cells (Ivie *et al.*, 2010). Receptor-type protein-tyrosine phosphatase (RPTP) enzymes are known to regulate host cell proliferation, differentiation and adhesion. VacA binds to RPTPα and RPTPβ proteins and to sphingomyelin on gastric epithelial cell surface (Isomoto *et al.*, 2010). Sphingomyelin is important for VacA association with cell membranes and VacA vacuolating activity. Moreover, a specific VacA (β_2-integrin subunit CD18) receptor on T-cell surface has been revealed (Isomoto *et al.*, 2010). VacA also targets fibronectin, epidermal growth factor (EGF) receptor and heparin sulfate, different lipids on T cells etc. (Backert *et al.*, 2010b.) Therefore, VacA can 'gun against' plenty of host surface molecules on both gastric and immune cells.

VacA protein has a key role in *H. pylori* pathogenesis. Unlike other bacterial pore-forming toxins, VacA toxin works intracellularly (Ivie *et al.*, 2010). This toxin (88 kDa monomer) is secreted by type V-secretion system of *H. pylori* and forms flower-like structures. These 'flowers' can be processed into two (p33 and p55) fragments, although the cleavage does not affect the VacA functions (Isomoto *et al.*, 2010). The 'flowers' exhibit a central ring surrounded by peripheral 'petals'. It has been shown that the p33 domain (33 kDa) exerts a pore-forming activity needed for vacuolation (Ivie *et al.*, 2010). The p55 subunit (55 kDa) undergoes acid activation and afterwards becomes resistant to acid and pepsin degradation. This subunit is important for binding to target cells and formation of anionic membrane channels (Atherton and Blaser, 2009; Costa *et al.*, 2009). The p55 domain has a β-helical arrangement with 13-coil structure. In case of deletions in the coils, the secretion of the toxin decreases (Ivie *et al.*, 2010). The p33 and p55 then can amass to produce single or double layered complexes, the latter representing the flower-like structures (Gangwer *et al.*, 2007; González-Rivera *et al.*, 2010).

According to VacA location, there are several types of activities of the toxin. When after oligomerization, the exotoxin is endocytosed by the GECs, the complex of six VacA subunits forms a hexameric anion-selective membrane channel in the endosomal membranes, interacting with their lipids (Isomoto *et al.*, 2010). In this way, the toxin induces nutrient leakage from the cells and forms large vacuoles in them (Isomoto *et al.*, 2010). SNAREs (soluble *N*-ethylmaleimide-sensitive factor attachment protein receptors) are small proteins, interacting with other SNAREs and present on cell membranes, including transport vesicles. Vacuolation results from alteration of protein trafficking in the intracellular vesicles and depends on the presence of vacuolar (V)-ATPase and SNARE proteins in the cells. SNAREs participate in last step of vacuolation, the late endosome–lysosome fusion (Mashima *et al.*, 2008). The vacuolation originates from the swollen late endosomes, produced during the late endocytotic pathway.

Both VacA subunits can associate at mitochondrial membranes (Foo *et al.*, 2010). When VacA reaches the mitochondria, it produces pores in their membranes and hence mitochondrial alterations. VacA also induces apoptosis of the gastric cells, activating proapoptotic Bcl-2 family proteins, which amass in the mitochondria, inducing cytochrome c release (Isomoto *et al.*, 2010).

When bound to cell membrane receptors, VacA causes changes in cytoskeleton-associated cell functions, disruption of endosomal/lysosomal activity and injury of gastric cells with erosions and loosening of tight junctions. It also induces development of pro-inflammatory response (Chung *et al.*, 2010). VacA increases the release of IL-8 and prostaglandin E_2 (PGE_2) by activation of p38 MAPK (mitogen-activated protein kinase)/ ATF-2 (activating transcription factor 2) cascade. The toxin modifies the integrin-receptor-induced cell signalling.

If the toxin passes through tight junctions, it immunomodulates T- and other immune cells (D'Elios *et al.*, 2007; Takeshima *et al.*, 2009; Isomoto *et al.*, 2010). On the T cells, VacA exerts both suppressive and pro-inflammatory effects (Takeshima *et al.*, 2009). The suppressive effect is due to inhibition of IL-2 release by interference with T-cell receptor/IL-2 signalling pathway at the level of Ca^{2+}–calmodulin-dependent phosphatase calcineurin (Atherton and Blaser, 2009). Thus, VacA diminishes T-cell proliferation and can induce cell cycle arrest. This can contribute to the infection chronicity. Pro-inflammatory effect is due to activation of NF-κB via the classical pathway (Takeshima *et al.*, 2009). VacA can inhibit B cell proliferation as well (Isomoto *et al.*, 2010).

Briefly, VacA forms membrane channels and nutrient leakage from GECs and cell vacuolation if the toxin is endocytosed, or initiates a pro-inflammatory response if VacA is cell-surface bound, or modulates T- and B-cell activation/proliferation if the toxin passes trough tight junctions, or induces cell apoptosis if VacA is transported to the mitochondria (Isomoto *et al.*, 2010; Fig. 4.2). VacA is a multifunctional virulence factor of *H. pylori* with activities against both GECs and immune cells.

The *vacA* signal region, encoding the signal peptide and mature protein *N*-terminus, has s1 (type s1a, s1b or s1c) and s2 alleles. The mid region, encoding part of the p55 epithelial cell-binding subunit, exists as m1 (m1a, m1b, and m1c) or m2 (m2a and m2b) alleles. Newly described *vacA* intermediate region, encoding part of the p33 subunit, has three allelic types, i1, i2 and the recently reported i3 (Rhead *et al.*, 2007; Chung *et al.*, 2010). The *vacA* deletion (d) region (between the i and m regions) exhibits either d1 type without the 81-bp deletion or d2 type with the deletion (Ogiwara *et al.*, 2008).

Differences between incidences of gastric cancer and geographic areas are associated, at least in part, by variations in *cagA* and *vacA* genotypes. Strains with Western type 3′ region *cagA* and *vacA* s2 and s1a/m1a genotype encompass the Western group, while those with East Asian-type 3′ *cagA* and *vacA* s1c/m1b genotype form the East Asian group, linked to common gastric cancer incidence (Tanaka *et al.*, 2009a). This can be associated with the *vacA* mosaic combinations within the three allelic types, each allele showing stronger or weaker activity.

The signal *vacA* region affects the vacuolating activity of the cytotoxin. Almost all s1 forms possess cytotoxic activity, while a 12-amino-acid hydrophilic segment on the s2 form blocks the activity (Dong *et al.*, 2009). The mid region is a determinant of binding specificity of VacA to different epithelial cells (Chung *et al.*, 2010). The *vacA* m1 type strains have been linked to stronger gastric damage than those of m2 type; the *vacA* s1a type has caused higher gastric infiltration by neutrophils and lymphocytes than s1b type, while the s2 type has shown no noticeable activity on the gastric cells (Podzorski *et al.*, 2003). Contrary examples exist, in Costa Rica, where *cagA*$^+$ status and *vacA* s1b allele have been associated with gastric cancer (Con *et al.*, 2009). While in Germany, the *vacA* s1a type has been 4.2 times more common in gastric cancer than in gastritis, in Taiwan, gastric cancer patients have shown more often m1T (modified primer) subtypes than other patients (Koehler *et al.*, 2003; Lin *et al.*, 2004). In Turkish patients with chronic gastritis, the corpus neutrophil infiltration has been higher in the m1 *H. pylori* group than in the m2 group (Umit *et al.*, 2009). As it has been suggested (Chung *et al.*, 2010), the m region seems to have a secondary importance as *H. pylori* virulence factor than *cagA*$^+$ or *vacA* s1 genotype. Nevertheless, the less virulent strains are also clinically important. In Germany, *vacA* m2 allele has been the most common type in strains from patients with MALT lymphoma (Koehler *et al.*, 2003).

The *vacA* i type can predict better the strain virulence than both s and m types (Douraghi *et al.*, 2009; Chung *et al.*, 2010). Association between strains of *vacA* i1 allele and gastric cancer has been reported in Iranian and Italian patients, and link between this allele and gastric ulcer has been observed in Iraqi and Italian patients (Rhead *et al.*, 2007; Basso *et al.*, 2008; Hussein *et al.*, 2008). For instance, in Italy, *vacA* s1 and i1 alleles have been associated with gastric cancer and *vacA* i1 allele has been associated with peptic ulcers (Basso *et al.*, 2008). The *vacA* i3 allele, which possibly has i2-like activity, has been found in more Turkish (in 25.7%) than Chinese strains (3.3%; Chung *et al.*, 2010).

However, like for many virulence genes, no link between *vacA* s1 and i1 alleles and gastric damage has been found in East Asia (Yamaoka *et al.*, 2008a). In Japan, China, South Korea and Thailand, *vacA* s1 prevalence has been up to >90%, like that of *cagA*$^+$ status (Park *et al.*, 2001; Yamazaki *et al.*, 2005; Chomvarin *et al.*, 2008; Hu *et al.*, 2009; Chung *et al.*, 2010). Most strains from East Asia have harboured the virulent *vacA* i1 allele, which frequency has reached 93.3% in Chinese strains (Ogiwara *et al.*, 2008; Chung *et al.*, 2010). The East Asian strains have shown also more often *vacA* s1, m1, and d1 genotypes than Western strains. In the Western countries, the *vacA* m1, i1, and d1 types have raised the risk of gastric cancer (Ogiwara *et al.*, 2008). The functions of *vacA* d1/d2 regions should be more clarified.

Prevalence of *vacA* genotypes varies between countries. Examples of prevalent subtypes are *vacA* s1c and m1b in East Asia, s1a and m1c in South Asia, and m1a in Europe and Africa (Yamaoka *et al.*, 2008a). The *vacA* s1a allele has been frequent in American, Northern and Eastern European strains, while s1b has been common in Latin American, Portuguese and Spanish strains (Ashour *et al.*, 2001). In Europe, strains of s1a subtype have been found in 61–89.8% in Northern and Eastern countries, while the prevalence of s1b type has been 72–89% in Spain and Portugal (Lin *et al.*, 2004; Kauser *et al.*, 2005a; Kauser *et al.*, 2005b; Boyanova *et al.*, 2009). In the Middle East, *vacA* s1, m1 and i1 genotypes have been found in 71.5%, 32.8% and 40.7% of strains, respectively, like in Europe and America (Sugimoto *et al.*, 2009). Overall prevalence of *vacA* s1a type has ranged from 0–14.8% in East Asia, Tunisia, Venezuela, Portugal and US children to >83% in Turkey and Bulgaria (Table 4.3). The i1 allele has been common (80%) in US African strains, unlike those from Turkey (22.9%) and Uruguay (46.7%; Chung *et al.*, 2010).

Differences in the prevalence of *vacA* alleles among regions of the same country and the age groups also have been found in some studies (Wang *et al.*, 2003; Lee *et al.*, 2004). Prevalence of *vacA* s1a has been 45.2% in South Korean patients aged <20 years versus 14.9% in those aged ≥50 years (Lee *et al.*, 2004).

Some *vacA*-allele double combinations are clinically important. Cytotoxin amount has been found to be large by strains of s1/m1 type, intermediate by s1/m2 type and low/absent by s2/m2 type strains (Yamazaki *et al.*, 2005). The s2/m1 type has been rare and non-toxic (Letley *et al.*, 2003). In Western countries, strains of s1/m1 type have been associated with peptic ulcers and gastric cancer (Kusters *et al.*, 2006). In countries with common gastric cancer such as Japan and Korea, almost all *H. pylori* strains have shown *vacA* s1/m1, whereas in countries with less frequent gastric cancer, such as Turkey and Thailand, <60% of strains have harboured the genotype (Nguyen *et al.*, 2008). Patients with gastric cancer have harboured usually strains of s1/m1 type (Yamaoka *et al.*, 2008a).

Highly virulent triple-allele combinations also should be considered. Most s1/m1 strains have exhibited the virulent i1 subtype and most s2/m2 strains have shown i2 subtype. However, *vacA* i1 prevalence among s1/m2 strains has varied according to the country, e.g. most s1/m2 strains have been i1 subtype in strains from China (93.3%) and U.S. Africans (71.4%), unlike those from Turkey (27.3%) despite the high CagA frequency there (Chung *et al.*, 2010). Of the s1/m2 strains, only few (8%) East Asian strains have been i2 but half of Western strains have shown i2 subtype (Yamaoka *et al.*, 2008a).

Western strains of *vacA* s1/i1/m1 type (producing the fully active VacA) have been more common in gastric cancer patients than those of s2/i2/m2 type, which produce inactive VacA (Rhead *et al.*, 2007). The s1/i1/m2 type has been active on less epithelial cell lines than the s1/i1/m1 type as m2 types bind to fewer epithelial cells (Atherton and Blaser, 2009). However, not only s1/i1/m1 but also s1/i1/m2 type strains have shown vacuolating activity and association with the most severe gastric diseases (Rhead *et al.*, 2007; Basso *et al*, 2008; Chung *et al.*, 2010). High prevalence of *vacA* s1/i1/m1 and s1/i1/m2 type strains in East Asia is a factor, contributing to the common gastric cancer there (Atherton and Blaser, 2009; Jang *et al.*, 2010).

Furthermore, VacA toxin can show evolution *in vivo* and can change its toxicity through recombination with other *H. pylori* strains in mixed infections. Mixed infections have been more common in some countries/regions, mainly in developing ones, e.g. Venezuela (55%) and Tunisia (31.4%), than in North America, Slovenia and Bulgaria (10.2–11.5%%; Ghose *et al.*, 2005; Homan *et al.*, 2009; Ben Mansour *et al.*, 2010; Boyanova *et al.*, 2010; Tanih *et al.*, 2010). Strains of the mixed infections have been more resistant to antibiotics and more virulent compared with those of single infections (Cellini *et al.*, 2006).

Usually, infections by type I ($CagA^+/VacA^+$) strains have correlated with the risk for peptic ulcers, gastric cancer and MALT lymphoma in Western countries (Apostolopoulos *et al.*, 2002). The s1 and s1/m1 isolates have been associated often with *cagA*$^+$ status and i1 allele (Boyanova *et al.*, 2009; Bindayna and Al Mahmeed, 2009; Chung *et al.*, 2010). In strains from China, Turkey, Uruguay and the USA, Chung *et al.* (2010) have found a coexistence of *cagA*$^+$, *vacA* s1, and *vacA* i1 all isolates except those from Turkey.

Importantly, virulent *cagA*$^+$/*vacA* s1a type strains have been eradicated easier than the rest (see Chapter 9). Fascinatingly, VacA and CagA can downregulate each other's activities on gastric mucosa to preserve it from excessive damage. NAT (nuclear factor of activated T cells) is activated by CagA but inhibited by VacA (Isomoto *et al.*, 2010). CagA can inhibit VacA-induced apoptosis by two mechanisms. Both earlier activity of the phosphorylated CagA, preventing the pinocytosed VacA to reach the mitochondria and the later activity of unphosphorylated CagA, blocking the VacA-induced cell death at the mitochondrial

Table 4.3 Prevalence of *vacA* alleles within *H. pylori* strains in some countries

Country	No. of strains	Prevalence (%) of *vacA* allele										Common types	Mixed	Association with PU/GC
		s1	s1a	s1b	s1c	s2	m1	m2	i1	i2	i3			
Europe														
Bulgaria	108	91.7	89.8	1.8	0	8.3	39.8	60.2	NA	NA	NA	s1/m2	6.9	s1, s1a
France	82 (43 ML, 39 CG)	58.1 (ML), 78.1 (CG)	NA	NA	NA	41.9 (ML), 28.2 (CG)	30.2 (ML), 56.4 (CG)	69.8 (ML), 43.6 (CG)	NA	NA	NA	NA	0	None
Germany	91	83.5	70.3	12.1	0	16.5	39.6	60.4	NA	NA	NA	s1	0	s1a (GC), m2 (ML)
Ireland	103	85.4	NA	NA	NA	14.6	18.4	81.6	NA	NA	NA	s1/m2	NA	NA
Italy	189	70.4	NA	NA	NA	29.6	43.6	56.4	60	40	NA	s1/m1 and s2/m2	NA	s1, i1, m1
Lithuania	81	56.8	NA	NA	NA	43.2	13.6	66.7	NA	NA	NA	s2/m2	NA	s1
Slovenia[1]	165	70	NA	NA	NA	24	42	52	NA	NA	NA	s1/m1	11.5	None
UK	65	87.8	NA	NA	NA	6	21.2 (m1a), 4.6 (m1c)	66.6	NA	NA	NA	s1/m2	NA	NA
North America														
Cuba	130	73.8	NA	NA	NA	26.2	57.7	40	NA	NA	NA	s1/m1	NA	s1/*cagA*⁺
Mexico (CG)	143	67.9	NA	NA	NA	32.2	63.6	36.4	NA	NA	NA	s1/m1	NA	NA
USA[1]	61	77.1	14.8	62.3	0	21.3	62.3	36.1	NA	NA	NA	s1b/m1	1.6	NA
USA[1]	37	73	NA	NA	NA	27	56.8	42.3	NA	NA	NA	s1/m1	0	NA
South America														
Chile	66 (41 patients)	63.6	42.4	21.2	0	25.8	31.8	43.9	NA	NA	NA	s1a, m2	22.1	s1/*cagA*⁺
Uruguay	30	56.7	NA	NA	NA	46.7	46.7	56.7	46.7	50	0	s2/m2 and s1/m1	3.3	NA
Venezuela	121	54	0	10–38	0–48	3–21	27–47	10–46	NA	NA	NA	NA	41–69	NA

Asia													
Bahrain	120	66	NA	NA	NA	34	47	53	NA	NA	NA	s1/m1	NA	s1 and s1/ m1
China	87	69	11.5	0	57.5	0	29.9	51.7	NA	NA	NA	s1c/m2a and m1	41	None
China	30	100	NA	NA	NA	0	6.7	93.3	93.3	3.3	3.3	s1/i1/m2	0	NA
China	150	96	NA	NA	NA	4	27	57	NA	NA	NA	s1	11	None
Japan	220	91.8	11.4	0.9	79.5	2.3	85	9.1	NA	NA	NA	s1c/m1b	NA	NA
Jordan	110	45.3	NA	NA	NA	54.7	48.9	51.1	NA	NA	NA	s1/m1 and s2/ m2	NA	NA
Pakistan	224	100	68.8	32.1	0	0	60.3	51.8	41.1	54	NA	s1a/m1	17	m1, s1a/ m1, s1bm1 and i1
South Korea	136	94.9	72.8	22.1	NA	5.1	86.8	13.2	NA	NA	NA	s1a/m1	NA	None
South Korea[1]	33	NA	27	NA	72	NA	82	15	NA	NA	NA	s1c/m1	NA	None
Taiwan	234		69.2	4.3	45.3	1.3	1.3 (m1), 23.9 (m1T)	49.6	NA	NA	NA	NA	>38	m1T in GC
Turkey	35	57.1	NA	NA	NA	42.8	22.9	88.6	22.9	42.9	25.7	s2/m2	11.4	NA
Africa														
Nigeria	41	97.6	NA	NA	NA	4.9	24.4	78.1	NA	NA	NA	s1/m2	NA	None
Tunisia	281	23.2	5.3	17.8	0	45.6	10.7	58.1	NA	NA	NA	s2/m2	31.4	s1/m1

[1]Children.

PU, peptic ulcers; GC, gastric cancer; NA, not available; CG, chronic gastritis; ML, mucosa-associated lymphoid tissue lymphoma.

Data for Bahrain (Bindayna and Al Mahmeed, 2009), Bulgaria (Boyanova *et al*., 2009), Chile (García *et al*., 2006), China (Wang *et al*., 2003; Zhang *et al*., 2006; Chung *et al*., 2010), Cuba (Torres *et al*., 2009), France (Lehours *et al*., 2004), Germany (Koehler *et al*., 2003), Ireland (Taneike *et al*., 2009), Italy (Basso *et al*., 2008), Japan (Yamazaki *et al*., 2005), Jordan (Nimri *et al*., 2006), Lithuania (Miciuleviciene *et al*., 2008); Mexico (Paniagua *et al*., 2009), Nigeria (Smith *et al*., 2002), Pakistan (Yakoob *et al*., 2009), Slovenia (Homan *et al*., 2009), South Korea (Park *et al*., 2001; Ko *et al*., 2008), Taiwan (Lin *et al*., 2004), Tunisia (Ben Mansour *et al*., 2010), Turkey (Chung *et al*., 2010), the UK (Kauser *et al*., 2005a); USA (Podzorski *et al*., 2003; Yamaoka *et al*., 2010), Uruguay (Chung *et al*., 2010) and Venezuela (Ghose *et al*., 2005).

level, protect the gastric cells against VacA activities (Oldani *et al.*, 2009). VacA suppresses EGF receptor signalling, which is needed for cell scattering and elongation (Costa *et al.*, 2009). In the GECs, *cagA* and *cagE* mutants have promoted increased VacA-induced vacuolation and *vacA* mutants have provoked enhanced CagA-induced hummingbird phenotype compared with wild-type strains. Thus, CagA reduces the vacuolation and VacA diminishes the hummingbird formation (Argent *et al.*, 2008a). By CagA, *H. pylori* defends the attached gastric cells against VacA toxin, generating simultaneously protected and unprotected cell populations (Akada *et al.*, 2010). Therefore, the opposite effects of CagA and VacA support the chronicity of the infection and cause the diversity of *H. pylori*-induced diseases.

Duodenal ulcer-promoting (*dupA*) gene

The *dupA* gene in *H. pylori* PZ encodes a homologue of VirB4 ATPase which participates in import and transfer of DNA and proteins. The *dupA* gene includes two overlapping genes (*jhp0917* and *jhp0918*) in the genome of strain J99. However, in most (≥97%) other strains, *dupA* exhibits 1 bp insertion after position 1385, which produces a frameshift, providing a continuous gene (Lu *et al.*, 2005b; Yamaoka, 2008a). Strains are read as $dupA^+$ if they are positive for both genes with the above-mentioned single base insertion (Yamaoka, 2008a).

The gene has been shown to induce IL-8 secretion from the gastric cells by activation in transcription factors responsible for IL-8 promoter activity as well as stability of the strains in low pH (Lu *et al.*, 2005b). Conversely, no association between *dupA* and IL-8 induction *in vitro* has been detected by Schmidt *et al.* (2009). Activation of transcription factors for *p53* gene for tumour suppression has been weak in *dupA*-deletion mutants (Lu *et al.*, 2005b), suggesting an anti-tumour effect of the gene. Hussein *et al.* (2010) have reported two types of *dupA* gene: a longer *dupA* (1884 bp; *dupA1*), present in most strains, and a truncated type (*dupA2*). Only *dupA1* type, unlike *dupA2* and *cag*PAI types, has been able to cause gastric inflammation by an increase in *H. pylori*-induced IL-12p40 and IL-12p70 release from $CD14^+$ mononuclear cells (Hussein *et al.*, 2010).

Evaluating 500 *H. pylori* strains in East Africa and South America, Lu *et al.* (2005b) have reported *dupA* gene to be a risk factor for duodenal ulcer, from where comes the name of the gene. The gene has been twice more common (42%) in duodenal ulcer patients than in those with gastritis (21%; Lu *et al.*, 2005b). The association between *dupA* gene and duodenal ulcer has not been universal. This association has not been found in studies from Belgium, Brazil, Iran and Sweden, but has been reported in studies from East Asia (China, South Korea and Japan), India and Colombia (Lu *et al.*, 2005b; Arachchi *et al.*, 2007; Argent *et al.*, 2007; Douraghi *et al.*, 2008; Zhang *et al.*, 2008; Schmidt *et al.*, 2009).

The *dupA* has been considered as protective gene against gastric cancer (Lu *et al.*, 2005b; Douraghi *et al.*, 2008). In a study of Argent *et al.* (2007) on strains from South Africa, Belgium, China and the USA, 71% of gastric cancer patients have harboured $dupA^+$ strains versus only 42.3% of those with gastritis. In Iran, Douraghi *et al.* (2008) have found no association between *dupA* gene and infection outcomes; however, *dupA*-negative strains have been linked to premalignant gastric lesions. Schmidt *et al.* (2009) have reported that although *dupA* has been a highly conserved gene, it has shown no steady association with duodenal ulcer or gastric cancer risk. Geographical differences have been reported on *dupA* activities. The $dupA^+$ strains have been inversely associated with gastric ulcer and gastric cancer in South America but have been linked to both diseases in China (Hussein, 2010).

Global prevalence of $dupA^+$ strains varies from <20% to 89%, showing an overall tendency to be less frequent in countries where gastric cancer is common (Hussein, 2010). In Japan, *dupA* positivity, duodenal ulcer and higher acid output have been inversely associated with the gastric cancer (Imagawa *et al.*, 2010). The prevalence of *dupA* has been reported to be 19.4–35.8% in East Asia (South Korea, Japan and China), Iraq, Columbia and India, higher (43.5–49.7%) in Belgium, the USA and Iran, but >84% in Brazil and South Africa (Table 4.4). A recent meta-analysis has shown that *dupA* is linked to duodenal ulcer in some countries and gastric ulcer and gastric cancer in other areas (Hussein, 2010). The authors of another meta-analysis have reported that *dupA* positivity increases the risk of duodenal ulcer 1.41-fold (1.57-fold in Asian countries and only 1.09-fold in Western countries), without association with gastric cancer and gastric ulcer (Shiota *et al.*, 2010). Briefly, in the different countries or regions, *dupA*$^+$ strains have been either cancer-promoting or peptic ulcer-promoting or insignificant for the outcome of *H. pylori* infection. This gene could be more important for the disease outcomes in East Asia than in the Western countries. The *dupA* subtypes possibly account for some differences in the results on the prevalence of the gene and should be further evaluated.

Frequencies of *jhp0917* and *jhp0918* genes are highly associated. Most strains have been either positive or negative for both genes. Zhang *et al.* (2008) and Douraghi *et al.* (2008) have found *jhp0917*$^+$/*jhp0918*-negative strains in only 2.5–5.1% of the strains. Lu *et al.* (2005b) have proved *jhp0917*-negative/*jhp0918*$^+$ strains in 2% of the cases. Slightly higher prevalence of *jhp0917*$^+$/*jhp0918*-negative and *jhp0917*-negative/*jhp0918*$^+$ strains has been reported (in 9% and 16% of all strains, respectively), by Arachchi *et al* (2007).

The *dupA* prevalence also may vary between patient groups according to the age. In Brazil, the gene has been more common in strains from children than in those from adults (Gomes *et al.*, 2008). This can suggest a loss of the gene throughout the infection and indirectly implies a negative association between *dupA* gene and the risk for gastric cancer (Yamaoka, 2008a).

Gomes *et al.* (2008) have observed frameshift mutations in 16% of *dupA*$^+$ sequenced samples. Single adenine insertion after position 1426 of *dupA* or at position 2998 of the *jhp0917–jhp0918* gene of the J99 strain, has been found to generate a premature stop codon, which could diminish the activities of the gene product. For diagnostic purpose, the identification of the intact DupA protein is important and therefore, immunoblot technique can be considered. Accurate *dupA* detection also can comprise PCR-sequencing or PCR with multiple primer pairs rather than with single primer pairs as it has been recommended (Yamaoka, 2008a; Matteo *et al.*, 2010). By these methods, it is important to detect frameshift mutations and insertions, which may alter the protein expression of DupA protein.

Another locus (*jhp940*) in *H. pylori* PZs also has been studied. Rizwan *et al.* (2008) have found *H. pylori* JHp940 to be a virulence factor, which could increase NF-κB translocation in macrophages, and could stimulate TNF-α and IL-8 release. The locus has been common (>90% of strains) in French, Indian and South African strains and rare (≤45%) in those from Spain, Costa Rica and Japan (Rizwan *et al.*, 2008). However, this locus has been found as stand-alone virulence factor owing to its independence on *cagA* status and the severe diseases (Rizwan *et al.*, 2008).

iceA (induced by contact with epithelium gene A)

The *iceA* gene has *iceA1* and *iceA2* alleles (Lehours *et al.* 2004). The *iceA1* allele encodes a CATG-specific restriction endonuclease, homologous to that (*nlaIIIR*) in *Neisseria lactamica,* and is induced by *H. pylori* contact with gastric cells. The *iceA1* and *iceA2* alleles are located upstream of the methyltransferase *hpy1M* gene and the expression of the methyltransferase gene accompanies *iceA1* thus, *iceA1-hpy1M* forms a CATG-specific type II R-M system (Proença Módena *et al.*, 2009). It is to be noted that *iceA* modulates IL-8 release by the host epithelial cells (Ma *et al.*, 2010). However, *iceA1* mutations, affecting the endonuclease activity of the protein, have been frequent (in about 80% of *iceA*1 strains; Gatti *et al.*, 2006).

While the more virulent *iceA1* allele has been prevalent in East Asia, *iceA2* allele has been common in America (Gong *et al.*, 2005). Prevalence of *iceA1* allele has been ≥70% in China, South Korea and Nigeria versus only <20% in the the USA and Jordan (Smith *et al.*, 2002; Podzorski *et al.*, 2003; Nimri *et al.*, 2006; Zhang *et al.*, 2006; Ko *et al.*, 2008). Mixed *iceA* alleles have been found as well, reaching 40% in South African strains (Kidd *et al.*, 2001). In a study, no Jordanian strain has been *iceA1*$^+$ but, conversely, the majority (90.2%) of the strains from Nigeria have shown *iceA1* type with no association with the diseases (Smith *et al.*, 2002; Nimri *et al.*, 2006).

Table 4.4 Prevalence of *dupA* gene in *H. pylori* strains in some countries

Country/ Region	No. of strains	Overall *dupA* prevalence (%)	*dupA* prevalence				Association with diseases	Reference
			DU	GU	GC	Other		
Europe								
Belgium	135	43.7	50	NA	52.6	38.2	None	Argent *et al.* (2007)
Belgium and Sweden	187	49.2	52.9	NA	57.5	43.8	None	Hussein *et al.* (2010)
Sweden	52	63.5	63.6	NA	61.9	65.0	None	Schmidt *et al.* (2009)
North America								
USA	46	43.5	42.9	66.7	NA	45	None	Argent *et al.* (2007)
South America								
Brazil	79	62	65	58.3	NA	62.1–66.7	None	Pacheco *et al.* (2008)
Colombia	165	35.8	55.6	40.0	12.0	40.0	DU and gastriris (vs. GC)	Lu *et al.* (2005b)
Colombia and Brazil	561	71.3	77.7	45.4	58.8	79.2	GU and GC (inverse)	Hussein *et al.* (2010)
Asia								
China	360	35.3	45.5	23.4	24.1	38.3	DU (vs. GC and GU)	Zhang *et al.* (2008)
East Asia (Japan, China and Korea)	1047	39.6	43.2	43.2	38.4	36.6	None	Hussein *et al.* (2010)
India	166	31.3	37.5	NA	NA	22.9	DU	Arachchi *et al.* (2007)
Iran	157	49.7	50	39.1	46.2	55.6	Dysplasia (inverse)	Douraghi *et al.* (2008)
Iraq	49	32.7	60.0	40.0	NA	17.2	Peptic ulcers	Hussein *et al.* (2008)
Iraq and Iran	206	45.6	53.3	39.3	55.6	40.2	None	Hussein *et al.* (2010)
Japan	160	21.3	36.7	26.0	10.0	14.0	DU (vs. GC and gastritis)	Lu *et al.* (2005b)
Japan	56	32.1	45.0	25.0	NA	NA	GC (inverse)	Imagawa *et al.* (2010)
South Korea	175	19.4	36.9	16.7	6.0	6.7	DU (vs. gastritis and GC)	Lu *et al.* (2005b)
Africa								
South Africa	46	84.8	92.3	NA	88.9	73.3	None	Argent *et al.* (2007)
Australia	45	37.4	NA	NA	NA	37.4	NA	Schmidt *et al.* (2009)

DU, duodenal ulcer; GC, gastric cancer.

Therefore, the significance of *iceA* alleles is still a matter of controversy. In some countries, *iceA1*⁺ strains have been prevalent in patients with severe diseases but in most studies, no such association has been found (Caner *et al.*, 2007; Proença-Modena *et al.*, 2009). The *iceA1* allele has been associated with gastric cancer in South Africa, Germany and Turkey and with peptic ulcers in Turkey. In Germany, the *iceA1* gene has been 3.6 times more frequent in gastric cancer patients than in those with gastritis (Koehler *et al.*, 2003). However, in many countries, e.g. China, France, Lithuania, Bulgaria, South Korea, etc., the association has not been proven (Kidd *et al.*, 2001; Ando *et al.*, 2002; Koehler *et al.*, 2003; Lehours *et al.*, 2004; Lin *et al.* 2004; Erzin *et al.* 2006; Caner *et al.*

2007; Miciuleviciene *et al.* 2008; Boyanova *et al.*, 2010; Table 4.5). Moreover, in Brazil, an inverse association of *iceA1* allele with peptic ulcer and gastric cancers has been found (Ashour *et al.*, 2001; Gatti *et al.*, 2006). Similarly, in Venezuela, *iceA2* allele has been linked to atrophic gastritis (Chiurillo *et al.*, 2010). In East Asia, no association has been found between the degree of gastric inflammation and *cagA*, *vacA* or *iceA* status (Ko *et al.*, 2008). In Turkey, *iceA1* allele has been associated with duodenal ulcers, while *iceA2* allele has been linked to chronic gastritis (Caner *et al.*,

Table 4.5 Prevalence of *iceA* alleles in *H. pylori* strains in some countries

Country	No. of strains	Prevalence (%) of *iceA1*	Prevalence (%) of *iceA2*	Mixed alleles (%)	Disease association with *iceA1*	Reference
Europe						
Bulgaria	176	69.3[1]	30.7[1]	10.2	None	Boyanova *et al.* (2010)
France	43 (ML) and 39 (CG)	55.8 (ML), 38.5 (CG)	44.2 (ML), 61.5 (CG)	0	None	Lehours *et al.* (2004)
Germany	91	69.2	30.8	0	GC	Koehler *et al.* (2003)
Lithuania	81	43.2	49.4	0	None	Miciuleviciene *et al.* (2008)
Slovenia[2]	165	62	31	11.5	None	Homan *et al.* (2009)
UK	64	57	36.3	3.1	NA	Kauser *et al.* (2005a)
North America						
USA[2]	61	16[3]	84[3]	4.9	NA	Podzorski *et al.* (2003)
South America						
Argentina	299	40.8	50.5	8.7	None	Vega *et al.* (2010)
Brazil	142	9.2	83.1	7.7	PU and GC (inverse)	Ashour *et al.* (2001)
Brazil	94	92.6	7.4	1.1	PU (inverse)	Gatti *et al.* (2006)
Chile	53 (CG)	15	60	21	None	Araya *et al.* (2004)
Asia						
China	150	75	19	NA	None	Zhang *et al.* (2006)
Jordan	110	0	73.6	NA	NA	Nimri *et al.* (2006)
South Korea[2]	33	76	24	NA	None	Ko *et al.* (2008)
Thailand	112	45.5	33.1	NA	None	Chomvarin *et al.* (2008)
Taiwan	234	70.1	18.8	3.4	None	Lin *et al.* (2004)
Turkey	91	74.7	25.3	NA	GC	Erzin *et al.* (2006)
Turkey	46	45.7	54.3	NA	DU	Caner *et al.* (2007)
Africa						
Nigeria	41	90.2	7.3	2.5	None	Smith *et al.* (2002)
South Africa	86	68[3]	80[3]	40	GC	Kidd *et al.* (2001)
Tunisia	281	60.2	16	23.8	None	Ben Mansour *et al.* (2010)

[1]In single strain infections.

[2]Children.

[3]Including single strain and mixed infections.

DU, duodenal ulcer; GU, gastric ulcer; PU, peptic ulcer; CG, chronic gastritis; ML, mucosa-associated lymphoid tissue lymphoma; GC, gastric cancer; NA, not available.

2007). This implies the importance of the coordinated effect of all *H. pylori* virulence genes rather than the activity of *iceA* alleles alone.

Combinations of *iceA1* allele with other virulence factors can be highly important. In Bulgaria, *iceA1* allele has been associated with *vacA* s1 and *cagA* status of strains (Boyanova *et al.*, 2010). In Argentina, no association has been found between gastric damage and *iceA1* or *cagA* status, while triple virulent genotypes (*cagA*$^+$, *vacA* s1/m1 and *iceA1* types) have been more common in ulcer patients than in the non-ulcer group (Vega *et al.*, 2010). In is known that the *cagA*$^+$ strains have shown inverse association with clarithromycin resistance in some studies. Similarly, in Bulgaria, *iceA1* allele has been more common (74.4%) within clarithromycin susceptible strains than within the resistant strains (55.3%; Megraud, 2004; Boyanova *et al.*, 2010). This draws attention to the statement (see above) that the virulent *H. pylori* strains could be eradicated more easily by clarithromycin-based regimens than the less virulent strains.

Adhesins as virulence factors of *H. pylori*

H. pylori adhesion to GECs is essential to establish the infection and to protect the bacteria from clearance by liquid flow, peristaltic and other unfavourable factors. About one-fifth of the bacteria are bound to the gastric mucosa in a dynamic process of attachment to and detachment from the cells (O'Brien *et al.*, 2008; de Reuse *et al.*, 2010). Larger part (4%) of *H. pylori* genome encodes genes for outer membrane proteins (OMPs) compared with that in most other bacteria (Dossumbekova *et al.*, 2006), showing that the adhesion is a highly important prerequisite for *H. pylori* infection. Many *H. pylori* adhesins such as BabA, SabA, OipA, AlpA or AlpB, belonging to the OMP family, are used by *H. pylori* (Dossumbekova *et al.*, 2006). Fascinatingly, many of these genes, e.g. *oipA, sabA, sabB, babB, babC* and *hopZ*, undergo phase variation by slipped-strand mispairing. This process involves deletion or insertion of nucleotides during DNA replication within repetitive sequences, usually near the 5′ end of genes, which change the reading frames of the genes (Odenbreit *et al.*, 2009). Thus, the genes are present in active ('on') or inactive ('off') status, i.e. the same gene produces/does not produce functional proteins, which is an adaptation mechanism.

Mucins are glycosylated high-molecular-weight glycoproteins which lubricate and protect GECs. *H. pylori* can bind to the mucins by different mechanisms, including BabA binding to fucosylated structures, SabA binding to sialylated structures and adhesin binding to salivary mucins (Lindén *et al.*, 2010). The specific adhesion to salivary mucins but not to gastric mucins in some *H. pylori* strains can lead to colonization of oral sites and possibly to transmission of infection (Lindén *et al.*, 2010). In the oral cavity, with neutral pH, the mucin may be bound by the saliva-specific adhesin, SabA and BabA (Lindén *et al.*, 2010).

Inflammatory outer membrane (*oipA*) gene functional status

Inflammatory outer membrane protein (OipA, 34 kDa) acts as and adhesin and induces, although to a lesser degree than CagA, mucosal IL-8 release, leading to the development of duodenal ulcer and gastric cancer; yet the association is not universal (Ando *et al.*, 2002; Yamaoka *et al.*, 2006; Table 4.6). The *oipA* (HopH) gene is important in carcinogenesis, as gene inactivation has been reported to decrease β-catenin nuclear localization *in vitro* and the incidence of gastric cancer in rodents (Franco *et al.*, 2008). OipA also induces a site-specific activation (phosphorylation) of FAK, leading to fibre formation in GECs and hence damaged cell morphology (Tabassam *et al.*, 2008). CagA$^+$/OipA$^+$ strains have produced the highest IL-8 level, which has fallen to ≤ 50% in CagA$^+$/OipA-negative strains and even more in double-negative strains (Odenbreit *et al.*, 2009). However, Dossumbekova *et al.* (2006) have found no influence of *oipA*$^+$ status on IL-8 release. According to another study, OipA induces the release of IL-8 and other pro-inflammatory cytokines such as IL-6, IL-18 and RANTES from the GECs (Yamaoka *et al.*, 2008a). Upregulation of IL-18 in gastric cells has been linked to both *cag*PAI and OipA, while the upregulation in monocytes has been found to depend only on OipA (Yamauchi *et al.*, 2008). In US- and Colombian patients, the

Table 4.6 Functional status of *oipA* gene in *H. pylori* strains in some countries

Country/ region	No. of strains	Common CT repeats	*oipA* status 'on'	*oipA* status 'off'	Clinical significance of 'on' status	Reference
Europe						
Italy	60	NA	60	40	NA	Zambon *et al.* (2003)
Germany (gastritis)	58	6	59	41	NA	Dossumbekova *et al.* (2006)
UK	65	6 and 7	70.5	29.5	NA	Kauser *et al.* (2005a)
Western countries	13	6	61.5	38.5	Severe diseases (PU, GC)	Ando *et al.* (2002)
North America						
USA[1]	37	NA	67.6	32.4	NA	Yamaoka *et al.* (2010)
USA (Caucasians)	28	6	57.1	42.9	Severe diseases	Ando *et al.* (2002)
USA (African Americans)	10	6	9 of 10	1 of 10	NA	Ando *et al.* (2002)
USA	80	NA	73.8	26.2	DU	Yamaoka *et al.* (2006)
South America						
Colombia	166	NA	74	26	None	Quiroga *et al.* (2005)
Colombia	120	NA	80	20	DU and GC	Yamaoka *et al.* (2006)
Asia						
India (gastritis)	10	2+1 and 3	8 of 10	2 of 10	NA	Kauser *et al.* (2005b)
Japan	38	3+1	100	0	None	Ando *et al.* (2002)
Malaysia and Singapore	159	NA	>85	<15	None	Schmidt *et al.* (2010)
Africa						
Tunisia	281	NA	90.8	9.2	Peptic ulcers	Ben Mansour *et al.* (2010)

[1]Children.

NA, not available or not applicable; PU, peptic ulcer; DU, duodenal ulcer; GC, gastric cancer.

infection with OipA$^+$ strains has exhibited high gastric *H. pylori* density and neutrophil infiltration (Yamaoka *et al.*, 2006). In brief, the release of pro-inflammatory cytokines in the gastric mucosa depends on *cag*PAI and, simultaneously, on OipA.

The OipA protein-encoding gene *Hp0638* is present in all *H. pylori* strains but its expression fluctuates owing to phase variation with different number of CT dinucleotides at the 5′ region (signal-peptide coding region) of the *oipA* gene (Kusters *et al.*, 2006). Frameshift mutation (loss or add of CT dinucleotides) can turn frameshift reading off and can inactivate the gene. The *cag*PAI$^+$ strains have shown *oipA* often in functional (in-frame) status. Within strains from nine countries, none of *cagA* negative strains but most (96%) *cagA*$^+$ strains have harboured *Hp0638* in frame (Ando *et al.*, 2002). However, as for other *H. pylori* virulence genes, the geographic variations in the prevalence of this gene have been very strong.

PCR-sequencing has been used to assess the *oipA* frame status by the number of CT dinucleotide repeats in signal encoding region (Kauser *et al.*, 2005a,b; Matteo *et al.*, 2010). Importantly, 5, 7 or 8 (less often 4, 9, 10 and 12) dinucleotide repeats in the signalling region shift the frame and turn the gene 'off', while with 6, 5+2 or 9 CT repeats, the gene is in 'on' status (Kauser *et al.*, 2005b; Dossumbekova *et al.*, 2006). Nevertheless, some strains with 9 CT repeats may have out-of-frame genes owing to deletions.

In Western strains, in-frame status of *Hp0638* has been associated with severe diseases (Ando *et al.*, 2002; Ben Mansour *et al.*, 2010). In US and Colombian strains, OipA status has been an independent risk factor for gastric cancer and duodenal ulcer. OipA has been detected in strains from more duodenal ulcer patients (88%) than in those with gastritis (61%; Yamaoka *et al.*, 2006).

Among strains with *Hp0638* in frame, often ≥5 repeats, e.g. 6-CT dinucleotide repeat pattern has been prevalent in 70% of strains from Western countries, while ≤5 CT repeats, most often 3 + 1-CT-repeat pattern has been common (found in 67%) in those from East Asia (Ando *et al.*, 2002). No association between *Hp0638* in frame and diseases has been found in East Asian countries, Malaysia and Singapore (Ando *et al.*, 2002; Schmidt *et al.*, 2010). In many European countries, the *oipA* gene in functional status has been found in 59–70.5% of strains in common association with *cagA*, *vacA* s1/m1 and *babA2* (Zambon *et al.*, 2003). Importantly, in Argentina, the *oipA* gene has shown variability in gastric antrum and corpus of about one-quarter of patients (Matteo *et al.*, 2010). In brief, both independent and the *cagA* supporting activities of the *oipA* gene strongly contribute to development of severe gastric diseases, which is evident in the Western countries.

Blood group antigen-binding adhesin (*babA2*) allele

Le^b is the main antigen in the gastric mucosa of most (secretor-positive) subjects, while secretor-negative subjects exhibit Le^a-dominant phenotype (Yamaoka, 2008b). *H. pylori* BabA adhesin (75 kDa, encoded by *babA2* gene) binds to fucosylated blood-group antigens H-type 1 and Le^b, expressed on the normal gastric mucosa of the secretor individuals, and thus participates in the early stage of the infection (O'Brien *et al.*, 2008; Magalhães and Reis, 2010). $BabA^+$ strains are also able to bind to mucins of saliva, gastric juice and gastric mucus layer (Lindén *et al.*, 2010). It is noteworthy that *H. pylori* strains can be BabA high producers with enough BabA protein to bind Le^b, BabA-low producers with small BabA amount, unable to bind Le^b and BabA-negative strains, producing no BabA. Using immunoblot assay, Yamaoka (2008b) has stated that of 80 strains studied, 85% have been BabA-high producers, 10% have been low producers and 5% have been non-producers (Yamaoka, 2008b).

Of the three babA alleles (*babA1, babA2* and *babB*), only *babA2* can code for active BabA adhesin. The very rare *babA1* allele is 'silent' as it lacks a sequence for translational start and signal peptide owing to a 10-bp deletion in the signal peptide sequence (Proença-Modena *et al.*, 2009). BabA non-producing strains often exhibit non-functional, silent *babA* gene sequences (Yamaoka, 2008b). Both antigenic and phase variations regulate BabA expression. The *babA/B* and *babB/A* chimeras have been detected, showing bp replacement in one allele by bp of the alternative allele (Yamaoka, 2008b). Therefore, the *babA* gene can carry frameshift mutations, causing inefficient translation or lack of translation initiation codon, or can produce chimeras (Yamaoka, 2008b). In all these cases, no BabA is produced by the strains.

The BabA protein plays a role in *H. pylori* adhesion. In animal models, the $BabA^+$ strains have induced higher inflammation and stronger IL-8 release from the GECs than the BabA negative strains (Atherton and Blaser, 2009). The BabA protein of the $babA2^+$ strains has been related to peptic ulcers and gastric cancer, although the association has not been universal (Fujimoto *et al.*, 2007; Yamaoka, 2008a; Con *et al.*, 2009). For instance, in Brazil, Mattar *et al.* (2005) have observed no association between *babA2* and *cagA* genotypes and peptic ulcer disease. Conversely, the $BabA^+$ status has been linked to severe intestinal metaplasia and gastric atrophy in US and Colombian patients (Yamaoka *et al.*, 2006). Moreover, Azevedo *et al.* (2008) have reported a noteworthy BabA association with intercellular localization of *H. pylori* and presence of intestinal metaplasia.

The proportion of $babA2^+$ *H. pylori* strains in Europe and Turkey has been reported to be 30–70%, while that in East Asia (Japan and South Korea), Thailand, Columbia and Cuba has been higher (70–100%; Zambon *et al.*, 2003; Lehours *et al.* 2004; Erzin *et al.* 2006; Chomvarin *et al.*,

2008; Chiarini *et al.*, 2009; Odenbreit *et al.*, 2009; Talarico *et al.*, 2009; Torres *et al.*, 2009; Boyanova *et al.*, 2010). The prevalence of *babA2* has varied from 46% to 82.3% in South America (Con *et al.*, 2009; Paniagua *et al.*, 2009). In duodenal ulcer patients from four European countries, however, some strains with good *in vitro* adhesion have been *babA2* negative by PCR owing to allelic variations (Olfat *et al.*, 2005). Like CagA, VacA and OipA, BabA has been independent on infection outcomes in East Asia (Yamaoka, 2008c). According to Yamaoka (2008b), immunoblot evaluation has shown all East Asian strains to express BabA protein. However, some BabA-negative/CagA$^+$ strains have shown higher association with peptic ulcers and gastric cancer than those with *in vitro* Le^b-binding activity; thus, *in vitro* Le^b-binding activity does not always indicate the degree of gastric damage (Yamaoka, 2008b).

The *babA2* allele has often been associated with *vacA* s1 and *cagA* status (Zambon *et al.*, 2003; Olfat *et al.*, 2005; Boyanova *et al.*, 2010). Co-expression of *vacA* s1 and *cagA* genotypes (type 1 strains), or that of triple-positive strains (with *vacA* s1, *cagA* and *babA2* genotypes), has been related to severe gastric inflammation, peptic ulcers, intestinal metaplasia and gastric cancer in Italy (Zambon *et al.*, 2003). Patients infected by strains with *cagA*$^+$, *vacA* s1/m1, and *babA2* genotypes have been at high risk for intestinal metaplasia (Zambon *et al.*, 2003). Triple-positive strains have been associated with duodenal ulcer in Portugal, Germany, Finland and Cuba, while no such association has been found in Sweden and Bulgaria (Olfat *et al.*, 2005; Wen *et al.*, 2007; Torres *et al.*, 2009; Boyanova *et al.*, 2010). In Bulgaria, neither *iceA1* nor *babA2* alone but only their combination has been more common (48.1%) in ulcer patients than in the rest (28.7%; Boyanova *et al.*, 2010). In brief, the *babA2* allele shows association, although not universal, with the disease outcomes of *H. pylori* infection, especially in combination with other virulence factors.

For diagnostic purpose, it is more useful to asses BabA by immunoblot rather than PCR methods as they detect strains with silent *babA* sequences and conversely, owing to sequence variations, using a single primer pair can give false-negativity (Yamaoka, 2008b).

Sialic acid-binding adhesin (*sabA*) gene status

BabA binding to Le^b antigen is important at the beginning of the infection when sialylated glycoconjugates are scarce in gastric mucosa. However, with growing inflammation, the expression of sialic acid containing glycans in the mucosa grows, resulting in *H. pylori* adhesion to the mucosa by sialic acid-binding adhesin (SabA; Yamaoka, 2008c). SabA$^+$ and BabA negative strains adhere to inflamed gastric mucosa but not to healthy tissues.

SabA adhesin, encoded by the *sabA* (JHp662) gene, binds to sialyl Le^x and sialyl Le^a antigens on inflamed gastric mucosa as well as to the extracellular matrix protein, laminin (Kusters *et al.*, 2006; Peek, 2008; Yamaoka, 2008c). SabA binds to sialyl-Le^x and sialyldimeric-Le^x structures and 43% of SabA$^+$ strains target sialyl-Le^a (Lindén *et al.*, 2010). Importantly, the sialyl Le^a is an established tumour antigen. In addition, SabA mediates *H. pylori* binding to sialylated structures on neutrophils and erythrocytes; therefore, SabA is important for non-opsonic activation of neutrophils and resultant oxidative gastric damage (Unemo *et al.*, 2005). SabB protein also exists but is inactive (Yamaoka, 2008c).

Variable SabA expression is a result of phase variation via slipped-strand mispairing during chromosomal replication (Goodwin *et al.*, 2008). The *sabA* gene can turn 'on' or 'off' owing to nucleotide sequence after the CT repeats (Yamaoka *et al.*, 2006). Double regulation of *sabA* locus by ArsRS TCS and generation of subpopulations with alternative *sabA* alleles by slipped-strand mispairing facilitate *H. pylori* adaptation to changing gastric conditions to avoid host immunity (Goodwin *et al.*, 2008). The 'off' status of SabA also can detach *H. pylori* from gastric cells to escape from the immunity (Yamaoka, 2008c). All these activities can contribute to the chronicity of the infection. SabA expression is dependent on the gastric pH, being low at pH 5 and higher at higher pH values (Yamaoka, 2008c).

SabA adhesin has been present in about 30–40% of *H. pylori* strains (Yamaoka, 2008c). SabA$^+$ status, being negatively related to duodenal ulcer, has been a risk factor for severe intestinal metaplasia, gastric atrophy and gastric cancer in the USA and Colombia (Yamaoka, 2008a). SabA$^+$ status has been less common in strains from duodenal ulcer patients (44%) compared with those with gastric cancer (70%; Yamaoka *et al.*, 2006).

However, like other virulence factors, SabA cannot indicate the outcomes of the infection in East Asia. In Japan, the *sabA* gene has been linked neither to anti Le^x antibodies in the host nor to severe gastric inflammation (Nakasato *et al.*, 2008). In addition, in Germany, SabA protein has been produced by only 5% of the strains but they have been unable to bind their substrate (Odenbreit *et al.*, 2009). Evaluation of SabA in patient groups from more countries is needed.

Interestingly, *H. pylori* binding to MUC7 and MUC5B salivary mucins has been reported (Costa *et al.*, 2009). This binding mode can favour the oral–oral transmission of *H. pylori* infection.

From diagnostic viewpoint, because *sabA* expression has not been associated with the number of CT dinucleotide repeats in the 5′ region of the gene, the SabA detection should be better performed by immunoblot than by PCR (Yamaoka, 2008c).

Intriguingly, the BabA and SabA show interactions (Nishioka *et al.*, 2010). Complexomes (protein complexes) of *H. pylori* strains from a MALT lymphoma- and a duodenal ulcer patient have been compared (Bernarde *et al.*, 2010). BabA and SabA have been shown as a complex in the strain associated with the MALT lymphoma. As SabA binds to tumour antigens and BabA$^+$ strains can point to intercellular location of the bacteria and intestinal metaplasia, BabA/SabA association can contribute to development of malignancy, facilitating the bacteria to persist inside the gastric cell (Bernarde *et al.*, 2010). Again, the cooperative and well-coordinated activity of *H. pylori* virulence factors is of primary importance for the clinical outcome of the infection.

H. pylori homB outer membrane protein gene status

Recently, *homB* gene of the *hom* family of *H. pylori* outer-membrane proteins has been reported as a virulence factor and a marker for gastric cancer versus duodenal ulcer (Jung *et al.*, 2009). Some *H. pylori* strains possess *homA* gene that has been more common in non-ulcer patients and in less virulent *H. pylori* genotypes (Oleastro *et al.*, 2009). Conversely, the *homB* status has been linked to gastric inflammation and only corpus atrophy, whereas *cagA*$^+$ status has been associated with inflammation and atrophy in all part of the stomach (Jung *et al.*, 2009). The *homB* inactivation, especially by double *homB* knockout mutant, has been shown to reduce strongly the IL-8 release and *H. pylori* binding to gastric cells. Prevalence of *homB* gene in USA and Colombian populations has been higher in gastric cancer patients (71.9%) than in those with duodenal ulcers (52.1%; Jung *et al.*, 2009). In another study, *homB* status has been associated with peptic ulcers in children and adults aged ≤40 years (Oleastro *et al.*, 2008). In Portugal, the presence of two *homB* copies in the same strain has been the most common genotype in ulcerogenic *H. pylori* strains (Oleastro *et al.*, 2008). Therefore, *homB* gene is associated with highly virulent strains.

Single or double-copy or one copy of *homB* or *homA* genes can be present in *H. pylori* DNA (Oleastro *et al.*, 2009). As a single copy, the gene has been found in *HP0710/jhp0649* locus, whilst as a double-copy, *homA* and *homB* have been detected in *HP0710/jhp0649* or *jhp0870* loci. While *homB* double-copy has been associated with peptic ulcers, the *homA* single copy has been linked to non-ulcer dyspepsia (Oleastro *et al.*, 2008, 2009). Both *homB* and *homA* are phase variable genes. Variations of 5′ and 3′ *homB* regions, which differ geographically, develop by accumulation of single nucleotide polymorphisms or by homologous recombinations (Oleastro *et al.*, 2010b).

In Western strains, both *homB/homA* single-copy and double-copy genotypes have been present but in East Asian strains, only single-copy genotype has been found (Oleastro *et al.*, 2009). Prevalence of *homB* and *homA* genes has been evaluated in 449 strains from many countries (Oleastro *et al.*, 2009). The *homB* (56%) and *homA* (60.4%) genes have shown similar proportions in Western countries with slightly higher *homA* prevalence in strains from Portugal, France, Sweden, the USA and Brazil, and higher *homB* prevalence in strains from Germany and Colombia. In East Asia (Japan and Korea), *homB* has been highly prevalent (>77–95.9% of strains;

Oleastro *et al.*, 2009). This confirms the marked difference between East Asian and Western *H. pylori* strains in numerous virulence genes.

H. pylori outer-membrane protein *hop* genes

The role of outer membrane proteins, *hopQ*, *hopZ* and *hopH* in *H. pylori* adhesion has been established (Table 4.1). *H. pylori hopQ* alleles belong to two families, type I and type II (Cao *et al.*, 2005). In some strains, HopQ protein from type I *hopQ* strains has been shown to reduce the bacterial adherence to AGS GECs (Loh *et al.*, 2008). Strong antiadhesive activity of the protein has been detected in strains, lacking other adhesins such as BabA. Type I *hopQ* alleles have been associated with gastric atrophy as well as with *vacA* s1, m1 and *cagA*$^+$ strains in Western countries (Ohno *et al.*, 2009). Type II *hopQ* alleles have been more common in *cag*PAI-negative than in *cag*PAI$^+$ strains (Loh *et al.*, 2008). Type II *hopQ* alleles have been prevalent in Western strains, while almost all East Asian strains have been *cagA*$^+$ and type I *hopQ* with higher risk for atrophic gastritis and gastric cancer there (Cao *et al.*, 2005). Although type I *hopQ* genotype has not been a major risk factor for ulceration or malignancy, it has shown the joint activity of *H. pylori* virulence factors.

Another *hop* gene, *hopZ*, modulates pH-regulated genes such as *hopP*, which binds sialylated glycans produced by gastric epithelial progenitors. Genetic inactivation of *hopZ* has led to decreased fitness of the strains in stomach of gnotobiotic mice (Giannakis *et al.*, 2009).

Adherence-associated lipoprotein (*alpA* and *alpB*) genes

The *alpA* and *alpB* (*alpAB*) genes are encoded by two adjacent homologous genes inside an operon and thus, are co-transcribed (Lu *et al.*, 2007). The *alpAB* genes are important for *H. pylori* adhesion, as their deletion has diminished strongly (by 60–70%) *H. pylori* binding to GECs (Lu *et al.*, 2007). Nearly all *H. pylori* strains produce the outer membrane adhesins AlpA and AlpB proteins, suggesting their importance for *H. pylori*.

Specific amino acid sequences for East Asian or Western strains have been found in 10 positions in AlpA and five positions in AlpB. AlpAB from East Asia has induced IL-8 release whereas AlpAB of all strains has induced IL-6 release in GECs (Lu *et al.*, 2007). As East Asian AlpAB proteins have more strongly stimulated NF-κB in the GECs and as NF-κB promotes cancer development, preventing apoptosis of pre-malignant cells, AlpAB proteins are associated with severe gastroduodenal diseases in East Asia (Lu *et al.*, 2007).

Other virulence factors

H. pylori helical shape depends on cell wall peptide cross-links. Lack of helical shape slightly alters *H. pylori* motility but reduces its stomach colonization (Sycuro *et al.*, 2010). To diminish host immune detection, *H. pylori* uses peptidoglycan deacetylation by PgdA peptidoglycan deacetylase. A *pgdA* mutant has shown low ability to colonize mouse stomach and strong release of macrophage inflammatory protein-2 (MIP-2), TNF-α and IL-10 (Wang *et al.*, 2010). *H. pylori* could acquire cholesterol from gastric cell membranes and the presence of cholesterol can render the strains more resistant to acid and oxidative stresses compared with those in the absence of cholesterol (Hildebrandt and McGee, 2009). *H. pylori* motility is a complex property associated with the products of approximately 40 genes (de Reuse *et al.*, 2010). For instance, *flaA* and *flaB*, and *flbA* mutants have exhibited non-flagellate and non-motile cells (Bernarde *et al.*, 2010). MotB protein is a motor protein for flagellar rotation; the flagella of mutants lacking MotB loose this ability (Li *et al.*, 2010).

Motility is an important *H. pylori* factor for both colonization and immune evasion. When *H. pylori* escapes pH stress and reaches a neutral pH environment, the motility diminishes or stops to favour the bacterial adherence to GECs. *H. pylori* mutants lacking functional flagella have lost their ability to colonize the gastric mucosa in animal models (Algood and Cover, 2006). Many genes, e.g. those for flagellin A (*flaA*), flagellar basal-body L-ring protein (*flgH*), flagellar basal-body rod protein (*flgG*), flagellar hook-associated protein 1 (*flgK*) and flagellar biosynthetic protein (*fliQ*) contribute to *H. pylori* colonization of the stomach (Duckworth *et al.*, 2009). Moreover, the Hp0703-Hp0244 (FlgRS) two-component system, which adjusts the intermediate flagellar gene expression, also modulates genes for *H. pylori* acid survival (see above). *H. pylori* flagellin is a poor activator of innate immunity as it is weakly recognized by TLR5 owing to substitution in key flagellar amino acids (Proença-Modena *et al.*, 2009). This contributes to long-term persistence of *H. pylori* in the host.

It is known that promoter hypermethylation of host CDKN2A leads to gene inactivation. Interestingly, the promoter hypermethylation may vary according to *H. pylori* type. Brazilian study has shown association between FlaA$^+$ *H. pylori* strains and tumours without methylation as well as between *cagA*$^+$ or *vacA* s1/m1 type strains without FlaA and tumours with methylation (Alves *et al.*, 2010). However, in the USA, another study has not proven the association (Vo *et al.*, 2002). Thus, the topic of tumour-suppressor gene inactivation should be further clarified.

Chemotaxis is important for the establishment of *H. pylori* infection. To sense external stimuli, *H. pylori* genome encodes a cytoplasmic (TlpD) and three membrane-bound (TlpA, TlpB, and TlpC) chemoreceptors (Alexandre, 2008). Methyl-accepting chemotaxis transducer (*tlpC*) and methyl-accepting chemotaxis receptor protein (*tlpA*) genes are essential for host colonization (Duckworth *et al.*, 2009). Chemotaxis receptor protein TlpA is a receptor for arginine and sodium bicarbonate and TlpB is essential for pH taxis (Israel and Peek, 2010). TlpB is responsible for a repellent response, permitting the bacteria to escape highly acid environments (Alexandre, 2008). Mutations in *tlpB* gene, encoding TlpB, have reduced the inflammation in animal models (Israel and Peek, 2010). TlpD detects perturbations of electron transport system (energy sensing), and starts a repellent response as well (Alexandre, 2008).

Response regulators are needed for chemotaxis. Chemoreceptors, a histidine kinase (CheA), a receptor-kinase adaptor (CheW) and a phosphorylatable response regulator (CheY) control flagellar rotation (Lowenthal *et al.*, 2009). *H. pylori* chemotactic system differs from that of *E. coli* by the presence of numerous response regulators: CheY1, one CheY-like-containing CheA protein (CheAY2) and three CheV proteins (Lam *et al.*, 2010). CheVs are bifunctional proteins and hybrids of CheW and CheY-like domain (Lowenthal *et al.*, 2009). *H. pylori* CheVs are in either CheA-activating or CheA-deactivating states, phosphorylation switching them towards the opposite form. Two CheVs probably activate CheA and one CheV possibly deactivates CheA (Lowenthal *et al.*, 2009).

CheW, CheA or CheY mutants have been non-chemotactic, although able to infect weakly mice (Israel and Peek, 2010). The CheW has been strongly needed for *H. pylori* colonization during the first months of the infection. In a mouse model, the non-chemotactic *H. pylori* mutants have colonized the gastric corpus unlike the wild-type strains that have been found in all gastric sites (Terry *et al.*, 2005). In mixed infections, the mutants have been replaced gradually by the wild-type strains. Terry *et al.* (2005) have stressed the importance of *H. pylori* chemotaxis to set up and maintain the infection, to favour survival of the best-fitted strain and to promote colonization in all gastric sites.

In addition, *H. pylori* possesses phospholipase A, which digests phospholipids of cell membranes and periplasmic deamidases, γ-glutamyltranspeptidase and L-asparaginase, which, together with Hp-DcuA (the aspartate importer) act as virulence factors, promoting *H. pylori* colonization. The γ-glutamyltranspeptidase produces glutamate and ammonia from glutamine and the L-asparaginase generates aspartate and ammonia from asparagine. They can draw off amino acids from GECs and can produce ammonia, which are toxic for the cells (Leduc *et al.*, 2010). Importantly, the *ggt* gene-encoded γ-glutamyltranspeptidase provokes apoptosis in GECs and inhibits T-cell proliferation (Kim *et al.*, 2007). This enzyme, a member of the *N*-terminal nucleophile hydrolase superfamily, is able to produce H_2O_2 in GECs, leading to NF-κB activation and upregulation of IL-8 release, and to oxidative damage of the DNA. Kim *et al.* (2010) have reported *H. pylori* γ-glutamyltranspeptidase to induce cell cycle arrest in GECs at the G1 phase of growth, downregulating the key regulators (cyclins E and A) of DNA synthesis and mitosis and upregulating cyclin-dependent kinase inhibitors (p27 and p21). The enzyme has been most active in strains from peptic ulcer patients (Gong *et al.*, 2010).

H. pylori arginase (RocF) participates in nitrogen metabolism, survival in the stomach and modulation of the gastric inflammation. The enzyme has been shown to increase 1000 times *H. pylori* acid resistance *in vitro* compared with that in *rocF* mutants. However, the antioxidant properties of the enzyme can reduce the gastric inflammation (Kusters *et al.*, 2006; Wang *et al.*, 2006). RocF diminishes the NO generation in host cells. This can lead to avoidance of host nitrosative stress and to decrease in NO-induced gastric damage, thereby facilitating *H. pylori* persistence (Das *et al.*, 2010; Romero-Adrián *et al.*, 2010). Moreover, *H. pylori* upregulates host arginase II in the stomach, inducing cell apoptosis (Das *et al.*, 2010).

TNF-α-inducing protein (Tipα) is newly discovered carcinogenic factor of *H. pylori*, encoded by *Tipa (Hp0596)* gene (Godlewska *et al.*, 2008). Tipα is unique for *H. pylori* and has been found in all *H. pylori* genomes sequenced (Tosi *et al.*, 2009). This protein has specific structure, different from those of the other *H. pylori* virulence factors. Receptor for Tipα is the nucleolin, present in high amounts on the cell surface of cancer cells. Tipα is secreted from *H. pylori* as a homodimer of 38 kDa assembled by disulfide bridges (Tosi *et al.*, 2009). The crystal structure of Tipα has revealed that its monomer includes a helical domain with four helices (α1 to α4) and hydrophobic region as well as a mixed domain (Jang *et al.*, 2009).

In the tumorigenesis, the nucleolin moves from the nucleoli to the cell surface of GECs and binds Tipα. Thereafter, Tipα attains the cell cytoplasm and nucleus. The protein shows a potent DNA-binding property with tumour-promoting potential (Jang *et al.*, 2009). Surface region at the Tipα bottom, across its monomers, has been found to be responsible for the DNA binding (Jang *et al.*, 2009). In addition, Tipα is able to bind to TNF-α promoter and to induce NF-kB activation and high expression of TNF-α and chemokine genes in the cells, which also results in carcinogenesis (Watanabe *et al.*, 2010). Production of TNF-α and IL-1α has been 15–50-time higher in Tipα stimulated macrophages compared with other macrophages (Godlewska *et al.*, 2008). Tipα amount has been larger in patients with gastric cancer compared with chronic gastritis (Suganuma *et al.*, 2008). Jang *et al.* (2009) have suggested that development of inhibitors of Tipα homodimerization could be of therapeutic value in the control of *H. pylori* infection. In brief, although Tipα is possibly present in all/most *H. pylori* strains and its dual activity sharply augments the carcinogenesis in case it is already initiated.

Conversely, the transmembrane cell–cell junction protein, E-cadherin, is known to exert tumour suppression activity. However, the E-cadherin–β-catenin complex can be affected by *H. pylori* in two ways i.e. the CagA-induced mechanism and that of high-temperature requirement A (HtrA) protein. HtrA is a highly active, multimeric serine protease and chaperone (Löwer *et al.*, 2008). Both ways can promote development of premalignant gastric changes. The HtrA protein has been reported to target and cut the E-cadherin ectodomain (Hoy *et al.*, 2010). This activity disrupts intercellular adhesions of the GECs and facilitates *H. pylori* entry into the intercellular space. The altered barrier of the GECs can promote gastric inflammation and possible tumorigenesis. Other *H. pylori* proteases, among them, collagenase and protease acting on PDGF (platelet-derived growth factor), also have shown activity in the pathogenesis of the infection (Löwer *et al.*, 2008).

To counteract oxidative stress and host immunity, *H. pylori* employs many enzymes, for example catalase (KatA), superoxide dismutase (SodB), alkyl hydroperoxide reductase, NADPH quinone reductase (MdaB), neutrophil-activating protein (Hp-NAP) etc. (Wang *et al.*, 2006.) The importance of the SodB, converting superoxide to H_2O_2, and the KatA, converting H_2O_2 into oxygen and H_2O, have been shown by the loss, to some degree, of *H. pylori* colonization by mutants if compared with wild-type strains (Israel and Peek, 2010).

Hp-NAP (150 kDa), encoded by *napA* gene, is a dodecameric cytosolic protein, fold into α-helix structure, which is present in almost all *H. pylori* strains (Wang *et al.*, 2008). Hp-NAP resembles bacterioferritins, showing a central cavity for iron-gathering purpose (Proença-Modena *et al.*, 2009). Reported Hp-NAP receptors have been sulfated oligosaccharides, including sulfo-Le^a, sulfogalactose and sulfo-*N*-acetyl-glucosamine on mucins (Magalhães and Reis, 2010). Cooksley *et al.* (2003) have reported Hp-NAP regulation by Fur and Hp-NAP co-localization with cellular DNA, implying their interplay.

Hp-NAP has multiple activities. It promotes bacterial adhesion to the mucin and neutrophils, attracts the neutrophils, inducing their adhesion to endothelial cells and provokes oxidative burst in the neutrophils with subsequent release of reactive oxygen species (ROS), thus intensifying the gastric inflammation (Polenghi *et al.*, 2007). Hp-NAP also mediates neutrophil production of chemokines such as CXCL8, CCL3, and CCL4 and following recruitment of monocytes, lymphocytes and dendritic cells in addition to the neutrophils (Polenghi *et al.*, 2007). Hp-NAP has been reported to stimulate the neutrophils to produce myeloperoxidase, which, on its turn, generates potent oxidants such as hypochlorous acid (Wang *et al.*, 2008). At the same time, Hp-NAP diminishes the activities of ROS inside the bacterial cell to protect it (Cooksley *et al.*, 2003; Wang *et al.*, 2006). Furthermore, Hp-NAP prolongs the lifetime of monocytes through indirect induction of antiapoptotic proteins (A1, Mcl-1, Bcl-2 and Bcl-XL), with participation of IL-1β (Bernarde *et al.*, 2010; Cappon *et al.*, 2010). This activity may contribute to development of malignancy. In addition, Hp-NAP stimulates neutrophils and monocytes to release IL-12, a key cytokine for Th1 bias of naive T cells, leading to tissue injury (D'Elios *et al.*, 2007). Complex of NapA and a supposed ribosomal protein L7/L12, found by complexomic study in a MALT lymphoma patient, may be important for the development of the disease (Bernarde *et al.*, 2010).

Besides, *H. pylori* has stress regulator proteins, e.g. Fur and post-transcriptional regulator CsrA (Wang *et al.*, 2006). To repair oxidative DNA damage, *H. pylori* employs many mechanisms, e.g. those by *AddA* gene product, DNA recombinations, Hp-NAP (mentioned above), oxoguanine-recognizing MutS2 and methionine sulphoxide reductase (Wang *et al.*, 2006). Heat-shock proteins (Hsps), found in both prokaryotic and eukaryotic cells, are induced when a cell undergoes environmental stresses. *H. pylori*-Hsp60 (a GroEL heat shock protein) induces IL-8 release by ERK and p38 mitogen-activated protein kinase (MAPK) signalling pathways and TLR receptor (TLR2). ERK and the p38 MAPK pathways activate NF-kB and the release of IL-8 (Zhao *et al.*, 2007). Lin *et al* (2010a) have suggested that the *H. pylori* Hsp60 participates in gastric carcinogenesis owing to induction of pro-inflammatory cytokine release, angiogenesis and metastasis. Hsp60 has been reported to be the main *H. pylori* factor that stimulates IL-6 release from murine macrophages after the NF-kB activation (Gobert *et al.*, 2004). In Japan, seropoprevalence to *H. pylori*-Hsp60 has been higher in *H. pylori*-positive (73.3%) versus negative patients (24.5%) and in patients with gastric cancer (87.7%) versus those without cancer (73.3%; Tanaka *et al.*, 2009b).

HspA is a nickel chaperone and homologue of the essential bacterial heat shock protein GroES (Schauer *et al.*, 2010). By immunoproteomics, GroES seropoprevalence has been associated with antral cancer in Taiwan. In the same study, GroES has been found to induce pro-inflammatory cytokine (IL-8, IL-6, GM-CSF, IL-1β and TNF-α) release in human mononuclear cells as well as IL-8 production in GECs (Lin *et al.*, 2006).

H. pylori often hides its antigens from the host immune mechanisms. Plasminogen-binding PgbA and PgbB proteins of *H. pylori* cover the bacterial surface with the host zymogen plasminogen (inactive form of the fibrinolytic enzyme plasmin) to avoid immune response (Algood and Cover, 2006). *H. pylori* can also change the gastric acid secretion by enhancing the release of acid-suppressive cytokine (IL-1β; Israel and Peek, 2010). Additionally, the *cag*PAI$^+$ strains can bind to and upregulate decay-accelerating factor (DAF) in the gastric mucosa and thus, can modulate the gastric inflammation (O'Brien *et al.*, 2008). DAF-protected human cells avoid phagocytosis and complement cytotoxicity through dissociation of complement C3 convertases (O'Brien *et al.*, 2008). Thus, DAF may enhance the growth and complement resistance in tumour cells.

In conclusion, *H. pylori* genetic diversity, molecular or functional mimicry, some virulence factors and the induced immunomodulation contribute to the chronicity of the infection. The overall combination of *H. pylori* virulence factors with host and environment factors determines the infection outcomes. Although severe outcomes have shown strong association with *cagA*, *cag*PAI and *vacA* status of the strains, mainly in Western countries, the combined and well-coordinated activity of all *H. pylori* pathogenicity factors, including also *Tipa*, *dupA*, *oipA*, *iceA*, *homB*, *babA*, *sabA*, *hopQ* and other genes, seems to be crucial for the infection pathogenesis. *H. pylori* genes with highest association with gastric cancer have been *cag*PAI, including *cagA* gene, and *vacA* mosaics, and *oipA* and *babA2* genes (Franco *et al.*, 2008; Wen and Moss, 2009). Combination of intact *cag*PAI, Eastern *cagA* type and *vacA* s1/i1/m1 in most strains from East Asia has been associated with the common gastric cancer there (Nguyen *et al.*, 2008). Strains harbouring *iceA1* allele, *sabA* 'on' and *hopZ* 'off' states, have increased 10-times the risk for MALT lymphoma and Le^y *cag*PAI-negative strains have contributed to the disease (Lehours *et al.*, 2004, 2009). Obviously, the multifaceted, cooperative activity of dozens of *H. pylori* virulence factors, acting either directly or through the chronic inflammation, can lead to development of peptic ulcers or gastric malignancy (Douraghi *et al.*, 2010) in some of the patients. Targeting virulent strains in more countries/regions and patient groups is strongly required to explain better the clinical significance and interplay among some virulence factors, to imply aggressive eradication strategy in the concerned patient groups and to provide new agents and improved regimens to control the infection.

References

Akada, J.K., Aoki, H., Torigoe, Y., Kitagawa, T., Kurazono, H., Hoshida, H., Nishikawa, J., Terai, S., Matsuzaki, M., Hirayama, T., *et al.* (2010). *Helicobacter pylori* CagA inhibits endocytosis of cytotoxin VacA in host cells. Dis. Model. Mech. *3*, 605–617.

Alexandre, G. (2008). A sense of self-worth: energy taxis provides insight into how *Helicobacter pylori* navigates through its environment. J. Bacteriol. *190*, 3095–3097.

Algood, H.M., and Cover, T.L. (2006). *Helicobacter pylori* persistence: an overview of interactions between *H. pylori* and host immune defenses. Clin. Microbiol. Rev. *19*, 597–613.

Ali, M., Khan, A. A., Tiwari, S. K., Ahmed, N., Rao, L. V., and Habibullah, C. M. (2005). Association between *cag*-pathogenicity island in *Helicobacter pylori* isolates from peptic ulcer, gastric carcinoma, and non-ulcer dyspepsia subjects with histological changes. World J. Gastroenterol. *11*, 6815–6868.

Alves, M.K., Lima, V.P., Ferrasi, A.C., Rodrigues, M.A., De Moura Campos Pardini, M.I., and Rabenhorst, S.H. (2010). CDKN2A promoter methylation is related to the tumor location and histological subtype and associated with *Helicobacter pylori flaA*(+) strains in gastric adenocarcinomas. APMIS *118*, 297–307.

Alvi, A., Devi, S.M., Ahmed, I., Hussain, M.A., Rizwan, M., Lamouliatte, H., Mégraud, F., and Ahmed, N. (2007). Microevolution of *Helicobacter pylori* type IV secretion systems in an ulcer disease patient over a ten-year period. J. Clin. Microbiol. *45*, 4039–4043.

Amundsen, S.K., Fero, J., Salama, N.R., and Smith, G.R. (2009). Dual nuclease and helicase activities of *Helicobacter pylori* AddAB are required for DNA repair, recombination, and mouse infectivity. J. Biol. Chem. *284*, 16759–16766.

Ando, T., Peek, R.M., Pride, D., Levine, S.M., Takata, T., Lee, Y.C., Kusugami, K., van der Ende, A., Kuipers, E.J., Kusters, J.G., and Blaser, M.J. (2002). Polymorphisms of *Helicobacter pylori* Hp0638 reflect geographic origin and correlate with *cagA* status. J. Clin. Microbiol. *40*, 239–246.

Apostolopoulos, P., Vafiadis-Zouboulis, I., Tzivras, M., Kourtessas, D., Katsilambros, N., Archimandritis, A. (2002). *Helicobacter pylori (H pylori)* infection in Greece: the changing prevalence during a ten-year period and its antigenic profile. BMC Gastroenterol. *16*, 11.

Arachchi, H.S., Kalra, V., Lal, B., Bhatia, V., Baba, C.S., Chakravarthy, S., Rohatgi, S., Sarma, P.M., Mishra, V., Das, B., *et al.* (2007). Prevalence of duodenal ulcer-promoting gene (*dupA*) of *Helicobacter pylori* in patients with duodenal ulcer in north Indian population. Helicobacter *12*, 591–597.

Araya, J.C., Anabal6n, L., Roá, I., Bravo, M., Villaseca, M.A., Guzmán, P., and Roa, J.C. (2004). Association between *Helicobacter pylori* genotype and the severity of gastritis in infected adults. Rev. Med. Chil. *132*, 1345–1354.

Argent, R.H., Burette, A., Miendje Deyi, V.Y., and Atherton, J.C. (2007). The presence of *dupA* in *Helicobacter pylori* is not significantly associated with duodenal ulceration in Belgium, South Africa, China, or North America. Clin. Infect. Dis. *45*, 1204–1206.

Argent, R.H., Thomas, R.J., Letley, D.P., Rittig, M.G., Hardie, K.R., and Atherton, J.C. (2008a). Functional association between the *Helicobacter pylori* virulence factors VacA and CagA. J. Med. Microbiol. *57*, 145–150.

Argent, R.H., Hale, J.L., El-Omar, E.M., and Atherton, J.C. (2008b). Differences in *Helicobacter pylori* CagA tyrosine phosphorylation motif patterns between western and East Asian strains, and influences on interleukin-8 secretion. J. Med. Microbiol. *57*, 1062–1067.

Ashour, A.A., Collares, G.B., Mendes, E.N., de Gusmão, V.R., Queiroz, D.M., Magalhães, P.P., de Carvalho, A.S., de Oliveira, C.A., Nogueira, A.M., Rocha, G.A., *et al.* (2001). *iceA* genotypes of *Helicobacter pylori* strains isolated from Brazilian children and adults. J. Clin. Microbiol. *39*, 1746–1750.

Atherton, J.C., and Blaser, M.J. (2009). Coadaptation of *Helicobacter pylori* and humans: ancient history, modern implications. Clin. Invest. *119*, 2475–2487.

Azevedo, M., Eriksson, S., Mendes, N., Serpa, J., Figueiredo, C., Resende, L. P., Ruvoen-Clouet, N., Haas, R., Boren, T., Le Pendu, J., *et al.* (2008). Infection by *Helicobacter pylori* expressing the BabA adhesin is influenced by the secretor phenotype. J. Pathol. *215*, 308–316.

Backert, S., Tegtmeyer, N., and Selbach, M. (2010a). The versatility of *Helicobacter pylori* CagA effector protein functions: The master key hypothesis. Helicobacter *15*, 163–176.

Backert, S., Mimuro, H., Israel, D.A., and Peek Jr, R.M. (2010b). Virulence factors of *Helicobacter pylori*. In *Helicobacter pylori* in the 21st Century, Sutton, P., and Mitchell, H., eds. (Cab International: Wallingford, UK), pp. 212–247.

Basso, D., Zambon, C.F., Letley, D.P., Stranges, A., Marchet, A., Rhead, J.L., Schiavon, S., Guariso, G., Ceroti, M., Nitti, D., *et al.* (2008). Clinical relevance of *Helicobacter pylori cagA* and *vacA* gene polymorphisms. Gastroenterology *135*, 91–99.

Ben Mansour, K., Fendri, C., Zribi, M., Masmoudi, A., Labbene, M., Filali, A., Ben Mami, N., Najjar, T., Meherzi, A., Sfar, T., *et al.* (2010). Prevalence of *Helicobacter pylori vacA, cagA, iceA* and *oipA* genotypes in Tunisian patients. Ann. Clin. Microbiol. Antimicrob. *9*, 10.

Benoit, S.L., Mehta, N., Weinberg, M.V., Maier, C., and Maier, R.J. (2007). Interaction between the *Helicobacter pylori* accessory proteins HypA and UreE is needed for urease maturation. Microbiology *15*,31474–14782.

Bernarde, C., Lehours, P., Lasserre, J.P., Castroviejo, M., Bonneu, M., Megraud, F., and Menard, A. (2010). Complexomics study of two *Helicobacter pylori* strains of two pathological origins: potential targets for vaccine development and new insight in bacteria metabolism. Mol. Cell. Proteomics *9*, 2796–2826.

Bindayna, K.M., and Al Mahmeed, A. (2009). *vacA* genotypes in *Helicobacter pylori* strains isolated from patients with and without duodenal ulcer in Bahrain. Indian J. Gastroenterol. *28*, 175–179.

Boyanova, L., Markovska, R., Yordanov, D., Marina, M., Ivanova, K., Panayotov, S., Gergova, G., and Mitov, I. (2009). High prevalence of virulent *Helicobacter pylori* strains in symptomatic Bulgarian patients. Diagn. Microbiol. Infect Dis. *64*, 374–380.

Boyanova, L., Yordanov, D., Gergova, G., Markovska, R., and Mitov, I. (2010). Association of *iceA* and *babA* genotypes in *Helicobacter pylori* strains with patient and strain characteristics. Antonie Van Leeuwenhoek *98*, 343–350.

Brandt, S., Kwok, T., Hartig, R., König, W., and Backert, S. (2005). NF-kappaB activation and potentiation of pro-inflammatory responses by the *Helicobacter pylori* CagA protein. Proc. Natl. Acad. Sci. U.S.A. *102*, 9300–9305.

Bury-Moné, S., Mendz, G.L., Ball, G.E., Thibonnier, M., Stingl, K., Ecobichon, C., Avé, P., Huerre, M., Labigne, A., Thiberge, J.M., *et al.* (2008). Roles of alpha and beta carbonic anhydrases of *Helicobacter pylori* in the urease-dependent response to acidity and in colonization of the murine gastric mucosa. Infect. Immun.*76*, 497–509.

Busler, V.J., Torres, V.J., McClain, M.S., Tirado, O., Friedman, D.B., and Cover, T.L. (2006). Protein–protein interactions among *Helicobacter pylori* cag proteins. J. Bacteriol. *188*, 4787–4800.

Camorlinga-Ponce, M., Romo, C., Gonzalez-Valencia, G., Munoz, O., and Torres, J. (2004). Topographical localisation of *cagA* positive and *cagA* negative *Helicobacter pylori* strains in the gastric mucosa; an *in situ* hybridisation study. J. Clin. Pathol. *57*, 822–828.

Caner, V., Yilmaz, M., Yonetci, N., Zencir, S., Karagenc, N., Kaleli, I., and Bagci, H. (2007). *H pylori iceA* alleles are disease-specific virulence factors. World J. Gastroenterol. *13*, 2581–2585.

Cao, P., Lee, K.J., Blaser, M.J., and Cover, T.L. (2005). Analysis of *hopQ* alleles in East Asian and Western strains of *Helicobacter pylori*. FEMS Microbiol. Lett. *251*, 37–43.

Cappon, A., Babolin, C., Segat, D., Cancian, L., Amedei, A., Calzetti, F., Cassatella, M.A., D'Elios, M.M., and de Bernard, M. (2010). *Helicobacter pylori*-derived neutrophil-activating protein increases the lifespan of monocytes and neutrophils. Cell Microbiol. *12*, 754–764.

Cellini, L., Grande, R., Di Campli, E., Di Bartolomeo, S., Capodicasa, S., and Marzio, L. (2006). Analysis of genetic variability, antimicrobial susceptibility and virulence markers in *Helicobacter pylori* identified in Central Italy. Scand. J. Gastroenterol. *41*, 280–287.

Cendron, L., Couturier, M., Angelini, A., Barison, N., Stein, M., and Zanotti, G. (2009). The *Helicobacter pylori* CagD (Hp0545, Cag24) protein is essential for CagA translocation and maximal induction of interleukin-8 secretion. J. Mol. Biol. *386*, 204–217.

Chiarini, A., Cala, C., Bonura, C., Gullo, A., Giuliana, G., Peralta, S., D'Arpa, F., and Giammanco, A. (2009). Prevalence of virulence-associated genotypes of *Helicobacter pylori* and correlation with severity of gastric pathology in patients from western Sicily, Italy. Eur. J. Clin. Microbiol. Infect. Dis. *28*, 437–446.

Chiurillo, M.A., Moran, Y., Cañas, M., Valderrama, E., Alvarez, A., and Armanie. E. (2010). Combination of *Helicobacter pylori-iceA2* and pro-inflammatory interleukin-1 polymorphisms is associated with the severity of histological changes in Venezuelan chronic gastritis patients. FEMS Immunol. Med. Microbiol. *59*, 170–176.

Chochi, K., Ichikura, T., Kinoshita, M., Majima, T., Shinomiya, N., Tsujimoto, H., Kawabata, T., Sugasawa, H., Ono, S., Seki, S., *et al.* (2008). *Helicobacter pylori* augments growth of gastric cancers via the lipopolysaccharide-toll-like receptor 4 pathway whereas its lipopolysaccharide attenuates antitumor activities of human mononuclear cells. Clin. Cancer Res. *14*, 2909–2917.

Chomvarin, C., Namwat, W., Chaicumpar, K., Mairiang, P., Sangchan, A., Sripa, B., Tor-Udom, S., and Vilaichone, R.K. (2008). Prevalence of *Helicobacter pylori vacA, cagA, cagE, iceA* and *babA2* genotypes in Thai dyspeptic patients. Int. J. Infect. Dis. *12*, 30–36.

Chung, C., Olivares, A., Torres, E., Yilmaz, O., Cohen, H., and Perez-Perez, G. (2010). Diversity of *VacA* intermediate region among *Helicobacter pylori* strains from several regions of the world. J. Clin. Microbiol. *48*, 690–696.

Cole, S.P., Harwood, J., Lee, R., She, R., and Guiney, D.G. (2004). Characterization of monospecies biofilm formation by *Helicobacter pylori*. J. Bacteriol. *186*, 3124–3132.

Con, S.A., Takeuchi, H., Con-Chin, G.R., Con-Chin, V.G., Yasuda, N., and Con-Wong, R. (2009). Role of bacterial and genetic factors in gastric cancer in Costa Rica. World J. Gastroenterol. *15*, 211–218.

Cooksley, C., Jenks, P.J., Green, A., Cockayne, A., Logan, R.P., and Hardie, K.R. (2003). NapA protects *Helicobacter pylori* from oxidative stress damage, and its production is influenced by the ferric uptake regulator. J. Med. Microbiol. *52*, 461–469.

Costa, A.C., Figueiredo, C., and Touati, E. (2009). Pathogenesis of *Helicobacter pylori* infection. Helicobacter *14* (Suppl. 1), 15–20.

Das, P., Lahiri, A., Lahiri, A., and Chakravortty, D. (2010). Modulation of the arginase pathway in the context of microbial pathogenesis: a metabolic enzyme moonlighting as an immune modulator. PLoS Pathog. *6*, e1000899.

Delahay, R.M., Balkwill, G.D., Bunting, K.A., Edwards, W., Atherton, J.C., and Searle, M.S. (2008). The highly repetitive region of the *Helicobacter pylori* CagY protein comprises tandem arrays of an alpha-helical repeat module. J. Mol. Biol. *377*, 956–971.

D'Elios, M.M., Montecucco, C., and de Bernard, M. (2007). VacA and Hp-NAP, Ying and Yang of *Helicobacter pylori*-associated gastric inflammation. Clin. Chim. Acta. *381*, 32–38.

Dong, Q.J., Wang, Q., Xin, Y.N., Li, N., and Xuan, S.Y. (2009). Comparative genomics of *Helicobacter pylori*. World J. Gastroenterol. *15*, 3984–3991.

Dorer, M.S., Fero, J., and Salama, N.R. (2010). DNA damage triggers genetic exchange in *Helicobacter pylori*. PLoS Pathog. *6*, e1001026.

Dossumbekova, A., Prinz, C., Mages, J., Lang, R., Kusters, J.G., Van Vliet, A.H., Reindl, W., Backert, S., Saur, D., Schmid, R.M., *et al.* (2006). *Helicobacter pylori* HopH (OipA) and bacterial pathogenicity: genetic and functional genomic analysis of *hopH* gene polymorphisms. J. Infect. Dis. *194*, 1346–1355.

Douraghi, M., Mohammadi, M., Oghalaie, A., Abdirad, A., Mohagheghi, M.A., Hosseini, M.E., Zeraati, H., Ghasemi, A., Esmaieli, M., and Mohajerani, N. (2008). *dupA* as a risk determinant in *Helicobacter pylori* infection. J. Med. Microbiol. *57*, 554–562.

Douraghi, M., Talebkhan, Y., Zeraati, H., Ebrahimzadeh, F., Nahvijoo, A., Morakabati, A., Ghafarpour, M., Esmaili, M., Bababeik, M., Oghalaie, A., *et al.* (2009). Multiple gene status in *Helicobacter pylori* strains and risk of gastric cancer development. Digestion *80*, 200–207.

Douraghi, M., Talebkhan, Y., Zeraati, H., and Mohammadi, M. (2010). Cooperative genotyping for *Helicobacter pylori* virulence determinants strengthens the predictive value of gastric cancer risk assessment. Dig. Liver Dis. *42*, 662–663.

Dubois, A., and Boren, T. (2007). *Helicobacter pylori* is invasive and it may be a facultative intracellular organism. Cell Microbiol. *9*, 1108–1116.

Duckworth, M.J., Okoli, A.S., and Mendz, G.L. (2009). Novel *Helicobacter pylori* therapeutic targets: the unusual suspects. Expert Rev. Anti Infect. Ther. *7*, 835–867.

Erzin, Y., Koksal, V., Altun, S., Dobrucali, A., Aslan, M., Erdamar, S., Dirican, A., and Kocazeybek, B. (2006). Prevalence of *Helicobacter pylori vacA, cagA, cagE, iceA, babA2* genotypes and correlation with clinical outcome in Turkish patients with dyspepsia. Helicobacter *11*, 574–580.

Fietta, P., and Delsante, G. (2009). The effector T helper cell triade. Riv. Biol. *102*, 61–74.

Fischer, W., Windhager, L., Rohrer, S., Zeiller, M., Karnholz, A., Hoffmann, R., Zimmer, R., and Haas, R. (2010). Strain-specific genes of *Helicobacter pylori*: genome evolution driven by a novel type IV secretion system and genomic island transfer. Nucleic Acids Res. *38*, 6089–6101.

Foo, J.H., Culvenor, J.G., Ferrero, R.L., Kwok, T., Lithgow, T., and Gabriel, K. (2010). Both the p33 and p55 subunits of the *Helicobacter pylori* VacA toxin are targeted to mammalian mitochondria. J. Mol. Biol. *401*, 792–798.

Franco, A.T., Johnston, E., Krishna, U., Yamaoka, Y., Israel, D.A., Nagy, T.A., Wroblewski, L.E., Piazuelo, M.B., Correa, P., and Peek, R.M., Jr. (2008). Regulation of gastric carcinogenesis by *Helicobacter pylori* virulence factors. Cancer Res. *68*, 379–387.

Fujimoto, S., Olaniyi Ojo, O., Arnqvist, A., Wu, J.Y., Odenbreit, S., Haas, R., Graham, D.Y., and Yamaoka, Y. (2007) *Helicobacter pylori BabA* expression, gastric mucosal injury, and clinical outcome. Clin. Gastroenterol. Hepatol. *5*, 49–58.

Gancz, H., Jones, K.R., and Merrell, D.S. (2008). Sodium chloride affects *Helicobacter pylori* growth and gene expression. J. Bacteriol. *190*, 4100–4105.

Gangwer, K.A., Mushrush, D.J., Stauff, D.L., Spiller, B., McClain, M.S., Cover, T.L., and Lacy, D.B. (2007). Crystal structure of the *Helicobacter pylori* vacuolating toxin p55 domain. Proc. Natl. Acad. Sci. U.S.A. *104*, 16293–16298.

García, A., Barra, R., Delgado, C., Kawaguchi, F., Trabal, N., Montenegro, S., and González, C. (2006). Genotypying of clinical isolates of *Helicobacter pylori* by *cagA, vacA* and *babA2* virulence associated genes. First detection of a *babA2* positive strain in Chilean patients. Rev. Med. Chil. *134*, 981–988.

Gatti, L.L., Módena, J.L., Payão, S.L., Smith Mde, A., Fukuhara, Y., Módena, J.L., de Oliveira, R.B., and Brocchi, M. (2006). Prevalence of *Helicobacter pylori cagA, iceA* and *babA2* alleles in Brazilian patients with upper gastrointestinal diseases. Acta Trop. *100*, 232–240.

Ghose, C., Perez-Perez, G.I., van Doorn, L.J., Domínguez-Bello, M.G., and Blaser, M.J. (2005). High frequency of gastric colonization with multiple *Helicobacter pylori* strains in Venezuelan subjects. J. Clin. Microbiol. *43*, 2635–2641.

Giannakis, M., Bäckhed, H.K., Chen, S.L., Faith, J.J., Wu, M., Guruge, J.L., Engstrand, L., and Gordon, J.I. (2009). Response of gastric epithelial progenitors to *Helicobacter pylori* Isolates obtained from Swedish patients with chronic atrophic gastritis. J. Biol. Chem. *284*, 30383–30394.

Gobert, A.P., Mersey, B.D., Cheng, Y., Blumberg, D.R., Newton, J.C., and Wilson, K.T. (2002). Cutting edge: urease release by *Helicobacter pylori* stimulates macrophage inducible nitric oxide synthase. J. Immunol. *168*, 6002–6006.

Gobert, A.P., Bambou, J.C., Werts, C., Balloy, V., Chignard, M., Moran, A.P., and Ferrero, R.L. (2004). *Helicobacter pylori* heat shock protein 60 mediates interleukin-6 production by macrophages via a toll-like receptor (TLR)-2-, TLR-4-, and myeloid differentiation factor 88-independent mechanism. J. Biol. Chem. *279*, 245–250.

Godlewska, R., Pawlowski, M., Dzwonek, A., Mikula, M., Ostrowski, J., Drela, N., and Jagusztyn-Krynicka, E.K. (2008). Tip-alpha (*hp0596* gene product) is a highly immunogenic *Helicobacter pylori* protein involved in colonization of mouse gastric mucosa. Curr. Microbiol. *56*, 279–286.

Gomes, L.I., Rocha, G.A., Rocha, A.M., Soares, T.F., Oliveira, C.A., Bittencourt, P.F., and Queiroz, D.M. (2008). Lack of association between *Helicobacter pylori* infection with *dupA*-positive strains and gastroduodenal diseases in Brazilian patients. Int. J. Med. Microbiol. *298*, 223–230.

Gong, Y.H., Wang, Y., and Yuan, Y. (2005). Distribution of *Helicobacter pylori* in north China. World J. Gastroenterol. *11*, 3523–3527.

Gong, M., Ling, S.S.M., Lui, S.Y., Yeoh, K.G., and Ho, B. (2010). *Helicobacter pylori* gamma-glutamyl transpeptidase is a pathogenic factor in the development of peptic ulcer disease. Gastroenterology *139*, 564–573.

González-Rivera, C., Gangwer, K.A., McClain, M.S., Eli, I.M., Chambers, M.G., Ohi, M.D., Lacy, D.B., and Cover, T.L. (2010). Reconstitution of *Helicobacter pylori* VacA toxin from purified components. Biochemistry *49*, 5743–5752.

Goodwin, A.C., Weinberger, D.M., Ford, C.B., Nelson, J.C., Snider, J.D., Hall, J.D., Paules, C.I., Peek, R.M., Jr., and Forsyth, M.H. (2008). Expression of the *Helicobacter pylori* adhesin SabA is controlled via phase variation and the ArsRS signal transduction system. Microbiology *154*, 2231–2240.

Graham, D.Y., Opekun, A.R., Osato, M.S., El-Zimaity, H.M., Lee, C.K., Yamaoka, Y., Qureshi, W.A., Cadoz, M., and Monath, T.P. (2004). Challenge model for *Helicobacter pylori* infection in human volunteers. Gut 53, 1235–1243.

Hatakeyama, M. (2009). *Helicobacter pylori* and gastric carcinogenesis. J. Gastroenterol. *44*, 239–248.

Hildebrandt, E., and McGee, D.J. (2009). *Helicobacter pylori* lipopolysaccharide modification, Lewis antigen expression, and gastric colonization are cholesterol-dependent. BMC Microbiol. *9*, 258.

Hilleringmann, M., Pansegrau, W., Doyle, M., Kaufman, S., MacKichan, M.L., Gianfaldoni, C., Ruggiero, P., and Covacci, A. (2006). Inhibitors of *Helicobacter pylori* ATPase Cagalpha block CagA transport and *cag* virulence. Microbiology *152*, 2919–2930.

Hirai, I., Sasaki, T., Kimoto, A., Yamamoto, Y., Azuma, T., Mahachai, V., Hansomburana, P., Lertkupinit, C., Luangjaru, S., Noophan, P., *et al.* (2010). Infection of less virulent *Helicobacter pylori* strains in asymptomatic healthy individuals in Thailand as a potential contributing factor to the Asian enigma. Microbes Infect. *12*, 227–230.

Homan, M., Luzar, B., Kocjan, B.J., Orel, R., Mocilnik, T., Shrestha, M., Kveder, M., and Poljak, M. (2009). Prevalence and clinical relevance of *cagA, vacA*, and *iceA* genotypes of *Helicobacter pylori* isolated from Slovenian children. J. Pediatr. Gastroenterol. Nutr. *49*, 289–296.

Hoy, B., Löwer, M., Weydig, C., Carra, G., Tegtmeyer, N., Geppert, T., Schröder, P., Sewald, N., Backert, S., Schneider, G., *et al.* (2010). *Helicobacter pylori* HtrA is a new secreted virulence factor that cleaves E-cadherin to disrupt intercellular adhesion. EMBO Rep. *11*, 798–804.

Hu, C.-T., Chiou, P.-Y., Wu, C.-C., Tseng, Y.-H., Chang, Y.-J., and Lin, N.-T. (2009). Analysis of resistance to clarithromycin and virulence markers in *Helicobacter pylori* clinical isolates from Eastern Taiwan. Tzu Chi Med. J. *21*, 123–128.

Hussein, N.R., Mohammadi, M., Talebkhan, Y., Doraghi, M., Letley, D.P., Muhammad, M.K., Argent, R.H., and Atherton, J.C. (2008). Differences in virulence markers between *Helicobacter pylori* strains from Iraq and those from Iran: potential importance of regional differences in *H. pylori*-associated disease. J. Clin. Microbiol. *46*, 1774–1779.

Hussein, N.R. (2010). The association of *dupA* and *Helicobacter pylori*-related gastroduodenal diseases. Eur. J. Clin. Microbiol. Infect. Dis. *29*, 817–821.

Hussein, N.R., Argent, R.H., Marx, C.K., Patel, S.R., Robinson, K., and Atherton, J.C. (2010). *Helicobacter pylori dupA* is polymorphic, and its active form induces pro-inflammatory cytokine secretion by mononuclear cells. J. Infect. Dis. *202*, 261–269.

Ikenoue, T., Maeda, S., Ogura, K., Akanuma, M., Mitsuno, Y., Imai, Y., Yoshida, H., Shiratori, Y., and Omata, M. (2001). Determination of *Helicobacter pylori* virulence by simple gene analysis of the *cag* pathogenicity island. Clin. Diagn. Lab. Immunol. *8*, 181–186.

Imagawa, S., Ito, M., Yoshihara, M., Eguchi, H., Tanaka, S., and Chayama, K. (2010). *Helicobacter pylori dupA* and gastric acid secretion are negatively associated with gastric cancer development. J. Med. Microbiol. *59*, 1484–1489.

Ishikawa, S., Ohta, T., and Hatakeyama, M. (2009). Stability of *Helicobacter pylori* CagA oncoprotein in human gastric epithelial cells. FEBS Lett. *583*, 2414–2418.

Isomoto, H., Moss, J., and Hirayama, T. (2010). Pleiotropic actions of *Helicobacter pylori* vacuolating cytotoxin, VacA. Tohoku J. Exp. Med. *220*, 3–14.

Israel, D.A., and Peek, R.M., Jr. (2010). Surreptitious manipulation of the human host by *Helicobacter pylori*. Gut Microbes *1*, 119–127.

Ivie, S.E., McClain, M.S., Algood, H.M., Lacy, D.B., and Cover, T.L. (2010). Analysis of a beta-helical region in the p55 domain of *Helicobacter pylori* vacuolating toxin. BMC Microbiol. *10*, 60.

Jang, J.Y., Yoon, H.J., Yoon, J.Y., Kim, H.S., Lee, S.J., Kim, K.H., Kim do, J., Jang, S., Han, B.G., Lee, B.I., *et al.* (2009). Crystal structure of the TNF-alpha-inducing protein (Tipalpha) from *Helicobacter pylori*: insights into its DNA-binding activity. J. Mol. Biol. *392*, 191–197.

Jang, S., Jones, K.R., Olsen, C.H., Joo, Y.M., Yoo, Y.J., Chung, I.S., Cha, J.H., and Merrell, D.S. (2010). Epidemiological link between gastric disease and polymorphisms in VacA and CagA. J. Clin. Microbiol. *48*, 559–567.

Jiménez-Soto, L.F., Kutter, S., Sewald, X., Ertl, C., Weiss, E., Kapp, U., Rohde, M., Pirch, T., Jung, K., Retta, S.F., *et al.* (2009). *Helicobacter pylori* type IV secretion apparatus exploits β1 Integrin in a novel RGD-independent manner. PLoS Pathog. *5*, e1000684.

Jones, K.R., Joo, Y.M., Jang, S., Yoo, Y.J., Lee, H.S., Chung, I.S., Olsen, C.H., Whitmire, J.M., Merrell, D.S., and Cha, J.H. (2009). Polymorphism in the CagA EPIYA motif impacts development of gastric cancer. J. Clin. Microbiol. *47*, 959–968.

Jung, S.W., Sugimoto, M., Graham, D.Y., and Yamaoka, Y. (2009). *homB* status of *Helicobacter pylori* as a novel marker to distinguish gastric cancer from duodenal ulcer. J. Clin. Microbiol. *47*, 3241–3245.

Jurik, A., Haußer, E., Kutter, S., Pattis, I., Praßl, S., Weiss, E., and Fischer, W. (2010). The coupling 0rotein Cag{beta} and its interaction partner CagZ are required for Type IV secretion of the *Helicobacter pylori* CagA protein. Infect. Immun. *78*, 5244–5251.

Kang, J., and Blaser, M.J. (2008). Repair and antirepair DNA helicases in *Helicobacter pylori*. J. Bacteriol. *190*, 4218–4224.

Kao, J.Y., Zhang, M., Miller, M.J., Mills, J.C., Wang, B., Liu, M., Eaton, K.A., Zou, W., Berndt, B.E., Cole, T.S., *et al.* (2010). *Helicobacter pylori* immune escape is mediated by dendritic cell-induced Treg skewing and Th17 suppression in mice. Gastroenterology *138*, 1046–1054.

Kato, S., Sugiyama, T., Kudo, M., Ohnuma, K., Ozawa, K., Iinuma, K., Asaka, M., and Blaser, M.J. (2000). CagA antibodies in Japanese children with nodular gastritis or peptic ulcer disease. J. Clin. Microbiol. *38*, 68–70.

Kauser, F., Khan, A. A., Hussain, M. A., Carroll, I. M., Ahmad, N., Tiwari, S., Shouche, Y., Das, B., Alam, M., Ali, S.M., *et al.* (2004). The *cag* pathogenicity island of *Helicobacter pylori* is disrupted in the majority of patient isolates from different human populations. J. Clin. Microbiol. *42*, 5302–5308.

Kauser, F., Hussain, M.A., Ahmed, I., Srinivas, S., Devi, S.M., Majeed, A.A., Rao, K.R., Khan, A.A., Sechi, L.A., and Ahmed, N. (2005a). Comparative genomics of *Helicobacter pylori* isolates recovered from ulcer disease patients in England. BMC Microbiol. *25*, 32.

Kauser, F., Hussain, M.A., Ahmed, I., Ahmad, N., Habeeb, A., Khan, A.A., and Ahmed, N. (2005b). Comparing genomes of *Helicobacter pylori* strains from the high-altitude desert of Ladakh, India. J. Clin. Microbiol. *43*, 1538–1545.

Kersulyte, D., Lee, W., Subramaniam, D., Anant, S., Herrera, P., Cabrera, L., Balqui, J., Barabas, O., Kalia, A., Gilman, R.H., *et al.* (2009). *Helicobacter pylori*'s plasticity zones are novel transposable elements. PLoS One *4*, e6859.

Khan, S., Karim, A., and Iqbal, S. (2009). *Helicobacter* urease: niche construction at the single molecule level. J. Biosci. *34*, 503–511.

Kidd, M., Peek, R.M., Lastovica, A.J., Israel, D.A., Kummer, A.F., and Louw, J.A. (2001). Analysis of *iceA* genotypes in South African *Helicobacter pylori* strains and relationship to clinically significant disease. Gut *49*, 629–635.

Kim, K.M., Lee, S.G., Park, M.G., Song, J.Y., Kang, H.L., Lee, W.K., Cho, M.J., Rhee, K.H., Youn, H.S., and Baikm, S.C. (2007). Gamma-glutamyltranspeptidase of *Helicobacter pylori* induces mitochondria-mediated apoptosis in AGS cells. Biochem. Biophys. Res. Commun. *355*, 562–567.

Kim, K.M., Lee, S.G., Kim, J.M., Kim, D.S., Song, J.Y., Kang, H.L., Lee, W.K., Cho, M.J., Rhee, K.H., Youn, H.S., and Baik, S.C. (2010). *Helicobacter pylori* gamma-glutamyltranspeptidase induces cell cycle arrest at the G1-S phase transition. J. Microbiol. *48*, 372–377.

Ko, J.S., Kim, K.M., Oh, Y.L., and Seo, J.K. (2008). *cagA*, *vacA*, and *iceA* genotypes of *Helicobacter pylori* in Korean children. Pediatr. Int. *50*, 628–631.

Koehler, C.I., Mues, M.B., Dienes, H.P., Kriegsmann, J., Schirmacher, P., and Odenthal, M. (2003). *Helicobacter pylori* genotyping in gastric adenocarcinoma and MALT lymphoma by multiplex PCR analyses of paraffin wax embedded tissues. Mol. Pathol. *56*, 36–42.

Kraft, C., Stack, A., Josenhans, C., Niehus, E., Dietrich, G., Correa, P., Fox, J.G., Falush, D., and Suerbaum, S. (2006). Genomic changes during chronic *Helicobacter pylori* infection. J. Bacteriol. *188*, 249–254.

Kulick, S., Moccia, C., Kraft, C., and Suerbaum, S. (2008). The *Helicobacter pylori mutY* homologue *Hp0142* is an antimutator gene that prevents specific C to A transversions. Arch. Microbiol. *189*, 263–270.

Kumar, S., Kumar, A., and Dixit, V.K. (2010). Diversity in the *cag* pathogenicity island of *Helicobacter pylori* isolates in populations from North and South India. J. Med. Microbiol. *59*, 32–40.

Kusters, J.G., van Vliet, A.H.M., and Kuipers, E.J. (2006). Pathogenesis of *Helicobacter pylori* infection. Clin. Microbiol. Rev. *19*, 449–490.

Kutter, S., Buhrdorf, R., Haas, J., Schneider-Brachert, W., Haas, R., and Fischer, W. (2008). Protein subassemblies of the *Helicobacter pylori* Cag type IV secretion system revealed by localization and interaction studies. J. Bacteriol. *190*, 2161–2171.

Kwok, T., Zabler, D., Urman, S., Rohde, M., Hartig, R., Wessler, S., Misselwitz, R., Berger, J., Sewald, N., Konig, W., *et al.* (2007). *Helicobacter* exploits integrin for type IV secretion and kinase activation. Nature *449*, 862–866.

Lam, K.H., Ling, T.K., and Au, S.W. (2010). Crystal structure of activated CheY1 from *Helicobacter pylori*. J. Bacteriol. *192*, 2324–2334.

Lamb, A., and Chen, L.-F. (2010). The many roads traveled by *Helicobacter pylori* to NFκB activation. Gut Microbes *1*, 109–113.

Leduc, D., Gallaud, J., Stingl, K., and de Reuse, H. (2010). Coupled amino acid deamidase-transport systems essential for *Helicobacter pylori* colonization. Infect. Immun. *78*, 2782–2792.

Lee, J.H., Choe, Y.H., Jeon, B.H., Oh, Y.J., Kim, J.J., Rhee, J.C., Kim, P.S., and Choi, M.S. (2004). Genotypes of the *Helicobacter pylori vacA* signal sequence differ with age in Korea. Helicobacter *9*, 54–58.

Lee, J.S., Choe, Y.H., Lee, J.H., Lee, H.J., Lee, J.H., and Choi, Y.O. (2010). *Helicobacter pylori* urease activity is influenced by ferric uptake regulator. Yonsei Med. J. *51*, 39–44.

Lehours, P., Ménard, A., Dupouy, S., Bergey, B., Richy, F., Zerbib, F., Ruskoné-Fourmestraux, A., Delchier, J.C., and Mégraud, F. (2004). Evaluation of the association of nine *Helicobacter pylori* virulence factors with strains involved in low-grade gastric mucosa-associated lymphoid tissue lymphoma. Infect. Immun. *72*, 880–888.

Lehours, P., Zheng, Z., Skoglund, A., Mégraud, F., and Engstrand, L. (2009). Is there a link between the lipopolysaccharide of *Helicobacter pylori* gastric MALT lymphoma associated strains and lymphoma pathogenesis? PLoS One *4*, e7297.

Letley, D.P., Rhead, J.L., Twells, R.J., Dove, B., and Atherton, J.C. (2003). Determinants of non-toxicity in the gastric pathogen *Helicobacter pylori*. J. Biol. Chem. *278*, 26734–26741.

Li, J., Ou, Z., Wang, F., Guo, Y., Zhang, R., Zhang, J., Li, P., Xu, W., and He, Y. (2009). Distinctiveness of the *cagA* genotype in children and adults with peptic symptoms in South China. Helicobacter *14*, 248–255.

Li, J., Go, A.C., Ward, M.J., and Ottemann, K.M. (2010). The chemical-in-plug bacterial chemotaxis assay is prone to false-positive responses. BMC Res. Notes. 3, 77.

Lima, V.P., Lima, M.A., Ferreira, M.V., Barros, M.A., and Rabenhorst, S.H. (2010). The relationship between *Helicobacter pylori* genes *cagE* and *virB11* and gastric cancer. Int. J. Infect. Dis. *14*, e613-e617.

Lin, H.J., Perng, C.L., Lo, W.C., Wu, C.W., Tseng, G.Y., Li, A.F., Sun, I.C., and Ou, Y.H. (2004). *Helicobacter pylori cagA, iceA* and *vacA* genotypes in patients with gastric cancer in Taiwan. World J. Gastroenterol. *10*, 2493–2497.

Lin, Y.F., Wu, M.S., Chang, C.C., Lin, S.W., Lin, J.T., Sun, Y.J., Chen, D.S., and Chow, L.P. (2006). Comparative immunoproteomics of identification and characterization of virulence factors from *Helicobacter pylori* related to gastric cancer. Mol. Cell Proteomics. *5*, 1484–1496.

Lin, Y.F., Chen, C.Y., Tsai, M.H., Wu, M.S., Wang, Y.C., Chuang, E.Y., Lin, J.T., Yang, P.C., and Chow, L.P. (2007). Duodenal ulcer-related antigens from *Helicobacter pylori*: immunoproteome and protein microarray approaches. Mol Cell Proteomics. *6*, 1018–1026.

Lin, C.S., He, P.J., Tsai, N.M., Li, C.H., Yang, S.C., Hsu, W.T., Wu, M.S., Wu, C.J., Cheng, T.L., and Liao, K.W. (2010a). A potential role for *Helicobacter pylori* heat shock protein 60 in gastric tumorigenesis. Biochem. Biophys. Res. Commun. *392*, 183–189.

Lin, W.C., Tsai, H.F., Kuo, S.H., Wu, M.S., Lin, C.W., Hsu, P.I., Cheng, A.L., and Hsu, P.N. (2010b). Translocation of *Helicobacter pylori* CagA into human B lymphocytes, the origin of mucosa-associated lymphoid tissue lymphoma. Cancer Res. *70*, 5740–5748.

Lindén, S.K., Arnqvist, A., Teneberg S., and Moran, A.P. (2010). *Helicobacter pylori* adhesion to the gastric surface. In *Helicobacter pylori* in the 21st Century, Sutton, P., and Mitchell, H., eds. (Cab International: Wallingford, UK), pp. 248–268.

Loh, J.T., Torres, V.J., and Cover, T.L. (2007). Regulation of *Helicobacter pylori cagA* expression in response to salt. Cancer Res. *67*, 4709–4715.

Loh, J.T., Torres, V.J., Algood, H.M., McClain, M.S., and Cover, T.L. (2008). *Helicobacter pylori* HopQ outer membrane protein attenuates bacterial adherence to gastric epithelial cells. FEMS Microbiol. Lett. *289*, 53–58.

Loh, J.T., Gupta, S.S., Friedman, D.B., Krezel, A.M., and Cover, T.L. (2010). Analysis of protein expression regulated by the *Helicobacter pylori* ArsRS two-component signal transduction system. J. Bacteriol. *192*, 2034–2043.

Loughlin, M.F., Barnard, F.M., Jenkins, D., Sharples, G.J., and Jenks, P.J. (2003). *Helicobacter pylori* mutants defective in RuvC Holliday junction resolvase display reduced macrophage survival and spontaneous clearance from the murine gastric mucosa. Infect. Immun. *71*, 2022–2031.

Lowenthal, A.C., Simon, C., Fair, A.S., Mehmood, K., Terry, K., Anastasia, S., and Ottemann, K.M. (2009). A fixed-time diffusion analysis method determines that the three *cheV* genes of *Helicobacter pylori* differentially affect motility. Microbiology *155*, 1181–1191.

Löwer, M., Weydig, C., Metzler, D., Reuter, A., Starzinski-Powitz, A., Wessler, S., and Schneider, G. (2008). Prediction of extracellular proteases of the human pathogen *Helicobacter pylori* reveals proteolytic activity of the Hp1018/19 protein HtrA. PLoS One 3, e3510.

Lu, H., Yamaoka, Y., and Graham, D.Y. (2005a). *Helicobacter pylori* virulence factors: facts and fantasies. Curr. Opin. Gastroenterol. *21*, 653–659.

Lu, H., Hsu, P.I., Graham, D.Y., and Yamaoka, Y. (2005b). Duodenal ulcer promoting gene of *Helicobacter pylori*. Gastroenterology *128*, 833–848.

Lu, H., Wu, J.Y., Beswick, E.J., Ohno, T., Odenbreit, S., Haas, R., Reyes, V.E., Kita, M., Graham, D.Y., and Yamaoka, Y. (2007). Functional and intracellular signaling differences associated with the *Helicobacter pylori* AlpAB adhesin from Western and East Asian strains. J. Biol. Chem. *282*, 6242–6254.

Lu, H., Murata-Kamiya, N., Saito, Y., and Hatakeyama, M. (2009). Role of partitioning-defective 1/microtubule affinity-regulating kinases in the morphogenetic activity of *Helicobacter pylori* CagA. J. Biol. Chem. *284*, 23024–23036.

Ma, Y.J., Duan, G.C., Zhang, R.G., Fan, Q.T., and Zhang, W.D. (2010). Mutation of *iceA* in *Helicobacter pylori* compromised IL-8 induction from human gastric epithelial cells. J. Basic Microbiol. *50 (Suppl. 1)*, S83–S88.

Magalhães, A., and Reis, C.A. (2010). *Helicobacter pylori* adhesion to gastric epithelial cells is mediated by glycan receptors. Braz. J. Med. Biol. Res. *43*, 611–618.

Maier, R.J., Benoit, S.L., and Seshadri, S. (2007). Nickel-binding and accessory proteins facilitating Ni-enzyme maturation in *Helicobacter pylori*. Biometals *20*, 655–664.

Mandell, L., Moran, A.P., Cocchiarella, A., Houghton, J., Taylor, N., Fox, J.G., Wang, T.C., and Kurt-Jones, E,A. (2004). Intact Gram-negative *Helicobacter pylori, Helicobacter felis,* and *Helicobacter hepaticus* bacteria activate innate immunity via toll-like receptor 2 but not toll-like receptor 4. Infect. Immun. *72*, 6446–6454.

Mashima, H., Suzuki, J., Hirayama, T., Yoshikumi, Y., Ohno, H., Ohnishi, H., Yasuda, H., Fujita, T., and Omata, M. (2008). Involvement of vesicle-associated membrane protein 7 in human gastric epithelial cell vacuolation induced by *Helicobacter pylori*-produced VacA. Infect. Immun. *76*, 2296–2303.

Matsumoto, Y., Marusawa, H., Kinoshita, K., Endo, Y., Kou, T., Morisawa, T., Azuma, T., Okazaki, I.M., Honjo, T., and Chiba, T. (2007). *Helicobacter pylori* infection triggers aberrant expression of activation-induced cytidine deaminase in gastric epithelium. Nat. Med. *13*, 470–476.

Mattar, R., dos Santos, A.F., Eisig, J.N., Rodrigues, T.N., Silva, F.M., Lupinacci, R.M., Iriya, K., and Carrilho, F.J. (2005). No correlation of *babA2* with *vacA* and *cagA* genotypes of *Helicobacter pylori* and grading of gastritis from peptic ulcer disease patients in Brazil. Helicobacter *10*, 601–608.

Mattar, R., Monteiro, M.S., Marques, S.B., Zilberstein, B., Hashimoto, C.L., and Carrilho, F.J. (2010). Association of *LEC* and *tnpA Helicobacter pylori* genes with gastric cancer in a Brazilian population Infect. Agent Cancer *5*, 1.

Matteo, M.J., Granados, G., Pérez, C.V., Olmos, M., Sanchez, C., and Catalano, M. (2007). *Helicobacter pylori cag* pathogenicity island genotype diversity within the gastric niche of a single host. J. Med. Microbiol. *56*, 664–669.

Matteo, M.J., Armitano, R.I., Granados, G., Wonaga, A.D., Sánches, C., Olmos, M., and Catalano, M. (2010). *Helicobacter pylori oipA, vacA* and *dupA* genetic diversity in individual hosts. J. Med. Microbiol. *59*, 89–95.

McClain, M.S., Shaffer, C.L., Israel, D.A., Peek, R.M., Jr., and Cover, T.L. (2009). Genome sequence analysis of *Helicobacter pylori* strains associated with gastric ulceration and gastric cancer. BMC Genomics *10*, 3.

Megraud, F. (2004). *H pylori* antibiotic resistance: prevalence, importance, and advances in testing. Gut *53*, 1374–1384.

Mégraud, F. (2008). *Helicobacter pylori*: bacteriological aspects, diagnostic methods, and susceptibility to antibiotics. Presse Med. *37*, 507–512.

Miciuleviciene, J., Calkauskas, H., Jonaitis, L., Kiudelis, G., Tamosiūnas, V., Praskevicius, A., Kupcinskas, L., and Berg, D. (2008). *Helicobacter pylori* genotypes in Lithuanian patients with chronic gastritis and duodenal ulcer. Medicina (Kaunas) *44*, 449–454.

Moran, A.P. (2008). Relevance of fucosylation and Lewis antigen expression in the bacterial gastroduodenal pathogen *Helicobacter pylori*. Carbohydr. Res. *343*, 1952–1965.

Nakasato, F., Shimoyama, T., Yoshimura, T., Mikami, T., Munakata, A., and Fukuda, S. (2008). Infection of *sabA*-positive *H. pylori* does not induce antiLewis X antibody in host. Hepatogastroenterology *55*, 1122–1125.

Ng, M.T., Van't Hof, R., Crockett, J.C., Hope, M.E., Berry, S., Thomson, J., McLean, M.H., McColl, K.E., El-Omar, E.M., and Hold, G.L. (2010). Increase in NF-{kappa}B binding affinity of the variant C allele of the toll-like receptor 9 - 1237T/C polymorphism is associated with *Helicobacter pylori*-induced gastric disease. Infect. Immun. *78*, 1345–1352.

Nguyen, L.T., Uchida, T., Murakami, K., Fujioka, T., and Moriyama, M. (2008). *Helicobacter pylori* virulence and the diversity of gastric cancer in Asia. J. Med. Microbiol. *57*, 1445–1453.

Nilsson, C., Sillen, A., Eriksson, L., Strand, M. L., Enroth, H., Normark, S., Falk, P., and Engstrand, L. (2003). Correlation between *cag* pathogenicity island composition and *Helicobacter pylori*-associated gastroduodenal disease. Infect. Immun. *71*, 6573–6581.

Nilsson, C., Skoglund, A., Moran, A.P., Annuk, H., Engstrand, L., and Normark, S. (2008). Lipopolysaccharide diversity evolving in *Helicobacter pylori* communities through genetic modifications in fucosyltransferases. PLoS One 3, e3811.

Nimri, L.F., Matalka, I., Bani Hani, K., and Ibrahim, M. (2006). *Helicobacter pylori* genotypes identified in gastric biopsy specimens from Jordanian patients. BMC Gastroenterol. *6*, 27.

O'Brien, D.P., Romero-Gallo, J., Schneider, B.G., Chaturvedi, R., Delgado, A., Harris, E.J., Krishna, U., Ogden, S.R., Israel, D.A., Wilson, K.T., *et al.* (2008). Regulation of the *Helicobacter pylori* cellular receptor decay-accelerating factor. J. Biol. Chem. *283*, 23922–23930.

Occhialini, A., Marais, A., Urdaci, M., Sierra, R., Muñoz, N., Covacci, A., and Mégraud, F. (2001). Composition and gene expression of the cag pathogenicity island in *Helicobacter pylori* strains isolated from gastric carcinoma and gastritis patients in Costa Rica. Infect. Immun. *69*, 1902–1908.

Odenbreit, S., Swoboda, K., Barwig, I., Ruhl, S., Borén, T., Koletzko, S., and Haas, R. (2009). Outer membrane protein expression profile in *Helicobacter pylori* clinical isolates. Infect. Immun. *77*, 3782–3790.

Ogiwara, H., Graham, D.Y., and Yamaoka, Y. (2008). *vacA* i-region subtyping. Gastroenterology *134*, 1267.

Ohno, T., Sugimoto, M., Nagashima, A., Ogiwara, H., Vilaichone, R.K., Mahachai, V., Graham, D.Y., and Yamaoka, Y. (2009). Relationship between *Helicobacter pylori hopQ* genotype and clinical outcome in Asian and Western populations. J. Gastroenterol. Hepatol. *24*, 462–468.

Olbermann, P., Josenhans, C., Moodley, Y., Uhr, M., Stamer, C., Vauterin, M., Suerbaum, S., Achtman, M., and Linz, B. (2010). A global overview of the genetic and functional diversity in the *Helicobacter pylori cag* pathogenicity island. PLoS Genet. *6*, e1001069.

Oldani, A., Cormont, M., Hofman, V., Chiozzi, V., Oregioni, O., Canonici, A., Sciullo, A., Sommi, P., Fabbri, A., Ricci, V., *et al.* (2009). *Helicobacter pylori* counteracts the apoptotic action of its VacA toxin by injecting the CagA protein into gastric epithelial cells. PLoS Pathog. *5*, e1000603.

Oleastro, M., Cordeiro, R., Ferrand, J., Nunes, B., Lehours, P., Carvalho-Oliveira, I., Mendes, A.I., Penque, D., Monteiro, L., Mégraud, F., *et al.* (2008). Evaluation of the clinical significance of *homB*, a novel candidate marker of *Helicobacter pylori* strains associated with peptic ulcer disease. J. Infect. Dis. *198*, 1379–1387.

Oleastro, M., Cordeiro, R., Ménard, A., Yamaoka, Y., Queiroz, D., Mégraud, F., and Monteiro, L. (2009). Allelic diversity and phylogeny of *homB*, a novel co-virulence marker *of Helicobacter pylori*. BMC Microbiol. *9*, 248.

Oleastro, M., Santos, A., Cordeiro, R., Nunes, B., Mégraud, F., and Ménard, A. (2010a). Clinical relevance and diversity of two homologous genes encoding glycosyltransferases in *Helicobacter pylori*. J. Clin. Microbiol. *48*, 2885–2891.

Oleastro, M., Cordeiro, R., Ménard, A., and Gomes, J.P. (2010b). Allelic diversity among *Helicobacter pylori* outer membrane protein genes *homB* and *homA* generated by recombination. J. Bacteriol. *192*, 3961–3968.

Olfat, F.O., Zheng, Q., Oleastro, M., Voland, P., Boren, T., Karttunen, R., Engstrand, L., Rad, R., Prinz, C., and Gerhard, M. (2005). Correlation of the *Helicobacter pylori* adherence factor BabA with duodenal ulcer disease in four European countries. FEMS Immunol. Med. Microbiol. *44*, 151–156.

Oyake, J., Otaka, M., Matsuhashi, T., Jin, M., Odashima, M., Komatsu, K., Wada, I., Horikawa, Y., Ohba, R., Hatakeyama, N., *et al.* (2006). Overexpression of 70-kDa heat shock protein confers protection against monochloramine-induced gastric mucosal cell injury. Life Sci. *79*, 300–305.

Pacheco, A.R., Proença-Módena, J.L., Sales, A.I.L., Fukuhara, Y., Da Silveira, W.D., Pimenta-Módena, J.L., De Oliveira, R.B., and Brocchi, M. (2008). Involvement of the *Helicobacter pylori* plasticity region and *cag* pathogenicity island genes in the development of gastroduodenal diseases. Eur. J. Clin. Microbiol. Infect. Dis. *27*, 1053–1059.

Paniagua, G.L., Monroy, E., Rodríguez, R., Arroniz, S., Rodríguez, C., Cortés, J.L., Camacho, A., Negrete, E., and Vaca, S. (2009). Frequency of *vacA, cagA* and *babA2* virulence markers in *Helicobacter pylori* strains isolated from Mexican patients with chronic gastritis. Ann. Clin. Microbiol. Antimicrob. *8*, 14.

Park, S.M., Park, J., Kim, J.G., and Yoo, B.C. (2001). Relevance of *vacA* genotypes of *Helicobacter pylori* to *cagA* status and its clinical outcome. Korean J. Intern. Med. *16*, 8–13.

Pattis, I., Weiss, E., Laugks, R., Haas, R., and Fischer, W. (2007). The *Helicobacter pylori* CagF protein is a type IV secretion chaperone-like molecule that binds close to the C-terminal secretion signal of the CagA effector protein. Microbiology *153*, 2896–2909.

Peek, R.M., Jr. (2008). Review: Prevention of gastric cancer: When is treatment of *Helicobacter pylori* warranted? Ther. Adv. Gastroenterol. *1*, 19–31.

Pflock, M., Kennard, S., Delany, I., Scarlato, V., and Beier, D. (2005). Acid-induced activation of the urease promoters is mediated directly by the ArsRS two-component system of *Helicobacter pylori*. Infect. Immun. *73*, 6437–6445.

Pflock, M., Finsterer, N., Joseph, B., Mollenkopf, H., Meyer, T.F., and Beier, D. (2006). Characterization of the ArsRS regulon of *Helicobacter pylori*, involved in acid adaptation. J. Bacteriol. *188*, 3449–3462.

Pinto-Santini, D.M., and Salama, N.R. (2009). Cag3 is a novel essential component of the *Helicobacter pylori* Cag type IV secretion system outer membrane subcomplex. J. Bacteriol. *191*, 7343–7352.

Podzorski, R.P., Podzorski, D.S., Wuerth, A., and Tolia, V. (2003). Analysis of the *vacA, cagA, cagE, iceA,* and *babA2* genes in *Helicobacter pylori* from sixty-one pediatric patients from the Midwestern United States. Diagn. Microbiol. Infect. Dis. *46*, 83–88.

Polenghi, A., Bossi, F., Fischetti, F., Durigutto, P., Cabrelle, A., Tamassia, N., Cassatella, M.A., Montecucco, C., Tedesco, F., and de Bernard, M. (2007). The neutrophil-activating protein of *Helicobacter pylori* crosses endothelia to promote neutrophil adhesion *in vivo*. J. Immunol. *178*, 1312–1320.

Proença Módena, J.L., Lopes Sales, A.I., Olszanski Acrani, G., Russo, R., Vilela Ribeiro, M.A., Fukuhara, Y., da Silveira, W.D., Módena, J.L., de Oliveira, R.B., and Brocchi, M. (2007). Association between *Helicobacter pylori* genotypes and gastric disorders in relation to the *cag* pathogenicity island. Diagn. Microbiol. Infect. Dis. *59*, 7–16.

Proença-Modena, J.L., Acrani, G.O., and Brocchi, M. (2009). *Helicobacter pylori*: phenotypes, genotypes and virulence genes. Future Microbiol. *4*, 223–240.

Quiroga, A.J., Cittelly, D.M., and Bravo, M.M. (2005). BabA2, *oipA* and *cagE Helicobacter pylori* genotypes in Colombian patients with gastroduodenal diseases. Biomedica 25, 325–234.

de Reuse, H., Heimesaat, M.M., and Bereswill, S. (2010). Helicobacteromics – genomics of a highly variable, well-adapted and persistent pathogen. In *Helicobacter pylori* in the 21st Century, Sutton, P., and Mitchell, H., eds. (CAB International: Wallingford, UK), pp. 269–291.

Rhead, J.L., Letley, D.P., Mohammadi, M., Hussein, N., Mohagheghi, M.A., Eshagh Hosseini, M., and Atherton, J.C. (2007). A new *Helicobacter pylori* vacuolating cytotoxin determinant, the intermediate region, is associated with gastric cancer. Gastroenterology *133*, 926–936.

Ribeiro, M.L., Godoy, A.P., Benvengo, Y.H., Mendonca, S., and Pedrazzoli, J., Jr. (2003). Clinical relevance of the *cagA, vacA* and *iceA* genotypes of *Helicobacter pylori* in Brazilian clinical isolates. FEMS Immunol. Med. Microbiol. *36*, 181–185.

Rizwan, M., Alvi, A., and Ahmed, N. (2008). Novel protein antigen (JHp940) from the genomic plasticity region of *Helicobacter pylori* induces tumor necrosis factor alpha and interleukin-8 secretion by human macrophages. J. Bacteriol. *190*, 1146–1151.

Robinson, K., Argent, R.H., and Atherton, J.C. (2007). The inflammatory and immune response to *Helicobacter pylori* infection. Best Pract. Res. Clin. Gastroenterol. *21*, 237–259.

Romero-Adrián, T.B., Leal-Montiel, J., Monsalve-Castillo, F., Mengual-Moreno, E., McGregor, e.g. Perini, L., and Antúnez, A. (2010). *Helicobacter pylori*: bacterial factors and the role of cytokines in the immune response. Curr. Microbiol. *60*, 143–155.

Romo-González, C., Salama, N.R., Burgeño-Ferreira, J., Ponce-Castañeda, V., Lazcano-Ponce, E., Camorlinga-Ponce, M., and Torres, J. (2009). Differences in genome content among *Helicobacter pylori* isolates from patients with gastritis, duodenal ulcer, or gastric cancer reveal novel disease-associated genes. Infect. Immun. *77*, 2201–2211.

Sachs, G., Kraut, J.A., Wen, Y., Feng, J., and Scott, D.R. (2006). Urea transport in bacteria: acid acclimation by gastric *Helicobacter* spp. J. Membr. Biol. *212*, 71–82.

Saha, A., Hammond, C.E., Beeson, C., Peek Jr., R.M., and Smolka, A.J. (2010). *Helicobacter pylori* represses proton pump expression and inhibits acid secretion in human gastric mucosa. Gut 59, 874–881.

Sakharkar, K.R., Sakharkar, M.K., and Chow, V.T. (2006). Gene fusion in *Helicobacter pylori*: making the ends meet. Antonie Van Leeuwenhoek *89*, 169–180.

Schmidt HM, Andres S, Kaakoush NO, Engstrand L, Eriksson L, Goh KL, Fock KM, Hilmi I, Dhamodaran S, Forman D, *et al.* (2009). The prevalence of the duodenal ulcer promoting gene (*dupA*) in *Helicobacter pylori* isolates varies by ethnic group and is not universally associated with disease development: a case-control study. Gut Pathog. *1*, 5.

Schmidt, H.M., Andres, S., Nilsson, C., Kovach, Z., Kaakoush, N.O., Engstrand, L., Goh, K.L., Fock, K.M., Forman, D., and Mitchell, H. (2010). The *cag* PAI is intact and functional but *Hp0521* varies significantly in *Helicobacter pylori* isolates from Malaysia and Singapore. Eur. J. Clin. Microbiol. Infect. Dis. *29*, 439–451.

Schneller, J., Gupta, R., Mustafa, J., Villanueva, R., Straus, E.W., and Raffaniello, R.D. (2006). *Helicobacter pylori* infection is associated with a high incidence of intestinal metaplasia in the gastric mucosa of patients at inner-city hospitals in New York. Dig. Dis. Sci. *51*, 1801–1809.

Scott, D.R., Marcus, E.A., Wen, Y., Oh, J., and Sachs, G. (2007). Gene expression *in vivo* shows that *Helicobacter pylori* colonizes an acidic niche on the gastric surface. Proc. Natl. Acad. Sci. U.S.A. *104*, 7235–7240.

Scott, D.R., Marcus, E.A., Wen, Y., Singh, S., Feng, J., and Sachs, G. (2010). Cytoplasmic histidine kinase (Hp0244)-regulated assembly of urease with UreI, a channel for urea and its metabolites, CO_2, NH_3, and NH_4^+, is necessary for acid survival of *Helicobacter pylori*. J. Bacteriol. *192*, 94–103.

Shak, J.R., Dick, J.J., Meinersmann, R.J., Perez-Perez, G.I., and Blaser, M.J. (2009). Repeat-associated plasticity in the *Helicobacter pylori* RD gene family. J. Bacteriol. *191*, 6900–6910.

Schauer, K., Muller, C., Carrière, M., Labigne, A., Cavazza, C., and De Reuse, H. (2010). The *Helicobacter pylori* GroES cochaperonin HspA functions as a specialized nickel chaperone and sequestration protein through its unique C-terminal extension. J. Bacteriol. *192*, 1231–1237.

Sheu, B.S., and Wu, J.J. (2008). Type 1 and 2 Lewis antigens of *Helicobacter pylori* – a potential marker of the human geographical distribution. J. Med. Microbiol. 57, 543–544.

Shiota, S., Matsunari, O., Watada, M., Hanada, K., and Yamaoka, Y. (2010). Systematic review and meta-analysis: the relationship between the *Helicobacter pylori dupA* gene and clinical outcomes. Gut Pathog. *2*, 13.

Shokrzadeh, L., Baghaei, K., Yamaoka, Y., Dabiri, H., Jafari, F., Sahebekhtiari, N., Tahami, A., Sugimoto, M., Zojaji, H., and Zali, M.R. (2010). Analysis of 3′-end variable region of the *cagA* gene in *Helicobacter pylori* isolated from Iranian population. J. Gastroenterol. Hepatol. *25*, 172–177.

Skoglund, A., Bäckhed, H.K., Nilsson, C., Björkholm, B., Normark, S., and Engstrand, L. (2009). A changing gastric environment leads to adaptation of lipopolysaccharide variants in *Helicobacter pylori* populations during colonization. PLoS One 4, e5885.

Slomiany, B.L., and Slomiany, A. (2010). Role of constitutive nitric oxide synthase S-nitrosylation in *Helicobacter pylori*-induced gastric mucosal cell apoptosis: effect of ghrelin. Inflammopharmacology *18*, 233–240.

Smith, S.I., Kirsch, C., Oyedeji, K.S., Arigbabu, A.O., Coker, A.O., Bayerdöffer, E., and Miehlke, S. (2002). Prevalence of *Helicobacter pylori vacA, cagA* and *iceA* genotypes in Nigerian patients with duodenal ulcer disease. J. Med. Microbiol. *51*, 851–854.

Sokic-Milutinovic, A., Wex, T., Todorovic, V., Milosavljevic, T., and Malfertheiner P. (2004). AntiCagA and antiVacA antibodies in *Helicobacter pylori*-infected patients with and without peptic ulcer disease in Serbia and Montenegro. Scand J. Gastroenterol. *39*, 222–226.

Stingl, K., Schauer, K., Ecobichon, C., Labigne, A., Lenormand, P., Rousselle, J.C., Namane, A., and de

Reuse, H. (2008). *In vivo* interactome of *Helicobacter pylori* urease revealed by tandem affinity purification. Mol. Cell Proteomics 7, 2429–2441.

Stingl, K., Müller, S., Scheidgen-Kleyboldt, G., Clausen, M., and Maier, B. (2010). Composite system mediates two-step DNA uptake into *Helicobacter pylori*. Proc. Natl. Acad. Sci. U.S.A. *107*, 1184–1189.

Su, B., Ceponis, P.J., and Sherman, P.M. (2003). Cytoskeletal rearrangements in gastric epithelial cells in response to *Helicobacter pylori* infection. J. Med. Microbiol. *52*, 861–867.

Suganuma, M., Yamaguchi, K., Ono, Y., Matsumoto, H., Hayashi, T., Ogawa, T., Imai, K., Kuzuhara, T., Nishizono, A., and Fujiki. H. (2008). TNF-alpha-inducing protein, a carcinogenic factor secreted from *H. pylori*, enters gastric cancer cells. Int. J. Cancer *123*, 117–122.

Sugimoto, M., Zali, M.R., and Yamaoka, Y. (2009). The association of vacA genotypes and *Helicobacter pylori*-related gastroduodenal diseases in the Middle East. Eur. J. Clin. Microbiol. Infect. Dis. *28*, 1227–1236.

Sycuro, L.K., Pincus, Z., Gutierrez, K.D., Biboy, J., Stern, C.A., Vollmer, W., and Salama, N.R. (2010). Peptidoglycan crosslinking relaxation promotes *Helicobacter pylori*'s helical shape and stomach colonization. Cell *141*, 822–833.

Tabassam, F.H., Graham, D.Y., and Yamaoka, Y. (2008). OipA plays a role in *Helicobacter pylori*-induced focal adhesion kinase activation and cytoskeletal re-organization. Cell Microbiol. *10*, 1008–1020.

Takeshima, E., Tomimori, K., Takamatsu, R., Ishikawa, C., Kinjo, F., Hirayama, T., Fujita, J., and Mori, N. (2009). *Helicobacter pylori* VacA activates NF-kappaB in T cells via the classical but not alternative pathway. Helicobacter *14*, 271–279.

Talarico, S., Gold, B.D., Fero, J., Thompson, D.T., Guarner, J., Czinn, S., and Salama, N.R. (2009). Pediatric *Helicobacter pylori* isolates display distinct gene coding capacities and virulence gene marker profiles. J. Clin. Microbiol. *47*, 1680–1688.

Tan, H.J., Rizal, A.M., Rosmadi, M.Y., and Goh, K.L. (2005). Distribution of *Helicobacter pylori cagA*, *cagE* and *vacA* in different ethnic groups in Kuala Lumpur, Malaysia. J. Gastroenterol. Hepatol. *20*, 589–594.

Tanaka, H., Yoshida, M., and Azuma, T. (2009a). The role of CagA in *H. pylori* infection. Nippon Rinsho *67*, 2245–2249.

Tanaka, A., Kamada, T., Yokota, K., Shiotani, A., Hata, J., Oguma, K., and Haruma, K. (2009b). *Helicobacter pylori* heat shock protein 60 antibodies are associated with gastric cancer. Pathol. Res. Pract. *205*, 690–694.

Tanaka, H., Yoshida, M., Nishiumi, S., Ohnishi, N., Kobayashi, K., Yamamoto, K., Fujita, T., Hatakeyama, M., and Azuma, T. (2010). The CagA protein of *Helicobacter pylori* suppresses the functions of dendritic cell in mice. Arch. Biochem. Biophys. *498*, 35–42.

Taneike, I., Nami, A., O'Connor, A., Fitzgerald, N., Murphy, P., Qasim, A., O'Connor, H., and O'Morain, C. (2009). Analysis of drug resistance and virulence-factor genotype of Irish *Helicobacter pylori* strains: is there any relationship between resistance to metronidazole and cagA status? Aliment. Pharmacol. Ther. *30*, 784–790.

Tanih, N.F., McMillan, M., Naidoo, N., Ndip, L.M., Weaver, L.T., and Ndip, R.N. (2010). Prevalence of *Helicobacter pylori vacA, cagA* and *iceA* genotypes in South African patients with upper gastrointestinal diseases. Acta Trop. *116*, 68–73.

Tegtmeyer, N., Hartig, R., Delahay, R.M., Rohde, M., Brandt, S., Conradi, J., Takahashi, S., Smolka, A.J., Sewald, N., and Backert, S. (2010). A small fibronectin-mimicking protein from bacteria induces cell spreading and focal adhesion formation. J. Biol. Chem. *285*, 23515–23526.

Terry, K., Williams, S.M., Connolly, L., and Ottemann, K.M. (2005). Chemotaxis plays multiple roles during *Helicobacter pylori* animal infection. Infect. Immun. *73*, 803–811.

Thiberge, J.M., Boursaux-Eude, C., Lehours, P., Dillies, M.A., Creno, S., Coppée, J.Y., Rouy, Z., Lajus, A., Ma, L., Burucoa, C., *et al.* (2010). From array-based hybridization of *Helicobacter pylori* isolates to the complete genome sequence of an isolate associated with MALT lymphoma. BMC Genomics *11*, 368.

Torres, J., and Backert, S. (2008). Pathogenesis of *Helicobacter pylori* infection. Helicobacter *13 (Suppl 1)*, 13–17.

Torres, L.E., Melian, K., Moreno, A., Alonso, J., Sabatier, C.A., Hernandez, M., Bermudez, L., and Rodriguez, B.L. (2009). Prevalence of *vacA, cagA* and *babA2* genes in Cuban *Helicobacter pylori* isolates. World J. Gastroenterol. *15*, 204–210.

Tosi, T., Cioci, G., Jouravleva, K., Dian, C., and Terradot, L. (2009). Structures of the tumor necrosis factor alpha inducing protein Tipalpha: a novel virulence factor from *Helicobacter pylori*. FEBS Lett. *583*, 1581–1585.

Tsang, Y.H., Lamb, A., Romero-Gallo, J., Huang, B., Ito, K., Peek, R.M., Jr., Ito, Y., and Chen, L.F. (2010). *Helicobacter pylori* CagA targets gastric tumor suppressor RUNX3 for proteasome-mediated degradation. Oncogene *29*, 5643–5650.

Uchida, T., Nguyen, L.T., Takayama, A., Okimoto, T., Kodama, M., Murakami, K., Matsuhisa, T., Trinh, T.D., Ta, L., Ho, D.Q., *et al.* (2009). Analysis of virulence factors of *Helicobacter pylori* isolated from a Vietnamese population. BMC Microbiol. *9*, 175.

Umeda, M., Murata-Kamiya, N., Saito, Y., Ohba, Y., Takahashi, M., and Hatakeyama, M. (2009). *Helicobacter pylori* CagA causes mitotic impairment and induces chromosomal instability. J. Biol. Chem. *284*, 22166–22172.

Umehara, S., Higashi, H., Ohnishi, N., Asaka, M., and Hatakeyama, M. (2003). Effects of *Helicobacter pylori* CagA protein on the growth and survival of B lymphocytes, the origin of MALT lymphoma. Oncogene *22*, 8337–8342.

Umit, H., Tezel, A., Bukavaz, S., Unsal, G., Otkun, M., Soylu, A.R., Tucer, D., Otkun, M., and Bilgi, S. (2009). The relationship between virulence factors of *Helicobacter pylori* and severity of gastritis in infected patients. Dig. Dis. Sci. *54*, 103–110.

Unemo, M., Aspholm-Hurtig, M., Ilver, D., Bergström, J., Borén, T., Danielsson, D., and Teneberg, S. (2005). The sialic acid binding SabA adhesin of *Helicobacter pylori* is essential for nonopsonic activation of human neutrophils. J. Biol. Chem. *280*, 15390–15397.

Vale, F.F., Mégraud, F., and Vítor, J.M. (2009). Geographic distribution of methyltransferases of *Helicobacter pylori*: evidence of human host population isolation and migration. BMC Microbiol. *9*, 193.

Vega, A.E., Cortiñas, T.I., Puig, O.N., and Silva, H.J. (2010). Molecular characterization and susceptibility testing of *Helicobacter pylori* strains isolated in western Argentina. Int. J. Infect. Dis. *14 (Suppl 3)*, e85–e92.

Vo, Q.N., Geradts, J., Boudreau, D.A., Bravo, J.C., and Schneider, B.G. (2002). CDKN2A promoter methylation in gastric adenocarcinomas: clinical variables. Hum. Pathol. 33, 1200–1204.

Wang, J., van Doorn, L.J., Robinson, P.A., Ji, X., Wang, D., Wang, Y., Ge, L., Telford, J.L., and Crabtree, J.E. (2003). Regional variation among *vacA* alleles of *Helicobacter pylori* in China. J.Clin. Microbiol. *41*, 1942–1945.

Wang, G., Alamuri, P., and Maier, R.J. (2006). The diverse antioxidant systems of *Helicobacter pylori*. Mol. Microbiol. *61*, 847–860.

Wang, C.A., Liu, Y.C., Du, S.Y., Lin, C.W., and Fu, H.W. (2008). *Helicobacter pylori* neutrophil-activating protein promotes myeloperoxidase release from human neutrophils. Biochem. Biophys. Res. Commun. *377*, 52–56.

Wang, G., Maier, S.E., Lo, L.F., Maier, G., Dosi, S., and Maier, R.J. (2010). Peptidoglycan deacetylation in *Helicobacter pylori* contributes to bacterial survival by mitigating host immune responses. Infect. Immun. *78*, 4660–4666.

Watanabe, T., Hirano, K., Takahashi, A., Yamaguchi, K., Beppu, M., Fujiki, H., and Suganuma, M. (2010). Nucleolin on the cell surface as a new molecular target for gastric cancer treatment. Biol Pharm Bull. *33*, 796–803.

Wen, Y., Marcus, E.A., Matrubutham, U., Gleeson, M.A., Scott, D.R., and Sachs, G. (2003). Acid-adaptive genes of *Helicobacter pylori*. Infect. Immun. *71*, 5921–5939.

Wen, S., Velin, D., Felley, C.P., Du, L., Michetti, P., and Pan-Hammarström, Q. (2007). Expression of *Helicobacter pylori* virulence factors and associated expression profiles of inflammatory genes in the human gastric mucosa. Infect. Immun. *75*, 5118–5126.

Wen, S., and Moss, S.F. (2009). *Helicobacter pylori* virulence factors in gastric carcinogenesis. Cancer Lett. *282*, 1–8.

Wen, Y., Feng, J., Scott, D.R., Marcus, E.A., and Sachs, G. (2009). The pH-responsive regulon of Hp0244 (FlgS), the cytoplasmic histidine kinase of *Helicobacter pylori*. J. Bacteriol. *191*, 449–460.

Wong, B.C., Lam, S.K., Wong, W.M., Chen, J.S., Zheng, T.T., Feng, R.E., Lai, K.C., Hu, W.H., Yuen, S.T., Leung, S.Y., *et al.* (2004). *Helicobacter pylori* eradication to prevent gastric cancer in a high-risk region of China: a randomized controlled trial. JAMA *291*, 187–194.

Yakoob, J., Abid, S., Abbas, Z., Jafri, W., Ahmad, Z., Ahmed, R., and Islam, M. (2009). Distribution of *Helicobacter pylori* virulence markers in patients with gastroduodenal diseases in Pakistan. BMC Gastroenterol. *9*, 87.

Yamaoka, Y., Ojo, O., Fujimoto, S., Odenbreit, S., Haas, R., Gutierrez, O., El-Zimaity, H.M., Reddy, R., Arnqvist, A., and Graham D.Y. (2006). *Helicobacter pylori* outer membrane proteins and gastroduodenal disease. Gut *55*, 775–781.

Yamaoka, Y. (2008a). Roles of the plasticity regions of *Helicobacter pylori* in gastroduodenal pathogenesis. J. Med. Microbiol. *57*, 545–553.

Yamaoka, Y. (2008b). Roles of *Helicobacter pylori* BabA in gastroduodenal pathogenesis. World J. Gastroenterol. *14*, 4265–4272.

Yamaoka, Y. (2008c). Increasing evidence of the role of *Helicobacter pylori* SabA in the pathogenesis of gastroduodenal disease. J. Infect. Dev. Countries *2*, 174–181.

Yamaoka, Y., Reddy, R., and Graham, D.Y. (2010). *Helicobacter pylori* virulence factor genotypes in children in the United States: clues about genotype and outcome relationships. J. Clin. Microbiol. *48*, 2550–2551.

Yamauchi, K., Choi, I.J., Lu, H., Ogiwara, H., Graham, D.Y., and Yamaoka, Y. (2008). Regulation of IL-18 in *Helicobacter pylori* infection. J. Immunol. *180*, 1207–1216.

Yamazaki, S., Yamakawa, A., Okuda, T., Ohtani, M., Suto, H., Ito, Y., Yamazaki, Y., Keida, Y., Higashi, H., Hatakeyama, M., *et al.* (2005). Distinct diversity of *vacA, cagA*, and *cagE* genes of *Helicobacter pylori* associated with peptic ulcer in Japan. J. Clin. Microbiol. *43*, 3906–3916.

Yokota, S., Okabayashi, T., Rehli, M., Fujii, N., and Amano, K. (2010). *Helicobacter pylori* lipopolysaccharides upregulate toll-like receptor 4 expression and proliferation of gastric epithelial cells via the MEK1/2-ERK1/2 mitogen-activated protein kinase pathway. Infect. Immun. *78*, 468–476.

Zambon, C.F., Navaglia, F., Basso, D., Rugge, M., and Plebani, M. (2003). *Helicobacter pylori babA2, cagA*, and s1 *vacA* genes work synergistically in causing intestinal metaplasia. J. Clin. Pathol. *56*, 287–291.

Zhang, M.J., He, L.H., Wong, B.C., Zhou, Z.F., and Zhang, J.Z. (2006). Distribution of *cagA* 3′ region, *iceA, vacA* and Hp0519 on *Helicobacter pylori* isolated from China. Zhonghua Liu Xing Bing Xue Za Zhi *27*, 508–512.

Zhang, Z., Zheng, Q., Chen, X., Xiao, S., Liu, W., and Lu, H. (2008). The *Helicobacter pylori* duodenal ulcer promoting gene, *dupA* in China. BMC Gastroenterol. *8*, 49.

Zhao, Y., Yokota, K., Ayada, K., Yamamoto, Y., Okada, T., Shen, L., and Oguma, K. (2007). *Helicobacter pylori* heat-shock protein 60 induces interleukin-8 via a Toll-like receptor (TLR)2 and mitogen-activated protein (MAP) kinase pathway in human monocytes. J. Med. Microbiol. *56*, 154–164.

Zhong, Q., Shao, S.H., Cui, L.L., Mu, R.H., Ju, X.L., and Dong, S.R. (2007). Type IV secretion system in *Helicobacter pylori*: a new insight into pathogenicity. Chin. Med. J. (Engl). *120*, 2138–2142.

Immunology of *Helicobacter pylori* Infection

5

Ivan Mitov

Abstract

Helicobacter pylori infection induces almost all mechanisms of innate and acquired immunity. Different bacterial, environmental and host factors may influence the balance between the protective role of the immune mechanisms and their role in gastric mucosal damage, respectively, the possibility of lifelong asymptomatic colonization of gastric mucosa or clinical manifestation of *H. pylori* infection. Bacterial virulence factors stimulate Toll-like and Nod-like receptors to induce innate and adaptive cell-mediated and humoral immune response. Balance of Th1/Th2 response is of great importance in host protection and in pathogenesis of *H. pylori*-mediated diseases. The polarized Th1 response is not sufficient to clear the bacteria. Moreover, a predominant activation of Th1 cells plays a key role in tissue damage. Th2 response appears to be protective against gastric inflammation. Cytotoxic activities of T cells are important for the outcome of *H. pylori* infection. Protection as a result of anti-*H. pylori*, local and systemic immune response is minimal. Furthermore, the antibodies may promote colonization of gastric mucosa.

Immune response to *H. pylori* infection

H. pylori colonization of the gastric mucosa is long term and often life-long despite the presence of innate and acquired humoral and cell-mediated immune response. *H. pylori* infection leads to development of a complex inflammatory and immune response in the gastric mucosa with participation of neutrophils, macrophages, plasma cells, eosinophils, and lymphocytes, including the major cells, $CD4^+$ and $CD8^+$ T cells (Algood and Cover, 2006; O'Keeffe and Moran, 2008).

The first line of host defence is the innate immune response. After entering the stomach, by means of its motility, *H. pylori* penetrates the mucus layer and reaches the gastric epithelial cells where the acidity is considerably less (pH 5–6) in comparison to the luminal pH of 1–2. Multiple antibacterial proteins are produced by the gastric mucosal cells, including α- and β-defensins and cathelicidin LL-37 (Hase *et al.*, 2003; Isomoto *et al.*, 2004, Joly *et al.*, 2005; Ouellette, 2005; Leszczyńska *et al.*, 2009). Lactoferrin in mucosal secretion suppresses bacterial growth by sequestering extracellular iron that does not support the growth of iron-requiring bacteria (Dial *et al.*, 1998) and through direct interaction with the bacterial cell surface (Miehlke *et al.*, 1996). O-glycans in gastric mucin with terminal α-1,4-linked *N*-acetylglucosamine have antimicrobial activity against *H. pylori*, inhibiting its biosynthesis of cholesteryl-α-D-glucopyranoside, a major cell wall component (Kawakubo *et al.*, 2004; Kobayashi *et al.*, 2009). Surfactant protein D, a component of innate immunity, expressed also in the human gastric mucosa during *H. pylori* infection, can cause *H. pylori* agglutination and inhibition of its motility (Khamri *et al.*, 2005; Moran *et al.*, 2005).

H. pylori infection triggers innate protective antibacterial mechanisms and stimulation of proinflammatory factors. Strong innate immune response is induced by several mechanisms that involve bacterial pathogen-associated molecular patterns (PAMPs: lipopolysaccharide- LPS, flagellin, peptidoglycan), various components of the innate immune system, including nucleotide-binding oligomerization domain protein I (Nod1), Toll-like receptors (TLRs) and secretion of antimicrobial peptides. *In vitro* PAMPs of *H. pylori* can be recognized by TLRs, expressed on antigen-presenting cells and gastric epithelial cells or by Nod1 receptor, expressed on epithelial cells, and this recognition can start the innate and acquired immune responses *in vivo* (Algood and Cover, 2006; Bergman and D'Elios, 2010). TLRs can be found on the surface of the gastric cells and are associated with the plasma membrane or lysosomal and/or endosomal vesicles, while Nod1 is expressed in the cytosol (Szczepanik, 2006).

Peptidoglycan fragments, delivered to the gastric epithelial cells by *H. pylori cag* pathogenicity island (*cag*PAI) are recognized by the cytosolic innate immune molecule, Nod1 (Grubman *et al.*, 2010). Moreover, Nod1 and human beta-defensin 2 (hBD-2) can kill directly the *H. pylori* bacteria by epithelial cells, which contribute to the host defence against *cag*PAI-positive bacteria (Grubman *et al.*, 2010). *H. pylori* induces proinflammatory gene expression in host cells via TLRs. TLRs are family of transmembrane proteins, which recognize conserved microbial products, such as LPS, recognized by TLR4, peptidoglycan (by TLR2), flagellin (by TLR5) and non-methylated bacterial DNA, to alert the immune system to the presence of pathogenic organisms (Smith *et al.*, 2006). Activation of TLRs by intact bacteria can lead to activation and translocation of nuclear factor kappa B (NF-κB) to the nucleus and the transcription of inflammatory cytokine genes. The major TLRs for whole *Helicobacter* bacteria are TLR2 and TLR5, although the *Helicobacter* LPS is recognized by TLR4 (Mandell *et al.*, 2004). *H. pylori* LPS and heat shock protein (Hsp) 60 are recognized by TLR2 instead of TLR4, moreover, *H. pylori* flagellin is not recognized by TLR5 (Lee *et al.*, 2003; Smith *et al.*, 2003). *In vitro* experiments with HEK293 cells, however, show that *H. pylori* LPS may be recognized by TLR2 instead of TLR4 (Smith *et al.*, 2003). Generally, to avoid activation of the innate immune response and inflammation, recognition of the main PAMPs, LPS and flagellin by TLRs is weakened (Andersen-Nissen *et al.*, 2005).

The innate immune response to *H. pylori* promotes development of gastritis and induces an influx of neutrophils and other immune cell by secretion of chemokines and cytokines. Most (99%) of *H. pylori* cells does not invade intact gastric mucosa and the contact between bacteria and phagocytic cells is rare (see Chapter 4). Chronic active gastritis usually exhibits neutrophils, located in lamina propria, epithelial layer and foveolar lumen, and also T and B cells and macrophages, recruiting in the lamina propria (Schmausser *et al.*, 2004). Infiltration of neutrophils is observed during initial colonization of the stomach by *H. pylori* as well as during persistent infection, especially in adults, while lymphocytic response with a small amount of neutrophils is observed mainly in children (Whitney *et al.*, 2000). Both neutrophil activation and migration are regulated by the neutrophil attractant chemokines, interleukin 8 (IL-8) and growth-related oncogene α (GROα), which release is induced by *H. pylori* (Schmausser *et al.*, 2004). Whole *H. pylori* bacteria and especially the vacuolating cytotoxin A (VacA) and the neutrophil activating protein of *H. pylori* (Hp-NAP), are able to activate the mast cells and to contribute to increased gastric inflammation (Algood and Cover, 2006). Hp-NAP acts as a TLR2 agonist, which can induce the expression of IL-12 and IL-23 by neutrophils and monocytes and can promote *H. pylori*-specific adaptive Th1 response in the gastric mucosa of *H. pylori*-infected patients (Amedei *et al.*, 2006). IL-12 is known to participate in the differentiation of naive Th cells into the Th1 phenotype and also acts as a powerful pro-inflammatory cytokine. The VacA, *cag* pathogenicity island (*cag*PAI), and the heat shock protein Hsp90 can upregulate the expression of IL-12 and the Th1 polarized immune response (D'Elios and Andersen, 2009). On the other hand, *H. pylori* can disturb the function of many immune cells *in vitro*. *H. pylori* can downregulate chemokine receptors CXCR1 and CXCR2 expression on neutrophils *in vitro* and can possibly inhibit the neutrophil migration *in vivo* (Schmausser *et al.*, 2004). In brief, some virulence factors (Hp-NAP, SabA) of *H. pylori* can activate the neutrophils, however, it can impede their proper functions. Virulent strains are able to escape phagocyte killing by delaying phagosome formation (Allen, 2007). However, Taiwanese clinical isolates have been able to transiently survive and replicate in human macrophage cell line THp-1 cells (Wang *et al.*, 2010). Therefore, *H. pylori* can behave as a facultative intracellular organism under some conditions (Dubois and Borén, 2007). *H. pylori* could multiply also in epithelial cells and in dendritic cells, as recently reported (Wang *et al.*, 2010).

In *H. pylori*-infected gastric mucosa, the levels of many cytokines such as interferon γ (IFN-γ), tumour necrosis factor-α (TNF-α) and interleukins (IL-1β, IL-6, IL-7, IL-8, IL-10 and IL-18) have been usually elevated (Lindholm *et al.*, 1998). Most of this cytokines possess proinflammatory activities, while IL-10 has immunoregulatory activity that may limit the inflammatory Th1 immune response (Algood and Cover, 2006). IL-10 has been found to inhibit *H. pylori*-induced IL-8 expression and activation of NF-kB, and to enhance *H. pylori* growth in a bacterial-cell co-culture model (Robinson *et al.*, 2008). There is still some controversy about the role of the suppressive cytokine IL-10 (Lundin *et al.*, 2007; Peek *et al.*, 2008). Another potent anti-inflammatory and Th2-type immunoregulatory cytokine, IL-4, has not been elevated in the gastric mucosa. It is not a major contributor to host resistance to *H. pylori* and enhanced IL-4 production has little if any effect on gastric colonization by this organism (Chen *et al.*, 1999).

H. pylori induces an intense humoral and cell mediated immune response, which can help for asymptomatic course of infection but can contribute to tissue damage as well. Bacteria interact with dendritic, T and B cells for development of acquired immune response. The dendritic cells can phagocytize and process bacteria to present antigenic peptides by major histocompatibility complex (MHC) class II molecules to specific T cells. Dysfunction of dendritic cells maturation could result in chronic gastric inflammation (Wang *et al.*, 2010). *H. pylori* stimulates dendritic cell expression of various cytokines such as IL-6, IL-8, IL-10 and IL-12. However, the dendritic cells incubated with Le antigen-positive strains have been found to express more IL-10, which downregulates inflammatory responses, compared with the dendritic cells incubated with Le antigen-negative strains (Bergman *et al.*, 2004). VacA can reduce the Ii-dependent pathway of antigen presentation, mediated by newly synthesized MHC class II molecules (Molinari *et al.*, 1998). In addition, the cytotoxin-associated protein (CagA) expression in the B cells has been shown to decrease IL-3-dependent B-cell proliferation by inhibiting JAK-STAT signalling (Umehara *et al.*, 2003). These events can lead to reduced antibody production and inhibited cytokine expression.

Cell-mediated immunity and balance of Th1/Th2 response are of great importance in host protection and in pathogenesis of *H. pylori*-mediated diseases. T-helper cells are activated by peptide fragments presented on antigen-presenting cells (O'Keeffe and Moran, 2008). T helper 1 cells produce IL-2, TNF-α and IFN-γ, participate in cell-mediated immune response, promote production of IgG2a, lyse antigen-loaded target cells by perforins or FAS (CD95) mediated mechanisms and lead to macrophage activation. Th2 cells release IL-4, IL-5 and IL-10, participate in downregulation of Th1 cell-mediated inflammation and facilitate IgG1, IgE and IgA (Wang *et al.*, 2000; Xie *et al.*, 2007; Bergman and D'Elios, 2010).

A number of studies report that bacterial load is inversely proportional to the degree of inflammation, which suggests the main role of cell-mediated Th1 proinflammatory response in bacterial eradication and *Helicobacter* immunity (Garhart *et al.*, 2002; Taylor *et al.*, 2008). Evidence for this has been obtained in immunodeficient mice: animals deficient in antibody production (Ermak *et al.*, 1998; Akhiani *et al.* 2004) or $CD4^+$ T cells cannot be protected by *H. pylori* immunization, whereas deficient $CD8^+$ T cells ones can be successfully immunized and protected against colonization (Pappo *et al.*, 1999).

Predominant Th1 response, showing high levels of expression of proinflammatory cytokines IFN-γ and IL-2 and lower levels of expression of IL-4 and IL-10, has been observed, however, in *H. pylori*-infected gastric mucosa from adults with gastritis and peptic ulcers (Lopes *et al.*, 2005). During *H. pylori*-mediated gastritis, there is a considerable amount of $CD4^+$ Th1 cells producing IFN-γ, but not of Th2 cells. In the presence of IFN-γ, macrophages are activated to secrete proinflammatory cytokines, including TNF-α and IL-12. The number of IFN-γ secreting cells in gastric mucosa correlates with the severity of gastric damage (Lehmann *et al.*, 2002). Since IFN-γ may induce binding of *H. pylori* to gastric epithelial cells followed by apoptosis of these cells, one may speculate that *H. pylori*-mediated diseases are in part autoimmune diseases initiated by *H. pylori*-specific Th1 cells infiltrating the gastric mucosa (Lohoff *et al.*, 2000). Although IFN-γ may play a protective role in *H. pylori* infection, in IFN-γ-deficient mice, *H. pylori* has displayed strongly reduced gastritis in the absence of IFN-γ (Sawai *et al.*, 1999). In children, the cytokine release is lower than in adults, and without clear Th1 dominance, suggesting downregulated gastric immune response (Lopes *et al.*, 2005).

Murine *H. felis* infection model resembles *H. pylori* infection in humans. Infection of B- and T-cell-deficient mice does not result in gastric pathology despite of heavy mucosal colonization, whereas infection of B-cell-deficient mice results in severe gastric pathology, identical to that in immunocompetent mice. The results indicate that T-cell but not B-cell response is a critical mediator of *Helicobacter*-induced gastric pathology (Roth *et al.*, 1999).

Peptic ulcer disease is associated with Th1 polarization of gastric T-cell responses, whereas uncomplicated chronic gastritis is associated with a mixed Th1/Th2 response. About half of the *H. pylori*-reactive Th cell clones from peptic ulcer patients have been specific for CagA, whereas about one-quarter of *H. pylori*-reactive Th clones from non-ulcer gastritis patients have been specific for *H. pylori* urease. Importantly, by *H. pylori* antigen-specific stimulation, >80% of T helper cells from ulcer patients have demonstrated a polarized Th1 profile, whereas about 2/3 of T helper cells from non-ulcer patients have a mixed Th1/Th2 cytokine profile with only one-third polarized to Th1 (D'Elios *et al.*, 1997). Interestingly, the high prevalence of parasitic infections, promoting predominant Th2 responses, which could reduce Th1-mediated gastric pathology and development of gastric cancer, can explain the infrequent peptic ulcers and gastric cancer in African countries (Fox *et al.*, 2000).

H. pylori can inhibit the T cells by several mechanisms. VacA decreases T-cell activity, reducing the release of IL-2 and inhibiting the T-cell proliferation, and both *H. pylori* arginase and a low-molecular-weight *H. pylori* factor inhibit the T-cell proliferation (Algood and Cover, 2006). The inhibition of the T-cell proliferation is important for the persistence of the infection and can lead to uncontrolled T-cell-dependent B-cell proliferation and risk for mucosa-associated lymphoid tissue (MALT) lymphoma in some patients (D'Elios *et al.*, 2005; Ferrand *et al.*, 2008).

Regulatory immune cells, mostly regulatory $Foxp^{3+}CD4^{+}CD25^{high}$ T cells (T_{reg} cells) are important regulators of the adaptive immune response and contribute to *H. pylori*-associated inflammation and persistence (Kandulski *et al.*, 2010). These T cells control and prevent the gastric mucosa from pathological changes during *H. pylori* infection but also lead to the chronicity of the infection, inhibiting the protective immune response. T_{reg} cells can inhibit the T-cell responses by direct cell–cell contact or by the secretion of cytokines such as the transforming growth factor (TGF)-β1 and IL-10 (D'Elios and Andersen, 2009). IL-10-producing T cells contribute to bacterial persistence in the gastric mucosa. Peptic ulcer patients have exhibited 2.4 times lower $CD4^{+}CD25^{high}IL10^{+}T_{reg}$ response but increased Th1 and Th2 responses compared with non-ulcer patients. It could be assumed that *H. pylori*-induced T_{reg} response leads to long-term coexistence of bacteria with the human host, whereas when this response is abnormal, it can contribute to development of peptic ulcers (Robinson *et al.*, 2008). In children, *H. pylori*-induced gastric pathology has been often limited and local T_{reg} cell responses have been elevated compared with those in adults, which implies that the gastric T_{reg} cell response reduces the inflammation and ulceration in children (Harris *et al.*, 2008).

Outcomes of infection depend on host immune response against *H. pylori*. Whereas Th1 cytokines contribute to development of gastritis, Th2 cytokines appear to be protective against gastric inflammation. Virulence factors that increase the dominance of Th1 responses involve

H. pylori neutrophil activating protein (Hp-NAP), which promotes IL-12 release by neutrophils and macrophages and induces Th1-skewing of antigen-specific CD4$^+$ T cells (Amedei *et al.*, 2006). When *H. pylori*-specific Th1 response prevails, it is associated with high IFN-γ, TNF-α, and IL-12 release and risk for peptic ulcer, conversely, when mixed release of both Th1 and Th2 cytokines prevails, it is often associated with uncomplicated gastritis (D'Elios *et al.*, 2005). In cases of chronic superficial gastritis, gastric tissues exhibit the presence of lymphocytes, macrophages, neutrophils and dendritic cells. CD4$^+$ T cells are more than CD8$^+$ T cells (Lundgren *et al.*, 2005; O'Keeffe and Moran, 2008). Briefly, it is now accepted that the development of gastric pathology is dependent mainly on the Th1 cells and Th1 cytokines (Kusters *et al.*, 2006). Some T-cell clones in *H. pylori*-positive patients are involved in the development of autoimmune gastritis (D'Elios *et al.*, 2007). Unconventional T helper cells, Th17 cells, can induce recruitment of polymorphs, including neutrophils and thus can contribute to *H. pylori*-associated inflammation (O'Keeffe and Moran, 2008).

Cytolytic effector activities of T cells are also important for the outcome of the infection. Cytolytic Th cells can cross-bind different epitopes of *H. pylori* proteins and H$^+$,K$^+$-ATPase autoantigen, which is also known to be a key autoantigen in autoimmune gastritis, and can promote gastric atrophy through long-lasting activation of Fas ligand-mediated apoptosis and perforin-induced cytotoxicity (Bergman and D'Elios, 2010). Moreover, gastric T cells from MALT lymphoma patients show defective perforin- and Fas-FasL-mediated killing of B cells, and resulting abnormal help for B-cell proliferation, therefore, the deregulated *H. pylori*-induced T-cell-dependent B-cell activation could promote the development of MALT lymphoma (Bergman and D'Elios, 2010). Protective role of cytotoxic pathways is proven in mice lacking perforin activity of CD8$^+$ T and NK cells, which developed B lymphomas after 1 year of age. These spontaneous lymphomas are rejected by CD8$^+$ T cells when transferred into wild-type mice (Smyth, *et al.*, 2000).

Humoral immune response is also developed in almost all *H. pylori* infected subjects (Perez-Perez *et al.*, 1988). Specific serum IgM antibodies appear usually during the first month of infection (Nurgalieva *et al.*, 2005). Serum IgA and IgG antibodies develop against different *H. pylori* epitopes and specific secretory IgA antibodies are present in gastric juice as a local immune response (Mattsson *et al.*, 1998) but their effect on bacterial colonization is controversial. Intragastric administration of specific monoclonal IgA antibodies protects against *H. felis* infection in mice (Czinn *et al.*, 1993). Secretory IgA (SIgA) and IgG antibodies however may promote colonization of gastric mucosa and may counteract resistance against infection (Akhiani *et al.*, 2004). Findings of cross-reactive antibodies to gastric epithelial cells and the parietal cell H$^+$,K$^+$-ATPase (Amedei *et al.*, 2003; D'Elios and Andersen, 2009) support the role of humoral immunity for local inflammation and damage. Manojlovic *et al.* (2008) have reported that gastric cancer patients have developed a different anti-*H. pylori* humoral immune response with low titter of IgG and IgA compared with patients with gastritis. As a whole, the role of anti-*H. pylori* humoral immune response as a protective mechanism is minimal (Kusters *et al.*, 2006).

Gastric inflammation in *H. pylori* infection can increase much by the bacterial virulence factors such as *cag*PAI, *vacA* s1/i1/m1 alleles, *dupA*, *babA2* and *oipA* status (see Chapter 4). Degree of gastric inflammation is also modulated by the host. In adults, the constant *H. pylori* colonization over decades can lead to development of atrophic gastritis as a preneoplastic lesion. The immunocompromised patients are not at higher risk for severe inflammation compared with the immunocompetent subjects (Algood and Cover, 2006). However, the immune response to *H. pylori* is complex and depends on host genetics. Functional polymorphisms in genes encoding cytokines such as IL-1β and IL-1 receptor antagonist secretion in response to microbes can modify the clinical expression of the infection and the gastric cancer risk. Polymorphisms in cytokine encoding genes result in increased proinflammatory or reduced anti-inflammatory cytokine release, and the genes encoding some TLRs also increase the risk for *H. pylori*-associated disease (Atherton and Blaser, 2009). Polymorphism, associated with increased level of IL-1β and TNF-α and decreased level of IL-10 is linked to elevated risk for atrophic gastritis and gastric cancer (El-Omar *et al.*, 2003; Wu *et al.*, 2003; Togawa *et al.*, 2005). Polymorphism in IL-1β gene promoter has been associated with increased expression of proinflammatory cytokine IL-1β, which inhibits the gastric acidity and increases the gastric cancer risk (Peek *et al.*, 2008). Patients with IL-1β polymorphism and *cagA*$^{+}$ *H. pylori* strains have been at 25–87 times higher risk for gastric cancer. Polymorphism in receptors for recognition such as TLR4, associated with altered reactivity to *H. pylori*, has been linked also to a risk for the cancer (El-Omar *et al.*, 2008).

Importantly, *H. pylori* infection usually does not result in development of protective immunity and reinfection is common after eradication therapy in regions where there is a high prevalence of the infection (Soto *et al.*, 2003). Balance of Th1/Th2 response is of great importance in host protection and in pathogenesis of *H. pylori*-mediated diseases. Polarized Th1 response plays a key role in tissue damage. Th2 response appears to be protective against gastric inflammation. Inadequate adaptive immune response is most probably a result of the inter-strain diversity, poor TLR activation, suppressive effects of the virulence and other *H. pylori* factors, and induction of T_{reg} cell response.

References

Akhiani, A.A., Schon, K., Franzen, L.E., Pappo, J. and Lycke, N. (2004). *Helicobacter pylori*-specific antibodies impair the development of gastritis, facilitate bacterial colonization, and counteract resistance against infection. J. Immunol. *172*, 5024–5033.

Algood, H.M., and Cover, T.L. (2006). *Helicobacter pylori* persistence: an overview of interactions between *H. pylori* and host immune defenses. Clin. Microbiol. Rev. *19*, 597–613.

Allen, L.A. (2007). Phagocytosis and persistence of *Helicobacter pylori*. Cell Microbiol. *9*, 817–828.

Amedei, A., Bergman, M.P., Appelmelk, B.J., Azzurri, A., Benagiano, M., Carlo Tamburini, C., Ruurd van der Zee, R., Telford, J.L., Vandenbroucke-Graulshttp, C.M.J.E., Mario M. D'Elios, M.M., *et al.* (2003). Molecular mimicry between *Helicobacter pylori* antigens and H+, K+ -adenosine triphosphatase in human gastric autoimmunity. J. Exp. Med., *198*, 1147–1156.

Amedei, A., Cappon, A., Codolo, G., Cabrelle, A., Polenghi, A., Benagiano, M., Tasca, E., Azzurri, A., D'Elios, M.M., Del Prete, G., *et al.* (2006). The neutrophil-activating protein of *Helicobacter pylori* promotes Th1 immune responses. J. Clin. Invest. *116*, 1092–1101.

Andersen-Nissen, E., Smith, K.D., Strobe, K.L., Barrett, S.L., Cookson, B.T., Logan, S.M., and Aderem, A. (2005). Evasion of Toll-like receptor 5 by flagellated bacteria. Proc. Natl. Acad. Sci. U.S.A. *102*, 9247–9252.

Atherton, J.C., and Blaser, M.J. (2009). Coadaptation of *Helicobacter pylori* and humans: ancient history, modern implications. Clin. Invest. *119*, 2475–2487.

Bergman, M.P., Engering, A., Smits, H.H., van Vliet, S.J., van Bodegraven A.A., Wirth, H.P., Kapsenberg, M.L., Vanderbroucke-Grauls, C.M.J.E., van Kooyk, I., and Appelmelk, B.J. (2004). *Helicobacter pylori* modulates the T helper cell 1/T helper cell 2 balance through phase-variable interaction between lipopolysaccharide and DC-SIGN. J. Exp. Med. *200*, 979–990.

Bergman, M.P., and D'Elios, M.M. (2010). Cytotoxic T cells in *H. pylori*-related gastric autoimmunity and gastric lymphoma. J. Biomed. Biotechnol. *2010*, 104918.

Chen, W., Shu, D., and Chadwick, V.S. (1999). *Helicobacter pylori* infection in interleukin-4-deficient and transgenic mice. Scand. J. Gastroenterol. *34*, 987–992.

Czinn, S.J., Cai, A., and Nedrud, J.G. (1993). Protection of germ-free mice from infection by *Helicobacter felis* after active oral or passive IgA immunization. Vaccine *11*, 637–642.

D'Elios, M. M., Manghetti, M., Almerigogna, F., Amedei, A., Costa, F., Burroni, D., Baldari, C.T., Romagnani, S., Telford, J.L., and Del Prete, G. (1997). Different cytokine profile and antigen-specificity repertoire in *Helicobacter pylori*-specific T cell clones from the antrum of chronic gastritis patients with or without peptic ulcer. Eur. J. Immunol. *27*, 1751–1755.

D'Elios, M.M., Amedei, A., Benagiano, M., Azzuri, A., and Del Prete, G. (2005). *Helicobacter pylori*, T cells and cytokines: the "dangerous liaisons". FEMS Immunol. Med. Microbiol. *44*, 113–119.

D'Elios, M.M., Montecucco, C., and de Bernard, M. (2007). VacA and Hp-NAP, Ying and Yang of *Helicobacter pylori*-associated gastric inflammation. Clin. Chim. Acta *381*, 32–38.

D'Elios, M.M., and Andersen, L.P. (2009). Inflammation, immunity, and vaccines for *Helicobacter pylori*. Helicobacter *14* (Suppl. 1), 21–28.

Dial, E.J., Hall, L.R., Serna, H., Romero, J.J., Fox, J.G., and Lichtenberger, L.M. (1998). Antibiotic properties of bovine lactoferrin on *Helicobacter pylori*. Dig. Dis. Sci. *43*, 2750–2756.

Dubois, A., and Borén, T. (2007). *Helicobacter pylori* is invasive and it may be a facultative intracellular organism. Cell Microbiol. *9*, 1108–1116.

El-Omar, E.M., Rabkin, C.S., Gammon, M.D., Vaughan, T.L., Risch, H.A., Schoenberg, J.B., Stanford, J.L., Mayne, S.T., Goedert, J., Blot, W.J., *et al.* (2003). Increased risk of non-cardia gastric cancer associated with pro-inflammatory cytokine gene polymorphisms. Gastroenterology *124*, 1193–1201.

El-Omar, E.M., Ng, M.T., and Hold, G.L. (2008). Polymorphisms in Toll-like receptor genes and risk of cancer. Oncogene *27*, 244–252.

Ermak, T.H., Giannasca, P.J., Nichols, R., Myers, G.A., Nedrud, J., Weltzin, R., Lee, C.K., Kleanthous, H., and Monath, T.P. (1998). Immunization of mice with urease vaccine affords protection against *Helicobacter pylori* infection in the absence of antibodies and is mediated by MHC class II-restricted responses. J. Exp. Med. *188*, 2277–2288.

Ferrand, J., Roumanes, D., Pitard, V., Moreau, J.F., Mégraud, F., and Lehours, P. (2008). Modulation of lymphocyte proliferation induced by gastric MALT lymphoma-associated *Helicobacter pylori* strains. Helicobacter *13*, 167–173.

Fox, J.G., Beck, P., Dangler, C.A., Whary, M.T., Wang, T.C., Shi, H.N., and Nagler-Anderson, C. (2000). Concurrent enteric helminth infection modulates inflammation and gastric immune responses and reduces helicobacter-induced gastric atrophy. Nat. Med. *6*, 536–542.

Garhart, C.A., Redline, R.W., Nedrud, J.G., and Czinn, S.J. (2002). Clearance of *Helicobacter pylori* infection and resolution of postimmunization gastritis in a kinetic study of prophylactically immunized mice. Infect. Immun. *70*, 3529–3538.

Grubman, A., Kaparakis, M., Viala, J., Allison, C., Badea, L., Karrar, A., Boneca, I.G., Le Bourhis, L., Reeve, S., Smith, I.A., *et al.* (2010). The innate immune molecule, NOD1, regulates direct killing of *Helicobacter pylori* by antimicrobial peptides. Cell Microbiol. *12*, 626–639.

Harris, P.R., Wright, S.W., Serrano, C., Riera, F., Duarte, I., Torres, J., Peña, A., Rollán, A., Viviani, P., Guiraldes, E., *et al.* (2008). *Helicobacter pylori* gastritis in children is associated with a regulatory T-cell response. Gastroenterology *134*, 491–499.

Hase, K., Murakami, M., Iimura, M., Cole, S.P., Horibe, Y., Ohtake, T., Obonyo, M., Gallo, R.L., Eckmann, L., and Kagnoff, M.F. (2003). Expression of LL-37 by human gastric epithelial cells as a potential host defense mechanism against *Helicobacter pylori*. Gastroenterology *125*, 1613–1625.

Isomoto, H., Mukae, H., and Mukae, H. (2004). Elevated concentrations of α-defensins in gastric juice of patients with *Helicobacter pylori* infection. Am. J. Gastroenterol. *99*, 1916–1923.

Joly, S., Compton, L.M., Pujol, C., Kurago, Z.B., and Guthmiller, J.M. (2009). Loss of human β-defensin 1, 2, and 3 expression in oral squamous cell carcinoma. Oral Microbiol. Immunol. *24*, 353–360.

Kandulski, A., Malfertheiner, P., and Wex, T. (2010). Role of regulatory T-cells in *H. pylori*-induced gastritis and gastric cancer. Anticancer Res. *30*, 1093–1103.

Kawakubo, M., Ito, Y., Okimura, Y., Kobayashi, M., Sakura, K., Kasama, S., Fukuda, M. N., Fukuda, M., Katsuyama, T., and Nakayama, J. (2004). Natural antibiotic function of a human gastric mucin against *Helicobacter pylori* Infection. Science *305*, 1003–1006.

Khamri, W., Moran, A.P., Worku, M.L., Karim, Q.N., Walker, M.M., Annuk, H., Ferris, J. A., Appelmelk, B. J., Eggleton, P., Reid, K.B.M, *et al.* (2005). Variations in *Helicobacter pylori* lipopolysaccharide to evade the innate immune component surfactant protein D. Infect. Immun. 73, 7677–7686.

Kobayashi, M., Lee, H., Nakayama, J., and Fukuda, M. (2009). Roles of gastric mucin-type O-glycans in the pathogenesis of *Helicobacter pylori* infection. Glycobiology *19*, 453–461.

Kusters, J.G., van Vliet, A.H.M., and Kuipers, E.J. (2006). Pathogenesis of *Helicobacter pylori* infection. Clin. Microbiol. Rev. *19*, 449–490.

Lee, S.K., Stack, A., Katzowitsch, E., Aizawa, S.I., Suerbaum, S., and Josenhans, C. (2003). *Helicobacter pylori* flagellins have very low intrinsic activity to stimulate human gastric epithelial cells via TLR5. Microbes Infect. *5*, 1345–1356.

Lehmann, F.S., Terracciano, L., Carena, I., Baeriswyl, C., Drewe, J., Tornillo, L., De Libero, G., and Beglinger, C. (2002). *In situ* correlation of cytokine secretion and apoptosis in *Helicobacter pylori*-associated gastritis. Am. J. Physiol. Gastrointest. Liver Physiol. *283*, G481–G488.

Leszczyńska, K., Namiot, A., Fein, D.E., Wen, Q., Namiotm Z., Savage, P.B., Diamond, S., Janmey, P.A., and Bucki, R. (2009). Bactericidal activities of the cationic steroid CSA-13 and the cathelicidin peptide LL-37 against *Helicobacter pylori* in simulated gastric juice. BMC Microbiol. *9*, 187.

Lindholm, C., Quiding-Jarbrink, M., Lonroth, H., Hamlet, A., and Svennerholm, A. M. (1998). Local cytokine response in *Helicobacter pylori*-infected subjects. Infect. Immun. *66*, 5964–5971.

Lohoff, M, Röllinghoff M, and Sommer F. (2000). *Helicobacter pylori* gastritis: a Th1 mediated disease? J. Biotechnol. *83*, 33–36.

Lopes, A.I., Quiding-Jarbrink, M., Palha, A., Ruivo, J., Monteiro, L., Oleastro, M., Santos, A., and Fernandes, A. (2005). Cytokine expression in pediatric *Helicobacter pylori* infection. Clin. Diagn. Lab. Immunol. *12*, 994–1002.

Lundgren, A., Trollmo, C., Edebo, A., Svennerholm, A.M., and Lundin, B.S. (2005). *Helicobacter pylori*-specific CD4+ T cells home to and accumulate in the human *Helicobacter pylori*-infected gastric mucosa. Infect. Immun. 73, 5612–5619.

Lundin, B.S., Enarsson, K., Kindlund, B., Lundgren, A., Johnsson, E., Quiding-Järbrink, M., and Svennerholm, A.M. (2007). The local and systemic T-cell response to *Helicobacter pylori* in gastric cancer patients is characterised by production of interleukin-10. Clin. Immunol. *125*, 205–213.

Mandell, L., Moran, A.P., Cocchiarella, A., Houghton, J., Taylor, N., Fox, J.G., Wang, T.C., and Kurt-Jones, E.A. (2004). Intact Gram-negative *Helicobacter pylori, Helicobacter felis,* and *Helicobacter hepaticus* bacteria activate innate immunity via Toll-like receptor 2 but not Toll-like receptor 4. Infect. Immun. 72, 6446–6454.

Manojlovic, N., Babic, D., Filipovic-Ljeshovic, I., and Pilcevic, D. (2008). Anti *Helicobacter pylori* IgG and IgA response in patients with gastric cancer and chronic gastritis. Hepatogastroenterology. 55, 807–813.

Mattsson, A., Quiding-Jarbrink, M., Lonroth, H., Hamlet, A., Ahlstedt, I., and Svennerholm, A. (1998). Antibody-secreting cells in the stomachs of symptomatic and asymptomatic *Helicobacter pylori*-infected subjects. Infect. Immun. *66*, 2705–2712.

Miehlke, S., Reddy, R., Osato, M. S., Ward, P. P., Conneely, M., and Graham D. Y. (1996). Direct activity of recombinant human lactoferrin against *Helicobacter pylori*. J. Clin. Microbiol. *34*, 2593–2594.

Molinari, M., Salio, M., Galli, C., Norais, N., Rappuoli, R., Lanzavecchia, A., and Montecucco C. (1998). Selective inhibition of Ii-dependent antigen presentation by *Helicobacter pylori* toxin VacA. J. Exp. Med. *187*, 135–140.

Moran, A.P., Khamri, W., Walker, M.M., and Thursz M.R. (2005). Role of surfactant protein D (SP-D) in innate immunity in the gastric mucosa: evidence of interaction with *Helicobacter pylori* lipopolysaccharide. J. Endotoxin Res. *11*, 357–362.

Nurgalieva, Z.Z., Conner, M.E., Opekun, A.R., Zheng, C.Q., Elliott, S.N., Ernst, P.B., Osato, M., Estes, M.K., and Graham, D.Y. (2005). B-cell and T-cell immune responses to experimental *Helicobacter pylori* infection in humans. Infect. Immun. 73, 2999–3006.

O'Keeffe, J., and Moran, A.P. (2008). Conventional, regulatory, and unconventional T cells in the immunologic response to *Helicobacter pylori*. Helicobacter *13*, 1–19.

Ouellette, A.J. (2005) Paneth cell α-defensins: peptide mediators of innate immunity in the small intestine. Springer Semin. Immunopathol. *27*, 133–146.

Peek, Jr. R.M. (2008). Review: Prevention of gastric cancer: When is treatment of *Helicobacter pylori* warranted? Ther. Adv. Gastroenterol. *1*, 19–31.

Perez-Perez, G.I., Dworkin, B.M., Chodos, J.E., and Blaser, M.J. (1988). *Campylobacter pylori* antibodies in humans. Ann. Intern. Med. *109*, 11–17.

Pappo, J., Torrey, D., Castriotta, L., Savinainen, A., Kabok, Z., and Ibraghimov, A. (1999). *Helicobacter pylori* infection in immunized mice lacking major histocompatibility complex class I and class II functions. Infect. Immun. *67*, 337–341.

Robinson, K., Kenefeck, R., Pidgeon, E., Shakib, S., Patel, S., Polson, R., Zaitoun, A.M., and Atherton, J.C. (2008). *Helicobacter pylori*-induced peptic ulcer disease is associated with inadequate regulatory T-cell responses. Gut 57, 1375–1385.

Roth, K.A., Kapadia, S.B., Martin, S.M., and Lorenz, R.G., (1999). Cellular immune responses are essential for the development of *Helicobacter felis*-associated gastric pathology. J. Immunol. *163*, 1490–1497.

Sawai, N., Kita, M., Kodama, T., Tanahashi, T., Yamaoka, Y., Tagawa, Y., Iwakura, Y., and Imanishi, J. (1999). Role of gamma interferon in *Helicobacter pylori*-induced gastric inflammatory responses in a mouse model. Infect. Immun. 67, 279–285.

Schmausser, B., Josenhans, C., Endrich, S., Suerbaum, S., Sitaru, C., Andrulis, M., Brandlein, S., Rieckmann, P., Muller-Hermelink, H.K., and Eck, M. (2004). Downregulation of CXCR1 and CXCR2 expression on human neutrophils by *Helicobacter pylori*: a new pathomechanism in *H. pylori* infection? Infect Immun. 72, 6773–6779.

Smith, M.F., Jr., Mitchell, A., Li, G., Ding, A., Fitzmaurice, A.M., Ryan, K., Crowe, S., and Goldberg, J.B. (2003). Toll-like receptor (TLR) 2 and TLR5, but not TLR4, are required for *Helicobacter pylori*-induced NF-kappa B activation and chemokine expression by epithelial cells. J. Biol. Chem. *278*, 32552–32560.

Smith, M.F., Jr., Novotny, J., Carl, V.S., and Comeau, L.D. (2006). *Helicobacter pylori* and toll-like receptor agonists induce syndecan-4 expression in an NF-kappaB-dependent manner. Glycobiology *16*, 221–229.

Smyth, M.J., Thia, K.Y.T., Street, S.E.A., MacGregor, D., Godfrey, D.I., and Trapani, J.A. (2000). Perforin-mediated cytotoxicity is critical for surveillance of spontaneous lymphoma. J. Exp. Med. *192*, 755–760.

Soto, G., Bautista, C.T., Roth, D.E., Gilman, R.H., Velapatino, B., Ogura, M., Dailide, G., Razuri, M., Meza, R., Katz, U., *et al.* (2003). *Helicobacter pylori* reinfection is common in Peruvian adults after antibiotic eradication therapy. J. Infect. Dis. *188*, 1263–1275.

Szczepanik, M. (2006). Interplay between *Helicobacter pylori* and the immune system. Clinical implications. J. Physiol. Pharmacol. 57 *(Suppl 3)*, 15–27.

Taylor, J.M., Ziman, M.E., Canfield, D.R., Vajdy, M., and Solnick, J.V. (2008). Effects of a Th1- versus a Th2-biased immune response in protection against *Helicobacter pylori* challenge in mice. Microb. Pathog. *44*, 20–27.

Togawa, S., Joh, T., Itoh, M., Katsuda, N., Ito, H., Matsuo, K., Tajima, K., and Hamajima, N. (2005). Interleukin-2 gene polymorphisms associated with increased risk of gastric atrophy from *Helicobacter pylori* infection. Helicobacter *10*, 172–178.

Umehara, S., H. Higashi, N. Ohnishi, M. Asaka, and M. Hatakeyama. (2003). Effects of *Helicobacter pylori* CagA protein on the growth and survival of B lymphocytes, the origin of MALT lymphoma. Oncogene 22, 8337–8342.

Wang, J., Fan, X., Lindholm, C., Bennett, M., O'Connoll, J., Shanahan, F., Brooks, e.g., Reyes,V.E., and Ernst, P.B. (2000). *Helicobacter pylori* modulates lymphoepithelial cell interactions leading to epithelial cell damage through Fas/Fas ligand interactions. Infect. Immun. *68*, 4303–4311.

Wang, Y.H., Gorvel, J.P., Chu, Y.T., Wu, J.J., and Lei, H.Y. (2010). *Helicobacter pylori* impairs murine dendritic cell responses to infection. PLoS One *5*, e10844.

Whitney, A.E., Guarner, J., Hutwagner, L., and Gold, B.D. (2000). *Helicobacter pylori* gastritis in children and adults: comparative histopathologic study. Ann. Diagn. Pathol. *4*, 279–285.

Wu, M.S., Wu, C.Y., Chen, C.J., Lin, M.T., Shun, C.T., and Lin, J.T. (2003). Interleukin-10 genotypes associate with the risk of gastric carcinoma in Taiwanese Chinese. Int. J. Cancer. *104*, 617–623.

Xie, Y., Zhou, N.J., Gong, Y.F., Zhou, X.J., Chen, J., Hu, S.J., Lu, N.H., and Hou, X.H. (2007). Th immune response induced by *H pylori* vaccine with chitosan as adjuvant and its relation to immune protection. World J. Gastroenterol. *13*, 1547–1553.

Vaccines

6

Ivan Mitov

Abstract

An effective vaccine is needed to improve the success of anti-*H. pylori* therapy. Cooperative action of cell-mediated, humoral and molecular responses is necessary for effective protection against *H. pylori*. Vaccines against *H. pylori* can be used as prophylactic vaccines to prevent the infection or as therapeutic vaccines to cure the infection, to improve the eradication success of standard regimens or to reduce the bacterial density in the gastric mucosa and the risk for emergence of antibiotic-resistant strains. In recent years, many attempts, using various *H. pylori* antigens such as urease, CagA, Hp-NAP, HspA or combinations, many adjuvants and different routes of immunization, have been made to create vaccines against *H. pylori* infection. Although some attempts are promising, no effective and safe vaccine against *H. pylori* is currently available for humans. New and innovative approaches for immunization using DNA, living vectors, microspheres etc. are currently under evaluation. The vaccination plan and the groups who should receive vaccination are still to be determined. The development of a successful vaccine will be important, especially in developing countries.

Development of *H. pylori* vaccines

H. pylori eradication faces several problems such as treatment failure due to increasing resistance, poor patient compliance, side-effects of the antibiotics and cost of the therapy (Aebischer *et al.*, 2005). High relapse and reinfection rates after antimicrobial eradication have been reported in countries with high prevalence of the infection (see Chapter 7). Effective vaccine is needed to improve significantly the management of *H. pylori* infection.

Vaccines against *H. pylori* can be used as prophylactic vaccines to prevent the infection or as therapeutic vaccines to eliminate the already established infection. Prophylactic vaccination depends on the ability of the immune system to recognize an antigen as 'foreign' and to develop a strong adaptive response to next exposures, which prevents the infection. Conversely, therapeutic vaccination depends on a pre-existing memory response against an antigen or can develop immune response against a target, to which the host shows reduced responsiveness (Chui *et al.*, 2005). Application of a prophylactic vaccine on infants in developed countries could almost eliminate *H. pylori* infection at the population level in about 10 years if the vaccine is only 50% effective. However, in the developing countries, a continuous vaccination could be needed (Rupnow *et al.*, 2001; Aebischer *et al.*, 2005). In chronically infected animals and possibly in humans, the vaccination could lead to cure of the infection or to improvement of the eradication success of standard regimens. Even without eradication, the therapeutic vaccination can be useful to reduce the bacterial number in the gastric mucosa and the risk for emergence of antibiotic-resistant strains.

During the last decades, many attempts have been made to create vaccines against *H. pylori* infection. Despite of efforts, no efficacious vaccine is currently available for humans. Gastric colonization by the bacteria usually results in development of strong but non-protective immune response, mainly associated with T helper 1 (Th1) cells (see Chapter 5). Mast cells also have been found to be important for the development of vaccine-induced, T cell-mediated protection from *H. felis* infection in mice (Velin *et al.*, 2005). Activated mast cells can release chemokines and cytokines to attract T cells, can upregulate MHC II molecules and surface molecules and, via histamine release, can promote Th1 responses (in an H1 receptor way) or Th2 responses (in an H2 receptor-dependent way; Aebischer *et al.*, 2008b). To obtain a full protection against *H. pylori* infection in humans, the vaccine should provide strong and specific cellular, molecular and humoral responses (D'Elios and Andersen, 2009).

Whole inactivated bacteria, cell lysates, purified *H. pylori* components and recombinant proteins have been evaluated as candidates for vaccine preparation. Antigens which are surface exposed and easily accessible to the immune response, those which are abundant (and possibly highly immunogenic), those which are well conserved in all strains and those which are important virulence factors have been chosen. Mostly murine models have been used to estimate immunogenicity of the antigen preparations and adjuvant/delivery systems.

Although the protective capacity of different *Helicobacter* vaccines is well documented, the protective immune mechanisms are not well clarified yet. It has been considered that the successful immunization depends on the induction of humoral and Th2 cell-mediated immune response. However, the mechanisms of the protection appear more complex. IL-18 could participate in the development of protective immunity against *H. pylori* by the induction of a Th1-type response and production of IFN-γ, while IL-10 could promote immune evasion of the bacteria (Akhiani *et al.*, 2004; Akhiani *et al.*, 2005). Increase in IL-12 has been suggested to participate in the protection of mice against *H. pylori* infection, while the significance of other cytokines such as IL-2, IL-4, IFN-γ, and TNF-α is still controversial (Taylor *et al.*, 2007). Finally, it is widely accepted that cooperative action of cell-mediated, humoral and molecular responses is needed for effective protection against *H. pylori*.

The initial studies demonstrating protective efficacy of mucosal vaccination against *H. pylori* were performed with inactivated bacteria and whole-cell lysates (Czinn *et al.* 1991; Chen *et al.*, 1993). An intranasal vaccine based on *H. pylori* Sydney strain 1 lysate and cholera toxin (CT) adjuvant induced high-level protection with a strong IFN-γ-producing T-cell response, associated with an increase in chemokines, which attract neutrophils to the stomach and are important for *H. pylori* eradication (DeLyria *et al.*, 2009). Using whole-cell lysates, it is possible to achieve complete protection in mice as determined by inability to grow *H. pylori* from gastric mucosa (Garhart *et al.*, 2002). Clinical trials have been performed with a formalin-inactivated *H. pylori* whole-cell vaccine, given orally in combination with the mucosal adjuvant mLT(R192G), a mutant of *Escherichia coli* heat-labile toxin (LT) (Summerton *et al.*, 2010).

H. pylori urease, a well-conserved enzyme which allows bacterial survival in the low pH in the stomach and contributes to *H. pylori* colonization, has appeared as an appropriate antigen for vaccines. However, although the urease has reduced *H. pylori* colonization in mice, antiurease antibodies have been present in only half of *H. pylori*-positive subjects and the urease-specific T helper (Th) cells have been relatively scarce in the gastric mucosa of patients with chronic gastritis and peptic ulcer (Del Giudice *et al.*, 2009). In any case, the urease is often used in the attempts to develop appropriate anti-*H. pylori* vaccine in animal models and in humans. Recombinant epitope, responsible for urease activity in Freund's complete adjuvant, has reduced bacterial load in the stomach after subcutaneous immunization of mice. This protection has been associated with antibody production and Th1 response (Morihara *et al.*, 2008). Recombinant fusion proteins of urease B epitope and cholera toxin B subunit (CT) in liposome, introduced orally as prophylactic and therapeutic vaccine, have induced urease IgA response, and have protected mice (Zhao *et al.*, 2007).

CagA is one of the highly immunogenic *H. pylori* antigens and the immune response to this virulence factor can predict severe infection. VacA possesses toxicity and enhances the gastric colonization of *H. pylori* (Del Giudice *et al.*, 2009). Neutrophil activating protein of *H. pylori* (Hp-NAP) induces the release of IL-12 and IL-23 by neutrophils and monocytes as Toll-like receptor 2 (TLR2) agonist and leads to polarization of the immune response towards a Th1-type functional phenotype (Del Giudice *et al.*, 2009). Immunogenicity of other *H. pylori* protein virulence factors such as BabA, SabA, HpaA and AlpA adhesins as well as OipA and HomB have been investigated as well. There are some problems with the use of certain proteins as vaccines. For example, urease is not immunodominant and induces a weak antibody response (Leal-Herrera *et al.*, 1999). One of the most immunogenic proteins, CagA, and other bacterial proteins are not present in all strains of *H. pylori.* Finally, the antibody response against separate proteins is not sufficient to develop intense and complete protection.

The high protection observed with the inactivated bacteria and whole-cell lysates led to the concept that multiple antigens may be necessary for the development of a truly protective vaccine. It has been shown that oral immunizations with a combination of recombinant *H. pylori* heat-shock protein A (HspA) and urease in mice have been 100% protective, whereas either component has induced only 80% protection against challenge with *H. felis* (Ferrero *et al.*, 1995). A multivalent subunit VacA, CagA and NAP *H. pylori* vaccine, currently in development for use in humans, has been shown to be protective in a therapeutic study using beagles (Rossi *et al.*, 2004). Another vaccine, consisting of urease B subunit (UreB), HspA and the surface protein HpaA, delivered either orally with a mutant LT or parenterally with aluminium hydroxide, has induced greater protection against *H. pylori* in Mongolian gerbils compared with mono- or bivalent vaccines (Wu *et al.*, 2008).

New types of vaccines such as live vector vaccines, DNA-, microsphere- and ghost vaccines have been recently evaluated. In animal studies, live vector vaccines have reduced *H. pylori* colonization in the stomach, while in humans, the vaccines have been less successful (Agarwal and Agarwal, 2008). DNA vaccines contain inserted DNA, encoding a bacterial antigen into a bacterial plasmid, which is then delivered as naked DNA or by an attenuated carrier, such as *Salmonella* species (Agarwal and Agarwal, 2008). Oral DNA vaccine from *H. pylori* Hp-NAP in live attenuated *Salmonella* serotype Typhimurium strain has induced humoral and mucosal immune responses in mice (Sun *et al.*, 2006). Microspheres induce humoral, mucosal and cell-mediated immunity. Poly(D,L-lactide)-polyethylene glycol microspheres loaded with *H. pylori* lysates have been found to be very immunogenic in mice (Ren *et al.*, 2002). Bacterial ghosts represent empty cell envelopes without cytoplasmic contents but with fully intact surface antigenic structures. *H. pylori* bacterial ghosts, loaded with recombinant Omp18 plus cholera toxin have induced serum anti-*H. pylori* and Omp18-specific antibodies and resultant reduction of gastric *H. pylori* colonization in mice (Talebkhan *et al.*, 2010). However, more studies, especially in humans, of all these new types of vaccines are needed.

Different routes of immunization against *H. pylori* have been used in mice, e.g. oral, intranasal, intrarectal, intradermal, intramuscular and subcutaneous routes of immunization, in combination with mucosal adjuvants or delivery vehicles. *H. pylori* causes local infection in the gastric mucosa, therefore vaccine strategies widely exploit the mucosal route of immunization developed for other gastrointestinal infections. It has been shown that the intranasal immunization with supernatant of a sonicate culture has been much more effective than oral immunization of mice. This prophylactic immunization of mice does not prevent colonization by *H. pylori* but enables mice to clear the infection or significantly reduces colonization. The reduction in bacterial load is associated with gastric inflammation that subsides over time (Garhart *et al.*, 2002). 'Sterilizing immunity' has been reported by rectal immunization with recombinant antigen and LT adjuvant (Kleanthous *et al.*, 2001), oral immunization with bacterial ghosts (Panthel *et al.*, 2003) and DNA vaccine applied intranasally with CT and CpG oligonucleotides as adjuvants (Jiang *et al.*, 2003).

The immune mechanisms against *H. pylori* infection are not only limited to the mucosal immune system. Protective immunity has been obtained by intraperitoneal immunization with live *H. felis* (Chen *et al.*, 1993) and by subcutaneous application of urease antigen with and without adjuvants (Ermak *et al.* 1998). Single injection of bacterial lysate and aluminium hydroxide has succeeded to induce protective immunity equivalent to oral vaccination against *H. pylori* (Gottwein *et al.*, 2001). Taylor *et al.* (2007) have obtained long-term (three months after immunization) protection against *H. pylori* by immunization either by the systemic route alone (intramuscular) or by a combination of mucosal followed by systemic routes (oral/intramuscular). Conversely, mice immunized orally or by intramuscular and then oral routes have not been protected from oral *H. pylori* challenge (Taylor *et al.*, 2007). It is important to assess the immunity during the resting memory phase rather than only during acute phase after vaccination (1–6 weeks; Taylor *et al.*, 2007).

There are several questions which should be considered when animal models are used for anti-*H. pylori* vaccine development. Immunization of mice against *H. pylori* seldom results in sterilizing immunity. Moreover, the infection in mice and other animals needs *H. pylori* strain adaptation to infect them with a reduced efficiency (Del Giudice *et al.*, 2009). Mice are not naturally infected by *H. pylori*; however, the bacteria can be adapted to colonize their gastric mucosa. Furthermore, observations on the vaccine effects in mice reflect the acute early infection but not the chronic infection over decades in humans. Type IV secretory system (T4SS), which transports CagA and other cell components into gastric cells, is generally unstable in *H. pylori*-infected murine stomach and mice usually do not develop severe expression of the infection such as gastric cancer (Hoffelner *et al.*, 2008). Better stability of T4SS-positive strains and development of severe diseases, as in the human infection, have been observed in Mongolian gerbils (*Meriones unguiculatus*); however, this model has disadvantages such as the lack of tools such as antibodies and the lack of a genome sequence and the option of genetic manipulation (Hoffelner *et al.*, 2008).

A difficulty in developing effective vaccines against *H. pylori* is the choice of adjuvants and delivery systems. Mucosal immunization by different antigens associated with mucosal adjuvants such as cholera toxin, heat-labile (LT) toxin of *E. coli*, CpG oligonucleotides or live attenuated bacterial vectors have been tested for prophylaxis or treatment of *Helicobacter* species infections. However, the cholera toxin and LT can cause diarrhoea in humans (Agarwal and Agarwal, 2008; Sijun and Yong, 2009). Less toxic mutant LT (R192G) and double mutant LT have been constructed and tested in clinical trials (Kotloff *et al.*, 2001; Losonsky *et al.*, 2003) and in mouse models (Summerton *et al.*, 2010). Whole-cell vaccines adjuvanted with mLT and dmLT have been well tolerated, have generated both mucosal and systemic immune responses and considerably weakened gastrointestinal side-effects.

Recently, new adjuvants have been evaluated and have given promising results for preparing anti-*H. pylori* vaccines. Chitosan, a non-toxic deacetylated product of chitin, used with bacterial whole-cell sonicates of *H. pylori*, has been able to induce both Th1 and Th2 protective responses against *H. pylori* infection (Xie *et al.*, 2007). Aluminium hydroxide, combined with multicomponent vaccine, has been safe and the vaccine formulation has been immunogenic in a phase I study (see below; Malfertheiner *et al.*, 2008).

Study on mouse model, vaccinated by intranasal or subdermal route by vaccines, containing HpaA and catalase antigens as well as ISCOMATRIX and ISCOM adjuvants, which have been shown to be safe and efficient, has demonstrated promising results (Skene *et al.*, 2008). ISCOMATRIX™ includes 40-nm cage-like structures, containing cholesterol, phospholipid and saponin from *Quillaja saponaria*, which possesses immunomodulatory properties. ISCOM™ vaccine is similar to ISCOMATRIX vaccine but the antigen is incorporated into the structure during formulation. Interestingly, HpaA ISCOM vaccine has induced protective anti-*H. pylori* immunity if delivered by the respiratory but not oral route and catalase ISCOMATRIX vaccine has induced protective immunity by subcutaneous administration.

Vaccination studies on humans have been performed but more such evaluations are still required (Johansson *et al.*, 2004; Keller and Michetti, 2001; Voland *et al.*, 2006). Recombinant vector vaccines, expressing *H. pylori* urease proteins, have employed avirulent strains of *Salmonella enterica* serovars (DiPetrillo *et al.*, 1999; Bumann *et al.*, 2001). Studies with *H. pylori*-negative volunteers, immunized orally by vaccinal strain *Salmonella* serovar Typhi Ty21a, expressing the urease of *H. pylori* or Hp0231 protein, or only Ty21a, have shown unsatisfactory results (Aebischer *et al.*, 2008a). Although the immunization by protein subunits of *H. pylori* in animals has been promising, studies by similar vaccines in humans have shown the adjuvant-associated side-effects and only a moderate protective efficacy (Agarwal and Agarwal, 2008). Formalin-inactivated whole-cell vaccine has been given orally with mutant low toxic LT in infected and uninfected subjects. Reduction of colonization levels has not been established (Kotloff *et al.* 2001). Several clinical trials in human volunteers, immunized by mucosal route, have been performed by recombinant urease (Michetti *et al.*, 1999; Sougioultzis *et al.*, 2002; Kabir, 2007). Reduction in bacterial load has been observed in studies of human therapeutic vaccine trial using recombinant urease with LT adjuvant (Michetti *et al.*, 1999) and cell-mediated immune response has been associated with clearance or reductions in bacterial load (Kabir, 2007; Aebischer *et al.*, 2008a). Interestingly, *H. pylori*-positive subjects with simultaneous helminthic infection have shown less frequent clinically expressed diseases induced by *H. pylori* (Elshal *et al.*, 2004; Du *et al.*, 2005). This can suggest that the administration of low dose-immunomodulating agents can be of clinical benefit.

Important success in vaccination has been recently reported. Malfertheiner *et al.* (2008) have evaluated the immunogenicity and safety of a vaccine, consisting of recombinant VacA, CagA, and Hp-NAP, administered intramuscularly with an aluminium hydroxide adjuvant to *H. pylori*-negative healthy volunteers. Both the vaccinated and placebo volunteers have exhibited only very mild adverse reactions. All vaccinated volunteers have developed specific IgG and cellular immune responses to one or two antigens and most of them (86%) have shown immune response to all three antigens. Importantly, booster dose 18–24 months later has induced antibody and cellular memory responses (Malfertheiner *et al.*, 2008).

Generally, although many efforts have been made during the last decades, no effective and safe vaccine against *H. pylori* is currently available for humans. New directions for immunization with use of DNA, living vectors, microspheres etc. are currently under evaluation. The question who and when should be vaccinated is still controversial but the vaccination will be useful, especially in the developing countries (Agarwal and Agarwal, 2008).

References

Aebischer, T., Schmitt, A., Walduck, A.K., and Meyer, T.F. (2005). *Helicobacter pylori* vaccine development: facing the challenge. Int. J. Med. Microbiol. *295*, 343–353.

Aebischer, T., Bumann, D., Epple, H.J., Metzger, W., Schneider, T., Cherepnev, G., Walduck, A.K., Kunkel, D., Moos, V., Loddenkemper, C., *et al.* (2008a). Correlation of T cell response and bacterial clearance in human volunteers challenged with *Helicobacter pylori* revealed by randomised controlled vaccination with Ty21a-based *Salmonella* vaccines. Gut *57*, 1065–1072.

Aebischer, T., Walduck, A., Schroeder, J., Wehrens, A., Chijioke, O., Schreiber, S., and Meyer, T.F. (2008b). A vaccine against *Helicobacter pylori*: towards understanding the mechanism of protection. Int. J. Med. Microbiol. *298*, 161–168.

Agarwal, K., and Agarwal, S. (2008). *Helicobacter pylori* vaccine: from past to future. Mayo Clin. Proc. *83*, 169–175.

Akhiani, A.A., Schoen, K., and Lycke, N. (2004). Vaccine-induced immunity against Helicobacter pylori infection is impaired in IL-18-deficient mice. J. Immunol. *173*, 3348–3356.

Akhiani, A.A., Stensson, A., Schoen, K., and Lycke, N. (2005). IgA antibodies impair resistance against *Helicobacter pylori* infection: studies on immune evasion in IL-10-deficient mice. J. Immunol. *174*, 8144–8153.

Bumann, D., Metzger, W.G., Mansouri, E., Palme, O., Wendland, M., Hurwitz, R., Haas, G., Aebischer, T., Von Specht, B.U., and Meyer, T.F. (2001). Safety and immunogenicity of live recombinant *Salmonella enterica* serovar Typhi Ty21a expressing urease A and B from *Helicobacter pylori* in human volunteers. Vaccine *20*, 845–852.

Chen, M., Lee, A., Hazell, S., Hu, P., and Li, Y. (1993). Immunisation against gastric infection with *Helicobacter* species: first step in the prophylaxis of gastric cancer? Zentralbl. Bakteriol. *280*, 155–165.

Chui, S.Y., Clay, T.M., Lyerly, H.K., and Morse, M.A. (2005). The development of therapeutic and preventive vaccines for gastric cancer and *Helicobacter pylori*. Cancer Epidemiol. Biomarkers Prev. *14*, 1883–1889.

Czinn, S.J., and Nedrud, J.G. (1991). Oral immunization against *Helicobacter pylori*. Infect. Immun. *59*, 2359-2363.

Del Giudice, G., Malfertheiner, P., and Rappuoli, R. (2009). Development of vaccines against *Helicobacter pylori*. Expert Rev. Vaccines *8*, 1037–1049.

D'Elios, M.M., and Andersen, L.P. (2009). Inflammation, immunity, and vaccines for *Helicobacter pylori*. Helicobacter *14* (Suppl.1), 21–28.

DeLyria, E.S., Redline, R.W., and Blanchard, T.G. (2009). Vaccination of mice against *H. pylori* induces a strong Th-17 response and immunity that is neutrophil dependent. Gastroenterology *136*, 247–256.

DiPetrillo, M.D., Tibbetts, T., Kleanthous, H., Killeen, K.P., and Hohmann, E.L. (1999). Safety and immunogenicity of phoP/phoQ-deleted *Salmonella typhi* expressing *Helicobacter pylori* urease in adult volunteers. Vaccine *18*, 449–459.

Du, Y., Agnew, A., Ye, X.P., Robinson, P.A., Forman, D., and Crabtree, J.E. (2005). *Helicobacter pylori* and *Schistosoma japonicum* co-infection in a Chinese population: helminth infection alters humoral responses to *H. pylori* and serum pepsinogen I/II ratio. Microbes Infect. *8*, 52–60.

Elshal, M.F., Elsayed, I.H., El Kady, I.M., Badra, G., El-Refaei, A., El-Batanony, M., and Hendy, O.M. (2004). Role of concurrent *S. mansoni* infection in *H. pylori*-associated gastritis: a flow cytometric DNA-analysis and oxyradicals correlations. Clin. Chim. Acta *346*, 191–198.

Ermak, T.H., Giannasca, P.J., Nichols, R., Myers, G.A., Nedrud, J., Weltzin, R., Lee, C.K., Kleanthous, H., and Monath, T.P. (1998). Immunization of mice with urease vaccine affords protection against *Helicobacter pylori* infection in the absence of antibodies and is mediated by MHC class II-restricted responses. J. Exp. Med. *188*, 2277–2288.

Ferrero, R.L., Thiberge, J.M., Kansau, I., Wuscher, N., Huerre, M., and Labigne, A. (1995). The GroES homolog of *Helicobacter pylori* confers protective immunity against mucosal infection in mice. Proc. Nat. Acad. Sci. U.S.A. *92*, 6499–6503.

Garhart, C.A., Redline, R.W., Nedrud, J.G., and Czinn, S.J. (2002). Clearance of *Helicobacter pylori* infection and resolution of postimmunization gastritis in a kinetic study of prophylactically immunized mice. Infect. Immun. *70*, 3529–3538.

Gottwein, J.M., Blanchard, T.G., Targoni, O.S., Eisenberg, J.C., Zagorski, B.M., Redline, R.V., Nedrud, J.G., Tary-Lehmann, M., Lehman, P.V., and Czinn, S.J. (2001). Protective anti*Helicobacter* immunity is induced with aluminum hydroxide or complete Freund's adjuvant by systemic immunization. J. Infect. Dis. *184*, 308–314.

Hoffelner, H., Rieder, G., and Haas, R. (2008). *Helicobacter pylori* vaccine development: optimisation of strategies and importance of challenging strain and animal model. Int. J. Med. Microbiol. *298*, 151–159.

Jiang, W., Baker, H.J., and Smith, B.F. (2003). Mucosal immunization with Helicobacter, CpG DNA, and cholera toxin is protective. Infect. Immun. *71*, 40–46.

Johansson, E.L., Bergquist, C., Edebo, A., Johansson, C., and Svennerholm, A.M. (2004). Comparison of different routes of vaccination for eliciting antibody responses in the human stomach. Vaccine *22*, 984–990.

Kabir, S. (2007). The current status of *Helicobacter pylori* vaccines: a review. Helicobacter *12*, 89–102.

Keller, W.C., and Michetti, P. (2001). Vaccination against *Helicobacter pylori* – an old companion of man. Expert Opin. Biol. Ther. *1*, 795–802.

Kleanthous, H., Tibbitts, T.J., Gray, H.L., Myers, G.A., Lee, C.K., Ermak, T.H., and Monath, T.P. (2001). Sterilizing immunity against experimental *Helicobacter pylori* infection is challenge-strain dependent. Vaccine *19*, 4883–4895.

Kotloff, K.L., Sztein, M.B., Wasserman, S.S., Losonsky, G.A., Dilorenzo, S.C., and Walker, R.I. (2001). Safety and immunogenicity of oral inactivated whole-cell *Helicobacter pylori* vaccine with adjuvant among volunteers with or without subclinical infection. Infect. Immun., *69*, 3581–3590.

Leal-Herrera, Y., Torres, J., Perez-Perez, G., Gomez, A., Monath, T., Tapia-Conyer, R., and Munoz, O. (1999). Serologic IgG response to urease in *Helicobacter pylori*-infected persons from Mexico. Am. J. Trop. Med. Hyg. *60*, 587–592.

Losonsky, G.A., Kotloff, K.L., and Walker, R.I. (2003). B cell responses in gastric antrum and duodenum following oral inactivated *Helicobacter pylori* whole cell (HWC) vaccine and LTR192G in *H. pylori* seronegative individuals. Vaccine *21*, 562–565.

Malfertheiner, P., Schultze, V., Rosenkranz, B., Kaufmann, S.H., Ulrichs, T., Novicki, D., Norelli, F., Contorni, M., Peppoloni, S., Berti, D., *et al.* (2008). Safety and immunogenicity of an intramuscular *Helicobacter pylori* vaccine in noninfected volunteers: a Phase I study. Gastroenterology *135*, 787–895.

Michetti, P., Kreiss, C., Kotloff, K.L., Porta, N., Blanco, J.L., Bachmann, D., Herranz, M., Saldinger, P.F., Corthesy-Theulaz, I., Losonsky, G., *et al.* (1999). Oral immunization with urease and *Escherichia coli* heat-labile enterotoxin is safe and immunogenic in *Helicobacter pylori*-infected adults. Gastroenterology *116*, 804–812.

Morihara, F., Hifumi, E., Yamada, M., Nishizono, A., and Uda, T. (2008). Therapeutic effects of molecularly designed antigen UREB138 for mice infected with *Helicobacter pylori*. Biotechnol. Bioeng. *100*, 634–643.

Panthel, K., Faller, G., and Haas, R. (2003). Colonization of C57BL/6J and BALB/c wild-type and knockout mice with *Helicobacter pylori*: effect of vaccination and implications for innate and acquired immunity. Infect. Immun. *71*, 794–800.

Ren, J.M., Zou, Q.M., Wang, F.K., He, Q., Chen, W., and Zen, W.K. (2002). PELA microspheres loaded *H. pylori* lysates and their mucosal immune response. World J. Gastroenterol. *8*, 1098–1102.

Rossi, G., Ruggiero, P., Peppoloni, S., Pancotto, L., Fortuna, D., Lauretti, L., Volpini, G., Mancianti, S., Corazza, M., Taccini, E., *et al.* (2004). Therapeutic vaccination against *Helicobacter pylori* in the beagle dog experimental model: safety, immunogenicity, and efficacy. Infect. Immun. *72*, 3252–3259.

Rupnow, M.F., Shachter, R.D., Owens, D.K., and Parsonnet, J. (2001). Quantifying the population impact of a prophylactic *Helicobacter pylori Vaccine* Vaccine *20*, 879–885.

Sijun, H, and Yong, X. (2009). *Helicobacter pylori* vaccine: mucosal adjuvant and delivery systems. Indian J. Med. Res. *130*, 115–124.

Skene, C.D., Doidge, C., and Sutton, P. (2008). Evaluation of ISCOMATRIX and ISCOM vaccines for immunisation against *Helicobacter pylori.* Vaccine *26*, 3880–3884.

Sougioultzis, S., Lee, C.K., Alsahli, M., Banerjee, S., Cadoz, M., Schrader, R., Guy, B., Bedford, P., Monath, T.P., Kelly, C.P., *et al.* (2002). Safety and efficacy of E. coli enterotoxin adjuvant for ureasebased rectal immunization against *Helicobacter pylori.* Vaccine *21*, 194–201.

Summerton, N.A., Welch, R.W., Bondoc, L., Yang, H.H., Pleune, B., Ramachandran, N., Harris, A.M., Bland, D., Jackson, W.J., Park, S., *et al.* (2010). Toward the development of a stable, freeze-dried formulation of *Helicobacter pylori* killed whole cell vaccine adjuvanted with a novel mutant of *Escherichia coli* heat-labile toxin. Vaccine *28*, 1404–1411.

Sun, B., Li, Z.S., Tu, Z.X., Xu, G.M., and Du, Y.Q. (2006). Construction of an oral recombinant DNA vaccine from *H pylori* neutrophil activating protein and its immunogenicity. World J. Gastroenterol. *12*, 7042–7046.

Talebkhan, Y., Bababeik, M., Esmaeili, M., Oghalaei, A., Saberi, S., Karimi, Z., Afkhami, N., and Mohammadi, M. (2010). *Helicobacter pylori* bacterial ghost containing recombinant Omp18 as a putative vaccine. J. Microbiol. Methods. *82*, 334–337.

Taylor, J.M., Ziman, M.E., Fong, J., Solnick, J.V., and Vajdy, M. (2007). Possible correlates of long-term protection against *Helicobacter pylori* following systemic or combinations of mucosal and systemic immunizations. Infect. Immun. *75*, 3462–3469.

Velin, D., Bachmann, D., Bouzourene, H., and Michetti, P. (2005). Mast cells are critical mediators of vaccine-induced *Helicobacter* clearance in the mouse model. Gastroenterology *129*, 142–155.

Voland, P., Zeitner, M., Hafsi, N., and Prinz, C. (2006). Human immune response towards recombinant *Helicobacter pylori* urease and cellular fractions. Vaccine *24*, 3832–3839.

Wu, C., Shi, Y., Guo, H., Zou, W.Y., Guo, G., Xie, Q.H., Mao, X.H., Tong, W.D., and Zou, Q.M. (2008). Protection against *Helicobacter pylori* infection in Mongolian gerbil by intragastric or intramuscular administration of *H. pylori* multicomponent vaccine. Helicobacter *13*, 191–199.

Xie, Y., Zhou, N.J., Gong, Y.F., Zhou, X.J., Chen, J., Hu, S.J., Lu, N.H., and Hou, X.H. (2007). Th immune response induced by *H pylori* vaccine with chitosan as adjuvant and its relation to immune protection. World J. Gastroenterol. *13*, 1547–1553.

Zhao, W., Wu, W., and Xu, X. (2007). Oral vaccination with liposome-encapsulated recombinant fusion peptide of urease B epitope and cholera toxin B subunit affords prophylactic and therapeutic effects against *H. pylori* infection in BALB/c mice. Vaccine *25*, 7664–7673.

Epidemiology of *Helicobacter pylori* Infection

7

Lyudmila Boyanova

Abstract

H. pylori causes the second most common chronic bacterial infection in humans. As a result of the childhood-acquired and usually life-long (unless eradicated) infection, about 12–24% of *H. pylori*-positive subjects develop severe diseases, e.g. peptic ulcers or gastric malignancy. In developed countries, the infection usually spreads intrafamilially; however, in developing countries or underdeveloped rural areas, it can be acquired often extrafamilially or via environmental contamination, leading to higher infection prevalence and greater intrafamilial diversity of the strains there. Infected mother and older siblings are important factors for *H. pylori* transmission to children. The transmission routes are oral–oral (by saliva), which prevails in the developed world, faecal–oral (person-to-person or by contaminated water, or maybe food), mainly in the developing countries or gastro–oral (by vomiting and regurgitation). The role of viable but not culturable coccoid forms and biofilms appears to be important. Oral *H. pylori* seems to be associated with combined oral and gastric infections, probably more often in the developing countries. However, the accuracy of PCR for detection of extragastric *H. pylori* needs improvement. As a whole, the infection prevalence is still high in countries/groups with poor socioeconomic status. In many developing countries, ≥50% of children and ≥70% of adults have been *H. pylori*-positive versus only <15% of children and ≤20–40% of adults in most developed countries. Detection of antiCagA antibodies has been used to reveal infections by virulent strains and the importance of East Asian CagA testing has been stressed. Both reinfection and infection clearance also have been reported, mainly in children, although in high-prevalence countries, the infection clearance is unimportant. Many risk factors for the infection mirror poor socioeconomic status and a strong birth cohort effect. Both dietary and environmental factors are likely to modify the infection prevalence and should be further evaluated. In brief, improved hygiene and living conditions, urbanization and growing antibiotic use for *H. pylori* and many other infections, have led to a decrease in both infection and reinfection rates in the developed countries and in some developing countries. However, the infection still affects every second person worldwide. Moreover, many important questions on the infection transmission routes and reservoirs still need elucidation.

Introduction

More than half of the people worldwide are *H. pylori* infected and some of them develop severe diseases, e.g. peptic ulcers (10–20%), gastric cancer (1–3%) and MALT lymphoma (<1%; Lee *et al.*, 2008; Malaty, 2010). *H. pylori* infection is usually life-long unless eradicated and has been considered to be the second most common chronic bacterial infection in humans (Tiwari *et al.*, 2005).

Sources of infection and transmission routes

H. pylori infection is an anthroponosis, affecting humans and, rarely, some primates. Occasionally,

H. pylori has been isolated from domestic animals (Dore *et al.*, 2001; Boomkens *et al.*, 2004) but the zoonotic transmission of the bacteria is still doubtful. *H. pylori* distribution patterns reflect the routes of human migrations, starting from East Africa, and confirm the interpersonal route of the infection transmission (Vale and Vítor, 2010). Epidemiology of the infection has been evaluated by ^{13}C urea breath test (UBT), stool antigen test (SAT), serology and many polymerase chain reaction (PCR)-based methods (Azevedo *et al.*, 2009). Among the non-invasive methods, UBT and SAT show the best performance, but the less expensive serological methods are used most often. However, serology cannot distinguish present from past infections and is suboptimal in children aged under 6 years because of possible lack of detectable antibodies. For instance, in Mexican children aged 2 years, the infection rate was found to be 19% by ^{13}C UBT but only 3% by serology (Nurgalieva *et al.*, 2008). Moreover, the transmission routes of *H. pylori* are still not completely elucidated.

Mechanisms of transmission of *H. pylori* infection are oral–oral through saliva (by kissing or touching contaminated oral secretions), faecal–oral (person-to-person or by contaminated water or perhaps food) and gastro-oral by vomiting and regurgitation (Leclerc, 2006; Mladenova *et al.*, 2006). Endoscopic transmission of *H. pylori* also has been reported in about 0.8–3% of patients in cases of manual disinfection but only occasionally in cases of automatic equipment (Buzás, 2010).

Importantly, *H. pylori* infection often spreads within families. In developed countries, *H. pylori* infection is usually acquired from an infected mother (most often) or other family member, highlighting the role of close contacts in oral–oral transmission (Leclerc, 2006; Weyermann *et al.*, 2006; Cartágenes *et al.*, 2009). Roma *et al.* (2009) have reported one or more infected family members in all *H. pylori*-positive Greek children tested. In the USA, exposure to infected and symptomatic family members has been found to increase the risk for infection 4.8-fold compared with controls (Perry *et al.*, 2006). In the USA, as a developed country, vomiting and hence the oral–oral route is a greater risk factor than only diarrhoea and hence faecal–oral transmission (Perry *et al.*, 2006). In Germany, the risk for child infection is 12.9-fold higher among those with an infected mother and only 1.4-fold higher among those with an infected father compared with controls (Weyermann *et al.*, 2006). In Japanese Brazilians, *H. pylori* prevalence has been higher in children with two positive parents (16%) and in those with only a positive mother (14.9%) than in those with a positive father (9.9%) or negative parents (3.5%; Ito *et al.*, 2006). Strains from mothers and their children have matched often (in 69% of the cases) in Japan but less frequently in Bangladesh (46%) and Peru (30%; Herrera *et al.*, 2008; Konno *et al.*, 2008; Nahar *et al.*, 2009). This can suggest other routes of transmission of the infection in addition to the oral–oral mechanism in some developing countries.

Spouse-to-spouse transmission also can occur (Brenner *et al.*, 2006). In Turkey, Sari *et al.* (2008) have obtained higher (92.9%) eradication of the infection, treating all infected family members versus only the examined patient (61.4%). Association of *H. pylori* infection for partners with non-German nationality has been strong, unlike that for partners born in Germany (Brenner *et al.*, 2006). Moreover, in Bangladesh, only 6% of strains from both partners have shared the same genotype (Nahar *et al.*, 2009). It should be considered, however, that both partners usually share the same risk factors for the infection, which can influence the interpretation of the results. As suggested by Brenner *et al.* (2006), the risk for spouse-to-spouse transmission of *H. pylori* is higher in countries/populations with high infection rates than elsewhere.

Whether healthcare personnel are at risk for *H. pylori* infection is still controversial. Higher *H. pylori* prevalence in endoscopy staff has been reported in some studies but not in others (Nelson and Muscarella, 2006; Loster *et al.*, 2009). It is, however, interesting that in Poland, the risk for the infection in dentists has been linked to male sex and prolonged (>15 years) professional exposure, suggesting *H. pylori* transmission by aerosol (Loster *et al.*, 2009). For healthcare staff, working in Belgian institutions for people with intellectual disability, faecal contact has been a risk factor for infection (De Schryver *et al.*, 2008).

Environmental factors also may be linked to the prevalence of *H. pylori* infection. Interestingly, in Germany, *H. pylori* risk has been elevated in subjects, living in homes at main roads (Strebel *et al.*, 2010). The authors have suggested the role of the long-term effect of pollutants to the immune system, which should be confirmed in further studies. Dietary factors also could influence the infection rates. Lower prevalence of *H. pylori* infection has been found in Mexican yoghurt consumers than in non-consumers (Ornelas *et al.*, 2007). In this study, the consumption of more than one yoghurt servings per week has decreased the seroprevalence twice in comparison with that in non-consumers. Consumption of fresh vegetables and fruits has prevented *H. pylori* reinfection (see Chapter 9). For example, in Pakistani children, *H. pylori* positivity was >2-fold lower in frequent consumers of vegetables and fruits than in others (Imanzadeh *et al.*, 2007). No such association has been found, however, in Canadian adults (Naja *et al.*, 2007). Alcohol intake, possibly owing to antimicrobial effects of wine, which contains the active substance resveratrol, has been associated with a decrease in *H. pylori* prevalence in adult women from Ontario (Canada), while an inverse association has been found in Ethiopian adults (Moges *et al.*, 2006; Naja *et al.*, 2007). In Canadian women, the alcohol consumption of ≥10 drinks per week has been associated with an odds ratio (OR) of only 0.6 for the infection, but even less frequent consumption has resulted in an OR of 0.5 compared with non-consumers (Naja *et al.*, 2007). Conversely, in Ethiopian adults, *H. pylori* seroprevalence has been higher (90.1%) in consumers of local alcoholic drinks and beer than in non-consumers (76.7%; Moges *et al.*, 2006). In brief, dietary and environmental factors may strongly modify the prevalence of *H. pylori* infection, but this association requires a proof by studies in more countries and larger patient groups of both children and adults.

While oral–oral transmission can be considered as the main *H. pylori* transmission route in the developed countries, additional extrafamilial or environmental contamination can occur, mostly in developing countries or underdeveloped rural regions (Herrera *et al.*, 2008). In developed countries such as Belgium, sewage exposure has not been found to be a risk factor for *H. pylori* acquisition (Van Hooste *et al.*, 2010). However, drinking water type has been a risk factor for the infection in South Africa and Brazil (Zaterka *et al.*, 2007; Dube *et al.*, 2009a). Especially in the developing countries, the following factors are usually more prevalent than those in the developed countries: contaminated water, enterocolitis, poor sanitation and intensive not only intrafamilial but also extrafamilial contacts. Faecal–oral transmission of the infection can be facilitated by diarrhoea, while gastro–oral transmission is usually facilitated by vomiting (Delport and van der Merwe, 2007). In rural South African families, transmission by contaminated water, food or frequent contacts between infants and caretakers has been suggested to be more common than intrafamilial transmission (Schwarz *et al.*, 2008). In brief, vertical transmission of the infection, e.g. from parents to children, is prevalent in the developed countries, while in the developing countries, horizontal transmission (outside the family) contributes to common infections and higher diversity, even within a family, of *H. pylori* strains. Both the prevalence and the transmission routes of *H. pylori* infection show great differences between the developed and developing countries.

Faecal–oral transmission of *H. pylori* has been proven in many studies. *H. pylori* has been isolated from vomit specimens, diarrhoeal stools and saliva (Suzuki *et al.*, 2007a; Vale and Vítor, 2010). Cimetidine-induced hypochlorhydria has increased the isolation of viable *H. pylori* from stool specimens, showing that *H. pylori* can survive, although perhaps only briefly, in the intestinal tract (Haggerty *et al.*, 2003). Faecal–oral transmission is also supported by acquisition of the infection after outbreaks of gastroenteritis in institutionalized young people (Laporte *et al.*, 2004). Institutionalized Czech children have also been reported to be at higher risk (OR, 6.3) for the infection than other children (Sykora *et al.*, 2009). In addition, the seroprevalence rate of the infection has been higher (50%) in Italian non-institutionalized children with mental retardation than in control subjects (19%; Luzza *et al.*, 2004). These studies highlight the role of hygiene in *H. pylori* transmission.

However, abundant intestinal flora and bile activity in the intestinal tract hamper *H. pylori* isolation from faecal specimens. Furthermore, faecal *H. pylori* is often in a viable but non-culturable form (VBNC; Vale and Vítor, 2010). PCR or PCR-based methods and SAT, used to detect faecal *H. pylori*, do not prove the viability of the bacteria and some PCR techniques could give false-positive results owing to cross reactions with DNA from non-*pylori* helicobacters (Sugimoto *et al.*, 2009). Conversely, low *H. pylori* number or presence of inhibitors in the extragastric samples could give false-negative results (Sugimoto *et al.*, 2009).

Different methods have been used to detect *H. pylori* in the environment, e.g. in water. Among them are the following: culture, PCR, real-time PCR (RT-PCR) and nested PCR and, for detection of *H. pylori* coccoids, scanning electron microscopy, epifluorescence microscopy and fluorescence *in situ* hybridization (FISH; Vale and Vítor, 2010). Many genes have been used for PCR detection of *H. pylori*, e.g. 16S rRNA, *glmM*, *ureA*, *ureB*, *vacA*, *cagA* and also 26K species-specific antigen gene but no gold standard has been established so far (Sugimoto *et al.*, 2009). To abolish the problem of high genetic polymorphism of *H. pylori* genes, Liu *et al.* (2008) have used a 76-bp sequence of 16S rRNA, which is common and highly specific to *H. pylori*, for RT-PCR and *in situ* hybridization for *H. pylori* detection. The methods can be helpful for detection of the bacteria in gastric and extragastric samples.

Indeed, *H. pylori* has been detected in the environment mostly by molecular methods. *H. pylori* DNA has been present in 0–23% of drinking water, 0–50% of river water and 14–66% of wastewater samples worldwide (Vale and Vítor, 2010). The choice of the detection method is of primary importance to obtain accurate results in similar studies. For instance, Nayak and Rose (2007) have detected *H. pylori* DNA at 2–28 cells/ml in sewage samples. Of all samples, the authors have proven presence of *vacA* gene in 84% by quantitative RT-PCR but not by conventional PCR (Nayak and Rose, 2007). Fascinatingly, although in few studies, *H. pylori* has been isolated from water such as fractionated water samples with planktonic organisms (by filtration, culture and PCR) and a Mexican wastewater canal (by immunomagnetic separation, subsequent culture and PCR; Lu *et al.*, 2002; Cellini *et al.*, 2005). Other data also suggest the water to be a source for the transmission of *H. pylori* infection. In sewage and water, *H. pylori* can remain viable and culturable for 24 h at 15°C and for as long as five days at 4°C (Nayak and Rose, 2007). In hostile conditions, *H. pylori* quickly produces VBNC forms, however, even in chlorinated drinking water, the VBNC forms have been FISH detected 3 h post inoculation (Moreno *et al.*, 2007).

The role of water has been additionally suggested by frequently reported risk factors for the infection such as drinking non-hygienic or well water in South Africa, Uganda, Brazil and Iran (Zaterka *et al.*, 2007; Dube *et al.*, 2009a; Nouraie *et al.*, 2009; Hestvik *et al.*, 2010). Water source can be linked to other risk factors for the infection, e.g. the place of residence, country and educational level of the subjects, therefore, the importance of VBNC in *H. pylori* transmission merits much more evaluation.

In addition, like other Gram-positive and Gram-negative bacteria, *H. pylori* can live in polysaccharide-containing biofilms on surfaces exposed to water (at the air–liquid border) or on stainless steel coupons as well as on glass surface and in batch cultures *in vitro* (Yonezawa *et al.*, 2009). Biofilms are bacterial communities, which form multi-layer microcolonies, producing thick extracellular matrix, where the bacteria communicate among them as well as with the host organism by quorum-sensing signals. Importantly, these bacterial communities strongly oppose hostile environment, including disinfection and antibiotics. The attached bacteria use biofilm channels for their nutrition. *H. pylori* biofilm formation has been a strain-specific property and has been associated with outer membrane vesicle production (Yonezawa *et al.*, 2009). By a two-stage chemostat model system, Gião *et al.* (2008) have observed numerous *H. pylori* cells with high rRNA levels in biofilms, suggesting cell viability, for more than one month. Biofilms could contribute to *H. pylori* transmission and survival as well as for strong colonization of the gastric mucosa and treatment failures.

Recently, *H. pylori* has been PCR detected also in 9% of soil samples from Spanish parks (Pérez *et al.*, 2010); however, the viability of the bacteria in soil is dubious.

The role of food, especially raw milk, for the infection transmission has been suggested. *H. pylori* DNA has been PCR detected in 26–72% of raw goat, sheep and cow milk samples, and the bacteria have been isolated, although seldom, from raw cow and sheep milk samples (Dore *et al.*, 2001; Vale and Vítor, 2010). It is noteworthy that in Japan, Fujimura *et al.* (2002) have detected *H. pylori* DNA by semi-nested PCR in 72.2% of 18 raw cow milk samples and have also isolated a *H. pylori* strain in 5.6% (one) of these samples. In addition, *H. pylori* has been able to survive for >9 days in artificially contaminated pasteurized and ultrahigh temperature milk samples (Quaglia *et al.*, 2007). Interestingly, after contamination of spinach, although *H. pylori* cells have lost their culturability, their mRNA transcripts, including *vacA* mRNA transcripts, have been found 6 days later (Buck and Oliver, 2010). Role of the food, where, most probably, *H. pylori* cannot grow, in the infection transmission, although still uncertain, appears possible. It may be of importance in case of consumption of food shortly after its contamination. Again, the role of VBNC forms in *H. pylori* transmission should be evaluated.

Oral *H. pylori*

The role of oral *H. pylori* for the gastric infection is controversial. However, *H. pylori* has been PCR detected and isolated in saliva, dental plaque and vomit specimens, and in patients with gastric *H. pylori*, the bacteria may reach the oral cavity by gastro-oesophageal reflux, vomiting and regurgitation (Loster *et al.*, 2006).

Successful *H. pylori* isolation from oral specimens has ranged widely from 0% to 10% in many countries to about 50% in Poland and 69.7% in Nigeria (Table 7.1). However, owing to the possible existence of VBNC forms and the large number (>300) of oral bacterial species, the isolation of oral *H. pylori* is difficult. In Bulgaria, *H. pylori* has been isolated in only 2.3% of dental specimens (Boyanova *et al.*, 2009). Similarly, Brazilian authors (Kignel *et al.*, 2005) have detected *H. pylori* in just one dental plaque but frequent (40.8%) gastric isolation in the dyspeptic subjects.

Methods for detection of oral *H. pylori* are important for the reliability of results. The rapid urease test (RUT) is inappropriate to detect oral *H. pylori* as other bacteria, e.g. *Staphylococcus, Streptococcus, Haemophilus* and *Actinomyces* spp., may give false-positive results. In India, false-positive RUT results have been found in two-thirds of the dental plaque specimens compared with culture (Sudhakar *et al.*, 2008). Direct Gram stain is also unreliable indicator for oral *H. pylori* since other curved bacteria, e.g. *Treponema denticola, Campylobacter curvus, Campylobacter showae*, etc., are also present in the oral cavity. Thus, gastric *H. pylori* can be easily identified by phenotypic tests, while extragastric *H. pylori* should be confirmed by molecular methods with primer pairs for over two different specific genes (Megraud and Lehours, 2007). In Pakistani patients, by simultaneous amplification of two *H. pylori* genes, 51.6% of the dental plaque samples have been *H. pylori* positive versus 73% when only one gene has been used (Chaudhry *et al.*, 2011). Nevertheless, in a study of Sugimoto *et al.* (2009), no combination of five primer pairs evaluated has increased the accuracy of PCR tests for *H. pylori* detection in saliva.

Although PCR detects minimal DNA amount, it cannot distinguish between live and dead bacteria. Choice of PCR primers, specimens and type of population also affects the results on the prevalence of oral *H. pylori*, which has varied from 0% to 53.2% in healthy subjects and from 0% to 72% in patients with gastritis (Table 7.1). According to the methods and subject characteristics, oral *H. pylori* detection by PCR has oscillated from 0% to 46% of cases in Iran and from 2% to 72% in Brazilian symptomatic patients (Kignel *et al.*, 2005; Nasrolahei *et al.*, 2008; Eskandari *et al.*, 2009; Assumpção *et al.*, 2010). In some studies, *H. pylori* has been PCR detected most often in gingival plaque than in saliva (Gebara *et al.*, 2004; Morales-Espinosa *et al.*, 2009), while other studies have not reported such a difference (Czesnikiewicz-Guzik *et al.*, 2005; Medina *et al.*, 2010). Importantly, 60% of the Brazilian symptomatic patients have been *H. pylori* positive in both oral and biopsy specimens and 26.6% of them have shown *cagA* positivity (Silva *et al.*, 2009). Although there are large variations in the prevalence of oral *H. pylori*

Table 7.1 Detection of *H. pylori* in oral cavity by culture or polymerase chain reaction (PCR) in some studies since 2000

Method	Country	Group	Age groups	No. of subjects	Type of oral specimen	% of positive specimens	Reference
Culture	Bulgaria	GDS and H	Total	43	DP, saliva	2.3	Boyanova *et al.* (2009)
	India	GDS	Adults	50	DP	10	Sudhakar *et al.* (2008)
	Nigeria	GDS	NA	66	DP	69.7	Ogunbodede *et al.* (2002)
	Poland	GDS	Adults	40	DP, saliva	50	Loster *et al.* (2006)
	Poland	GDS and H	Adults	49	DP, saliva	42.9, 53.2 of H, 56.9, 54.9 of GDS	Czesnikiewicz-Guzik *et al.* (2005)
	Turkey	GDS	Adults	295	DP, denture plaque	5.4, 0 in dentate patients, 9.1, 0.3 in edentulous patients	Agüloğlu *et al.* (2006)
	Turkey	Chronic pharyngitis	NA	70	Pharyngeal mucosa	5.8 (pharyngitis), 0 (healthy)	Kaptan *et al.* (2009)
	UK	GDS	Children	100	DP	0	Allaker *et al.* (2002)
	Venezuela	GDS and H	NA	97	DP, saliva	0	De Sousa *et al.* (2006)
PCR	Argentina	GDS and H	Total	98	DP, saliva	18.4, 34.9 in both gastric and oral sites	Medina *et al.* (2010)
	Brazil	GDS and periodontitis	Adults	30	DP, saliva	20–26.6, 10	Gebara *et al.* (2004)
	Brazil	GDS	NA	49	DP	2	Kignel *et al.* (2005)
	Brazil	GDS	Adults	99	DP	72	Assumpção *et al.* (2010)
	Brazil	GDS and H	Adults	62	DP, saliva	36.6, 53.3 in GDS and 0, 0 in H; *cagA*$^+$ strains: 27.3, 43.8 in GDS and 0, 0 in H	Silva *et al.* (2009)
	China	GDS	NA	443	DP	59.4	Liu *et al.* (2009)
	Germany	GDS and H	NA	94	DP	10.6 in oral sites only, 6.4 in oral and gastric sites	Bürgers *et al.* (2008)
	Iran	GDS	Adults	260	DP	46	Nasrolahei *et al.* (2008)
	Iran	GDS and periodontitis	Adults	23	DP	17.4	Eskandari *et al.* (2009)
	Iran	Periodontitis	Adults	44	DP	0	Eskandari *et al.* (2009)
	Israel	H	NA	47	DP	24	Al-Hawajri *et al.* (2004)
	Japan	GDS	NA	57	DP	35.1	Umeda *et al.* (2003)
	Japan	H, periodontitis	Total	326	Saliva	6.4 (total), 15.7 (periodontitis)	Suzuki *et al.* (2008)
	South Africa	H	NA	58	DP, saliva	3.4 (total): 1.7, 1.7	Goosen *et al.* (2002)
	South Africa	H	Total	79	DP	0	Olivier *et al.* (2006)
	Turkey	GDS	NA	67	DP	28.4	Teoman *et al.* (2007)
	Turkey	Chronic pharyngitis	NA	70	Pharyngeal mucosa	34.3 in the pharynx and stomach, 20 in the pharynx	Kaptan *et al.* (2009)

Table 7.1 continued

Method	Country	Group	Age groups	No. of subjects	Type of oral specimen	% of positive specimens	Reference
	UK	GDS	Children	100	DP	36	Allaker *et al.* (2002)
	USA	GDS	NA	37	Saliva	35.1 (by ≥1 primer pair, 54% of them false-positive[1])	Sugimoto *et al.* (2009)

[1]Positive result in a patient whose gastric mucosa was negative for *H. pylori* by multiple tests.

GDS, gastroduodenal symptoms; H, healthy; DP, dental plaque; NA, not available.

among the studies, it could be speculated that the highest prevalence might be observed mostly in some developing countries.

Choice of primers for detection of the few *H. pylori* cells in the oral specimens with abundant normal flora is highly important. EHC-L/EHC-U primers (for 860-bp *H. pylori* DNA) have been more accurate for oral *H. pylori* than HpU1/HpU2 (for *ureA* gene) and Hp1/Hp2 primers (for 16S rRNA gene) according to Song *et al.* (2000). Importantly, false-positive PCR results (in patients with *H. pylori* negative gastric mucosa) have been frequent in the saliva. Using five primer pairs (Hp1/Hp2, Hp64-f/Hp64-r, EHC-U/EHC-L, VAG-F/VAG-R and ICT37/ICT38), 54% of positive saliva samples have been false-positive in the study of Sugimoto *et al.* (2009). The PCR sensitivity in saliva has been only 18–29% and specificity has ranged from 75% to 100% (Sugimoto *et al.*, 2009). In other studies, hemi-nested and nested PCR techniques have been used to improve specificity (Umeda *et al.*, 2003; Olivier *et al.*, 2006). Hemi-nested PCR assay based on *H. pylori* phosphoglucosamine mutase gene (*glmM*) has been able to detect 5×10^2 cells of oral *H. pylori* (Goosen *et al.*, 2002). However, Sugimoto *et al.* (2009) have recommended the use of magnetic beads or other techniques to separate and concentrate bacteria as well as primers for *H. pylori* virulence genes to detect the extragastric bacteria.

Risk factors for the presence of oral *H. pylori* can be gastric infection, poor oral hygiene and periodontitis. Medina *et al.* (2010) found 6.5-fold higher prevalence of oral *H. pylori* in Argentinean patients with gastroduodenal diseases (34.9%) than in healthy subjects (5.4%). Similar results have been reported in Mexico (Morales-Espinosa *et al.*, 2009). In Iran, 17.4% of patients with both gastritis and periodontitis harboured oral *H. pylori* compared with none of those with periodontitis without gastritis (Eskandari *et al.*, 2009).

Interestingly, subjects with good oral hygiene have harboured oral *H. pylori* less often, possibly owing to *H. pylori* inhibition by oral streptococci and other bacteria, which participate in the early colonization of dental biofilm (Suzuki *et al.*, 2008). Conversely, patients with oral *H. pylori* have more often shown occult blood and periodontopathogenic anaerobes such as *Prevotella intermedia* (Suzuki *et al.*, 2008) or *Tannerella forsythia* (Umeda *et al.*, 2003) in the saliva compared with the *H. pylori*-negative subjects. The high inflammation in the periodontitis patients has been suggested to enhance the colonization by *H. pylori* (Suzuki *et al.*, 2008).

However, the role of oral cavity as a reservoir of *H. pylori* is not convincing as the oral and the gastric strains may differ or the oral strains may be absent in the stomach (Leclerc, 2006; Bürgers *et al.*, 2008). Nevertheless, in Brazil, 89% of *H. pylori*-positive dental plaques harboured the same *vacA* and *cagA* genotypes as in gastric mucosa (Assumpção *et al.*, 2010). In China, the prevalence of gastric infection has been higher in patients with oral *H. pylori* than in those with *H. pylori*-negative dental plaques (Liu *et al.*, 2009).

There is still a controversy about the association of oral *H. pylori* and the eradication of the gastric infection. In Poland, the eradication of the gastric bacteria has not influenced the oral *H. pylori* in the study of Czesnikiewicz-Guzik *et al.* (2007) who have proven no association between the PCR profiles of the bacteria from both sites. Conversely, Japanese authors (Miyabayashi *et al.*, 2000) have reported the oral *H. pylori* to be associated with a risk for infection relapse. Two years after treatment by lansoprazole, metronidazole and clarithromycin, the authors have found eradication in only 69.5% of the patients with oral *H. pylori* but in much more (95.8%) patients without the oral bacteria. The importance of oral *H. pylori* has been stressed also by Zaric *et al.* (2009). The authors have found that more (77.3%) Serbian patients treated with combined periodontal and triple therapy have shown successful eradication of the gastric *H. pylori* compared with those on triple therapy only (47.6%). Similarly, in China, Jia *et al.* (2009) reported the prevalence of gastric *H. pylori* to be 4.3-fold lower (19.6%) in patients with dental plaque control than in those without the control (84.3%). The possible protective role of the long-term professional dental plaque control against *H. pylori* infection or reinfection has been suggested by the authors. In brief, the importance of oral *H. pylori* has been suggested by many but not all studies. This can imply that the bacteria inhabiting the oral cavity may be important in some patients and in some countries.

Other sites of the oral cavity also have been evaluated as possible reservoirs of *H. pylori* infection with so far controversial results. In Finnish children prone to respiratory infections, cultures of adenoid and middle ear fluid samples yielded no growth of *H. pylori* (Pitkäranta *et al.*, 2005). Conversely, in Turkish patients with chronic pharyngitis, *H. pylori* has been detected by PCR in the pharyngeal mucosa in 34.3% of patients with gastric infection and in 20% of uninfected patients, and, although only in 5.8% of all patients, *H. pylori* has been successfully isolated by Kaptan *et al.* (2009). The detection of *H. pylori* has been more common in the patients with chronic pharyngitis than in the control group. This association merits evaluation in more studies. In addition, controversial data have been reported on *H. pylori* symbiosis with *Candida* or *Acanthamoeba* species (Goldman *et al.*, 2006; Azevedo *et al.*, 2009; Vale and Vítor, 2010; Vilarinho *et al.*, 2010). Possible significance of the recently suggested reservoirs of the infection is interesting but should be further clarified.

Briefly, the oral *H. pylori* could be of importance for part of the patients with gastric or chronic oral infections, possibly more often in the developing countries. Detection of oral *H. pylori* strongly depends on the methods, specimens and population type. Diagnosis of oral *H. pylori*, however, should not be used to identify the gastric infection because of the low sensitivity of PCR detection of oral *H. pylori* and its variable presence in this site (Megraud and Lehours, 2007; Sugimoto *et al.*, 2009). Improved PCR accuracy for *H. pylori* detection in extragastric samples is required. Significance of oral *H. pylori* for periodontal and gastric diseases, reinfection and treatment success as well as the presence of additional reservoirs of the infection merit extensive evaluation in more countries and larger patient groups and, not to forget, by highly accurate tests.

Prevalence and risk factors for the infection

Prevalence of *H. pylori* infection has shown large geographic variations (Tables 7.2 and 7.3). The present chapter focuses mostly on the prevalence of the infection in asymptomatic subjects because that in symptomatic patients, reasonably, is much higher (Sedlackova *et al.*, 2003) and dependent on the disease.

In countries or groups with poor socioeconomic status, the infection prevalence has been higher than elsewhere. In many developing countries, ≥ 50% of children and ≥ 70% of adults have been infected versus only < 15% of children and ≤ 20–40% of adults from the developed countries, (Fig. 7.1). Highest (≥ 70% for children or ≥80% for adults) infection rates have been reported in studies from Africa (Egypt, Nigeria and South Africa) and South America (Brazil and Venezuela, in Amerindians; Fritz *et al.*, 2006; Marini *et al.*, 2007; Mohammad *et al.*, 2008; Cartágenes *et al.*, 2009; Dube *et al.*, 2009a; Tijjani and Umar, 2009). In European adults, the infection rates have ranged from 11% in Sweden to > 70% in Albania and Poland (Celiński *et al.*, 2006; Thjodleifsson *et al.*, 2007; Monno *et al.*, 2008; Table 7.3).

Fascinatingly, some populations in developing countries have shown a strangely low prevalence of the infection. Low prevalence (< 18%) of the infection has been reported in Javanese subjects, Malaysian blood donors and pregnant women in Zanzibar (Tokudome *et al.*, 2005; Graham *et al.*, 2007; Sasidharan and Uyub, 2009). A role of

Table 7.2 Prevalence of *H. pylori* infection in asymptomatic children in some countries during the last 10 years

Country (subjects)	No. tested	Age (years)	Methods	Prevalence (%) of the infection	Main risk factors	Reference
Europe						
Czech Republic	1545	0–15	MAb-SAT	7.1	Age[1], SeS (number of children, poor paternal education, institutionalized children)	Sykora *et al.* (2009)
Germany (mostly children)	1083	0–30	ELISA IgG	13.1 (Germans); 30.4 (Turks in Germany)	Age, ethnicity	Porsch-Ozcürümez *et al.* (2003)
Germany (review)	NA	5–7	NA	5–7; 36–44 (immigrants)	Age, ethnicity (Turkish)	Fischbach *et al.* (2009)
Italy	172	0.1	SAT	2.9	Formula feeding	Baldassarre *et al.* (2009)
Russia	370	<5–19	ELISA IgG	13	Study year, age, SeS (poor parent education), anti-*H. pylori* therapy (inverse)	Tkachenko *et al.* (2007)
Serbia	283	7–18	ELISA IgG/IgA	36.4	Age	Dinić *et al.* (2007)
Spain (primary care)	59	0–19	ELISA IgG	11.9	Age	Baena Diez *et al.* (2002)
Switzerland	196	15–16	ELISA IgG	9.7	SeS (immigrants)	Heuberger *et al.* (2003)
The Netherlands	1258	2–4	ELISA IgG	1.2	SeS (non-Dutch parent)	Mourad-Baars *et al.* (2007)
North America						
Mexico	472	0.5–2	UBT	7 (0.5 years), 19 (2 years)	Age	Nurgalieva *et al.* (2008)
USA	1227	3–11	ELISA IgG	5.5 (3–5 years), 9.2 (6–11 years)	Age, SeS (racial/ethnic minorities, birthplace outside the USA), non-breastfed children	Cardenas *et al.* (2006)
South America						
Brazil	217	0.2–6	UBT, ELISA IgG	40	Age, infected mother	Braga *et al.* (2007)
Brazil	100	1–12	PAb-SAT, ELISA IgG/CagA	54.4 (SAT); 50 (ELISA), of them, 78 CagA$^+$	Infected mother, age	Cartágenes *et al.* (2009)
Ecuador	257	0.5–16	ELISA	63	Geographic region (Andes mountains), symptoms	Gómez *et al.* (2004)
Venezuela (Amerindians)	56	2–20	UBT	46–47	Age (versus adults), nutritional status (better)	Marini *et al.* (2007)
Asia						
Bangladesh	238	NA	ELISA IgG/IgA, SAT	50 (SAT), 60 (serology)	Blood group (A)	Bhuiyan *et al.* (2009)
China	1036	8–15	PAb-SAT	31.7 (25.6–37.7)	Age, geographic region	Zhang *et al.* (2009a)

Table 7.2 continued

Country (subjects)	No. tested	Age (years)	Methods	Prevalence (%) of the infection	Main risk factors	Reference
India (healthy)	137	0.7–16	PCR on stool	4.2 (<5 years)–50 (11–16 years)	Age, SeS (poor)	Mishra *et al.* (2008)
Iran	386	1–15	ELISA IgG/ CagA	46.6, of them, 72.8 CagA$^+$	Age, sex (male) for antiCagA positivity	Jafarzadeh *et al.* (2007)
Israel (daycare)	316	0.25–5	MAb-SAT	24.7	Age (>1 year)	Kori *et al.* (2009)
Japan	454	0–15	ELISA IgG	12.2	Age	Kato *et al.* (2003)
Japan	452	4–10	Urinary ELISA IgG	4–6.7	NA	Naito *et al.* (2008)
Malaysia (Penan)	114	0.6–10	PAb-SAT	37.7	SeS (lack of a flushing toilet), geographic region	Huang *et al.* (2004)
Pakistan	1976	1–15	ELISA IgG	47	Age, SeS (low/middle; poor father's education)	Jafri *et al.* (2010)
Turkey	275	1–15	ELISA IgG, SAT	23.6	Age, mother (infected), place of residence (urban)	Ceylan *et al.* (2007)
Turkey	327	3–12	UBT	52.9–56.6	SeS (poor, household crowding), antibiotic use (no use during last six months)	Ozen *et al.* (2006)
Turkey	529	0–19	ELISA IgG/ CagA	30, of them, 50.8–54 CagA$^+$	Age, sex (male) for antiCagA positivity	Apan *et al.* (2008)
Vietnam	824	0.5–15	ELISA IgG	34	Breastfeeding under 6 months	Nguyen *et al.* (2006)
Vietnam	224	0–19	ELISA IgG	62.5 (total), 33.3 (0–4 years)–78.3 (15–19 years)	Age, place of residence (urban)	Hoang *et al.* (2005)
Africa						
Cameroon	176	0–10	PAb-SAT	52.3	Age, SeS (low), sex (male)	Ndip *et al.* (2004)
Egypt	286	6–15	UBT	72.4	SeS (household crowding)	Mohammad *et al.* (2008)
Libya	100	1–19	ELISA IgG	67 (total): 50 (1–9 years), 84 (10–19 years)	Age, SeS (low), education (poor)	Bakka and Salih (2002)
Kenya	195	0–3	R-SAT	45.6	SeS (low, household crowding, poor sanitation, low family income)	Langat *et al.* (2006)
South Africa	NA	<12	MAb-SAT	75.9	Age, SeS (drinking water type, sanitation)	Dube *et al.* (2009a)
Tunisia	1055	6–7	ELISA IgG	51.4	SeS (low, household crowding, bed-sharing, cup-sharing), late bottle-weaning, abdominal pain	Siai *et al.* (2008)
Uganda	427	0–12	R-SAT	44.3 (total), 28.7 (<1 year), 40 (9–12 years)	Age, sex (male), SeS (sanitation – sharing toilet; type of housing –semi-permanent; drinking water type	Hestvik *et al.* (2010)

Table 7.2 continued

Country (subjects)	No. tested	Age (years)	Methods	Prevalence (%) of the infection	Main risk factors	Reference
Australia						
Australia	902	1–19	ELISA IgG	4 (1–4 years), 10 (15–19 years)	Age	Moujaber *et al.* (2008)
Australia (Africans)	193	<16	MAb-SAT	81.9	Age	Cherian *et al.* (2008)

[1]Age-increased age.

MAb SAT, monoclonal antibody EIA-based stool antigen test (SAT); PAb-SAT, polyclonal antibody EIA-based SAT; R-SAT, monoclonal antibody-based rapid SAT; UBT, urea breath test; ELISA, enzyme-linked immunosorbent assay; SeS, socioeconomic factors.

Table 7.3 Prevalence of *H. pylori* infection in asymptomatic adults in some countries during the last 10 years

Country (subjects)	No. tested	Age (years)	Methods	Prevalence (%) of the infection	Main risk factors	Reference
Europe						
Albania (mainly adults)	1088	16–64	ELISA IgG	70.7	Age, sex (female)	Monno *et al.* (2008)
Belgium	567	NA	ELISA IgG	16.7 (sewage workers), 13.6 (control group)	None	Van Hooste *et al.* (2010)
Estonia	240	25–50	ELISA IgG/ CagA	69, of them, 62 CagA$^+$	SeS, possibly smoking	Thjodleifsson *et al.* (2007)
Germany (review)	NA	<30–45	NA	24–30, 52–86 (immigrants)	Age, ethnicity	Fischbach *et al.* (2009)
Iceland	447	25–50	ELISA IgG/ CagA	36, of them, 36 CagA$^+$	Age, SeS, possibly smoking	Thjodleifsson *et al.* (2007)
Poland	585	20->80	ELISA IgG	78.5	Age, SeS (poor sanitation), place of residence (small towns), birthplace (in the country)	Celiński *et al.* (2006)
Poland (shepherds and farmers)	131	NA	UBT, serology	97.6 (shepherds), 65.1 (farmers)	SeS (shepherds)	Papiez *et al.* (2003)
Spain (primary care)	208	20->69	ELISA IgG	63.9	Age	Baena Diez *et al.* (2002)
Sweden	359	25–50	ELISA IgG/ CagA	11, of them, 69 CagA$^+$	Age, SeS, possibly smoking	Thjodleifsson *et al.* (2007)

Table 7.3 continued

Country (subjects)	No. tested	Age (years)	Methods	Prevalence (%) of the infection	Main risk factors	Reference
North America						
Canada (Ontario)	1306	50–80	ELISA IgG	23.1	SeS (non-whites, immigrants, birthplace; poor education), age, number of siblings, alcohol non-consumption	Naja *et al.* (2007)
USA	5153	16->70	ELISA IgG	15–39.5 (16–69 years)–45 (≥70 years)	Age, SeS (racial/ethnic minorities, birthplace outside the USA; poverty), low vitamin C intake	Cardenas *et al.* (2006)
South America						
Brazil	993	NA	ELISA	63.2–66.5	Age, prior gastrointestinal endoscopy, SeS (household crowding, drinking water type, poor sanitation, low family income, poor education)	Zaterka *et al.* (2007)
Brazil (women)	100	Young	ELISA IgG/ CagA	86, of them, 82.6 $CagA^+$	NA	Cartágenes *et al.* (2009)
Chile	2615	17–>65	ELISA IgG	73	Age, sex (male), SeS (residence in areas with high gastric cancer mortality), hepatitis A	Ferreccio *et al.* (2007)
Venezuela (Amerindians)	34	20–70	UBT	82	Age (versus children), nutritional status (better)	Marini *et al.* (2007)
Asia						
China	1029	40–79	PAb-SAT, histology	41.4–51	Geographic region	Zhang *et al.* (2009a)
Georgia	94	NA	ELISA IgG/ CagA	70.2, 57.4 $CagA^+$	Age	Olivares *et al.* (2006)
India	108	17–60	PCR on stool	58.6–64	SeS (poor)	Mishra *et al.* (2008)
Iran	200	20–60	ELISA IgG/ CagA	67.5, of them, 67.4 $CagA^+$	Age, sex (male) for $CagA^+$	Jafarzadeh *et al.* (2007)
Iran	2561	18–65	ELISA IgG	69	Age, SeS (low education, household crowding, drinking water type, sanitation in childhood)	Nouraie *et al.* (2009)
Japan	186	NA	R-SAT, RT-PCR on faeces, PCR for *cagA*	40.3, of them 86.7 by PCR, of them 86.2 *cagA*$^+$ and 63.1 East Asian type	NA	Hirai *et al.* (2010)

Table 7.3 continued

Country (subjects)	No. tested	Age (years)	Methods	Prevalence (%) of the infection	Main risk factors	Reference
South Korea	8020	≥16	ELISA IgG	59.6	Age, SeS (low income, province), sex (male), year of study	Yim *et al.* (2007)
Thailand	179	40–80	R-SAT, RT-PCR	43.6, of them, 88.5 by PCR, of them 43.5 *cagA*⁺ and 7.2 East Asian type	NA	Hirai *et al.* (2010)
Turkey	351	20->60	ELISA IgG/ CagA	58.1, of them, 53.9 CagA⁺	Age, sex (female for IgG, male for CagA⁺)	Apan *et al.* (2008)
United Arab Emirates (workers)	426	NA	ELISA IgG	64.3–78.4	SeS (poor living conditions, education, sanitation)	Bener *et al.* (2002)
Vietnam	738	20->80	ELISA IgG	78	Place of residence (urban)	Hoang *et al.* (2005)
Africa						
Ethiopia (dyspeptic)	215	<20–79	ELISA IgG	85.6	Age, alcohol consumption (local drinks)	Moges *et al.* (2006)
Libya	260	20–>70	ELISA IgG	80	Age, SeS, education (illiterates), marital status (married)	Bakka and Salih (2002)
Nigeria (normal endoscopy)	134	35.9[1]	Histology	80.6	NA	Tijjani and Umar (2009)
South Africa	90	NA	PCR on biopsies	83.3	NA	Fritz *et al.* (2006)
South Africa	356	31[1]	MAb-SAT	100 (25–47 and ≥60 years)	Age, SeS (drinking water type, sanitation)	Dube *et al.* (2009a)
Tunisia	250	NA	ELISA IgG/ CagA	64, 11 CagA⁺	NA	Mansour *et al.* (2010)
Australia						
Australia	1511	20–59	ELISA IgG	19.4	Age	Moujaber *et al.* (2008)

[1]Mean age.

RT-PCR, real-time polymerase chain reaction; UBT, urea breath test; MAb-/PAb-/R-SAT, monoclonal-/polyclonal antibody-/rapid-stool antigen test; SeS, socioeconomic factors.

founding populations and dietary factors has been suggested but is not well elucidated so far.

In many countries, *H. pylori* seroconversion often peaks at 2–5 years of age (Delport and van der Merwe, 2007; Salih, 2009). In developing countries, e.g. India and Brazil, the infection rate increases sharply after 5 years of age and remains high thereafter, confirming that the acquisition of infection occurs mostly in early childhood (Braga *et al.*, 2007; Mishra *et al.*, 2008; Nurgalieva *et al.*, 2008). Conversely, in developed countries such as Germany and Spain, the infection rates have

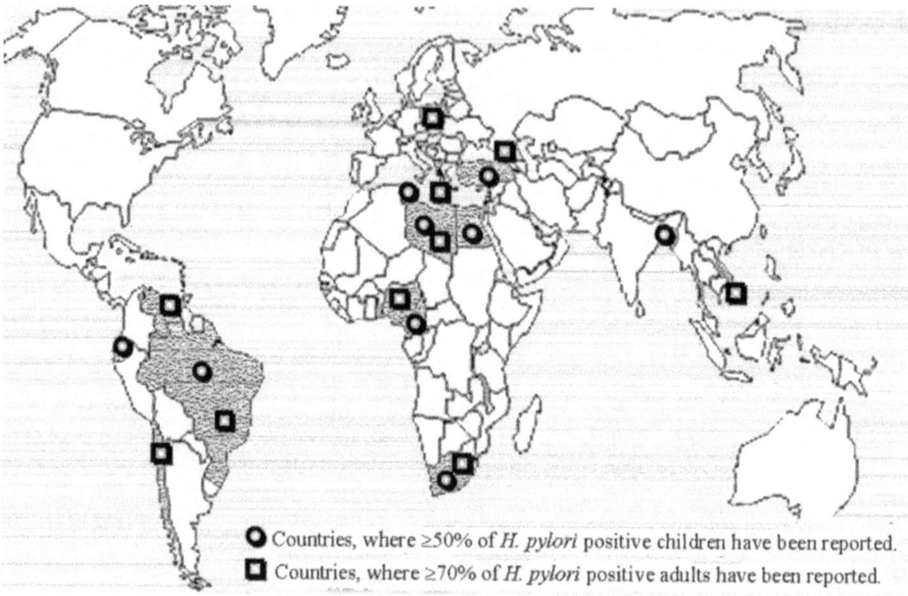

Figure 7.1 Countries in which ≥50% of *H. pylori*-positive children and ≥70% of *H. pylori*-positive adults have been reported.

been low in children, increasing slowly afterwards (Baena Diez *et al.*, 2002; Fischbach *et al.*, 2009). In Germany, the infection has affected 30% of adults and 5–13.1% of children versus 55.9% of adolescents and 68.9% of persons aged 31–45 years in India (Porsch-Ozcürümez *et al.*, 2003; Mishra *et al.*, 2008; Fischbach *et al.*, 2009). In some developing countries, mainly in Africa, half to three-quarters of children have been infected (Ndip *et al.*, 2004; Mohammad *et al.*, 2008; Dube *et al.*, 2009a) and infection is common even in youngest children. In Brazil, Vietnam and Kenya, 22.6–45.6% of children aged under 3 years have been reported positive, compared with 2.4–8.6% of those in Ireland and in Germany (Huang *et al.*, 2004; Langat *et al.*, 2006; Nguyen *et al.*, 2006; Salih, 2009).

Local prevalence of *H. pylori* infection within a country also should be considered. The infection is more common in rural developing regions than in urban developed ones (Vale and Vítor, 2010). In Poland, almost all (95%) adults in small towns have been infected compared with 72% of those in Lublin Province; similarly, 87% of adults born in the rural areas have been positive compared with 64% of those born in large towns (Celiński *et al.*, 2006). In Egypt, *H. pylori* seroprevalence has been twice as high in rural as in urban groups (Awdalla *et al.*, 2010). Similar results have been found in China (Cheng *et al.*, 2009). Only occasional studies, such as that of Hoang *et al.* (2005), have reported the opposite association. These authors stated that in Vietnam the overall seroprevalence was higher in urban Hanoi (78.8%) than in rural Hatay (69.2%).

Unsurprisingly, in Chile, *H. pylori* prevalence (79.7%) has been reported to be higher in regions with high gastric cancer mortality than elsewhere (62.3%; Ferreccio *et al.*, 2007). Conversely, the 'African enigma' (high infection prevalence with low gastric cancer incidence) can be explained by T-helper 2 (Th2)-biased profile after bacterial/parasitic co-infections, diet, poor cancer registry or low strain virulence (Mbulaiteye *et al.*, 2006). Patient sources also can influence the results. Of Swedish people aged over 70 years, fewer (<8%) blood donors have been seropositive than those (>40%) in the population, possibly because few elderly people have been blood donors owing to comorbidity or bleeding risk (Sörberg *et al.*, 2003).

Ethnicity, reflecting different exposure to *H. pylori* infection rather than genetic factors, has been a commonly reported risk factor for the infection. In developed countries such as Australia, Germany and Canada, the infection frequency is higher in immigrants from developing countries, probably because of poor socioeconomic conditions (Perez-Perez *et al.*, 2005; Naja *et al.*, 2007; Sanz-Pelaez *et al.*, 2008; De Vries *et al.*, 2008). In the USA, the infection rate in non-Hispanic white subjects has been 2-fold lower than in other ethnic groups (Cardenas *et al.*, 2006). Similarly, Swiss adolescents have shown a 4.1-fold lower seroprevalence than immigrants (Heuberger *et al.*, 2003). Only 19.6% of German children aged 10–15 years have been seropositive compared with 42.2% of Turkish children in Germany and 71.9% of Turkish children in Turkey (Porsch-Ozcürümez *et al.*, 2003). In the study of Cherian *et al.* (2008), using the stool antigen test, most African refugee children in Australia (63.2% of children aged under 2 years and 95.2% of those aged over 14 years) have been found to be *H. pylori* infected. However, the authors discourage the use of serology in refugee children. Briefly, the higher prevalence of *H. pylori* infection in immigrants compared with the other inhabitants of the developed countries shows again the difference between developed and developing countries.

Porsch-Ozcürümez *et al.* (2003) have reported that the prevalence of seropositive children aged under 1 year was higher (39.5%) in Turkish children in Germany than in German children (14.3%). The authors found a loss of maternal anti-*H. pylori* IgG antibodies within the first year in both paediatric groups. Similarly, among Kenyan children, stool prevalence in those aged under 6 months was 2-fold higher (61.8%) than that of the children aged 13–18 months (27.3%; Langat *et al.*, 2006). This can be due to passive immunity (the maternal antibodies), lactoferrin and other protective factors in the human milk (Chak *et al.*, 2009). Breast-feeding can provide protection against *H. pylori* infection, mainly in middle/low-income countries and groups (Nguyen *et al.*, 2006; Chak *et al.*, 2009). However, the protection lasts less than 1 year and does not seem to influence the prevalence of the infection later in life.

The main risk factors for *H. pylori* infection, especially if present during childhood, have been associated with socioeconomic status. Most reported risk factors, e.g. subject age (older subjects according to the cohort effect), ethnicity (immigrants from developing countries), education or parent education (poor), geographic areas (developing countries), sanitation and living conditions (deprived), household overcrowding, family members (infected mothers or older siblings), family income (low) and water or food (contaminated), mirror well the poor socioeconomic conditions (Moreira *et al.*, 2004; Tam *et al.*, 2008; Dube *et al.*, 2009a; Sykora *et al.*, 2009). For example, in Islamabad suburbs (Pakistan), 73.6% of children aged 3–8 years have been *H. pylori* positive by ^{13}C UBT compared with only 1.2% of Dutch children aged 2–4 years by ELISA (Mourad-Baars *et al.*, 2007; Ahmad *et al.*, 2008). Poor hygiene and sanitation have been reported by many authors as risk factors for the infection, facilitating the faecal–oral transmission (Bakka and Salih, 2002; Bener *et al.*, 2002; Huang *et al.*, 2004; Celiński *et al.*, 2006; Cartágenes *et al.*, 2009; Dube *et al.*, 2009a; Nouraie *et al.*, 2009). In Poland, >90% of adults with poor hygiene habits have been *H. pylori* seropositive compared with 65% of other subjects (Celiński *et al.*, 2006). The high infection rate in Africa has been associated with multifarious risk factors, e.g. household overcrowding, bed-sharing, poor hygiene such as use of unprotected water (by ≤33% of subjects in some countries); unpasteurized milk or unwashed vegetables/fruits; sharing cups, premastication of food for infants, sharing water for bathing/washing hands, limited sanitary facilities, bathing/swimming in contaminated water, etc. (Siai *et al.*, 2008; Dube *et al.*, 2009b.)

Increasing age has been a risk factor for the infection in most studies (Tables 7.2 and 7.3). This is due to the birth cohort phenomenon rather than a late acquisition of infection. A birth cohort phenomenon implies that different exposure to the infection occurs during childhood; thus, a high infection rate in childhood establishes a high infection rate in the concerned age group later in life. Rate of new infections in adults from Western countries has been only <0.5% yearly, therefore, the higher infection rate in the elderly subjects has reflected the higher infection prevalence in nineteenth century and the first half of twentieth century than afterwards (Sonnenberg *et al.*, 2002).

An additional risk factor for *H. pylori* infection is antibiotic use. Rising antibiotic use has been associated with decreased *H. pylori* prevalence. In Turkish children, a reduced infection rate has been associated with antibiotic consumption during the last 6 months (Ozen *et al.*, 2006). Although *H. pylori* eradication requires regimens with several antibiotics, monotherapy with macrolides or β-lactams can lead to eradication in 10–50% (Blaser *et al.*, 2008). In Russia, the antibiotic therapy for *H. pylori* eradication has been reported also as a factor, leading to decrease in the infection rate (Tkachenko *et al.*, 2007). The antibiotic consumption can explain some cases of *H. pylori* clearance in children (see below). Diabetes, smoking and sex (often male) have been additional risk factors for the infection or for antiCagA antibodies in some studies (Ndip *et al.*, 2004; Ferreccio *et al.*, 2007; Jafarzadeh *et al.*, 2007; Thjodleifsson *et al.*, 2007; Apan *et al.*, 2008; Hestvik *et al.*, 2010).

The relationship between diabetes mellitus and *H. pylori* infection is a topic that needs further elucidation. In some studies, the infection has been found to be more common in diabetic patients, possibly owing to impaired gastrointestinal motility and weakened non-specific immunity, although the association has not been proven in other studies (Ariizumi *et al.*, 2008; Hamed *et al.*, 2008). In Pakistan, *H. pylori* positivity, as determined by a stool antigen test, was significantly higher (73%) in diabetic patients with dyspepsia than in those without diabetes (51.3%; Devrajani *et al.*, 2010). However, no link between *H. pylori* and diabetes was found by Czech authors (Bures *et al.*, 2004). On the other hand, Japanese and Iranian studies on asymptomatic subjects have revealed an increase in insulin resistance in *H. pylori*-infected subjects (Gunji *et al.*, 2009; Ojetti *et al.*, 2010). Moreover, the reinfection rates appear to be higher in diabetic than in other patients, possibly owing to the increased susceptibility to infections and immunosuppression in many of these patients. For instance, in Italy, *H. pylori* reinfection incidence was found to be higher (27%) in type 1 diabetic patients than in control subjects (4%; Ojetti *et al.*, 2007).

Importantly, the significance of antiCagA antibodies has been highlighted in many studies. Among *H. pylori*-positive patients, the presence of antiCagA antibodies, predicting a strong risk for intestinal metaplasia and severe diseases (Huang and Hunt (2003), varies widely, e.g. from 36% in Iceland to >86% in Japan, a country with a high incidence of gastric cancer (Table 7.2). In Portugal, anti-CagA IgG antibody-positive subjects have had an 11.3-fold higher risk of gastric cancer than antibody-negative subjects (Peleteiro *et al.*, 2010). Importantly, some subjects infected by *cagA*$^{+}$ strains may remain antiCagA seronegative and, conversely, some *H. pylori*-seronegative subjects may exhibit antiCagA seropositivity (see Chapter 3).

In 2007, an important advance was made in understanding the role of antiCagA antibody titre. In Japan, Suzuki *et al.* (2007b) have found low CagA IgG titre to be highly associated with risk for non-cardia gastric cancer and to be superior to antiCagA and anti-*H. pylori* testing. The authors have suggested two possible reasons, related to either immune response gene control (HLA polymorphisms) or changed *cagA* gene expression in the cases with hypochlorhydria. This suggestion is worth extensive evaluation worldwide.

In children, unlike ELISA IgG assays with low sensitivity for *H. pylori* infection, the anti-CagA ELISA IgG showed higher accuracy as CagA is highly immunogenic and the related antibodies are long-lasting (Nurgalieva *et al.*, 2008). In Brazilian children aged 1– 4 years, only half of SAT-positive children have been ELISA positive (Cartágenes *et al.*, 2009). However, similar overall antiCagA seroprevalence has been found in *H. pylori*-positive children and adults from Iran (72.8% and 67.4%), Turkey (50.4–58% and 46.6–62.5%) and Brazil (78% and 82.6%, respectively; Jafarzadeh *et al.*, 2007; Apan *et al.*, 2008; Cartágenes *et al.*, 2009). Nevertheless, Jafarzadeh *et al.* (2007) have stated that in the Iranian groups, the antiCagA seropositivity in children aged under 5 years (59.6%) has grown to 83.3% in adults aged 20–30 years and has dropped to the first value in those aged 51–60 years. The authors also have found the prevalence of antiCagA antibodies to be 1.2 times higher in boys than in girls and 1.4 times higher in men than in women. A similar result has been reported in Turkey (Apan *et al.*, 2008). Jafarzadeh *et al.* (2007) have associated the difference with the subsequent higher frequency of severe diseases in men than in women. Nevertheless, more studies are needed in relation to the sex and age as possible risk factors for antiCagA seropositivity. As a whole, the antiCagA ELISA is reliable to detect infections by virulent strains in both children and adults, although this method has been inferior to antiCagA immunoblot in terms of sensitivity (Malaty, 2010).

In addition to the prevalence of *cagA*$^+$-positive strains, the type of CagA is very important, especially in East Asia, as infections by East Asian CagA strains result in more severe gastric diseases than those by Western strains. In a study by Hirai *et al.* (2010), similar rates of *H. pylori* positivity have been found by rapid SAT (TestMate Rapid Pylori Antigen) in Thai (43.6%) and Japanese (40.3%) asymptomatic adults. Prevalence and type of *cagA* gene have been evaluated by nested PCR on DNA from stool specimens. Importantly, strains from more Japanese adults have exhibited the most virulent types than those from the Thai adults (Hirai *et al.*, 2010). The 2-fold lower *cagA*$^+$ prevalence in *H. pylori* strains and the 8.8-fold lower prevalence of East Asian genotype *cagA* gene in the Thai than in Japanese adults is one of the reasons for the higher gastric cancer incidence in Japan than in Thailand, i.e. for the 'Asian enigma'.

H. pylori infection is decreasing in the developed world and, already, in developing countries, although the infection prevalence is still high in most developing countries. Presently, in the Western countries, <10% of children aged under 10 years are *H. pylori* positive compared with ≥70% in the first half of the twentieth century (Blaser *et al.*, 2008). In symptomatic US veterans, the seroprevalence has been reported to be >3-fold higher among those born before 1920 then in those born after 1980 (Carmack and Genta, 2009). From 1990 to 2006, a decline in infection rate (by 22.8% in a high gastric cancer region and by 14.1% in a low-incidence region) was found in China (Zhang *et al.*, 2009a). *H. pylori* seroprevalence has decreased from 44% to 13% in Russian children and from 60.4% to 30.4% in Brazilian children over 10 years (Tkachenko *et al.*, 2007; Kawakami *et al.*, 2008). In Italy, the gastric cancer mortality rates decreased more than 3-fold from 1971–1974 to 2000–2004 (Gorini *et al.*, 2008).

In brief, the declining infection rates and the ensuing decrease in associated diseases both have resulted from the improved hygiene and living conditions in the developed and, already, in many developing countries and from the growing antibiotic use for *H. pylori* and many other infections.

Spontaneous clearance of the infection and reinfection

The incidence of spontaneous clearance of *H. pylori* infection and reinfection should be carefully considered. Recurrence of *H. pylori* infection encompasses reinfection and recrudescence. Reinfection is an infection by a new strain >12 months after successful eradication, whereas recrudescence is the reappearance of the same strain within 1 year after eradication (Niv, 2008). To determine a reinfection case, it is necessary to prove the genetic difference between the isolates before and after treatment. In Japan, 37.5% of the strains isolated 6 months after therapy have shown the same PCR-based RFLP type as the initial isolates, suggesting recrudescence, compared with 0% of those isolated 1 year after treatment (Okimoto *et al.*, 2003). Recrudescence, especially in developed countries, results from unsuccessful eradication, decreasing by 46% after the first year after the eradication; thus in these countries, improvement of both treatment regimens and patient compliance are of primary importance (Niv, 2008). Unfortunately, recrudescence increases as the success of eradication therapy declines in many countries (see Chapter 10). Conversely, reinfection requires a strong improvement in the prophylaxis of the infection. Generally, the observed recurrence of *H. pylori* infection depends on treatment regimen, country, patient age and diseases, number of patients in the study, diagnostic method used and time point after the infection to determine the eradication. The eradication success is easier to be obtained in patients infected by the most virulent strains than in the rest (see Chapter 9). It is usually accepted to control the eradication at least 1 month after the end of therapy; however, the benefit from a longer period (after 2.5–3.5 months), large numbers of patients to examine and more common use of molecular techniques have been suggested

(Zhang *et al.*, 2009b). The reinfection has been observed more often in children than in adults in correlation with the usual infection acquisition in the early childhood.

H. pylori annual recurrence has ranged from <1–2.7% in most developed and some developing countries to >14% in other developing countries (Niv, 2008; Zhang *et al.*, 2009b; Najafi *et al.*, 2010). Reinfection rates (per patient-year) have been only <0.5–2% in the UK, Japanese, Israeli and Brazilian adults but 5.4–6% in French children and adolescents and US adults, and 9.1% in South Korean adults (Okimoto *et al.*, 2003; Cameron *et al.*, 2006; Niv *et al.*, 2008; Ryu *et al.*, 2010; Silva *et al.*, 2010). In Iranian children, reinfection has been only <0.6–1.8% over 1 year (Najafi *et al.*, 2010), while in Iranian adults, the reinfection rate has reached 20.4% 3 years after eradication (Zendehdel *et al.*, 2005).

It is noteworthy that during the last years, the global reinfection incidence has been rather low worldwide except (>10%) in several developing countries such as Bangladesh and Iran (Hildebrand *et al.*, 2001; Zendehdel *et al.*, 2005; Zhang *et al.*, 2009b). For example, In Brazil, the annual reinfection rate was only 1.8% in the recent study of Silva *et al.* (2010) compared with 13.6% in an earlier study of Coelho *et al.* (1992). Close contact between young children has been stressed as an important risk factor for reinfections. Halitim *et al.* (2006) have reported that among 45 French children, evaluated 1–9 years after eradication, 24% have been reinfected. The authors suggested that having siblings aged under 5 years is associated with a reinfection rate that is 2.7-fold higher than in other children. Common risk factors for reinfection are patient age and close contact with infected persons, poor sanitation, host genetic factors but not obligatory the overall prevalence of the infection in the country (Zhang *et al.*, 2009b). In brief, it is obvious that the gradual fall in *H. pylori* infection rates worldwide has reduced the reinfection incidence as well.

Spontaneous clearance of *H. pylori* infection (loss of the infection in earlier positive subject without eradication) is usually uncommon and can be due to different immune responses to *H. pylori* antigens in children and adults or to preceding use of antibiotics with activity against *H. pylori*. Initial antibody response has been found to be directed mainly against low-molecular-weight antigens in children versus both high- and low-molecular-weight proteins in adults, which could explain cases of spontaneous clearance of the infection in some children (Torres *et al.*, 2002). As for the past antibiotic use, the antibiotics active against *H. pylori* should be considered, although overall use of systemic antibiotic has been reported as well. In 13% of US and Mexican children, single or multiple use of antibiotics (active against *H. pylori*) for other diseases has been linked to clearance of infection (Broussard *et al.*, 2009). However, in youngest (aged 6 years and under and especially 2 years and under) children, it is important to detect the spontaneous clearance by suitable tests, e.g. UBT rather than serology and by appropriate cut-off points (Nurgalieva *et al.*, 2008). It is not to forget that even UBT results can be false positive in children aged 6 years and under (see Chapter 3).

The balance between the recurrence and the spontaneous clearance of *H. pylori* infection indicates the overall infection rate in an age group. Low clearance and slightly higher recurrence rates have been observed in Japanese children over 2 years and in US adults with preneoplastic gastric lesions over 12 years. The clearance and the recurrence rates have been 1.1% and 1.5% per year in the Japanese children as well as 2.9% and 5.4% in the US adults, respectively (Mera *et al.*, 2007; Naito *et al.*, 2008). In asymptomatic Czech children, the infection rate has decreased from 11% in children aged under 6 years to 6% in those aged over 10 years (Sedlackova *et al.*, 2003). Spontaneous clearance of the infection has been reported even in newborns (Baldassarre *et al.*, 2009). Ozen *et al.* (2006) have reported 2.5-fold higher rate (14%) of *H. pylori* infection acquisition in Turkish children aged 3–12 years compared with the loss of infection (5.5%) over 6 years, suggesting that the spontaneous clearance is not of primary importance in the developing countries.

Briefly, the improved sanitation and overall living conditions, urbanization and rising antibiotic use for *H. pylori* or, mainly, for other infections have led to global decrease in infection and reinfection rates in the developed and, already, in some developing countries. However, the infection still affects more than half of the global human population. *H. pylori* prevalence of ≥70% in adults or ≥50% in children has been reported, mostly in the developing countries, in Africa (Cameroon, Libya, Nigeria, South Africa and Tunisia), Asia (Bangladesh, Georgia and Turkey), South America (Brazil, Chile and Venezuela) and Europe (Albania and Poland). This implies that about 500 millions of people, mainly in the developing countries, would develop peptic ulcers and about 30 millions would develop gastric cancer (Salih, 2009). Therefore, both the prophylaxis and therapy of the infection need considerable improvement. For this purpose, the further elucidation of many important topics on the infection transmission and reservoirs can be of significant help.

References

Agüloğlu, S., Turhanoğlu, M., Eskimez, S., and Tacir, I. (2006). Detection of *Helicobacter pylori* colonization in human dental plaques and saliva of patients with chronic gastritis. Biotechnol. Biotechnol. Eq. *20*, 173–178.

Ahmad, T., Bilal, R., and Khanum, A. (2008). Prevalence of *Helicobacter pylori* infection in asymptomatic children of Islamabad suburbs (Pakistan) Int. J. Agr. Biol. *10*, 685–688.

Al-Hawajri, A.A., Keret, D., Simhon, A., Zlotkin, A., Fishman, Y., Bercovier, H., and Rahav, G. (2004). *Helicobacter pylori* DNA in dental plaques, gastroscopy, and dental devices. Dig. Dis Sci. *49*, 1091–1094.

Allaker, R.P., Young, K.A., Hardie, J.M., Domizio, P., and Meadows, N.J. (2002). Prevalence of *Helicobacter pylori* at oral and gastrointestinal sites in children, evidence for possible oral-to-oral transmission. J. Med. Microbiol. *51*, 312–317.

Apan, T.Z., Iseri, L., Aksoy, A., and Güliter, S. (2008). The antibody response to *Helicobacter pylori* in the sera from a rural population in the central Anatolia region of Turkey. J. Health Sci. *54*, 671–674.

Ariizumi, K., Koike, T., Ohara, S., Inomata, Y., Abe, Y., Iijima, K., Imatani, A., Oka, T., and Shimosegawa, T. (2008). Incidence of reflux esophagitis and *H. pylori* infection in diabetic patients. World J. Gastroenterol. *14*, 3212–3217.

Assumpção, M.B., Martins, L.C., Melo Barbosa, H.P., Barile, K.A., de Almeida, S.S., Assumpção, P.P., and Corvelo, T.C. (2010). *Helicobacter pylori* in dental plaque and stomach of patients from Northern Brazil. World J. Gastroenterol. *16*, 3033–3039.

Awdalla, H.I., Ragab, M.H., and Hanna, L.N. (2010). Environmental risk factors affecting transmission of *Helicobacter pylori* infection in Egypt. J. Public Health. *18*, 237–224.

Azevedo, N.F., Huntington, J., and Goodman, K.J. (2009). The epidemiology of *Helicobacter pylori* and public health implications. Helicobacter *14* (*Suppl 1*), 1–7.

Baena Diez, J.M., García Lareo, M., Martí Fernández, J., León Marín, I., Muñiz Llama, D., Teruel Gila, J., Baena Díez, J.M., Rams Rams, F., and Hernández Ibáñez, M.R. (2002). Prevalence of *Helicobacter pylori* infection in primary care: sero-epidemiological study. Aten. Primaria. *29*, 553–557.

Bakka, A.S., and Salih, B.A. (2002). Prevalence of *Helicobacter pylori* infection in asymptomatic subjects in Libya. Diagn. Microbiol. Infect. Dis. *43*, 265–268.

Baldassarre, M.E., Monno, R., Laforgia, N., Fumarola, L., Fanelli, M., Sgobba, C., Hassan, C., Panella, C., and Ierardi, E. (2009). The source of *Helicobacter pylori* infection in the neonatal period. J. Perinat. Med. 37, 288–292.

Bener, A., Uduman, S.A., Ameen, A., Alwash, R., Pasha, M.A., Usmani, M.A., Al-Naili, S.R., and Amiri, K.M. (2002). Prevalence of *Helicobacter pylori* infection among low socioeconomic workers. J. Commun. Dis. *34*, 179–184.

Bhuiyan, T.R., Qadri, F., Saha, A., and Svennerholm, A.M. (2009). Infection by *Helicobacter pylori* in Bangladeshi children from birth to two years: relation to blood group, nutritional status, and seasonality. Pediatr. Infect. Dis. J. *28*, 79–85.

Blaser, M.J., Chen, Y., and Reibman, J. (2008). Does *Helicobacter pylori* protect against asthma and allergy? Gut *57*, 561–567.

Boomkens, S.Y., Kusters, J.G., Hoffmann, G., Pot, R.G., Spee, B., Penning, L.C., Egberink, H.F., van den Ingh, T.S., and Rothuizen, J. (2004). Detection of *Helicobacter pylori* in bile of cats. FEMS Immunol. Med. Microbiol. *42*, 307–311.

Boyanova, L., Panov, Vl., Yordanov, D., Markovska, R., Gergova, G., and Mitov, I. (2009). Study on the extraoral localization of *Helicobacter pylori*. Paediatrics. *XLIX*, 31–35.

Braga, A.B., Fialho, A.M., Rodrigues, M.N., Queiroz, D.M., Rocha, A.M., and Braga, L.L. (2007). *Helicobacter pylori* colonization among children up to 6 years, results of a community-based study from Northeastern Brazil. J. Trop. Pediatr. *53*, 393–397.

Brenner, H., Weyermann, M., and Rothenbacher, D. (2006). Clustering of *Helicobacter pylori* infection in couples, differences between high- and low-prevalence population groups. Ann. Epidemiol. *16*, 516–520.

Broussard, C.S., Goodman, K.J., Phillips, C.V., Smith, M.A., Fischbach, L.A., Day, R.S., and Aragaki, C.C. (2009). Antibiotics taken for other illnesses and spontaneous clearance of *Helicobacter pylori* infection in children. Pharmacoepidemiol. Drug Saf. *18*, 722–729.

Buck, A., and Oliver, J.D. (2010). Survival of spinach-associated *Helicobacter pylori* in the viable but nonculturable state. Food Control *21*, 1150–1154.

Bures, J., Smahelová, A., Kopácová, M., and Rejchrt, S. (2004). Clinical importance of *Helicobacter pylori* infection in patients with diabetes mellitus. Vnitr Lek. *50*, 350–353.

Bürgers, R., Schneider-Brachert, W., Reischl, U., Behr, A., Hiller, K.A., Lehn, N., Schmalz, G., and Ruhl, S. (2008). *Helicobacter pylori* in human oral cavity and stomach. Eur. J. Oral Sci. *116*, 297–304.

Buzás, G.M. (2010). Gastric tubes as vectors of *Helicobacter pylori* transmission. Med. Hypotheses. *75*, 47–49.

Cameron, E.A., Bell, G.D., Baldwin, L., Powell, K.U., and Williams, S.G. (2006). Long-term study of re-infection following successful eradication of *Helicobacter pylori* infection. Aliment. Pharmacol. Ther. *23*, 1355–1358.

Cardenas, V.M., Mulla, Z.D., Ortiz, M., and Graham, D.Y. (2006). Iron deficiency and *Helicobacter pylori* infection in the United States. Am. J. Epidemiol. *163*, 127–134.

Carmack, S.W., and Genta, R.M. (2009). *Helicobacter pylori* seroprevalence in symptomatic veterans: a study of 7310 patients over 11 years. Helicobacter *14*, 298–302.

Cartágenes, V.D., Martins, L.C., Carneiro, L.M., Barile, K.A., and Corvelo, T.C. (2009). *Helicobacter pylori* in children and association with CagA strains in mother-child transmission in the Brazilian Amazon region. Rev. Soc. Bras. Med. Trop. *42*, 298–302.

Celiński, K., Kurzeja-Mirosław, A., Słomka, M., Cichoz-Lach, H., Madro, A., and Kasztelan-Szczerbińska, B. (2006). The effects of environmental factors on the prevalence of *Helicobacter pylori* infection in inhabitants of Lublin province. Ann. Agric. Environ. Med. *13*, 185–191.

Cellini, L., Di, C.E., Grande, R., Di, B.S., Prenna, M., Pasquantonio, M., and Pane, L. (2005). Detection of *Helicobacter pylori* associated with zooplankton. Aquat. Microb. Ecol. *40*, 115–120.

Ceylan, A., Kirimi, E., Tuncer, O., Türkdoğan, K., Ariyuca, S., and Ceylan, N. (2007). Prevalence of *Helicobacter pylori* in children and their family members in a district in Turkey. J. Health Popul. Nutr. *25*, 422–427.

Chak, E., Rutherford, G.W., and Steinmaus, C. (2009). The role of breast-feeding in the prevention of *Helicobacter pylori* infection: a systematic review. Clin. Infect. Dis. *48*, 430–437.

Chaudhry, S., Idrees, M., Izhar, M., Butt, A.K., and Khan, A.A. (2011). Simultaneous amplification of two bacterial genes: more reliable method of *Helicobacter pylori* detection in microbial rich dental plaque samples. Curr. Microbiol. *62*, 78–83.

Cheng, H., Hu, F., Zhang, L., Yang, G., Ma, J., Hu, J., Wang, W., Gao, W., and Dong, X. (2009). Prevalence of *Helicobacter pylori* infection and identification of risk factors in rural and urban Beijing, China. Helicobacter *14*, 128–133.

Cherian, S., Burgner, D.P., Carson, C.F., Sanfilippo, F.M., Cook, A.G., and Forbes, D.A. (2008). Diagnosis of *Helicobacter pylori* infection in a high-prevalence pediatric population: a comparison of 2 fecal antigen testing methods and serology. J. Pediatr. Gastroenterol. Nutr. *47*, 130–135.

Coelho, L.G., Passos, M.C., Chausson, Y., Costa, E.L., Maia, A.F., Brandao, M.J., Rodrigues, D.C., and Castro LP. (1992). Duodenal ulcer and eradication of *Helicobacter pylori* in a developing country. An 18-month follow-up study. Scand. J. Gastroenterol. *27*, 362–366.

Czesnikiewicz-Guzik, M., Bielanski, W., Guzik, T.J., Loster, B., and Konturek, S.J. (2005). *Helicobacter pylori* in the oral cavity and its implications for gastric infection, periodontal health, immunology and dyspepsia. J. Physiol. Pharmacol. *56 (Suppl 6)*, 77–89.

Czesnikiewicz-Guzik, M., Loster, B., Bielanski, W., Guzik, T.J., Konturek, P.C., Zapala, J., and Konturek, S.J. (2007). Implications of oral *Helicobacter pylori* for the outcome of its gastric eradication therapy. J. Clin. Gastroenterol. *41*, 145–151.

Delport, W., and van der Merwe, S.W. (2007). The transmission of *Helicobacter pylori*, the effects of analysis method and study population on inference. Best Pract. Res. Clin. Gastroenterol. *21*, 215–236.

De Schryver, A., Cornelis, K., Van Winckel, M., Moens, G., Devlies, G., Derthoo, D., and van Sprundel, M. (2008). The occupational risk of *Helicobacter pylori* infection among workers in institutions for people with intellectual disability. Occup. Environ. Med. *65*, 587–591.

De Sousa, L., Vásquez, L., Velasco, J., and Parlapiano, D. (2006). Isolation of *Helicobacter pylori* in gastric mucosa, dental plaque and saliva in a population from the Venezuelan Andes. Invest. Clin. *47*, 109–116.

Devrajani, B.R., Shah, S.Z., Soomro, A.A., and Devrajani, T. (2010). Type 2 diabetes mellitus: A risk factor for *Helicobacter pylori* infection: A hospital based case-control study. Int. J. Diabetes Dev. Ctries. *30*, 22–26.

De Vries, A.C., Van Driel, H.F., Richardus, J.H., Ouwendijk, M., Van Vuuren, A.J., De Man, R.A., and Kuipers, E.J. (2008). Migrant communities constitute a possible target population for primary prevention of *Helicobacter pylori*-related complications in low incidence countries. Scand. J. Gastroenterol. *43*, 403–409.

Dinić, M., Tasić, G., Stanković-Dordević, D., Otasević, L., Tasić, M., and Karanikolic, A. (2007). Serum anti*Helicobacter pylori* IgA and IgG antibodies in asymptomatic children in Serbia. Scand. J. Infect. Dis. *39*, 303–307.

Dore, M.P., Sepulveda, A.R., El-Zimaity, H., Yamaoka, Y., Osato, M.S., Mototsugu, K., Nieddu, A.M., Realdi, G., and Graham, D.Y. (2001). Isolation of *Helicobacter pylori* from sheep-implications for transmission to humans. Am. J. Gastroenterol. *96*, 1396–1401.

Dube, C., Nkosi, T.C., Clarke, A.M., Mkwetshana, N., Green, E., and Ndip, R.N. (2009a). *Helicobacter pylori* antigenemia in an asymptomatic population of Eastern Cape Province, South Africa: public health implications. Rev. Environ. Health *24*, 249–255.

Dube, C., Tanih, N.F., Clarke, A.M., Mkwetshana, N., Green, E., and Ndip, R.N. (2009b). *Helicobacter pylori* infection and transmission in Africa: Household hygiene and water sources are plausible factors exacerbating spread. Afr. J. Biotechnol. *8*, 6028–6035.

Eskandari, A., Mahmoudpour, A., Abolfazli, N., and Lafzi, A. (2009). Detection of *Helicobacter pylori* using PCR in dental plaque of patients with and without gastritis. Med. Oral Patol. Oral. Cir. Bucal. *15*, e28–e31.

Ferreccio, C., Rollán, A., Harris, P.R., Serrano, C., Gederlini, A., Margozzini, P., Gonzalez, C., Aguilera,

X., Venegas, A., and Jara, A. (2007). Gastric cancer is related to early *Helicobacter pylori* infection in a high-prevalence country. Cancer Epidemiol. Biomarkers Prev. *16*, 662–667.

Fischbach, W., Malfertheiner, P., Hoffmann, J.C., Bolten, W., Kist, M., and Koletzko, S. (2009). *Helicobacter pylori* and gastroduodenal ulcer disease. Dtsch. Arztebl. Int. *106*, 801–808.

Fritz, E.L., Slavik, T., Delport, W., Olivier, B., and van der Merwe, S.W. (2006). Incidence of *Helicobacter felis* and the effect of coinfection with *Helicobacter pylori* on the gastric mucosa in the African population. J. Clin. Microbiol. *44*, 1692–1696.

Fujimura, S., Kawamura, T., Kato, S., Tateno, H., and Watanabe, A. (2002). Detection of *Helicobacter pylori* in cow's milk. Lett. Appl. Microbiol. *35*, 504–507.

Gebara, E.C., Pannuti, C., Faria, C.M., Chehter, L., Mayer, M.P., and Lima, L.A. (2004). Prevalence of *Helicobacter pylori* detected by polymerase chain reaction in the oral cavity of periodontitis patients. Oral. Microbiol. Immunol. *19*, 277–280.

Gião, M.S., Azevedo, N.F., Wilks, S.A., Vieira, M.J., and Keevil, C.W. (2008). Persistence of *Helicobacter pylori* in heterotrophic drinking-water biofilms. Appl. Environ. Microbiol. *74*, 5898–5904.

Goldman, C., Barrado, A., Janjetic, M., Balcarce, N., Cueto Rua, E., Oshiro, M., Calcagno, M.L., Sarrasague, M.M., Fuda, J., Weill, R., *et al.* (2006). Factors associated with *H. pylori* epidemiology in symptomatic children in Buenos Aires, Argentina. World J. Gastroenterol. *12*, 5384–5388.

Gómez, N.A., Salvador, A., Vargas, P.E., Zapatier, J.A., and Alvarez, J. (2004). Seroprevalence of *Helicobacter pylori* among the child population of Ecuador. Rev. Gastroenterol. Peru *24*, 230–233.

Goosen, C., Theron, J., Ntsala, M., Maree, F.F., Olckers, A., Botha, S.J., Lastovica, A.J., and van der Merwe, S.W. (2002). Evaluation of a novel heminested PCR assay based on the phosphoglucosamine mutase gene for detection of *Helicobacter pylori* in saliva and dental plaque. J. Clin. Microbiol. *40*, 205–209.

Gorini, G., Giovannetti, L., Masala, G., Chellini, E., Martini, A., Mallone, S., and Costantini, A.S. (2008). Gastric cancer mortality trends in Tuscany, Italy, 1971–2004. Tumori *94*, 787–792.

Graham, D.Y., Yamaoka, Y., and Malaty, H.M. (2007). Thoughts about populations with unexpected low prevalences of *Helicobacter pylori* infection. Trans. R. Soc. Trop. Med. Hyg. *101*, 849–851.

Gunji, T., Matsuhashi, N., Sato, H., Fujibayashi, K., Okumura, M., Sasabe, N., and Urabe, A. (2009). *Helicobacter pylori* infection significantly increases insulin resistance in the asymptomatic Japanese population. Helicobacter *14*, 144–150.

Haggerty, T., Shmuely, H., and Parsonnet, J. (2003). *Helicobacter pylori* in cathartic stools of subjects with and without cimetidine-induced hypochlorhydria. J. Med. Microbiol. *52*, 189–191.

Halitim, F., Vincent, P., Michaud, L., Kalach, N., Guimber, D., Boman, F., Turck, D., and Gottrand, F. (2006). High rate of *Helicobacter pylori* reinfection in children and adolescents. Helicobacter *11*, 168–172.

Hamed, S.A., Amine, N.F., Galal, G.M., Helal, S.R., Tag El-Din, L.M., Shawky, O.A., Ahmed, E.A., and Abdel Rahman, M.S. (2008). Vascular risks and complications in diabetes mellitus: the role of *Helicobacter pylori* infection. J. Stroke Cerebrovasc. Dis. *17*, 86–94.

Herrera, P.M., Mendez, M., Velapatiño, B., Santivañez, L., Balqui, J., Finger, S.A., Sherman, J., Zimic, M., Cabrera, L., Watanabe, J., *et al.* (2008). DNA-level diversity and relatedness of *Helicobacter pylori* strains in shantytown families in Peru and transmission in a developing-country setting. J. Clin. Microbiol. *46*, 3912–3918.

Hestvik, E., Tylleskar, T., Kaddu-Mulindwa, D.H., Ndeezi, G., Grahnquist, L., Olafsdottir, E., and Tumwine, J.K. (2010). *Helicobacter pylori* in apparently healthy children aged 0–12 years in urban Kampala, Uganda: a community-based cross sectional survey. BMC Gastroenterol. *10*, 62.

Heuberger, F., Pantoflickova, D., Gassner, M., Oneta, C., Grehn, M., Blum, A. L., and Dorta, G. (2003). *Helicobacter pylori* infection in Swiss adolescents: prevalence and risk factors. Eur. J. Gastroenterol. Hepatol. *15*, 179–183.

Hildebrand, P., Bardhan, P., Rossi, L., Parvin, S., Rahman, A., Arefin, M.S., Hasan, M., Ahmad, M.M., Glatz-Krieger, K., Terracciano, L., *et al.* (2001). Recrudescence and reinfection with *Helicobacter pylori* after eradication therapy in Bangladeshi adults. Gastroenterology *121*, 792–798.

Hirai, I., Sasaki, T., Kimoto, A., Yamamoto, Y., Azuma, T., Mahachai, V., Hansomburana, P., Lertkupinit, C., Luangjaru, S., Noophan, P., *et al.* (2010). Infection of less virulent *Helicobacter pylori* strains in asymptomatic healthy individuals in Thailand as a potential contributing factor to the Asian enigma. Microbes Infect. *12*, 227–230.

Hoang, T.T., Bengtsson, C., Phung, D.C., Sörberg, M., and Granström, M. (2005). Seroprevalence of *Helicobacter pylori* infection in urban and rural Vietnam. Clin. Diagn. Lab. Immunol. *12*, 81–85.

Huang, J.Q., and Hunt, R.H. (2003). The evolving epidemiology of *Helicobacter pylori* infection and gastric cancer. Can. J. Gastroenterol. *17* (*Suppl B*), 18B–20B.

Huang, S.S., Hassan, A.K., Choo, K.E., Ibrahim, M.I., and Davis, T.M. (2004). Prevalence and predictors of *Helicobacter pylori* infection in children and adults from the Penan ethnic minority of Malaysian Borneo. Am. J. Trop. Med. Hyg. *71*, 444–450.

Imanzadeh, F., Sayyari, A.A., Ighbali, A., Javaherizadeh, H., and Valaie, N. (2007). Study of the risk factors related with *Helicobacter pylori* infections in children with peptic ulcer disease. Pak. J. Med. Sci. *23*, 446–448.

Ito, L.S., Oba-Shinjo, S.M., Shinjo, S.K., Uno, M., Marie, S.K., and Hamajima, N. (2006). Community-based familial study of *Helicobacter pylori* infection among healthy Japanese Brazilians. Gastric Cancer *9*, 208–216.

Jafarzadeh, A., Rezayati, M.T., and Nemati, M. (2007). Specific serum immunoglobulin G to *H pylori* and CagA in healthy children and adults (south-east of Iran). World J. Gastroenterol. *13*, 3117–3121.

Jafri, W., Yakoob, J., Abid, S., Siddiqui, S., Awan, S., and Nizami, S. (2010). *Helicobacter pylori* infection in children: population-based age-specific prevalence and risk factors in a developing country. Acta Paediatr. *99*, 279–82.

Jia, C.L., Jiang, G.S., Li, C.H., and Li, C.R. (2009). Effect of dental plaque control on infection of *Helicobacter pylori* in gastric mucosa. J. Periodontol. *80*, 1606–1609.

Kaptan, Z.K., Emir, H., Uzunkulaoğlu, H., Yücel, M., Karakoç, E., Koca, G., Tüzüner, A., Samim, E., and Korkmaz, M. (2009). Determination of *Helicobacter pylori* in patients with chronic nonspecific pharyngitis. Laryngoscope *119*, 1479–1483.

Kato, S., Okamoto, H., Nishino, Y., Oyake, Y., Nakazato, Y., Okuda, M., Fujisawa, T., Iinuma, K., and Blaser, M.J. (2003). *Helicobacter pylori* and TT virus prevalence in Japanese children. J. Gastroenterol. *38*, 1126–1130.

Kawakami, E., Machado, R.S., Ogata, S.K., and Langner, M. (2008). Decrease in prevalence of *Helicobacter pylori* infection during a 10-year period in Brazilian children. Arq. Gastroenterol. *45*, 147–151.

Kignel, S., de Almeida Pina, F., André, E.A., Alves Mayer, M.P., and Birman, E.G. (2005). Occurrence of *Helicobacter pylori* in dental plaque and saliva of dyspeptic patients. Oral Dis. *11*, 17–21.

Konno, M., Yokota, S., Suga, T., Takahashi, M., Sato, K., and Fujii, N. (2008). Predominance of mother-to-child transmission of *Helicobacter pylori* infection detected by random amplified polymorphic DNA fingerprinting analysis in Japanese families. Pediatr. Infect. Dis. J. *27*, 999–1003.

Kori, M., Goldstein, E., and Granot, E. (2009). *Helicobacter pylori* infection in young children detected by a monoclonal stool antigen immunoassay. Pediatr. Infect. Dis. J. *28*, 157–159.

Langat, A.C., Ogutu, E., Kamenwa, R., and Simiyu, D.E. (2006). Prevalence of *Helicobacter pylori* in children less than three years of age in health facilities in Nairobi Province. East Afr. Med. J. *83*, 471–477.

Laporte, R., Pernes, P., Pronnier, P., Gottrand, F., and Vincent, P. (2004). Acquisition of *Helicobacter pylori* infection after outbreaks of gastroenteritis, prospective cohort survey in institutionalised young people. B.M.J. *329*, 204–205.

Leclerc, H. (2006). Epidemiological aspects of *Helicobacter pylori* infection. Bull. Acad. Natl. Med. *190*, 949–962.

Lee, Y.-C., Liou, J.-M., Wu, C.-Y., and Lin, J.-T. (2008). Review: Eradication of *Helicobacter pylori* to prevent gastroduodenal diseases: Hitting more than one bird with the same stone. Ther. Adv. Gastroenterol. *1*, 111–120.

Liu, H., Rahman, A., Semino-Mora, C., Doi, S.Q., and Dubois, A. (2008). Specific and sensitive detection of *H. pylori* in biological specimens by real-time RT-PCR and *in situ* hybridization. PLoS ONE 3, e2689.

Liu, Y., Yue, H., Li, A., Wang, J., Jiang, B., Zhang, Y., and Bai, Y. (2009). An epidemiologic study on the correlation between oral *Helicobacter pylori* and gastric *H. pylori*. Curr. Microbiol. *58*, 449–453.

Loster, B.W., Majewski, S.W., Cześnikiewicz-Guzik, M., Bielanski, W., Pierzchalski, P., and Konturek, S.J. (2006). The relationship between the presence of *Helicobacter pylori* in the oral cavity and gastric in the stomach. J. Physiol. Pharmacol. *57* (*Suppl* 3), 91–100.

Loster, B.W., Czesnikiewicz-Guzik, M., Bielanski, W., Karczewska, E., Loster, J.E., Kalukin, J., Guzik, T.J., Majewski, S., and Konturek, S.J. (2009). Prevalence and characterization of *Helicobacter pylori (H. pylori)* infection and colonization in dentists. J. Physiol. Pharmacol. *60* (Suppl. 8), 13–18.

Lu, Y., Redlinger, T.E., Avitia, R., Galindo, A., and Goodman, K. (2002). Isolation and genotyping of *Helicobacter pylori* from untreated municipal wastewater. Appl. Environ. Microbiol. *68*, 1436–1439.

Luzza, F., Concolino, D., Imeneo, M., Pascuzzi, A., Pietragalla, E., Pallone, F., and Strisciuglio, P. (2004). High seroprevalence of *Helicobacter pylori* infection in non-institutionalised children with mental retardation. Clin. Microbiol. Infect. *10*, 670–673.

Malaty, H.M. (2010). Epidemiology of *Helicobacter pylori* infection. In *Helicobacter pylori* in the 21st Century, Sutton, P., and Mitchell, H., eds. (Cab International: Wallingford, UK), pp. 1–12.

Mansour, K.B., Keita, A., Zribi, M., Masmoudi, A., Zarrouk, S., Labbene, M., Kallel, L., Karoui, S., Fekih, M., Matri, S., *et al.* (2010). Seroprevalence of *Helicobacter pylori* among Tunisian blood donors (outpatients), symptomatic patients and control subjects. Gastroenterol. Clin. Biol. *34*, 75–82.

Marini, E., Maldonado-Contreras, A.L., Cabras, S., Hidalgo, G., Buffa, R., Marin, A., Floris, G., Racugno, W., Pericchi, L.R., Castellanos, M.E., *et al.* (2007). *Helicobacter pylori* and intestinal parasites are not detrimental to the nutritional status of Amerindians. Am. J. Trop. Med. Hyg. *76*, 534–540.

Mbulaiteye, S.M., Gold, B.D., Pfeiffer, R.M., Brubaker, G.R., Shao, J., Biggar, R.J., and Hisada, M. (2006). *H. pylori*-infection and antibody immune response in a rural Tanzanian population. Infect. Agent Cancer. *1*, 3.

Medina, M.L., Medina, M.G., Martín, G.T., Picón, S.O., Bancalari, A., and Merino, L.A. (2010). Molecular detection of *Helicobacter pylori* in oral samples from patients suffering digestive pathologies. Med. Oral Patol. Oral. Cir. Bucal. *15*, e38-e42.

Megraud, F., and Lehours, P. (2007). *Helicobacter pylori* detection and antimicrobial susceptibility testing. Clin. Microbiol. Rev. *20*, 280–322.

Mera, R., Fontham, E.T., Bravo, L.E., Bravo, J.C., Piazuelo, M.B., Camargo, M.C., and Correa, P. (2007). Re: long term follow up of patients treated for *Helicobacter pylori* infection. Gut *56*, 436.

Mishra, S., Singh, V., Rao, G.R., Dixit, V.K., Gulati, A.K., and Nath, G. (2008). Prevalence of *Helicobacter pylori* in asymptomatic subjects–a nested PCR based study. Infect. Genet. Evol. *8*, 815–819.

Miyabayashi, H., Furihata, K., Shimizu, T., Ueno, I., and Akamatsu, T. (2000). Influence of oral *Helicobacter pylori* on the success of eradication therapy against gastric *Helicobacter pylori*. Helicobacter *5*, 30–37.

Mladenova, I., Durazzo, M., and Pellicano, R. (2006). Transmission of *Helicobacter pylori*, are there evidences for a fecal–oral route? Minerva Med. *97*, 15–18.

Moges, F., Kassu, A., Mengistu, G., Adugna, S., Andualem, B., Nishikawa, T., and Ota, F. (2006). Seroprevalence of *Helicobacter pylori* in dyspeptic patients and its relationship with HIV infection, ABO blood groups and life style in a university hospital, Northwest Ethiopia. World J. Gastroenterol. *12*, 1957–1961.

Mohammad, M.A., Hussein, L., Coward, A., and Jackson, S.J. (2008). Prevalence of *Helicobacter pylori* infection among Egyptian children: impact of social background and effect on growth. Public Health Nutr. *11*, 230–236.

Monno, R., Volpe, A., Basho, M., Fumarola, L., Trerotoli, P., Kondili, L.A., Bino, S., Schinaia, N., Dentico, P., and Albanian-Italian collaborating group. (2008). *Helicobacter pylori* seroprevalence in selected groups of Albanian volunteers. Infection. *36*, 345–350.

Morales-Espinosa, R., Fernandez-Presas, A., Gonzalez-Valencia, G., Flores-Hernandez, S., Delgado-Sapien, G., Mendez-Sanchez, J.L., Sanchez-Quezada, E., Muñoz-Pérez, L., Leon-Aguilar, R., Hernandez-Guerrero, J., and Cravioto, A. (2009). *Helicobacter pylori* in the oral cavity is associated with gastroesophageal disease. Oral Microbiol. Immunol. *24*, 464–468.

Moreira, E.D., Jr., Santos, R.S., Nassri, V.B., Reis, A.T., Guerra, A.L., Alcantara, A.P., Matos, J.F., Carvalho, W.A., Moura, C.G., Silvani, C.S., *et al.* (2004). Risk factors for *Helicobacter pylori* infection in children, is education a main determinant? Epidemiol. Infect. *132*, 327–335.

Moreno, Y., Piqueres, P., Alonso, J.L., Jimenez, A., Gonzalez, A., and Ferrus, M.A. (2007). Survival and viability of *Helicobacter pylori* after inoculation into chlorinated drinking water. Water Res. *41*, 3490–3496.

Moujaber, T., MacIntyre, C.R., Backhouse, J., Gidding, H., Quinn, H., and Gilbert, G.L. (2008). The seroepidemiology of *Helicobacter pylori* infection in Australia. Int. J. Infect. Dis. *12*, 500–504.

Mourad-Baars, P.E., Verspaget, H.W., Mertens, B.J., and Mearin, M.L. (2007). Low prevalence of *Helicobacter pylori* infection in young children in the Netherlands. Eur. J. Gastroenterol. Hepatol. *19*, 213–216.

Nahar, S., Kibria, K.M., Hossain, M.E., Sultana, J., Sarker, S.A., Engstrand, L., Bardhan, P.K., Rahman, M., and Endtz, H.P. (2009). Evidence of intra-familial transmission of *Helicobacter pylori* by PCR-based RAPD fingerprinting in Bangladesh. Eur. J. Clin. Microbiol. Infect. Dis. *28*, 767–773.

Naito, Y., Shimizu, T., Haruna, H., Fujii, T., Kudo, T., Shoji, H., and Suzuki, K. (2008). Changes in the presence of urine *Helicobacter pylori* antibody in Japanese children in three different age groups. Pediatr. Int. *50*, 291–294.

Naja, F., Kreiger, N., and Sullivan, T. (2007). *Helicobacter pylori* infection in Ontario: prevalence and risk factors. Can. J. Gastroenterol. *21*, 501–506.

Najafi, M., Sobhani, M., Khodadad, A., Farahmand, F., and Motamed, F. (2010). Reinfection rate after successful *Helicobacter pylori* eradication in children. Iran. J. Ped. *20*, 58–62.

Nasrolahei, M., Sharif, M., Ahanjan, M., and Daryani, A. (2008). Prevalence of *Helicobacter pylori* in gastric mucosa and dental plaque in Sari, northern Iran. J. Chin. Clin. Med. 3, 383–388.

Nayak, A.K., and Rose, J.B. (2007). Detection of *Helicobacter pylori* in sewage and water using a new quantitative PCR method with SYBR green. J. Appl. Microbiol. *103*, 1931–1941.

Ndip, R.N., Malange, A.E., Akoachere, J.F., MacKay, W.G., Titanji, V.P., and Weaver, L.T. (2004). *Helicobacter pylori* antigens in the faeces of asymptomatic children in the Buea and Limbe health districts of Cameroon: a pilot study. Trop. Med. Int. Health. *9*, 1036–1040.

Nelson, D.B., and Muscarella, L.F. (2006). Current issues in endoscope reprocessing and infection control during gastrointestinal endoscopy. World J. Gastroenterol. *12*, 3953–3964.

Nguyen, B.V., Nguyen, K.G., Phung, C.D., Kremp, O., Kalach, N., Dupont, C., Raymond, J., and Vidal-Trecan, G. (2006). Prevalence of and factors associated with *Helicobacter pylori* infection in children in the north of Vietnam. Am. J. Trop. Med. Hyg. *74*, 536–539.

Niv, Y. (2008). *H pylori* recurrence after successful eradication. World J. Gastroenterol. *14*, 1477–1478.

Niv, Y., Hazazi, R., Waked, A., Lederfein, T., and Achiel, K. (2008). *Helicobacter pylori* recurrence and infection rate in Israeli adults. Dig. Dis. Sci. *53*, 1211–1214.

Nouraie, M., Latifi-Navid, S., Rezvan, H., Radmard, A.R., Maghsudlu, M., Zaer-Rezaii, H., *et al.* (2009). Childhood hygienic practice and family education status determine the prevalence of *Helicobacter pylori* infection in Iran. Helicobacter *14*, 40–46.

Nurgalieva, Z., Goodman, K.J., Phillips, C.V., Fischbach, L., de la Rosa, J.M., and Gold, B.D. (2008). Correspondence between *Helicobacter pylori* antibodies and urea breath test results in a US-Mexico birth cohort. Paediatr. Perinat. Epidemiol. *22*, 302–312.

Ogunbodede, E.O., Lawal, O.O., Lamikanra, A., Okeke, I.N., Rotimi, O., and Rasheed, A.A. (2002). *Helicobacter pylori* in the dental plaque and gastric mucosa of dyspeptic Nigerian patients. Trop. Gastroenterol. *23*, 127–133.

Ojetti, V., Migneco, A., Nista, E.C., Gasbarrini, G., Gasbarrini, A., Pitocco, D., and Ghirlanda, G. (2007). *H. pylori* re-infection in type 1 diabetes: A 5 years follow-up. Dig. Liver Dis. *39*, 286–287.

Ojetti, V., Pellicano, R., Fagoonee, S., Migneco, A., Berrutti, M., and Gasbarrini, A. (2010). *Helicobacter pylori* infection and diabetes. Minerva Med. *101*, 115–119.

Okimoto, T., Murakami, K., Sato, R., Miyajima, H., Nasu, M., Kagawa, J., Kodama, M., and Fujioka, T. (2003). Is the recurrence of *Helicobacter pylori* infection after eradication therapy resultant from recrudescence or reinfection, in Japan. Helicobacter *8*, 186–191.

Olivares, A., Buadze, M., Kutubidze, T., Lobjanidze, M., Labauri, L., Kutubidze, R., Chikviladze, D., Zhvania, M., Kharzeishvili, O., Lomidze, N., *et al.* (2006). Prevalence of *Helicobacter pylori* in Georgian patients with dyspepsia. Helicobacter *11*, 81–85.

Olivier, B.J., Bond, R.P., van Zyl, W.B., Delport, M., Slavik, T., Ziady, C., Terhaar Sive Droste, J.S., Lastovica, A., and van der Merwe, S.W. (2006). Absence of *Helicobacter pylori* within the oral cavities of members of a healthy

South African community. J. Clin. Microbiol. *44*, 635–636.

Ornelas, I.J., Galvan-Potrillo, M., and López-Carrillo, L. (2007). Protective effect of yoghurt consumption on *Helicobacter pylori* seropositivity in a Mexican population. Public Health Nutr. *10*, 1283–1287.

Ozen, A., Ertem, D., and Pehlivanoglu, E. (2006). Natural history and symptomatology of *Helicobacter pylori* in childhood and factors determining the epidemiology of infection. J. Pediatr. Gastroenterol. Nutr. *42*, 398–404.

Papiez, D., Konturek, P.C., Bielanski, W., Plonka, M., Dobrzanska, M., Kaminska, A., Szczyrk, U., Bochenek, A., and Wierzchos, E. (2003). Prevalence of *Helicobacter pylori* infection in Polish shepherds and their families. Dig. Liver Dis. *35*, 10–15.

Peleteiro, B., Lunet, N., Barros, R., La Vecchia, C., and Barros, H. (2010). Factors contributing to the underestimation of *Helicobacter pylori*-associated gastric cancer risk in a high-prevalence population. Cancer Causes Control *21*, 1257–1264.

Pérez, L.M., Codony, F., López Leyton, D., Fittipaldi, M., Adrados, B., and Morató, J. (2010). Quantification of *Helicobacter pylori* levels in soil samples from public playgrounds in Spain. J. Zhejiang. Univ. Sci. B. *11*, 27–29.

Perez-Perez, G.I., Olivares, A.Z., Foo, F.Y., Foo, S., Neusy, A.J., Ng, C., Holzman, R.S., Marmor, M., and Blaser, M.J. (2005). Seroprevalence of *Helicobacter pylori* in New York City populations originating in East Asia. J. Urban Health. *82*, 510–516.

Perry, S., de la Luz Sanchez, M., Yang, S., Haggerty, T.D., Hurst, P., Perez-Perez, G., and Parsonnet, J. (2006). Gastroenteritis and transmission of *Helicobacter pylori* infection in households. Emerg. Infect. Dis *12*, 1701–1708.

Pitkäranta, A., Kolho, K.-L., and Rautelin, H. (2005). *Helicobacter pylori* in children who are prone to upper respiratory tract infections. Arch. Otolaryngol. Head Neck Surg. *131*, 256–258.

Porsch-Ozcürümez, M., Doppl, W., Hardt, P.D., Schnell-Kretschmer, H., Tuncay, M., Akinci, A., Bilgin, Y., and Klör, H.U. (2003). Impact of migration on *Helicobacter pylori* seroprevalence in the offspring of Turkish immigrants in Germany. Turk. J. Pediatr. *45*, 203–208.

Quaglia, N.C., Dambrosio, A., Normanno, G., Parisi, A., Firinu, A., Lorusso, V., and Celano, G.V. (2007). Survival of *Helicobacter pylori* in artificially contaminated ultrahigh temperature and pasteurized milk. Food Microbiol. *24*, 296–300.

Roma, E., Panayiotou, J., Pachoula, J., Kafritsa, Y., Constantinidou, C., Mentis, A., and Syriopoulou, V. (2009). Intrafamilial spread of *Helicobacter pylori* infection in Greece. J. Clin. Gastroenterol. *43*, 711–715.

Ryu, K.H., Yi, S.Y., Na, Y.J., Baik, S.J., Yoon, S.J., Jung, H.S., and Song HJ. (2010). Reinfection rate and endoscopic changes after successful eradication of *Helicobacter pylori*. World J. Gastroenterol. *16*, 251–255.

Salih, B.A. (2009). *Helicobacter pylori* infection in developing countries: The burden for how long? Saudi J. Gastroenterol. *15*, 201–207.

Sanz-Pelaez, O., Santana-Rodriguez, E., Maroto, A.A., Carranza-Rodriguez, C., Pisos-Alamo, E., and Perez-Arellano, J.L. (2008). *Helicobacter pylori* and *cagA* seroprevalence in sub-Saharan inmigrants recently arrived to Gran Canaria (Spain). Scand. J. Infect. Dis. *7*, 1–3.

Sari, Y.S., Can, D., Tunali, V., Sahin, O., Koc, O., and Bender, O. (2008). *H pylori*, treatment for the patient only or the whole family? World J. Gastroenterol. *14*, 1244–1247.

Sasidharan, S., and Uyub, A.M. (2009). Prevalence of *Helicobacter pylori* infection among asymptomatic healthy blood donors in Northern Peninsular Malaysia. Trans. R. Soc. Trop. Med. Hyg. *103*, 395–398.

Schwarz, S., Morelli, G., Kusecek, B., Manica, A., Balloux, F., Owen, R.J., Graham, D.Y., van der Merwe, S., Achtman, M., and Suerbaum, S. (2008). Horizontal versus familial transmission of *Helicobacter pylori*. PLoS Pathog. *4*, e1000180.

Sedlackova, M., Malaty, H., Volf, V., Fruhauf, P., Marx, D., Soucek, A., and Graham, D.Y. (2003). *Helicobacter pylori* infection in a group of symptomatic and asymptomatic children and adolescents in the Czech Republic. Cas. Lek. Cesk. *142*, 102–105.

Siai, K., Ghozzi, M., Ezzine, H., Medjahed, N., and Azzouz, M.M. (2008). Prevalence and risk factors of *Helicobacter pylori* infection in Tunisian children: 1055 children in Cap-Bon (northeastern Tunisia). Gastroenterol. Clin. Biol. *32*, 881–886.

Silva, D.G., Stevens, R.H., Macedo, J.M., Albano, R.M., Falabella, M.E., Veerman, E.C., and Tinoco, E.M. (2009). Detection of cytotoxin genotypes of *Helicobacter pylori* in stomach, saliva and dental plaque. Arch. Oral Biol. *54*, 684–688.

Silva, F.M., Navarro-Rodriguez, T., Barbuti, R.C., Mattar, R., Hashimoto, C.L., and Eisig, J.N. (2010). *Helicobacter pylori* reinfection in Brazilian patients with peptic ulcer disease: A 5-year follow-up. Helicobacter *15*, 46–52.

Song, Q., Lange, T., Spahr, A., Adler, G., and Bode, G. (2000). Characteristic distribution pattern of *Helicobacter pylori* in dental plaque and saliva detected with nested PCR. J. Med. Microbiol. *49*, 349–353.

Sonnenberg, A., Cucino, C., and Bauerfeind, P. (2002). Commentary: the unresolved mystery of birth-cohort phenomena inGastroenterology Int. J. Epidemiol. *31*, 23–26.

Sörberg, M., Nyrén, O., and Granström, M. (2003). Unexpected decrease with age of *Helicobacter pylori* seroprevalence among Swedish blood donors. J. Clin. Microbiol. *41*, 4038–4042.

Strebel, K., Rolle-Kampczyk, U., Richter, M., Kindler, A., Richter, T., and Schlink, U. (2010). A rigorous small area modelling-study for the *Helicobacter pylori* epidemiology. Sci. Total. Environ. *408*, 3931–3942.

Sudhakar, U., Anusuya, C.N., Ramakrishnan, T., and Vijayalakshmi, R. (2008). Isolation of *Helicobacter pylori* from dental plaque: A microbiological study. J. Indian Soc. Periodontol. *12*, 67–72.

Sugimoto, M., Wu, J.Y., Abudayyeh, S., Hoffman, J., Brahem, H., Al-Khatib, K., Yamaoka, Y., and Graham,

D.Y. (2009). Unreliability of results of PCR detection of *Helicobacter pylori* in clinical or environmental samples. J. Clin. Microbiol. *47*, 738–742.

Suzuki, H., Hibi, T., and Marshall, B.J. (2007a). *Helicobacter pylori*: present status and future prospects in Japan. J. Gastroenterol. *42*, 1–15.

Suzuki, G., Cullings, H., Fujiwara, S., Hattori, N., Matsuura, S., Hakoda, M., Akahoshi, M., Kodama, K., and Tahara, E. (2007b). Low-positive antibody titer against *Helicobacter pylori* cytotoxin-associated gene A (CagA) may predict future gastric cancer better than simple seropositivity against *H. pylori* CagA or against *H. pylori*. Cancer Epidemiol. Biomarkers Prev. *16*, 1224–1228.

Suzuki, N., Yoneda, M., Naito, T., Iwamoto, T., Masuo, Y., Yamada, K., Hisama, K., Okada, I., and Hirofuji, T. (2008). Detection of *Helicobacter pylori* DNA in the saliva of patients complaining of halitosis. J. Med. Microbiol. *57*, 1553–1559.

Sykora, J., Siala, K., Varvarovska, J., Pazdiora, P., Pomahacova, R., and Huml, M. (2009). Epidemiology of *Helicobacter pylori* infection in asymptomatic children, a prospective population-based study from the Czech Republic. Application of a monoclonal-based antigen-in-stool enzyme immunoassay. Helicobacter *14*, 286–297.

Tam, Y.H., Yeung, C.K., Lee, K.H., Sihoe, J.D., Chan, K.W., Cheung, S.T., and Mou, J.W. (2008). A population-based study of *Helicobacter pylori* infection in Chinese children resident in Hong Kong, prevalence and potential risk factors. Helicobacter *13*, 219–224.

Teoman, I., Ozmeriç, N., Ozcan, G., Alaaddinoğlu, E., Dumlu, S., Akyön, Y., and Baloş, K. (2007). Comparison of different methods to detect *Helicobacter pylori* in the dental plaque of dyspeptic patients. Clin. Oral Investig. *11*, 201–205.

Thjodleifsson, B., Asbjörnsdottir, H., Sigurjonsdottir, R.B., Gíslason, D., Olafsson, I., Cook, E., Gíslason, T., Jogi, R., and Janson, C. (2007). Seroprevalence of *Helicobacter pylori* and *cagA* antibodies in Iceland, Estonia and Sweden. Scand. J. Infect. Dis. *39*, 683–689.

Tijjani, B., and Umar, A. (2009). Peptic ulcer disease and *Helicobacter pylori* infection at Kano, Nigeria. Internet J. Gastroenterol. *8*.

Tiwari, S.K., Khan, A.A., Ahmed, K.S., Ahmed, I., Kauser, F., Hussain, M.A., Ali, S.M., Alvi, A., Habeeb, A., Abid, Z., *et al.* (2005). Rapid diagnosis of *Helicobacter pylori* infection in dyspeptic patients using salivary secretion: a non-invasive approach. Singapore Med. J. *46*, 224–228.

Tkachenko, M.A., Zhannat, N.Z., Erman, L.V., Blashenkova, E.L., Isachenko, S.V., Isachenko, O.B., Graham, D.Y., and Malaty, H.M. (2007). Dramatic changes in the prevalence of *Helicobacter pylori* infection during childhood: a 10-year follow-up study in Russia. J. Pediatr. Gastroenterol. Nutr. *45*, 428–432.

Tokudome, S., Soeripto, Triningsih, F.X., Ananta, I., Suzuki, S., Kuriki, K., Akasaka, S., Kosaka, H., Ishikawa, H., Azuma, T., *et al.* (2005). Rare *Helicobacter pylori* infection as a factor for the very low stomach cancer incidence in Yogyakarta, Indonesia. Cancer Lett. *219*, 57–61.

Torres, J., Camorlinga-Ponce, M., Perez-Perez, G., Muñoz, L., and Muñoz, O. (2002). Specific serum immunoglobulin G response to urease and CagA antigens of *Helicobacter pylori* in infected children and adults in a country with high prevalence of infection. Clin. Diagn. Lab. Immunol. *9*, 97–100.

Umeda, M., Kobayashi, H., Takeuchi, Y., Hayashi, J., Morotome-Hayashi, Y., Yano, K., Aoki, A., Ohkusa, T., and Ishikawa, I. (2003). High prevalence of *Helicobacter pylori* detected by PCR in the oral cavities of periodontitis patients. J. Periodontol. *74*, 129–134.

Vale, F.F., and Vítor, J.M. (2010). Transmission pathway of *Helicobacter pylori:* does food play a role in rural and urban areas? Int. J. Food Microbiol. *138*, 1–12.

Van Hooste, W., Charlier, A.-M., Rotsaert, P., Bulterys, S., Moens, G., Van Sprundel, M., and De Schryver, A. (2010).Work-related *Helicobacter pylori* infection among sewage workers in municipal wastewater treatment plants in Belgium. Occup. Environ. Med. *67*, 91–97.

Vilarinho, S., Guimarães, N.M., Ferreira, R.M., Gomes, B., Wen, X., Vieira, M.J., Carneiro, F., Godinho, T., and Figueiredo, C. (2010). *Helicobacter pylori* colonization of the adenotonsillar tissue: Fact or fiction? Int. J. Pediatr. Otorhinolaryngol. *74*, 807–811.

Weyermann, M., Adler, G., Brenner, H., and Rothenbacher, D. (2006). The mother as source of *Helicobacter pylori* infection. Epidemiology *17*, 332–334.

Yim, J.Y., Kim, N., Choi, S.H., Kim, Y.S., Cho, K.R., Kim, S.S., Seo, G.S., Kim, H.U., Baik, G.H., Sin, C.S., Cho, S.H., and Oh, B.H. (2007). Seroprevalence of *Helicobacter pylori* in South Korea. Helicobacter *12*, 333–340.

Yonezawa, H., Osaki, T., Kurata, S., Fukuda, M., Kawakami, H., Ochiai, K., Hanawa, T., and Kamiya, S. (2009). Outer membrane vesicles of *Helicobacter pylori* TK1402 are involved in biofilm formation. BMC Microbiol. *9*, 197.

Zaric, S., Bojic, B., Jankovic, L.J., Dapcevic, B., Popovic, B., Cakic, S., and Milasin, J. (2009). Periodontal therapy improves gastric *Helicobacter pylori* eradication. J. Dent. Res. *88*, 946–950.

Zaterka, S., Eisig, J.N., Chinzon, D., and Rothstein, W. (2007). Factors related to *Helicobacter pylori* prevalence in an adult population in Brazil. Helicobacter *12*, 82–88.

Zhang, D.H., Zhou, L.Y., Lin, S.R., Ding, S.G., Huang, Y.H., Gu, F., Zhang, L., Li, Y., Cui, R.L., Meng, L.M., *et al.* (2009a). Recent changes in the prevalence of *Helicobacter pylori* infection among children and adults in high- or low-incidence regions of gastric cancer in China. Chin. Med. J. (Engl.). *122*, 1759–1763.

Zhang, Y.Y., Xia, H.H., Zhuang, Z.H., and Zhong, J. (2009b). Review article: 'true' re-infection of *Helicobacter pylori* after successful eradication – worldwide annual rates, risk factors and clinical implications. Aliment. Pharmacol. Ther. *29*, 145–160.

Zendehdel, N., Nasseri-Moghaddam, S., Malekzadeh, R., Massarrat, S., Sotoudeh, M., and Siavoshi, F. (2005). *Helicobacter pylori* reinfection rate 3 years after successful eradication. J. Gastroenterol. Hepatol. *20*, 401–404.

Helicobacter pylori-associated Diseases

8

Borislav Vladimirov

Abstract

Helicobacter pylori plays a main role in the development of gastritis all over the world. In addition, it is well known that *H. pylori* infection is associated with many non-malignant and malignant gastrointestinal and extragastric diseases. *H. pylori* remains one of the most common causes of peptic gastro-duodenal ulcers, gastric mucosa-associated lymphoid tissue (MALT) lymphoma and gastric cancer. In recent years, many clinical data have been collected about the relationship between *H. pylori* infection and gastro-oesophageal reflux disease (GORD), non-steroidal anti-inflammatory drug/acetylsalicylic acid-induced gastric injury and functional dyspepsia as well as about pathogenetic mechanisms of these correlations. There is also evidence confirming the role of genetic differences in host and bacterial factors and the role of environmental factors. Recent data have shown a decline in incidence and prevalence rate of peptic ulcer related to *H. pylori*. For patients with functional dyspepsia, eradication of *H. pylori* offers a modest but significant benefit. An inverse relationship between *H. pylori* infection and reflux oesophagitis/Barrett's oesophagus has been also confirmed. Despite of the controversial results, eradication of *H. pylori* infection has been recommended for non-steroidal anti-inflammatory drug- and acetylsalicylic acid users as well as for patients treated with antiplatelet therapy. The beneficial effects of *H. pylori* eradication on MALT lymphoma and on the prophylaxis of gastric cancer have been proven. On the other hand, an increasing amount of evidence for extragastric manifestation of *H. pylori* infection has been shown.

Chronic gastritis and non-ulcer dyspepsia

H. pylori is the commonest cause of gastritis and gastritis-associated diseases such as duodenal ulcer (DU), gastric ulcer (GU), low grade gastric mucosa-associated lymphoid tissue (MALT) lymphoma and gastric cancer. Gastritis due to *H. pylori* is to go through a cascade of events that defines the natural history of *H. pylori* infection. It is characterized by progressive inflammation, starting from the antrum and propagating into the corpus, resulting firstly in non-atrophic gastritis and following corpus injury with a loss of parietal cells, reduction in acid secretion and development of corpus atrophy (atrophic gastritis) with or without intestinal metaplasia, and finally dysplasia (El-Zimaity *et al.*, 2002). Atrophy begins at borderline mucosa (corpus-antrum border) and shifts proximally to the antrum, replacing fundic gland mucosa with advancing atrophic gastritis. Corpus atrophy progresses proximally to variably sized regions of the adjacent greater curve, proximal half of the lesser curve and neighbouring anterior and posterior walls of the corpus (El-Zimaity *et al.*, 2002).

Three phenotypes of *H. pylori* gastritis have been described. The 'mild pangastritis phenotype' is characterized with little disturbances of gastric acid secretion. The 'duodenal ulcer phenotype' includes predominantly antral pattern of gastritis with high gastrin release and high acid secretion. This phenotype is associated with duodenal and pre-pyloric ulcers. The third one, 'gastric cancer phenotype' is defined as corpus-predominant pattern of gastritis, atrophy with multifocal intestinal metaplasia, hypochlorhydria or achlorhydria, and is associated with ulcers of the gastric body and fundus and gastric cancer (El-Zimaity, 2007; de Vries *et al.*, 2008b; Harvey *et al.*, 2010). The rate of progression of gastritis, gastritis pattern and patient's acid secretory status are important to determine the *H. pylori* gastritis-associated diseases (El-Zimaity, 2007). In early stages of *H. pylori* infection, gastritis is associated with an exaggerated gastrin response to meals and other stimuli, resulting in an increase in acid secretion that is sufficient in some patients to cause DU. *H. pylori*-induced inflammation is often mild and superficial, localized mainly in the antrum and may absent in the gastric corpus, therefore, *H. pylori* is easily identified in the antrum but less frequently in the corpus (Elitsur *et al.*, 2002). Usually, gastritis does not advance or spreads slowly to involve the gastric corpus. Atrophic pangastritis with hypochlorhydria is rare or develops in later life. The risk of gastric cancer for DU patients remains low.

GU and the intestinal type of gastric cancer are typically associated with extensive gastritis with widespread intestinal metaplasia (El-Zimaity, 2007). The incidence of gastric cancer is highest in countries/regions with early development of atrophic corpus gastritis. Prolonged inflammation and antral atrophy cause destruction and loss of gastrin-producing cells, resulting in a decrease in acid secretion of parietal cells (Sipponen *et al.*, 2002a,b). This facilitates proximal migration of *H. pylori* with possible development of corpus gastritis and eventually corpus atrophy. These are the predisposing factors for the development of GU (El-Zimaity, 2007). *H. pylori,* acute and chronic inflammatory cells and intestinal metaplasia are easily identified in the antrum and the corpus. Pseudopyloric metaplasia is also identified in the corpus (Graham *et al.*, 2003). In intestinal type of gastric cancer, *H. pylori* is more difficult to identify in the antrum and corpus owing to low *H. pylori* prevalence and density (Tokunaga *et al.*, 2000).

In the recent years, it has been suggested that *H. pylori* infection could trigger the development of autoimmune gastritis and pernicious anaemia (D'Elios *et al.*, 2004; Lahner and Annibale, 2009; de Vries and Kuipers, 2010). Chronic autoimmune gastritis is a rare type of chronic gastritis that is common in Caucasians. El-Zimaity (2007) has described the histological picture of autoimmune gastritis. Atrophic corpus gastritis has been associated with a relatively normal antral mucosa. In earlier stage, chronic inflammation has developed primarily in the corpus of the stomach. In advanced cases, the degree of inflammation is minimal and there is loss of specialized glands of the upper stomach. In addition, pseudopyloric and intestinal metaplasia are often present in the late stages of both primary autoimmune and environmental *H. pylori*-associated gastritis (El-Zimaity *et al.*, 2002). Autoimmune response may be accountable for different clinical outcomes in *H. pylori* gastritis. In this process, molecular mimicry is involved by cross-activation between *H. pylori*-derived antigens and autoantigens of the gastric mucosa. *H. pylori* lipopolysaccharide contains Lewis X (Le^x) and Lewis Y (Le^y) blood group antigens that are expressed on gastric H^+,K^+,-ATPase. This H^+,K^+-ATPase is well recognized as a target of gastric auto-immunity (D'Elios *et al.*, 2004). As in classical autoimmune gastritis, anti-H^+,K^+-ATPase antibodies are significant indicators for the development of chronic atrophic corpus gastritis. El-Zimaity (2007) has stated that countries with a high incidence of gastric cancer have not shown high incidence of pernicious anaemia. *H. pylori* may be responsible for initiating the process; nonetheless, only patients with a genetic predisposition will develop early destruction of corpus glandular components. This process occurs late in *H. pylori*-associated gastritis. In one study, *H. pylori* eradication has been followed by healing of active autoimmune corpus gastritis in 80% of cases (Müller *et al.*, 2001).

Longstanding *H. pylori* infection is needed to develop pernicious anaemia. The low grade of colonization and disappearance of *H. pylori* in gastric mucosa with atrophy are associated with pernicious anaemia. De Vries and Kuipers (2010) have reported that a negative serology several years after clearance of *H. pylori* infection, the low incidence of autoimmune gastritis and the asymptomatic onset could explain why autoimmune gastritis is rarely diagnosed at an early stage. Large cohort studies of *H. pylori*-positive subjects should investigate this association.

In the recent years, the role of *H. pylori* infection in the development of gastric hyperplastic polyps has been discussed because some gastric polyps have disappeared after *H. pylori* eradication (El-Zimaity, 2007; Furuta and Delchier, 2009). Role of *cagA* gene for the development of gastric epithelial hyperplasia and gastric polyps of the stomach and small intestine has been confirmed on mouse models (Ohnishi *et al.*, 2008b).

In clinical practice, the diagnosis of gastritis is based on the Sydney system (Dixon *et al.*, 1996). Updated classification of chronic gastritis is based on topography, morphology, and aetiology. According this classification, gastritis due to *H. pylori* infection has included the following subtypes: pangastritis, antral predominant, atrophic, lymphocytic and other gastritis. While *H. pylori* related chronic non-atrophic gastritis could be associated with asymptomatic *H. pylori* infection, peptic ulcers or MALT lymphoma, the atrophic gastritis is associated with gastric cancer. As gastric cancer development is a slow and unpredictable process, recognizing the presence and extent of atrophy can help in the identification of patients at different risks for gastric cancer (El-Zimaity, 2007). To regret, the Sydney system fails to establish the risk of malignancy. For better correlation with the risk of malignant progression, an international group of pathologists (Operative Link for Gastritis Assessment) has proposed a new staging system for evaluation of gastritis, the so-called the OLGA staging system (see Chapter 3). Recent data have shown the interobserver agreement of OLGA classification to increase by grading of intestinal metaplasia of atrophic gastritis (Capelle *et al.*, 2010). On the clinical viewpoint, this combined classification system improves evaluation of the risk for neoplastic progression. This has been supported by a study on cancer risk in patients with different grades of premalignant gastric changes (de Vries *et al.*, 2008b).

Serum pepsinogen I and II levels and PI/PII ratio are good non-invasive surrogate markers to distinguish corpus atrophic gastritis versus non-atrophic or normal mucosa (El-Zimaithy, 2007). Very low pepsinogens levels as well as PI/PII ratio may indicate severe corpus atrophy. Our data also have confirmed the diagnostic value of these non-invasive markers for the selection of persons at high risk for gastric cancer (Vladimirov *et al.*, 2007).

Polymorphisms from both bacterial and host sides have been associated with the severity and/or the type of gastritis. Some markers may predict the individual response to *H. pylori* infection and *H. pylori*-induced gastritis (Tahara *et al.*, 2008). For example, polymorphisms of cytokines such as interleukins (ILs) and tumour necrosis factor (TNF) are associated with different patterns of gastritis among different individuals. Toll-like receptors (TLRs) as TLR2–196 to –174ins alleles are associated with severity of gastritis, gastric mucosal atrophy and intestinal metaplasia. In addition, TLR4 polymorphism is associated with inflammatory cytokines and chemokines in the gastric mucosa infected with *H. pylori* (Trejo-de la *et al.*, 2008). Polymorphism of regulated upon activation, normal T-cell expressed and secreted RANTES-28G allele decreases the risk of intestinal metaplasia (Tahara *et al.*, 2009). Tumour necrosis factor-alpha (TNF-α–308G/A) and interleukin 10 (IL-10 819C/T) gene polymorphisms are associated with increased risk of gastritis and lymphoid follicle formation (Achyut *et al.*, 2008). The possible importance of regulatory T cells (T_{reg}) and Treg cytokines such as transforming growth factor (TGF) β1 and interleukin 10 in the reduction of inflammation and ulceration induced by *H. pylori* has been investigated (Robinson *et al.*, 2008). The role of IL-10 in this process has been confirmed. The key role of *H. pylori* virulence factors is discussed in Chapter 4.

Kim *et al.* (2008b) have evaluated risk factors for atrophic gastritis and intestinal metaplasia within *H. pylori* virulence factors (i.e. *cagA*, *vacA* m1, and *oipA*) and environmental factors (i.e. smoking and alcohol) and host polymorphisms (i.e. IL-1b-511, IL-1RN, TNFA-308, IL-10-592, IL-10-819, IL-10-1082, IL-8-251, IL-6-572, GSTP1, p53 codon 72, and ALDH2). They have found that the bacterial factors are important risk factors for atrophic gastritis but that environmental and host factors are more important for intestinal metaplasia. Therefore, useful marker that predicts the individual response to *H. pylori* infection remains to be elucidated in relation to environmental factors (Furuta and Delchier, 2009).

Functional dyspepsia

According to Rome III diagnostic criteria for functional gastrointestinal disorders, functional dyspepsia is defined as ≥1 of the following symptoms: bothersome postprandial fullness; early satiation; epigastric pain or epigastric burning in patients without structural gastrointestinal disease, including at upper endoscopy (Drossman, 2006). Diagnosis of functional dyspepsia is based on presence of the symptoms for 3–6 months. The role of *H. pylori* infection in functional dyspepsia has not been fully elucidated and the effect of *H. pylori* eradication is still controversial. The prevalence of *H. pylori* in cases with functional dyspepsia is various in different countries. On the other hand, *H. pylori* prevalence in patients with dyspeptic symptoms is usually low compared with peptic ulcer disease (PUD). For example, *H. pylori* positivity in dyspeptic patients has been found in 30–40% of the Bulgarian population (Mateva *et al.*, 2001). This rate is about twice lower than that in peptic ulcer. Kivistö *et al.* (2010) have detected *Helicobacter bizzozeronii* from the gastric mucosa of a Finish patient with severe dyspeptic symptoms. The relevance of potentially zoonotic gastric *Helicobacter* spp. remains to be determined.

In primary care, a test and treat strategy using a non-invasive test is recommended in adult patients with unexplored persistent dyspepsia under the age of 45 (the age cut off point may vary between countries according to gastric cancer prevalence) and without alarm symptoms as anaemia or signs of gastrointestinal bleeding, unexplainable weight loss, dysphagia, persistent severe abdominal pain, palpable mass, malabsorption etc. (Malfertheiner *et al.*, 2007; de Vries *et al.*, 2008b). The Maastricht 3–2005 Consensus Report has involved recommendation based on the evidence of the meta-analysis and Cochrane collaboration systematic review updated in 2005 (Delaney *et al.*, 2005; Moayyedi *et al.*, 2006). According to these evidences, *H. pylori* eradication has a modest but significant beneficial effect on *H. pylori*-positive functional dyspepsia and leads to long-term symptom improvement (Moayyedi *et al.*, 2006). In addition to the symptomatic improvement, *H. pylori* eradication also reduces the risk of developing PUD, atrophic gastritis, and gastric cancer (Delaney *et al.*, 2005; Malfertheiner *et al.*, 2007). Moayyedi *et al.* (2000) have shown the test-and-treat strategy to be cost-effective for patients with uninvestigated dyspepsia, while Jarbol *et al.* (2006) have not confirmed the suggestion. In addition, in areas of low *H. pylori* prevalence (<20%), proton pump inhibitor (PPI) empirical treatment or a test and treat strategy have been considered to be equivalent options (Malfertheiner *et al.*, 2007; Talley *et al.*, 2008; de Vries *et al.*, 2008a). A recent 'screen-and-threat' study from Finland on 3326 persons has shown that a low prevalence of *H. pylori* infection might limit the cost efficacy of the programme (Salomaa-Räsänen *et al.*, 2010). Patients with functional dyspepsia in Asia would benefit from treatment for *H. pylori* infection (Gwee *et al.*, 2009). *H. pylori* eradication can improve the dyspeptic symptoms not only in Asian population but also in other areas of high *H. pylori* prevalence. In populations at high risk for gastric cancer, endoscopy before *H. pylori* eradication has been also recommended for patients with dyspepsia (Chey and Wong, 2007; Hermansson *et al.*, 2009). A population-based randomized controlled study has shown that *H. pylori* eradication has reduced consultations with a general practitioner for dyspepsia with 25% in a 2- to 7-year follow-up compared with treatment with placebo (Harvey *et al.*, 2010). *H. pylori* eradication probably prevents PUD development on long-term and gastric cancer on population level. Thus, this strategy warrants further evaluation and should be compared with long-term acid suppression, preferably by end-points of PUD and gastric cancer (de Vries and Kuipers, 2010).

Peptic ulcers

It is well known that *H. pylori* infection and non-steroidal anti-inflammatory drugs (NSAIDs)/acetylsalicylic acid (ASA) are the major factors for gastroduodenal ulcers. *H. pylori* plays a different role in GU and DU pathogenesis. It is well established that >80–90% of DUs but only up to 70% of GUs are associated with *H. pylori* infection (Vu and Ng, 2000; Ford *et al.*, 2004). There has been a reduction of relative risk of 54% in the recurrence of DU after *H. pylori* eradication, but smaller reduction of 37% for GU. In some countries, *H. pylori* prevalence has been reported to be over 90% in DU and over 80% in GU (Boyanova, 2007). The prevalence of *H. pylori* in bleeding ulcers is still not well defined, ranging from 40% to 90% (Vaira *et al.*, 1997).

Declining *H. pylori* infection rate among peptic ulcer patients has been observed. Nervi *et al.* (2006) have reported a decrease in ulcer prevalence: from 12.7% to 6.3% in Padova and from 15.6% to 12% in Parma, Italy, over 10 years. The decrease has been significant for DU and GU in Padova and for DU in Parma. Authors have stated that 10-year extensive *H. pylori* eradication in symptomatic patients has led to a significant reduction in peptic ulcer prevalence.

Lifetime risk for development of PUD in *H. pylori*-positive subjects ranges between 10% and 20% only (de Vries and Kuiper, 2010). Many data have shown a decreasing prevalence and incidence of *H. pylori* related gastroduodenal ulcer in Western countries over the past decades. A large epidemiological population-based prospective cohort study in Denmark enrolled 2416 subjects without history of PUD and has reported an overall 11-year (from 1983 to 1994 year) cumulative incidence of PUD of 2.9%, i.e. 1.6% for DU, and 1.3% for GU (Rosenstock *et al.*, 2003). Poor socioeconomic status has increased the risk of PUD. Similarly, in the USA, the prevalence of PUD decreased slowly from 1995–1999 to 2000–2004 (Smith *et al.*, 2005). Conversely, increasing frequency of *H. pylori*-negative idiopathic ulcers was found in patients with bleeding gastroduodenal ulcers in Hong Kong for the period of 1997–1998 to 2000 (Hung *et al.*, 2005). Other authors also have confirmed a rising in *H. pylori*-negative PUD (Sung *et al.*, 2009). Wu *et al.* (2009b) reported a dramatic decrease in incidence of PUD in Taiwan from 1997 to 2006 following an increase in anti-*H. pylori* therapy and use of PPIs.

According to Shirin *et al.* (2008), the ulcer-related mortality is continuing to decline all over the world. The dynamics of mortality due to ulcers has been demonstrated in Europe and in eight countries outside Europe (Sonnenberg, 2007a,b). The increasing consumption of anti-inflammatory and antisecretory drugs has not changed the trend over the last decades, and therefore, *H. pylori* has been considered the key factor in causing ulcer related mortality. Sung *et al.* (2009) have analysed the incidence of PUD in Western countries and have found PUD to remain a common condition despite decreasing prevalence in *H. pylori* infection. In contrast, in developing countries, a decline of PUD prevalence might be the consequence of the global decreasing prevalence of *H. pylori* infection. Meta-analysis of recent studies shows a decline of both incidence and prevalence of PUD. Current population-based incidences of PUD diagnoses have ranged between 100/100,000 and 190/100,000 per year in Western countries, and current population-based prevalences have ranged between 120/100,000 and 4700/100,000. Prevalence of PUD in patients undergoing upper gastrointestinal endoscopy has ranged between 5% and 16%.

Data on the incidence of PUD complications are controversial (Hermansson *et al.*, 2009). Upper gastrointestinal bleeding occurs in about 10–20% of patients with *H. pylori*-associated PUD. Barkun *et al.* (2010) have strongly recommended to always test the patents with peptic ulcer for *H. pylori* infection (Barkun *et al.*, 2010). *H. pylori*-positive patients need eradication therapy followed by testing to confirm eradication (Kamburov *et al.*, 2007). A negative *H. pylori* test in the acute setting should be repeated during follow-up. The recommendation for eradication in patients with PUD includes also those with bleeding ulcers (Hermansson *et al.*, 2009; Barkun *et al.*, 2010). *H. pylori* eradication rates can achieves >80% eradication on intention to treat basis by orally administered drugs but not during the bleeding phase. Moreover, drug pharmacokinetics at the gastric mucosal level is less predictable in the acute bleeding phase. Pre-treatment with PPIs does not influence the success of PPI-based triple therapy in eradicating *H. pylori*.

It is now well established that *H. pylori* eradication can significantly modify the natural history of PUD. Non-recurrence of GU and DU is strictly dependent on the success of *H. pylori* eradication and the persistence of infection is a negative prognostic marker. This is also true for relapses of ulcer complications such as bleeding. PUD (active or not, including complicated disease and following gastric surgery for peptic ulcer), is one of the strongly recommended indications for *H. pylori* eradication therapy (Malfertheiner *et al.*, 2007).

More than 15 years of experience was summarized in a comprehensive Cochrane collaboration systematic review, updated by Ford *et al.* (2004). In DU healing, eradication therapy has been superior to ulcer-healing drugs or no treatment. In GU healing, no significant difference has been detected between eradication therapy and ulcer-healing drugs. In preventing DU recurrence, no significant differences have been found between eradication therapy and maintenance therapy with ulcer-healing drugs but eradication therapy has been superior to no treatment. In preventing GU recurrence, eradication therapy has been superior to no treatment. The authors concluded that a 1- to 2-week course of *H. pylori* eradication therapy is an effective treatment for *H. pylori*-positive PUD. Gisbert and Pajares (2005) have performed a meta-analysis of randomized clinical trials and have concluded that in patients with PUD and *H. pylori* infection, prolonging the therapy with PPI after a triple therapy for 7 days with a PPI and two antibiotics has not been necessary to induce ulcer healing. The mean ulcer-healing rate with a 7-day treatment has been 91% versus 92% when PPI was prolonged for 2–4 more weeks. Murakami *et al.* (2006) have suggested that post-eradication maintenance therapy with H_2-receptor antagonist could reduce greatly the ulcer recurrence rate without affecting the *H. pylori* eradication.

The role of virulent, especially *cagA*$^+$ *H. pylori* strains as a predictive factor for *H. pylori* susceptibility to macrolides and for the outcome of eradication is discussed in Chapter 9. Treatment of *H. pylori* infection has been more effective than antisecretory therapy alone, with or without long-term maintenance antisecretory therapy, in preventing recurrent bleeding from PU, thus, all patients with PU bleeding should be tested for *H. pylori* infection, and eradication therapy should be prescribed to *H. pylori*-positive patients (van Leerdam, 2008). Gisbert *et al.* (2007) found that rebleeding did not occur in patients with complicated ulcers after *H. pylori* eradication and, therefore, the maintenance antisecretory therapy is not necessary in these patients if eradication is achieved.

The outcome of a *H. pylori* infection to developed PUD depends of the virulence of the infecting strain, host genetics and environmental factors. *H. pylori* virulence factors, which can predict the outcome of infection, are presented in Chapter 4. Host factors are important in the pathogenesis of *H. pylori*–induced PUD and the host immune response probably plays a key role in the outcome of the infection (de Vries and Kuipers, 2010). The important role of regulatory T cells (T_{reg} cells) in the development of PUD has been reported (see Chapter 5).

Over the past years, evidence is expanding largely on the role of specific genetic polymorphisms involved in the outcome of *H. pylori* infection, especially progression towards gastric cancer. Cytokine polymorphisms have previously been shown to modulate host response to *H. pylori* infection and to determine the occurrence of PUD (Furuta and Delchier, 2009). Several studies have confirmed an association between IL-1β and TNF-α gene polymorphism and risk of DU. In one Korean study, the IL-1β –511T-carrier polymorphism has had a preventative effect on the development of GU (Kim *et al.*, 2006). A case control study of *H. pylori*-infected subjects from India has shown higher frequency of IL-1β-511TT and -31CC genotypes in patients with DU compared with those with normal mucosa (Chakravorty *et al.*, 2006). In contrast, Garcia-Gonzalez *et al.* (2005) have not found any association between allelic variants of IL-1 and TNF gene and the susceptibility to DU. An association between polymorphisms in IL-10, IL-8 and IL-6 and both gastric and DUs has been demonstrated in a Korean population (Kang *et al.*, 2009). Lu *et al.* (2005) have assessed the role of host genotypes of TNF-α promoter single nucleotide polymorphism in clinical and histological outcomes in patients with *H. pylori* infection. Both TNF-α-1031C and -863A carriers have shown increased DU and GU risk in the presence of *H. pylori* infection. The authors have shown TNF-α-1031 and -863 promoter single nucleotide polymorphism to be novel host factors to determine the risk of PUD upon *H. pylori* infection. Vascular factors could also play a role in the pathogenesis of PUD regardless of the main ulcerogenic agent involved, *H. pylori* or NSAIDs. Kim *et al.* (2008c) have found vascular endothelial growth factor (VEGF) polymorphism to be associated with peptic ulcer risk. T-1676C polymorphism in COX-1 gene promoter has been linked only to GU but not DU, and frequency of the A-842G polymorphism has been rare in Japanese population (Arisawa *et al.*, 2007).

Gastric cancer and MALT lymphoma

It is well established that *H. pylori* is the most important risk factor for gastric adenocarcinomas and gastric MALT lymphomas. About 75% of non-cardia gastric cancers and gastric MALT lymphomas and 65% of all stomach cancers worldwide are associated with *H. pylori*, especially with CagA antibodies. In contrast to Western populations, in China and Japan, *H. pylori* infection has been associated with increased risk of both non-cardia and cardia cancers (Moss and Malfertheiner, 2007; Kamangar *et al.*, 2007). *H. pylori* eradication therapy has been recommended as prevention of gastric cancer prior to the development of gastric mucosal atrophy, intestinal metaplasia and dysplasia, but it is less efficacious when administered later. *H. pylori* eradication alone has only a moderate effect on gastric cancer incidence worldwide. *H. pylori* eradication is also recommended for low-grade gastric B-cell MALT lymphoma and may lead to complete remission in grade I MALT lymphomas. The pathogenetic mechanisms, in which *H. pylori* induce malignancy, continue to be investigated.

Gastric cancer

Gastric cancer is still the second most frequent cancer worldwide with around one million new diagnoses each year, and the second most common cause of cancer related deaths (Kamangar *et al.*, 2006). There is little improvement of long-term survival during the past decades. In the Western world, most patients are diagnosed in advanced stage of the disease. In 1994, the International Agency for Research on Cancer categorized *H. pylori* infection as a class I human carcinogen. Both the intestinal and diffuse histological types of gastric cancer are associated with *H. pylori* infection. *H. pylori*-induced chronic gastritis may pass through atrophic gastritis, metaplasia and dysplasia. All of them may result in development of gastric cancer. In addition, environmental and host factors modulate its outcome (Correia *et al.*, 2009).

It is also well established that infection with *cagA+/vacA* i1 type *H. pylori* strains increases the risk for gastric cancer (see Chapter 4). All bacterial virulence factors, host genetic factors, and environmental factors shape up the individual risk of developing gastric cancer (Moss and Malfertheiner, 2007; Ferreira *et al.*, 2008; Correia *et al.*, 2009). *H. pylori*-infected subjects are at higher risk of developing non-cardia gastric cancer than non-infected subjects or those in whom the infection has been eradicated (Uemura *et al.*, 2001). *H. pylori* eradication has been shown to prevent development of pre-neoplastic changes of the gastric mucosa, on the one hand, and to stop/reduce precancerous lesions as compared with controls (Asaka *et al.*, 2001) In contrast, other authors have found only significant reduction of development of gastric cancer in the group without pre-neoplastic lesions (Wong *et al.*, 2004). Thus, the optimal time to eradicate *H. pylori* is before pre-neoplastic lesions as atrophy and intestinal metaplasia, are present. Unfortunately, the potential for gastric cancer prevention globally is restricted by currently available treatments (Wong *et al.*, 2004). Therefore, new treatments are needed for a global strategy of gastric cancer prevention. In a review, Fuccio *et al.* (2007) observed that in the large and randomized studies, *H. pylori* eradication has reduced the gastric cancer rate only slightly. Thus, *H. pylori* eradication seems to be relevant in only a subset of subjects. Based on the relation between *H. pylori* and gastric cancer, Maastricht III Consensus Report has made the following statements:

- The global burden of gastric cancer is increasing, predominantly in developing countries (level of evidence 1, grade of recommendation A).

- *H. pylori* infection is the most common proven risk factor for human non-cardia gastric cancer (grade of recommendation A).
- The risk for gastric cancer development depends on bacterial virulence factors (grade of recommendation A), host genetic factors (grade of recommendation B), and on environmental factors, which contribute to the risk (grade of recommendation A).
- Evidence for *H. pylori* as an important factor for gastric cancer development is shown by experimental animal models (grade of recommendation B).
- Eradication of *H. pylori* prevents development of pre-neoplastic changes of the gastric mucosa (level of evidence 1b, grade of recommendation A).
- Eradication of *H. pylori* has the potential to reduce the risk of gastric cancer development (level of evidence 1c, grade of recommendation B).
- The optimal time to eradicate *H. pylori* is before pre-neoplastic conditions (atrophy, intestinal metaplasia) are present, probably in early adulthood (level of evidence 1b, grade of recommendation A).
- *H. pylori* eradication for gastric cancer prevention is cost effective in economic analyses. Feasibility studies are required to evaluate further the benefits and risks of this strategy (level of evidence 2, grade of recommendation B).
- The potential for gastric cancer prevention on a global scale is restricted by currently available treatments (level of evidence 1b, grade of recommendation A).
- New treatments are required for a global strategy of eradication to prevent gastric cancer (grade of recommendation A).
- *H. pylori* eradication for gastric cancer prevention in populations at risk should be considered (level of evidence 2a, grade of recommendation B; Malfertheiner *et al.*, 2007).

Although the evidence confirmed the close relationship between *H. pylori* and environmental factors and gastric cancer, only a small proportion of infected individuals (<1%) develop gastric adenocarcinoma (Yamaoka *et al.*, 2008; Mizuno *et al.*, 2010). The molecular nature of this disease and its precursor lesions has been under intense investigation. Correia *et al.* (2009) have suggested the role of host-related genetic susceptibility included genes involved in inflammation and carcinogen metabolism, the role of overexpression of a pro-inflammatory cytokine (IL-1β) and deletion of a cell–cell adhesion molecule (E-cadherin), as well as the role of stem cells in experimental mouse models of gastric carcinogenesis. Polymorphisms in several genes involved in the inflammatory response is critical for gastric carcinogenesis; in the mucosal protection towards *H. pylori* infection; in the protection of DNA to oxidative damage; and in the detoxification enhancement (Crucius *et al.*, 2008; Tripathi *et al.*, 2008; Zhang *et al.*, 2008; Canedo *et al.*, 2009; Gao *et al.*, 2009a,b; Shibata *et al.*, 2009a,b). The polymorphisms include IL-1β and IL-1 receptor antagonist genes (IL-1RN) and TNF-α IL-16 genes (Crucius *et al.*, 2008; McNamara *et al.*, 2008; Tu *et al.*, 2008). For instance, the procarcinogenic effect of IL-1-β has increased in the presence of *H. pylori* (McNamara *et al.*, 2008; Tu *et al.*, 2008). Inactivating mutations of the p53 gene can be found in 38–71% of gastric cancer (Kountouras *et al.*, 2008). The p53 mutations may be an early event in gastric carcinogenesis (Zhou *et al.*, 2007). Moreover, alterations in p73 gene may play a role in gastric carcinogenesis (Wei *et al.*, 2008).

At the molecular level, diffuse type gastric cancer can be distinguished from intestinal-type gastric cancer on the basis of the cell–cell adhesion molecule, E-cadherin (Correia *et al.*, 2009). E-cadherin is downregulated very early during diffuse gastric cancer development, suggesting a role in the initiation of this disease. In fact, a causal relationship between E-cadherin deficiency and the initiation of diffuse gastric cancer has been established. Inhibition of TGF-b signalling has also been reported to prevent progression and metastasis onset in diffuse gastric cancer, mainly because of its ability to enhance angiogenesis (Ferreira *et al.*, 2008; Komuro *et al.*, 2009). Recent data suggest an important role not only of genetic susceptibility, but also of accumulation of genetic and epigenetic alteration in gastric carcinogenesis. Genetic variation in the genes DNMT3A, PSCA, VEGF, and XRCCI has been reported to modify the risk of developing gastric cancer. Several genes have been newly associated with gastric carcinogenesis, through both oncogenic activation (MYC, SEMA5A, BCL2L12, RBP2 and BUBR1) and a tumour suppressor gene inactivation mechanism (KLF6, RELN, PTCH1A, CLDN11, and SFRP5). At the level of gastric cancer treatment, the HER-2 tyrosine kinase receptor has been demonstrated to be a molecular target of therapy.

Patients presenting with clinical manifestations of gastric cancer have limited options for treatment. Because of that, the diagnosis of premalignant lesions and early gastric cancer is very important. Prevention plays a key role in the fight against gastric cancer. For adequate individualized risk assessment in planning prevention strategies, the question of whether gastric atrophy and intestinal metaplasia (IM) are premalignant rather than paramalignant lesions is still under debate. A glandular atrophy, intestinal metaplasia and dysplasia are selected as precancerous lesions (Humar *et al.*, 2007, 2009). Bornschein *et al.* (2009) have reported that the cancer risk in patients with mild/moderate dysplasia has been comparable to the risk for the development of colorectal cancer after removal of colonic adenomas. In a large retrospective analysis, de Vries *et al.* (2008a) have found that the annual incidence of gastric cancer in Dutch patients has increased according to the severity of the mucosal alteration present in 0.1% for atrophic gastritis, 0.25% for intestinal metaplasia, 0.6% for mild-to-moderate dysplasia and 6% for severe dysplasia within 5 years after initial diagnosis. However, no recommendation for surveillance has been proposed. There is an important need for guidelines to determine at what intervals patients will require endoscopic control. So far, the best estimate for the regression in histopathology scores can be calculated as a function of the square of the time the patient is *H. pylori* negative after eradication therapy (Komuro *et al.*, 2009).

Population-based screening and treatment of *H. pylori* infection most probably represent the current best option for primary prevention of gastric cancer but several aspects need to be considered, such as timing, methods and cost-effectiveness in various regions of the world (Bornschein *et al.*, 2009). In high-risk populations as East Asia, the prevalent *H. pylori* strain type is CagA compared with populations in Africa, South Asia or Europe (Houghton *et al.*, 2004). Screening and treatment in individuals at the age of 20 in a high risk region in China, has resulted in adequate reduction of the lifetime risk for gastric cancer with costs below US$1500 per life year saved (McDonald *et al.*, 2008). Assessing persons at older age or rescreening of negative individuals has not been not cost-effective. Results of prospective trials on a global scale are needed to support these theoretical estimations (Bornschein *et al.*, 2009).

Serological-based tests as screening markers for pre-neoplastic changes of the gastric mucosa have the potential to select subjects with high risk of gastric cancer that require a close clinical and endoscopic follow-up (Selgrad *et al.*, 2010). Because of large differences in incidence between various populations, population-based screening is not applicable all over the world. During the last decades, serological screening has been applied in countries at high risk of gastric cancer such as Japan. The combination of pepsinogen I and II, gastrin and *H. pylori* antibodies have been the most accurate to diagnose atrophic gastritis, but when compared the pepsinogen tests with the conventional X-ray examination, the serum pepsinogen test alone has not been appropriate for gastric cancer screening (Mizuno *et al.*, 2009; Mizuno *et al.*, 2010). On the other hand, long-term results of gastric cancer screening using the serum pepsinogen test have shown that 80% of the newly diagnosed cancers have been early stage (Miki *et al.*, 2009). In addition, according to the results of prospective case cohort study in China with a follow-up period of 15 years, a low pepsinogen I/II ratio may identify subjects at increased risk of non-cardia and cardia gastric adenocarcinomas (Ren *et al.*, 2009). When old age, alcohol, and smoking habits have been added in 'atrophic' pepsinogen profile, selection of cases with high cancer risk has increased (Yamaji *et al.*, 2009). However, the sensitivity of the pepsinogen test is low. According to Selgrad *et al.* (2010), the combination of *H. pylori* infection and atrophic gastritis determined by serological examination is of value to predict the risk of gastric cancer and might be suitable for population-based gastric cancer screening in high-risk regions.

H. pylori eradication is effective to prevent gastric cancer, especially in Asia (Takenaka *et al.*, 2007; McNamara and El-Omar, 2008). On the other hand, some studies have not confirmed this beneficial effect (Fall *et al.*, 2006; Fuccio *et al.*, 2009; de Vries *et al.*, 2009). These controversial results may be explained by various degree and extent of pre-neoplastic changes as gastric atrophy and intestinal metaplasia at the time of eradication. The optimal time of *H. pylori* eradication is still controversial (Selgrad *et al.*, 2010). Some authors have suggested that patients receiving early *H. pylori* eradication, as well as those with frequent use of non-steroidal anti-inflammatory drugs or acetylsalicylic acid have lower risk for development of gastric cancer (Wu *et al.*, 2009a). The so-called 'point of no return' when mucosal alterations (i.e. intestinal metaplasia, glandular atrophy) are no longer reversible has been identified to be critical for an effective prevention of gastric cancer incidence or recurrence. According to Watari *et al.* (2008), *H. pylori* eradication changes the cellular phenotype of gastric intestinal metaplasia, which might be an important factor in the reduction of gastric cancer incidence after successful eradication. Yanaoka *et al.* (2009) have proposed the individuals with high risk of gastric cancer to be identified by the serum PG test. The risk of gastric cancer development after eradication therapy has been related to the presence of extensive atrophy before the eradication. Because of these data, a population-based screen-and-eradicate strategy for *H. pylori* infection can now be recommended only in high incidence regions (Correia *et al.*, 2009). Positive experience with endoscopic mucosal resection and sub-mucosal dissection for treatment of early gastric cancer has been confirmed and extended. New studies have found the incidence of recurrence and development meta-chronic gastric cancer post endoscopic resection of early adenocarcinomas to be significantly reduced after *H. pylori* eradication (Zhou *et al.*, 2007; Fukase *et al.*, 2008; Wei *et al.*, 2008). Moreover, *H. pylori* eradication can improve furthermore healing rates of treatment-induced GUs. Based on the multifactorial process of gastric carcinogenesis, including genetic polymorphisms of the host, virulence determinants of *H. pylori* and environmental factors, Selgrad *et al.* (2010) have suggested further diagnostic tools to be necessary to predict individual risk of gastric cancer in an *H. pylori*-infected patient. According to the authors, *H. pylori* eradication prevents gastric cancer development, and it seems that the earlier the bacteria are eradicated, the more significant is the decrease in gastric cancer risk. A prophylactic vaccine as primary prevention, especially with infant vaccination, would be an ideal strategy to eradicate *H. pylori* and prevent gastric cancer.

Gastric mucosa-associated lymphoid tissue lymphoma

H. pylori infection plays a critical role in the development and progression of MALT lymphoma. Owing to the relationship between MALT lymphoma and *H. pylori* factors, the presence of antiCagA IgG antibodies is suggested to be a risk factor for lymphoid follicle development in patients with gastritis (Achyut *et al.*, 2008). Moreover, cytoplasmic heat shock protein B (HspB) immunostaining has been observed in groups of neoplastic cells of MALT lymphoma, suggesting a possible involvement of HspB in the pathogenesis of *H. pylori*-related development of MALT lymphoma (De Luca *et al.*, 2008). On the other hand, *Helicobacter heilmannii* is known to induce gastric low-grade MALT lymphoma in mouse model (Nakamura *et al.*, 2007a). Epidemiological studies have also revealed that there is a high rate of MALT lymphoma in patients with *H. heilmannii* infection compared with those infected with *H. pylori* (Ferreira *et al.*, 2008).

H. pylori eradication has been strongly recommended as a first line treatment in *H. pylori*-positive patients with low-grade MALT lymphoma in the 1st and 2nd Maastricht Consensus Report, and has been reinforced as strongly recommended at Maastricht 3rd Consensus Report, 2005. (Malfertheiner *et al*, 2007). According to the data which were published after the 2nd Maastricht Consensus Report, 62% of patients with low-grade gastric MALT lymphoma achieve complete remission after *H. pylori* eradication within 12 months (Fischbach *et al.*, 2000, 2004). *H. pylori* eradication is the treatment of first choice for *H. pylori*-infected patients with localized stage I (according to Lugano classification) gastric MALT lymphoma (Inagaki,

2007; Malfertheiner *et al.*, 2007; Morgner *et al.*, 2007). Our unpublished data also have showed a high remission rate after *H. pylori* eradication in patients with low-grade gastric MALT lymphoma. Study from Korea has reported a long-term complete remission in 94% of patients with low grade MALT lymphoma, especially of cases with tumours in the distal stomach (Kim *et al.*, 2007). In another study, including patients with low-grade MALT lymphoma after eradication therapy for 14 days, 94% of them have reached complete remission. Only 10% of patients have been found with recurrence of MALT lymphoma. Cumulative recurrence rates are 2.7%, 11.5%, and 12.2% at 1, 2, and 3 years respectively. Persistence of *H. pylori* has been identified as the most important risk factor involved in recurrence and, therefore, an adequate eradication regimen and accurate regular evaluations for *H. pylori* status are needed during follow-up of primary gastric low-grade MALT lymphoma (Hong *et al.*, 2006). A recent Japanese study found a high rate (76%) of complete remission among patients with localized gastric MALT lymphoma 12 months after eradication therapy (Terai *et al.*, 2008). In a large European international study, 94% of patients with stage I gastric MALT lymphoma had a favourable disease course based on a follow-up of a median of 42 months (Fischbach *et al.*, 2007). The authors observed complete remission in 32% of participants and, in 62%, they found the minimal histological residuals to remain stable. These reports have confirmed that *H. pylori* eradication provides a favourable long-term prognosis, with a genuine chance of a cure or at least of long-lasting complete remission in the majority of patients (Ferreira *et al.*, 2008). Thus, a watch-and-wait strategy with regular endoscopies and biopsies is a justified approach for treating patients with localized gastric MALT lymphoma, even those with minimal residual disease. Nevertheless, careful monitoring for the development of gastric cancer is recommended during follow-up, given recent case reports on meta-chronic stomach cancers after eradication therapy for gastric MALT lymphoma (Isaka *et al.*, 2007; Seo *et al.*, 2007). It is also not strange for *H. pylori* to be associated with extragastric locations such as reported for primary orbital lymphoma (Chan *et al.*, 2006). Chemotherapy, radiotherapy and as surgical resection are currently applied with good results if the lymphoma does not reveal regression after eradication therapies or if the lymphoma is in stage II (Inagaki, 2007; Morgner *et al.*, 2007; Kuo *et al.*, 2008). Some authors have suggested that some patients respond to therapy even in the absence of detectable *H. pylori* (Grünberger *et al.*, 2006; Raderer *et al.*, 2006).

MALT lymphoma is genetically characterized by four chromosome translocations. The t(11;18)(q21;q21) generates a functional API2 (apoptosis inhibitor 2)-MALT1 fusion transcript, whereas the other three translocations involve translocation to immunoglobulin gene (IGH) loci and consequently increased expression of BCL10 (1p22), MALT1 (18q21) and fork head box protein P1 gene (FOXP1) (3p14; Ferreira *et al.*, 2008). Gastric MALT lymphomas with gene t(11; 18) (q21; q21), translocation with fusion of API2 and MALT1 are usually unresponsive to *H. pylori* eradication, and often present with advanced disease (Du, 2007; Fukuhara *et al.*, 2007; Inagaki, 2007; Malfertheiner *et al.*, 2007; Nakamura *et al.*, 2007b,c; Sakuma *et al.*, 2007). Fusion of both leads to suppression of apoptosis and strongly predicts failure to respond to eradication therapy (Malfertheiner *et al.*, 2007). A careful monitoring for disease progression is necessary for gastric MALT lymphoma carrying API2-MALT1 (Nakamura *et al.*, 2008.)

Array-based comparative genomic hybridization has demonstrated no significant genetic alterations in the t(11;18)(q21;q21)-positive MALT lymphoma, thus it is considered to have a good prognosis. In contrast, gastric t(11;18)(q21;q21)-negative MALT lymphoma with *H. pylori* independency has exhibited numerous genomic abnormalities similar to those in diffuse large B-cell lymphoma, which is characterized by a clinically advanced form of the disease and shows an increase in the large cell component (Fukuhara *et al.*, 2007; Inagaki, 2007; Nakamura *et al.*, 2008). Overall, these data suggest that acquired genomic imbalances may play a more important role in the *H. pylori*-independent than the *H. pylori*-dependent stage of gastric MALT lymphoma (Fukuhara *et al.*, 2007). In contrast to t(11;18)(q21;q21), there is little information on the clinical features of gastric MALT lymphomas bearing other translocations or numerical changes, but during recent years these genetic abnormalities have also been intensively studied (Nakamura *et al.*, 2007b,c; Bernasconi *et al.*, 2008). Cases with t(14;18)/IGH-involved translocation-BCL2 have responded well to *H. pylori* eradication therapy. The presence of extra copies of *MALT1* has been significantly associated with progression or relapse of the disease and has been an independent adverse prognostic factor for event-free survival. Moreover, longer follow-up studies with a large number of cases are necessary to clarify whether this numerical change has a major influence on the prognosis of gastric MALT lymphoma (Ferreira *et al.*, 2008). Toll-like receptor (CTLR) 4 expression has been localized on the surface of MALT lymphoma cells but not other lymphomas, suggesting a pathogenic role and a therapeutic target of the receptor for gastric MALT lymphoma (Adam *et al.*, 2008). Low-grade MALT lymphoma with superficial gastric involvement and BCL2-positive and p53-negative immunophenotype, disappears after successful *H. pylori* eradication (Venizelos *et al.*, 2007).

Gastro-oesophageal reflux disease

The role of *H. pylori* in the pathogenesis of GORD is not completely understood (de Vries and Kuipers, 2010). According to the Maastricht III Consensus Report, there is a negative association between the prevalence of *H. pylori* and GORD, but the reason of this relationship is unclear (Malfertheiner *et al.*, 2007). The prevalence of *H. pylori* infection is lower in GORD patients than in the general population (Axon, 2004). Therefore, *H. pylori* infection has been thought to be possibly protective against the development of GORD (Furuta and Delchier, 2009). Moreover, most countries with a high prevalence of *H. pylori* also have a low prevalence of GORD (Franchini and Veneri, 2004). Boyanova (2007) has found about 20% lower prevalence of *H. pylori* infection in Bulgarian untreated patients with reflux oesophagitis with or without Barrett's oesophagus than that in PUD. With the decreasing prevalence of *H. pylori* infection and related PUD and gastric cancer, an increase in GORD and its complications has been observed (Kinoshita, 2007; Malfertheiner *et al.*, 2007). The data from meta-analysis on the role of the CagA$^+$ *H. pylori* strains are contradictory. Several studies have supported the negative association between CagA$^+$ *H. pylori* strains against GORD, which has not been confirmed by other authors (Kupcinskas and Malfertheiner, 2005; Somi *et al.*, 2008).

Prevalence of *H. pylori* is higher in non-erosive disease and gradually decreases in more severe grades of erosive oesophagitis and Barrett's oesophagus (Abe *et al.*, 2004; Jonaitis *et al.*, 2004; Labenz *et al.*, 2004; Vakil *et al.*, 2004; Weston *et al.*, 2004). Somi *et al.* (2008) have found inverse association between CagA$^+$ *H. pylori* strains and reflux oesophagitis. Other studies, including a recent meta-analysis, have confirmed that *H. pylori* is also negatively associated with Barrett's oesophagus and oesophageal adenocarcinoma (Rokkas *et al.*, 2007). The trend is the same in the countries with a high prevalence of *H. pylori* (Kupcinskas and Malfertheiner, 2005; Koike *et al.*, 2007; Kim *et al.*, 2008a). In different studies on *H. pylori* infection, especially by CagA$^+$ strains, the prevalence of *H. pylori* infection has been reported to be lower in patients with Barrett's oesophagus and adenocarcinoma of the cardia (Jakobs *et al.*, 2004; Anderson *et al.*, 2008; Corley *et al.*, 2008a,b). This association has been confirmed in most but not all studies (Ye *et al.*, 2004; Ramus *et al.*, 2007; Fass *et al.*, 2008; Grande *et al.*, 2008). The possible protective role of *H. pylori* infection in development of GORD and its complications are still controversial (Shirin *et al.*, 2008). Several studies have been performed to clarify the relationship between *H. pylori* status, gastric inflammation or atrophy, and GORD. *H. pylori*-associated severe inflammation including the fundus of the stomach is associated with reduced gastric acid secretion and is inversely correlated with GORD and its complications (Kudo *et al.*, 2005; Malfertheiner *et al.*, 2007). A Spanish study also has provided evidence supporting the independent protective role of CagA$^+$ *H. pylori* status and IL-1B and ILRN allele polymorphisms against GORD (Queiroz *et al.*, 2004).

Maastricht III Consensus Report and many other studies have stated that eradication of *H. pylori* does not cause GORD and does not exacerbate symptoms in patients with GORD when untreated or in those receiving PPI maintenance treatment (Shimatani *et al.*, 2006; Vakil *et al.*, 2006; Delaney *et al.*, 2008; Harvey *et al.*, 2010). *H. pylori* eradication does not affect the outcome of PPI treatment in patients with GORD in Western populations (De Boer *et al.*, 2006; Malfertheiner *et al.*, 2007). Screening for *H. pylori* in patients with GORD is not recommended. The impact of long-term acid suppression on the gastric mucosa remains controversial (Rokkas *et al.*, 2007). *H. pylori* testing should be considered in patients receiving long-term maintenance treatment with PPIs because continuous administration of PPI for GORD patients carries the risk for exacerbating of the gastritis. The profound acid suppression with PPI or histamine 2 receptor antagonists in GORD patients with *H. pylori*-positive corpus gastritis may accelerate the process of loss of specialized glands, leading to atrophic gastritis and potentially to gastric cancer. *H. pylori* eradication stops the progression of atrophic gastritis and may lead to regression of atrophy. The effect on intestinal metaplasia is uncertain. Kuipers *et al.* (2004) have stated that *H. pylori*-positive GORD patients should undergo eradication therapy for prevention of gastric cancer, especially in Asian populations with a high gastric cancer incidence (Wu, 2008).

The possible role of *H. pylori* eradication for improvement or exacerbation of symptoms in patients with GORD remains also controversial (Peitz *et al.*, 2004; Raghunath *et al.*, 2004; Ramus *et al.*, 2007; Anderson *et al.*, 2008; Furuta and Delchier, 2009). Kim *et al.* (2008a) have found that erosive oesophagitis occurs after eradication of *H. pylori*, but non-erosive reflux disease does not. The results of the cohort studies and randomized controlled trials in 1983–2007 showed that *H. pylori* eradication did not increase the overall risk of subsequent development of GORD (Yaghoobi *et al.*, 2010). Within subgroup analysis of patients with PUD, a significant increased risk of GORD has been observed in cohort studies. A resolution of antral gastritis, associated with DU patients, reduces gastric acid secretion to normal levels by a reduction in gastrin production, leading to a reduced risk of GORD after *H. pylori* eradication. In corpus-predominant gastritis without atrophic changes, associated with GU disease, *H. pylori* eradication may increase gastric acid secretion and thus, may lead to an increased risk of GORD (de Vries and Kuipers, 2010). Gastric atrophy, associated with *H. pylori* infection, protects against oesophageal adenocarcinoma and its precursor Barrett's oesophagus as well as reflux oesophagitis (Anderson *et al.*, 2008; Kwon *et al.*, 2008). Whether reduction of GORD is only mediated through *H. pylori*-induced corpus atrophy is still being debated (Furuta and Delchier, 2009).

The lack of association between *H. pylori* eradication and development of GORD may differ for Asian populations, who more often have a pangastritis with reduced acid output. The East-West theory of McColl (2007) suggests that *H. pylori* protects against GORD in the East but not in the West. In a recent cohort of Japanese patients with ulcer disease, the risk of oesophagitis after *H. pylori* eradication has been significantly higher when compared with PUD patients with persistent infection (Take *et al.*, 2009).

H. pylori infection and NSAID

During recent years, the relation and interaction of non-steroidal anti-inflammatory drugs (NSAIDs) and acetylsalicylic acid (ASA) with *H. pylori* infection in causing upper gastrointestinal damage remain a controversial and complex issue (Kupcinskas and Malfertheiner, 2005; Malfertheiner *et al*, 2007; Venerito and Malfertheiner, 2010). From a clinical point of view, very important question is whether *H. pylori* infected patients receiving NSAIDs or ASA have an increased risk of gastroduodenal mucosal lesions, and whether *H. pylori* eradication will be needed (Shirin *et al.*, 2008). Maastricht Consensus Reports, the consensus of the American College Guidelines for the management of *H. pylori*, the Guidelines of the American College of Cardiology on antiplatelet therapy and NSAID use and the Guidelines of the American College of Gastroenterology for prevention of NSAID-related ulcer complications, all have discussed the association between NSAIDs/ASA-induced gastric injury and *H. pylori* infection (Chey and Wong, 2007; Malfertheiner *et al*, 2002, 2007). Many authors have investigated the interaction between NSAIDs/ASA and *H. pylori* infection as similarity and differences in mechanisms of mucosal damage, synergistic or opposite effects, naïve patients and chronic NSAIDs/ASA users, GU or DU, primary and secondary prophylaxis etc. It is well known that *H. pylori* as well as NSAID independently can cause PUD. The risk of peptic ulcer bleeding is about twice as high in patients with *H. pylori* infection, and five times higher in NSAIDs users. This risk increases 6-fold when both factors are present, but it is not clear yet if their ulcerogenic effect is additive (Huang *et al.*, 2002).

Many data have shown that there are some differences in pathogenetic mechanisms of mucosal damage and development of peptic ulcer for *H. pylori* and NSAIDs/ASA. The development of *H. pylori*-induced peptic ulcer depends on the severity and phenotype of the *H. pylori*-induced gastritis and related alteration in gastric acid secretion and degree of infiltration. The localization of ulcers also strongly depends on the type of gastritis and gastric acid output (Tummala *et al.*, 2004; Malfertheiner *et al.*, 2009). As inhibitors of cyclooxygenase (COX) enzymes (COX-1 and/or COX-2,), NSAIDs induce peptic ulcer as a result of suppression of gastric prostaglandin E_2 and by their direct toxic effect on gastroduodenal mucosa, epithelial and microvascular injuries, resulting in increase of acid secretion and loss of mechanism for mucosal protective barrier (Bhatt *et al.*, 2008a). NSAIDs have caused more frequently GU than DUs (Malfertheiner *et al.*, 2009). As seen on endoscopy, roughly, 15–20% of endoscopic lesions are GUs, and 5–8% are DUs. NSAID-induced GU develops independently of the presence of *H. pylori*, and may occur in patients on acid-suppressing therapy (Gillen *et al.*, 2004). On the other hand, development of DU is more frequent in NSAID users and *H. pylori* infection (Venerito and Malfertheiner, 2010). ASA has shown the same effects as other NSAIDs, but COX inhibition is irreversible and dose- related. Venerito *et al.* (2006) have found that in healthy subjects, low-dose ASA given for 1 week does not affect COX expression or mucosal prostaglandin E_2 levels. The mechanisms by which *H. pylori* and low dose ASA induce gastric damage are not completely elucidated.

In murine models, *H. pylori* has increased gastric prostaglandin E_2 levels, neutrophil activity and the degree of chronic inflammation, whereas indomethacin has decreased significantly all these effects. NSAIDs can also reverse the increased gastric epithelial cell apoptosis and cell proliferation induced by *H. pylori* (Shirin *et al.*, 2008). However, many other studies have found opposing data. Venerito and Malfertheiner (2010) have pointed out that *H. pylori*, as well as NSAIDs, increases acid secretion, and may have an additive effect. Moreover, both *H. pylori* and NSAIDs directly influence the increased apoptosis and proliferation in epithelial cells but without additive effect. According to the results of Leung *et al.* (2000), there is evidence that *H. pylori* eradication may reduce NSAID-associated gastric mucosal injury by lowering NSAID-induced apoptosis. Synergistic effect has been found in increasing of polymorphonuclear cells. *H. pylori* increases gastroduodenal blood flow because of increasing of mucosal prostaglandins. On the other hand, NSAIDs inhibit the synthesis of mucosal prostaglandins and decrease the blood flow. Data from Elizalde *et al.* (2003) have shown that chronic NSAID medication induces a decrease in gastric blood flow irrespective of the *H. pylori* status. *H. pylori* infection induces vasodilatation and chemotaxis of activated neutrophils in the lamina propria, resulting in mucosal damage, while NSAIDs induce vasoconstriction and intravascular activation of neutrophils, resulting in vasoconstriction and induction of ischaemic lesions (Venerito and Malfertheiner, 2010). One study has examined the effect of eradication of *H. pylori* prior to NSAID use, in *H. pylori*-infected and *H. pylori* eradicated gerbils, followed by administration of indomethacin and rofecoxib (Chang *et al.*, 2005). The authors have found reduction of gastric damage in Mongolian gerbils. Indeed, rofecoxib has caused less severe gastric damage than indomethacin in *H. pylori* eradicated gerbils. A prospective cohort study has evaluated the interaction of *H. pylori* and NSAIDs and their effect on gastric expression of COX-2 mRNA (Wu *et al.*, 2005). *H. pylori* infection has been associated with increased COX-2 mRNA expression in gastric antral mucosa for both NSAID users and non-users, but not in GU. The authors have suggested that *H. pylori* eradication does not interfere with GU healing in NSAID users.

Epidemiological studies have revealed that the incidence of gastric cancer is lower in subjects receiving NSAID and/or ASA compared with non-users. The studies on Mongolian gerbils have shown that administration of aspirin has not changed *H. pylori*-induced gastritis, but has alleviated *H. pylori*-induced hyperplasia and the development of heterotopic proliferative glands (Li *et al.*, 2008). Moreover, ASA has accelerated *H. pylori*-associated apoptosis but has decreased *H. pylori*-associated cell proliferation. It is suggested that the combination of increased apoptosis and decreased *H. pylori*-associated proliferation may be involved in the antineoplastic activities of ASA in *H. pylori*-related gastric carcinogenesis. Dysregulation in the balance between hyperproliferative response and apoptosis may ultimately be responsible for the pathological reaction to *H. pylori*, which may, in some individuals, result in the development of peptic ulcer or even gastric cancer. Many publications have assessed the relationship between NSAIDs/ASA-induced gastric injury and *H. pylori* infection regarding clinical practice, especially *H. pylori* eradication. Some of them have confirmed the beneficial effect of *H. pylori* eradication, but others have not. Lai *et al.* (2003) have stressed has the crucial role of the *H. pylori* infection in the development of DU in NSAIDs users. Other authors have found that the *H. pylori* infection is one of selected factors associated with the risk of DU in patients taking low doses of ASA for prevention of cardiovascular events (Yeomans *et al.*, 2005).

H. pylori eradication is recommended in NSAID users with a history of PUD (Bhatt *et al.*, 2008a,b) and in naïve patients who start a treatment with NSAID (Kiltz *et al.*, 2008). NSAID-naive users have benefited from testing for *H. pylori* infection and subsequent eradication therapy prior to the initiation of NSAID because *H. pylori* eradication has been better than placebo in preventing peptic ulcer and upper gastrointestinal bleeding (Chan *et al.*, 2002; Labenz *et al.*, 2002; Vergana *et al.*, 2005). In addition, *H. pylori* eradication may provide a long-term beneficial effect on NSAID-related symptoms (Schaeverbeke *et al.*, 2005). Venetito and Malfertheiner (2010) also have studied the role of *H. pylori* eradication for primary ulcer prevention. They have stated that *H. pylori* eradication before starting NSAID therapy reduces the risk for NSAID-induced GU and abolishes the risk for DU. The patients who should be tested for *H. pylori* and eventually treated are also these, who received long term ASA and have a history of PUD or active PUD or a history of significant bleeding (Chan *et al.*, 2001; Lai *et al.*, 2002; Bhatt *et al.*, 2008a,b). Some authors have suggested that *H. pylori* eradication may also prevent medium-dose aspirin (300 mg)-induced gastroduodenal mucosal injury, may heal peptic ulcers and may prevent bleeding relapse in patients on long-term low-dose aspirin (Giral *et al.*, 2004; Venetito and Malfertheiner, 2010). In contrast, in chronic NSAID users with peptic ulcer or ulcer bleeding, *H. pylori* eradication gives no benefit compared with placebo. In addition, the effect of PPI in these cases is better than that by *H. pylori* eradication alone in preventing ulcer recurrence and/or gastrointestinal bleeding (Pilotto *et al.*, 2000; Kiltz *et al.*, 2008). In a study by Chan *et al.* (2001), *H. pylori* eradication and PPI treatment have been equivalent in patients on low-dose aspirin. For secondary prophylaxis of NSAID-induced peptic ulcer, PPI co-medication is mandatory.

Despite these results, patients who receive long-term PPI treatment for prevention of NSAID or ASA ulcers also have to be tested and treated to reduce the PPI-*H. pylori* interaction, leading to loss of specialized glands and atrophic gastritis (Malfertheiner *et al.*, 2007, Bhatt *et al.*, 2008a; Bhatt *et al.*, 2008b). Venetito and Malfertheiner (2010) also have confirmed that *H. pylori* eradication alone is not sufficient for secondary prevention of NSAID-induced GU and DU. *H. pylori* infection appears to increase furthermore the protective effects of PPIs to reduce the risk of ulcer relapse. *H. pylori* eradication does not influence the healing of both GU and DU if NSAID intake is discontinued. PPI therapy is mandatory for secondary prevention of PUD and may reduce further the risk of ulcer relapse in the presence of *H. pylori*. *H. pylori* infection appears to positively influence omeprazole-induced GU but not DU healing (Chan *et al.*, 2001). The test-and-treat strategy for *H. pylori* infection is recommended in all patients, starting a long-term therapy with

NSAIDs if *H. pylori* is the main risk factor for DU (Venetito and Malfertheiner, 2010). After successful *H. pylori* eradication and gradual discontinuation of PPI, DU is unlikely to occur in patients on NSAIDs without a previous history of symptomatic or asymptomatic DU. Low-dose aspirin may be less ulcerogenic than NSAIDs and induces bleeding only in pre-existing *H. pylori*-related ulcers.

Despite all, the data about the utility of *H. pylori* eradication in ASA-naive patients are not available yet. *H. pylori* eradication alone may not be a sufficient risk reduction strategy for patients with a history of ulcer complication who require subsequent therapy with aspirin (Shirin *et al.*, 2008). Co-therapy with a PPI in such patients at high risk has been recommended. *H. pylori* eradication is important to prevent gastroduodenal injury due to antiplatelet and NSAID therapy. For antiplatelet therapy, the recommendation is to examine *H. pylori* infection in patients with a history of PUD and to eradicate the infection when present (Furuta and Delchier, 2009). Kiltz *et al.* (2008) have reviewed the literature from a rheumatological viewpoint and have given similar recommendations. Marinchev *et al.* (2007) have reported the prevalence of *H. pylori* infection in patients with rheumatoid arthritis and other connective tissue diseases under different continuous therapeutic regimes to be 75%, corresponding to that of the Bulgarian population. To relieve patients with recent ulcer complications from further gastrointestinal events, *H. pylori* eradication alone is insufficient and long-term acid inhibition is required (Vergara *et al.*, 2005).

H. pylori infection and extragastric diseases

H. pylori may induce gastric mucosal damage as well as chronic gastric and systemic inflammation that in many cases might be lifelong. In addition, the immunological response may cause not only local damage but also systemic effects. Therefore, it is logical to evaluate the extragastric manifestations of *H. pylori* infection. Today, there is evidence that *H. pylori* has some role in variety of extragastric diseases. The pathogenic role of *H. pylori* infection in idiopathic thrombocytopenic purpura (ITP) and iron-deficiency anaemia (IDA) appears to be strong. There is also a growing evidence for the role of *H. pylori* in other diseases, including ischaemic heart disease even though results are not conclusive (Pellicano *et al.*, 2009; Mitchell *et al.*, 2010). It is very difficult to explain the pathogenic mechanism in multifactorial diseases by only one cause. Goodman *et al.* (2006) have reviewed many studies on the association between *H. pylori* infection and extragastric disorders. They have concluded that there are many limitations concerning the experimental design and statistical power, e.g. limited number and selection of investigated subjects and controls. On the other hand, *H. pylori* infection is strongly associated with age, sex, ethnic background and socioeconomic status. Failure to match patients and controls for such factors is likely to confound these studies (Goodman *et al.*, 2006). Randomized, long-term, and large studies on the follow-up of *H. pylori* eradication are rare.

Haematological diseases

The exact mechanisms by which *H. pylori* affects ITP have not yet been fully elucidated. *H. pylori* infection could induce cross-reactive antibodies by molecular mimicry between bacterial and platelet antigens, for example, between amino acids sequence of CagA or VacA and human platelet peptides (Asensio-Sánchez *et al.*, 2008; Emilia *et al.*, 2007). One study has shown a cross reaction of *H. pylori* urease B monoclonal antibody with platelet glycoprotein IIIa and has suggested that the immune response to UreB could be involved in the pathogenesis of ITP (Bai *et al.*, 2009). Asahi *et al.* (2007) have suggested that activation of macrophages in *H. pylori*-infected ITP patients has a role in platelet destruction and in the generation of autoimmune antibodies.

The role of *H. pylori* in ITP is now well established (Malfertheiner *et al.*, 2007). Several studies have demonstrated higher prevalence of *H. pylori* in patients with ITP than in controls (Malfertheiner *et al.*, 2007; Moyaert *et al.*, 2008, Mitchell *et al.*, 2010). Because of that, the Maastricht III Consensus Report has recommended eradication therapy of *H. pylori* infection in patients in with chronic ITP (Malfertheiner *et al.*, 2007). Long-term follow-up studies have shown an overall platelet response in >50% of the patients successfully treated for *H. pylori* infection and increased response rates in countries with a high prevalence of the infection and in patients with less severe degrees of thrombocytopenia and also in those with short disease duration (Asahi *et al.*, 2008; Jackson *et al.*, 2008; Stasi *et al.*, 2009). In the meta-analysis of Arnold *et al.* (2009), the platelet count response following eradication therapy has been found to increase 14.5-fold. Not all new studies have revealed beneficial effect of eradication, especially in childhood ITP (Moyaert *et al*, 2008; Treepongkaruna *et al.*, 2009). Small improvement of ITP after healing of the infection has been found also in other studies (Arnold *et al.*, 2009; Ferrara *et al.*, 2009). Reduction in antiCagA antibody titre after eradication therapy has been greater in responders than in non-responders, thus confirming a possible role of *H. pylori* CagA antigen of in ITP pathogenesis (Kodama *et al.*, 2007). AntiCagA serum antibodies have been present in 83% versus 12.5% of cases with positive and negative response to *H. pylori* eradication, respectively (Scandellari *et al.*, 2009). The authors have stated than antibiotic treatment of the infection should be considered in ITP cases, particularly for patients who have CagA antibodies. Suzuki *et al.* (2008) have found no difference in the frequencies of polymorphisms in IL-1β, IL-1RN, and TNF-α genes between responders and non-responders. In contrast, the frequency of responders has been higher in ITP patients with the TNFβ G/G or G/A genotype than in those with the TNFβA/A genotype. Therefore, the TNFβ (+252) G/G or G/A genotype may be considered as a good predictor of platelet recovery in ITP patients after *H. pylori* eradication. Asahi *et al.* (2008) have observed that the recovery in platelet numbers has been mediated through a change in FccR balance towards the inhibitory FccRIIB. According to these data, the best pathogenic model postulated is based on the antibody cross-reaction between *H. pylori* ureB and human platelet GP IIIa (Bai *et al.*, 2009).

IDA may be another haematological disease associated with *H. pylori* infection. The Maastricht III Consensus Report has recommended eradication therapy for *H. pylori* infection in patients with IDA (Malfertheiner *et al.*, 2007). Several mechanisms of induction of IDA by *H. pylori* have been discussed. The most important could be occult bleeding of gastro-duodenal erosions and ulcers and impaired iron absorption as a result of chronic corpus gastritis with hypochlorhydria or achlorhydria (Malfertheiner *et al.*, 2007; Franceschi and Gasparrini, 2007; Muhsen and Cohen, 2008). A third possible mechanism by which *H. pylori* leads to IDA is the increased capacity of the bacteria to uptake and use the iron in gastric mucosa (Lee, 2007; Yokota *et al.*, 2008; Lee *et al.*, 2009). On the other hand, autoimmune atrophic gastritis has been shown to be responsible for refractory IDA in >20% (Konno, 2009). Figura *et al.* (2002) have discussed mechanisms of molecular mimicry for the development of anaemia. They have found a linear homology with numerous bacterial proteins, for example the bacterial catalase, an enzyme expressed by erythrocytes. Since the late 1980s, many studies have suggested a link between *H. pylori* infection and IDA both in adults and in children (Malfertheiner *et al.*, 2007; Carter *et al.*, 2008). The levels of ferritin and iron absorption have been lower in *H. pylori*-positive versus *H. pylori*-negative subjects, especially in CagA seropositive (Seo *et al.*, 2002; Muhsen and Cohen, 2008). In addition, patients with chronic hepatitis C with coexistent *H. pylori* infection have had lesser serum ferritin levels and hepatic iron than subjects without *H. pylori* infection, suggesting that the infection could affect hepatic iron accumulation in HCV-related liver disease (Sumida *et al.*, 2006). Choi (2006) have shown the serum-soluble transferrin receptor (sTfR) to be significantly higher in IDA *H. pylori*-positive than in negative patients, and have concluded that serum sTfR reflects more accurately body iron status than serum iron or ferritin in *H. pylori*-positive adolescents. Most investigations have found increasing of ferritin values and iron absorption after clearance of *H. pylori* infection and correction of IDA (Cardamone *et al.*, 2008). A systematic review and a meta-analysis have shown that *H. pylori* is risk factor for reduction of body iron, iron deficiency and IDA (Baggett *et al.*, 2006; Cardenas *et al.*, 2006; Goodman *et al.*, 2006; Muhsen and Cohen, 2008). The results have shown a 2.8-time increased risk for developing IDA in *H. pylori*-infected patients. Additionally, the risk of appearance of iron deficiency alone has been approximately 40% higher in *H. pylori*-infected patients (Baggett *et al.*, 2006; Cardenas *et al.*, 2006). Kaye *et al.* (2008) have found the gastric atrophy to be strongly associated with IDA. In addition, pregnant women with IDA have had a high prevalence of active *H. pylori* infection (Mulayim *et al.*, 2008). Regardless of these supporting results, some authors have not confirmed *H. pylori* role in the development of IDA (Fayed *et al.*, 2008). In addition, in some studies, no evidence has been found to confirm that the eradication of infection resolved IDA (Gessner *et al.*, 2006; Cherian *et al.*, 2008; Sarker *et al.*, 2008). In a study conducted in six Latin American countries, no association has been found between *H. pylori* eradication and correction of IDA (Santos *et al.*, 2009). A major problem in some countries, however, is that only approximately 50% of anaemia cases can be attributed to iron deficiency. Other causes, as malaria, schistosomiasis and dietary vitamin deficiency may play role in the pathogenesis of anaemia (Duque *et al.*, 2010). In contrast, one large epidemiological study has shown that during a follow-up of 40 months of children in Alaska, *H. pylori* eradication has reduced the prevalence of iron deficiency and IDA (Fagan *et al.*, 2009). To assess the real role of *H. pylori* infection in IDA as well as its pathogenic mechanisms, we need large new controlled trials.

Some authors suggest a protective role of active *H. pylori* infection in the leukaemic transformation of myelodysplastic syndromes (Diamantidis *et al.*, 2008). Increase of TNF-α in peripheral blood, followed by increased apoptosis of bone marrow cells by increasing free radical production, may lead to oxidative damage, possibly delaying the leukaemic evolution of myelodysplastic syndromes (Moyaert *et al.*, 2008). Significant improvement has been reported in about 30% of patients with monoclonal gammopathy of unknown significance after the eradication of *H. pylori*; however, another study, conducted on a larger number of patients, failed to find any evidence of remission of monoclonal gammopathy of unknown significance after antibacterial therapy (Rajkumar *et al.*, 2002).

Heart and vascular diseases (cardiovascular and cerebrovascular diseases)

The role of chronic inflammation in the pathogenesis of atherosclerosis is well established. The associations between a range of chronic viral and bacterial agents including *H. pylori* and cardiovascular disease have been investigated. Many studies have been published in this field in recent years, though with controversial results. *H. pylori* causes not only gastric, but also systemic and vascular, inflammation (Bunch *et al.*, 2008; Tamer *et al.*, 2009). For example, a significant correlation has been found between CagA seropositivity and CRP levels, which have been recognized to be among the main risk factors for atherosclerosis, ischaemic heart disease (IHD) and atrial fibrillation (AF; Badran and Mahfouz, 2008; Ayada *et al.*, 2009a,b).

Many studies have investigated the possibility of a direct role of *H. pylori* in atherosclerosis (Mitchell *et al.*, 2010). *H. pylori* DNA or immunohistochemical evidence of *H. pylori* infection in higher detection rate have been found in atherosclerotic plaques compared with none in controls (Kaplan *et al.*, 2006). Most detectable *H. pylori* DNAs in atherosclerotic tissue samples have been *cagA*$^{+}$ (Arias *et al.*, 2006; Berrutti *et al.*, 2008; Ciervo *et al.*, 2008). However, many other studies have failed to detect *H. pylori* in atherosclerotic plaques (Reszka *et al.* 2008; Honda *et al.*, 2008; Szklo *et al.*, 2009). Specimens of aortic wall, obtained from patients with atherosclerotic three-vessel stable coronary artery disease (CAD) referred to surgical revascularization (coronary group) and controls referred to aortic valve replacement (valve group) have been studied for the presence of *H. pylori* DNA. *H. pylori* has been detected in macroscopically healthy aortic wall of either coronary or valve patients, thus denying any possible pathogenic role in atherosclerotic disease (Iriz *et al.*, 2008) There has been an observation that *H. pylori* infection may cause endothelial dysfunction (Evrengul *et al.*, 2007). A number of studies have examined the relationship between *H. pylori* and carotid intima–media thickness. The majority of studies have found no association between carotid intima–media thickness and *H. pylori* positivity (Altintas *et al.*, 2007; Mitchell *et *al.*, 2010). Other trials have reported a link with CagA+ strains. Polyclonal antiCagA antibodies have cross-reacted with antigens of both normal and atherosclerotic blood vessels, suggesting that molecular mimicry may exist between CagA and vascular wall peptides (Franceschi *et al.*, 2002, 2009). Nyberg *et al.* (2008) has observed virulent CagA+ *H. pylori* strains could be related to abdominal aortic aneurysm rupture with negative findings. Regarding other arteriopathies, Sawayama *et al.* (2008) have showed that *H. pylori* prevalence has been higher in patients with peripheral arterial disease than in controls.

The involvement of *H. pylori* in metabolic disturbance or inflammatory processes potentially related to atherosclerotic, cardiovascular dysfunction, and IHD has been investigated (Pellicano *et al.*, 2009). A large number of studies with conflicting results have focused on the effect of *H. pylori* on a range of classical and non-classical risk factors, including the cytokines (Yoshikawa *et al.* 2007; Kucukazman *et al.*, 2009). In some of them, no strong correlations have been found between *H. pylori* seropositivity and any of these vascular risk factors (Liu *et al.*, 2007; Mitchell *et al.*, 2010). In other studies, a higher prevalence of *H. pylori* has been observed in patients with metabolic syndrome compared with controls (Eskandarian *et al.*, 2006; Assadi *et al.*, 2009). The authors have proposed a possible causative effect of *H. pylori* infection in the pathogenesis of metabolic syndrome. Prospective clinical investigation in an African population has revealed a significant association between *H. pylori* seropositivity and higher cardiometabolic risk (Longo-Mbenza *et al.*, 2007). After adjusting for age and sex, *H. pylori*-seropositive males have shown higher mean values for body weight, waist circumference, waist-to-hip ratio, blood pressure, haematocrit, uric acid, triglycerides, and total cholesterol. The levels of uric acid, plasma glucose, total cholesterol, fibrinogen, and blood pressure after 3 weeks' antibiotics therapy have been lower than their baseline levels. Gunji *et al.* (2008) also have reported that high systolic blood pressure, low HDL-cholesterol, and high LDL-cholesterol levels have been significantly associated with *H. pylori* seropositivity by multiple linear regression analysis. Other studies have shown the C-reactive protein level to be higher in *H. pylori*-positive subjects but to decrease 2 months after the eradication (Jha *et al.*, 2008; Siddiqui, 2009).

H. pylori has also been shown to affect the vascular risks and complications in patients with diabetes mellitus although data on the prevalence of *H. pylori* infection among these patients are insufficient and controversial (Hamed *et al.*, 2008). The prevalence of *H. pylori* infection as well as carotid artery intima–media thickness and inflammatory biomarkers such as IL-6 and TNF-α have been higher in infected patients (Ohnishi *et al.*, 2008a). In the multivariate analysis, blood glucose, triglycerides, erythrocytic sedimentation rate, IL-6 and TNF-α are risk factors for atherothrombotic cause of cerebral ischaemia in patients with *H. pylori* infection (Hamed *et al.*, 2008). These data have suggested a causal role of *H. pylori* infection in atherosclerosis and ischaemic cerebrovascular disturbances in diabetic patients. This effect could be mediated by increasing cytokine levels (Moyaert *et al.*, 2008). In contrast, other authors have not supported the role of cytokines, the overexpression of TNF-α and the vascular cell adhesion molecule-1 in the pathophysiology of IHD (Nikolopoulou *et al.*, 2008).

Several studies have confirmed the association between CagA seropositivity and ischaemic heart disease (IHD; Moyaert *et al.*, 2008; Pellicano *et al.*, 2009). However, the relationship between *H. pylori* serological status and coronary heart disease (CHD) or ischaemic stroke remains controversial. A more recent systematic meta-analysis also has confirmed the role of CagA+ *H. pylori* strains as a risk of ischaemic stroke and CHD (Pasceri *et al.*, 2006; Zhang *et al.*, 2008). The positive association has been small but significant. No significant difference between Chinese and Caucasian populations has been observed. Further prospective studies examining the effect of the eradication of CagA-positive strains of *H. pylori* are required. A weak association was found between CagA+ *H. pylori* and the occurrence of fatal and non-fatal myocardial infarction in follow-up. *H. pylori* seems to be more important in unstable angina and acute coronary events than in stable angina (Pasceri *et al.*, 2006; Jha *et al.*, 2008; Franceschi *et al.*, 2009). On the other hand, Ozdogru *et al.* (2007) have found the seropositivity rates for *H. pylori* to be similar between patients and controls (angiographically proven normal coronary arteries) and among patients with myocardial infarction, unstable angina and stable angina. The findings support the potential role of virulent *H. pylori* strains in the acute phase of IHD, which may be a trigger factor in the pathophysiology of IHD by inducing an inflammatory cascade on gastric mucosa.

Infection with CagA$^+$ *H. pylori* strains may lead to development of CAD as high levels of antiheat-shock protein 60 (Hsp60) antibodies have been shown to constitute a marker and/or a concomitant pathogenic factor of the disease (Ayada *et al.*, 2009a). Heat-shock proteins (Hsp) are a class of functionally related proteins whose expression is increased when cells are exposed to environmental stress. Hsp can be expressed/and or secreted by a range of pathogens, including *H. pylori*, by endogenous cells within atherosclerotic plaques. By T1 immune response, microbial Hsp may cross-react with homologous host proteins in a form of molecular mimicry (Ayada *et al*, 2009a). The possible association between *H. pylori* infection and cardiac arrhythmias, such as atrial fibrillation (AF), related to chronic inflammation has been published (Bunch *et al.*, 2008). A significant correlation has been found between CagA seropositivity and either left atrial dimension or C-reactive protein levels (Badran and Mahfouz, 2007). On the opposite, Lunetta *et al.* (2009) have not found the correlation. In conclusion, the relationship between *H. pylori* and cardiovascular disease remains controversial, despite the numerous studies on the topic.

Respiratory diseases

In the last years, several studies have evaluated the potential relation between *H. pylori* infection and atrophy (asthma, allergic rhinitis, and other allergy symptoms) and various benign and malignant respiratory, ear, nose, and throat diseases. (Pellicano *et al.*, 2009; Figura *et al.*, 2010). Regarding ear, nose, and throat (ENT) diseases presence of *H. pylori*, confirmed by PCR and histochemical analysis with Giemsa staining, has been found in nasal polyps, adenotonsillar, pharyngeal or laryngeal tissue (Cvorovic *et al.*, 2008; Kaptan *et al.*, 2009), while other authors have not found the association (Ozcan *et al.*, 2009; Ozyurt *et al.*, 2009; Vayisoglu *et al.*, 2008). However, Kaptan *et al.* (2009) have found a correlation between *H. pylori* infection and non-specific pharyngitis, and have suggested that eradication therapy could be also a suitable treatment of chronic pharyngitis. Another group has shown that *H. pylori* eradication leads to chronic non-specific pharyngeal symptom relief (Aladag *et al.*, 2009). The mechanism probably depends on acid inhibition rather than *H. pylori* eradication (Oridate *et al.*, 2008; Toros *et al.*, 2009). In adenoids, Fancy *et al.* (2009) have observed more *H. pylori* DNA in children with otitis media than in controls. Rezaii *et al.* (2008) have shown the seropositivity for antibodies to *H. pylori* to be associated with laryngohypopharyngeal carcinoma. *H. pylori* DNA has been found in both normal and laryngeal squamous cell carcinoma tissue (76.2 vs. 42.9%). However, Grbesa *et al.* (2008) have found seldom *H. pylori* DNA (16%) in samples from patients with laryngeal squamous cell carcinoma. In an evidence-based meta-analysis, including a case–control study, laryngeal cancer risk for patients with *H. pylori* infection was twice higher (Zhuo *et al.*, 2008).

According to 'hygiene hypothesis' lack of exposure to microorganisms in the early years of life, predisposes to later development of allergic diseases, including asthma. Some clinical studies have found inverse association between anti-*H. pylori* and antiCagA IgG antibody titres and atopic asthma but other studies have not confirmed the link (Fullerton *et al.*, 2009; Pellicano *et al.*, 2009). Results of first major study on 7663 adults has shown that *H. pylori* infection with more virulent strains (CagA$^+$) is inversely associated with asthma or rhinitis, thus suggesting a possible protective role (Chen and Blaser, 2007; Janhson *et al.*, 2007). Data from another U.S. study have confirmed the inverse association between *H. pylori cagA*$^+$ strains and asthma (Reibman *et al.*, 2008). CagA positivity has been associated with older age of asthma onset. In a study of Chen and Blaser (2008), in children aged 3–13 years, *H. pylori* positivity has been inversely associated with current asthma or ever having had asthma (Chen and Blaser, 2007). However, a recent study by Dowd *et al.* (2009), in American children aged 6 years or over, has failed to show a protective effect of *H. pylori* and other infections. *H. pylori* infection primarily involves a Th1 immune response that may suppress the Th2 pathway, which is mainly associated with atrophy (Cam *et al.*, 2009). In an animal model, *H. pylori* neutrophil-activating protein (Hp-NAP) has reduced eosinophilia, IgE, and Th2 cytokine levels in bronchoalveolar lavage and has stimulated the production of IL-12 and IL-23 via TLR2 pathway (Amedei *et al.*, 2006). Whether Hp-NAP is a candidate for novel strategies of management of allergic diseases remains to be elucidated.

The role of *H. pylori* has been discussed also in other lung disease. Gencer *et al.* (2007) have found association between serum levels of anti-*H. pylori* IgG and chronic obstructive pulmonary disease (COPD). Another study has failed to demonstrate the association (Fullerton *et al.*, 2009). Presence of *H. pylori* DNA in bronchial specimens from patients with bronchiectasis could not be demonstrated (Angrill *et al.*, 2006; Gulhan *et al.*, 2007). According to these results, there is no direct association between *H. pylori* and bronchiectasis, even though an indirect role of soluble products of *H. pylori* could not be excluded. Yahav *et al.* (2006) have found the patients with cystic fibrosis with pancreatic sufficiency or distal intestinal obstruction and those carrying mutations associated with a severe phenotype to be protected against *H. pylori* infection. Zycinska *et al.* (2008) have suggested that pulmonary Wegener's granulomatosis disease severity, prevalence of gastroduodenal lesions, type and duration of treatment are dependent on *H. pylori* infection. Another important disease has evaluated a possible association with *H. pylori* infection is lung cancer (Moyaert *et al.*, 2008). Some studies have reported that *H. pylori* is two to three times more prevalent in patients with lung cancer than in healthy subjects and may play a pathogenic role in this disease. The following mechanisms have been discussed: gastrin induced increased bronchial epithelium cell proliferation and angiogenesis; bacterial toxins and direct local oncogenic effects into the respiratory epithelium (Ishikawa *et al.*, 2009; Zhuo *et al.*, 2009).

Hepatobiliary and pancreatic diseases

Studies on the association between *H. pylori* infection and hepatobiliary and pancreatic diseases have mainly focused on enterohepatic *Helicobacter* species (see Chapter 2). The results, based predominantly on animal models are still inconclusive (Pellicano *et al.*, 2008). There is an increasing evidence that *H. pylori*, and *Helicobacter* species toxin and constituents, reaching liver by venous blood, can damage liver cells (Figura *et al.*, 2010). In addition, bile and biliary epithelium can be colonized by duodenal bacteria. *H. pylori* DNA and *vacA* gene-specific markers have been found in liver cells, bile and bile duct epithelium in patients with various hepatobiliary diseases such as liver cirrhosis, chronic hepatitis C, primary sclerosing cholangitis and primary biliary cirrhosis, non-alcoholic fatty liver disease, hepatocellular carcinoma and cholangiocarcinoma, non-calculous cholecystitis and gallstone disease (Krasinskas *et al.*, 2007; Isaeva *et al.*, 2009; Pirouz *et al.*, 2009). However, some studies have failed to find a difference in *H. pylori* DNA detection rates between patients and controls (Vivekanandan and Torbenson, 2008). These controversial data suggested that the role of *H. pylori* in hepatobiliary diseases remains unclear.

Several mechanisms have been proposed to explain how *H. pylori* infection could lead to hepatobiliary injuries (Mitchell *et al.*, 2010). For example: easier adhesion and internalization of *H. pylori* into hepatocytes; production of vacuoles into hepatocytes (endocytotic vesicles) and disturbance of apoptosis and DNA synthesis. In addition, infection with virulent strains can result in cell arrest and increase DNA fragmentation (Ito *et al.*, 2008). Goo *et al.* (2008) have induced lesions resembling those of human primary biliary cirrhosis in an infected mouse and have stated that *H. pylori* VacA toxin can play a role in the development of cirrhosis. The mechanism of molecular mimicry has been also discussed. An *in vitro* study has shown that *H. pylori* induces different gallbladder epithelial cell morphological damages (Chen *et al.*, 2008). To investigate the effect of *H. pylori* on human hepatic cells, comparative proteome analysis of untreated and *H. pylori*-treated hepatic cell line HepG2 has been performed by Zhang *et al.* (2005). Seven proteins, which have been upregulated in *H. pylori*-treated cells, have been identified including oncogene-encoded products, regulatory proteins of transcription and signal transduction. The result might be a basis for further studies on the role of *Helicobacter* spp. in liver diseases. A role of *H. pylori* in the pathogenesis of hepatic encephalopathy in cirrhotic patients has been also reported (Abdel-Hady *et al.*, 2007; Yang *et al.*, 2007). *H. pylori* is a source of ammonia, stimulates the release of cytokines in gastric mucosa and increases the levels of cytokines and endotoxins in the peripheral blood as well as circulating levels of ammonia. These data suggest that *H. pylori* eradication can be helpful in treatment and prevention of hepatic encephalopathy in patients with cirrhosis. The possible role of non-*pylori Helicobacter* spp. in human hepatobiliary tract is discussed Chapter 2. The results of studies, investigating the relationship between *H. pylori* infection and pancreatic diseases have been controversial. Kountouras *et al.* (2005) have suggested that *H. pylori* might trigger autoimmune pancreatitis by induction of autoimmunity and apoptosis, but Chang *et al.* (2009) have not confirmed the suggestion. Lindkvist *et al.* (2008) have observed that subjects with healthier lifestyles (non-smokers) have had an increased risk of pancreatic cancer if they *H. pylori* positive.

Intestinal diseases

An interest in the association between *H. pylori* or enterohepatic helicobacters and intestinal diseases has been increased in the last years (Moyaert *et al.*, 2008; Pellicano *et al.*, 2009, Figura *et al.*, 2010; see also Chapter 2). The pathogenic role of *Helicobacter* spp. has been discussed for colon tumours and inflammatory bowel disease. *H. pylori* may be protective against Crohn's disease. This inverse association between *H. pylori* and Crohn's disease has been confirmed (Feeney *et al.*, 2002; Song *et al.*, 2009). Recent systematic review and meta-analysis has shown that *H. pylori* infection can have a protective effect for inflammatory bowel disease development, especially Crohn's disease (Luther *et al.*, 2010). A number of factors can influence the results, including also previous eradication therapy. Some studies suggested that T_{reg} cells can play a role in this protective effect. Expect that, further investigations are required to explain the mechanism of *H. pylori* protection on inflammatory bowel disease.

Presence of *H. pylori*, investigated by immunohistochemistry has been observed in colon adenomas and carcinomas, but correlation with polyp size, localization or histology has not been found (Soylu *et al.*, 2008). Robertson *et al.* (2009) have found that both *H. pylori* infection and hypergastrinaemia have been associated with increased risk of recurrence of colon adenomas. Two studies from Japan have suggested an association between *H. pylori* infection and colorectal neoplasia (Fujimori *et al.*, 2005; Mizuno *et al.*, 2005). The first study has shown a positive relationship between *H. pylori* infection and the risk of adenoma and carcinoma, especially in women. The second study has found a high increase in the incidence of adenomas in the seropositive group (44.3 vs. 18.9%).

Metabolic diseases (diabetes mellitus and obesity)

Many studies have investigated the association between *H. pylori* infection and type 1 or type 2 diabetes mellitus but the published data remain controversial. Some authors have found higher prevalence of *H. pylori* infection among diabetic patients, especially those with obesity, compared with healthy controls (Bener *et al.*, 2007). In contrast, other studies have not confirmed the association (Mateva *et al.*, 2001; Demir *et al.*, 2008; Lutsey *et al.*, 2009). However, *H. pylori*-infected diabetic patients have had a significantly higher incidence of neuropathy. Delayed gastric clearance, result of autonomic neuropathy, leads to bacterial colonization and bacterial overgrowth in the gastrointestinal tract. Most investigators have concluded that there is no relationship between glycaemic control in patients with type 1 or type 2 diabetes mellitus and *H. pylori* infection (Fernandini-Paredes *et al.*, 2008; Ciortescu *et al.*, 2009; Krause *et al.*, 2009). On the contrary, other authors have found correlation between *H. pylori* infection and blood glucose levels and daily glucose fluctuations or glycosylated haemoglobin levels (Wang *et al.*, 2009). So *et al.* (2009) have shown that anti-*H. pylori* antibodies are an independent risk factor for hyperglycaemia and insulin resistance in Chinese subjects. Other authors also have confirmed relation between *H. pylori* infection and surrogate markers of insulin resistance (Gunji *et al.*, 2009; Eshraghian *et al.*, 2009).

Higher prevalence (>45%) of *H. pylori* infection in obese patients compared with controls has been found in some studies in Turkey and Italy (Arslan *et al.*, 2009; Figura *et al.*, 2010). In contrast, the role of *H. pylori* infection in obesity has not been confirmed in a large population-based cohort study in the USA (Cho *et al.*, 2005). No association has been observed between *H. pylori* status and body mass index. The controversial results may be explained by the multifactorial pathogenesis of obesity. In the recent years, a potential link between gut microflora and appetite control has been discussed (Fetissov *et al.*, 2008; Figura *et al.*, 2010). The levels of ghrelin and leptin, hormones, which regulate body weight and food intake, may be related to *H. pylori* status (Pacifico *et al.*, 2008; Chuang *et al.*, 2009). Ghrelin is mainly secreted by cells, situated close to the gastric parietal cells and by cells of intestine, pancreas and adipose tissue. The role of *H. pylori* infection in ghrelin secretion and the development of obesity is still controversial (Figura *et al.*, 2010). Some authors observed that ghrelin levels have been lower in *H. pylori* infected- than in uninfected subjects, especially obese patients with CagA serum antibodies but other authors have not confirm that (Gao *et al.*, 2009c; Stec-Michalska *et al.*, 2009; Figura *et al.*, 2010). Roper *et al.* (2008) have shown that *H. pylori* colonization has been associated with reduced circulating leptin levels, independent of body mass index. On the other hand, levels of ghrelin in infected and uninfected patients have been similar. Ghrelin has been strongly correlated with gastric pH. A recent study has showed that gastric and systemic ghrelin levels depend on the presence and extension of gastric mucosal atrophy (Kawashima *et al.*, 2009). Figura *et al.* (2010) have suggested that in countries with a higher prevalence of CagA$^+$ *H. pylori* and atrophic gastritis, the corpus atrophy is associated with decreased ghrelin secretion and decreased appetite but no obesity. Conversely, the increased or normal levels of ghrelin and obesity may be observed in countries with low or moderate prevalence of *H. pylori*, especially CagA$^+$ strains, and antral-predominant gastritis with absence of gastric atrophy.

Neurological diseases

Some authors have assessed the association between *H. pylori* and Alzheimer's disease. *H. pylori* infection can modulate humoral and cellular immune responses by molecular mimicry cross-reaction with components of nerves, and can lead to neural tissue damage. Kountouras *et al.* (2009) have observed higher prevalence of *H. pylori* infection in patients with Alzheimer's disease than in controls and benefit of *H. pylori* eradication on cognitive and functional status parameters. Regarding Parkinson's disease, *H. pylori* systemic inflammation can induce neuroinflammation and some authors have proposed that *H. pylori* is one of the prognostic factors of this disease (Charlett *et al.*, 2009). Lee *et al.* (2008) have showed that *H. pylori* infection with active gastroduodenitis could interfere with the absorption of L-dopa to provoke motor fluctuations and that *H. pylori* eradication may improve the clinical status.

In Japanese patients with multiple sclerosis, *H. pylori* seropositivity has been lower compared with that in healthy controls (Fiddian-Green, 2007). Moreover, *H. pylori* seropositivity has shown an inverse association with disease severity. The data have suggested a potential protective role of *H. pylori* infection. Attallah *et al.* (2007) have reported that the 58-kDa *H. pylori* antigen, which may cross the blood–brain barrier and enter the cerebrospinal fluid of patients with meningitis, may play a role in that disease. Persistent *H. pylori* infection might be a significant risk factor for migraine in patients without other predisposing factors, probably as a result of prolonged oxidative injury caused regional cerebral flow changes (Yiannopoulou *et al*, 2007). However, Tunca *et al.* (2007) have not proven the role of oxidative stress in patients with *H. pylori* infection and migraine.

Gynaecological diseases and fertility disorders

An increased interest in relationship between *H. pylori* infection and some gynaecological and fertility disorders has been observed. Some investigations found a higher prevalence of *H. pylori* infection and higher antiCagA seropositivity in pregnant women with pre-eclampsia than in pregnant controls (Pugliese *et al.*, 2008). Only some seroepidemiological studies have shown that *H. pylori* infection is associated with hyperemesis gravidarum (Figura *et al.*, 2010). A meta-analysis of case–control studies has shown that the presence of *H. pylori* increases more than twice the risk of hyperemesis gravidarum (Sandven *et al.*, 2009). The association has been more evident among Africans than in non-Africans. Yavasoglu *et al.* (2009) have observed an association between *H. pylori* infection and polycystic ovary syndrome, but Kaya *et al.* (2009) have not found the presence of *H. pylori* in cervical mucosa and cervicovaginal secretions.

Regarding infertility, Collodel *et al.* (2010) have shown that infertile patients infected with *H. pylori* have had a low sperm compared with uninfected ones and reduced quality of sperm motility and fertility, particularly in CagA$^+$ patients (Collodel *et al.*, 2010). Kurotsuchi *et al.* (2008) have suggested the presence of an antigenic mimicry between the flagella of *H. pylori* and spermatozoa. The antibodies produced increasing the risk of infertility.

Other diseases

In the last years, researchers have investigated the correlation between *H. pylori* infection and several other diseases, including different organs and systems. Significance of oral *H. pylori* is discussed in Chapter 7. As for the ophthalmology, some authors have reported an association of *H. pylori* infection and glaucoma. *H. pylori* infection has also been linked also to central serous chorioretinopathy, uveitis and blepharitis through different pathogenic mechanisms (Misiuk-Hojfo *et al.*, 2009; Izzotti *et al.*, 2009; Figura *et al.*, 2010). With regard to dermatology, some studies have reported no relationship with skin diseases (Abdel-Hafez *et al.*, 2008; Ali and Whitehead, 2008; Hellmig *et al.*, 2008). A possible pathogenic role of *H. pylori* infection has been discussed with chronic urticaria and there exacerbation and with lichen planus (Yadav *et al.*, 2008; Abdou *et al.* 2009; Attia *et al.*, 2010).

A relationship between *H. pylori* infection and proteinuria also has been suggested (Kanbay *et al.*, 2007; Kusano *et al.*, 2007; Sugimoto *et al.*, 2006). Nevertheless, whether the improvement of nephrotic syndrome is due to eradication of this infection or to spontaneous remission of membranous nephropathy is still unclear. Some authors have shown that CagA$^+$ strains of *H. pylori* may be a risk factor for microalbuminuria in type 2 diabetes. Strain antigenic sequences and those in endothelial cells can show an antigenic mimicry (Pietroiusti *et al.*, 2006). Afsar *et al.* (2007) have reported that *H. pylori* infection could increase renal resistive index. *H. pylori* eradication has been recommended in haemodialysed patients or those before kidney transplantation with aim of reducing the incidence of gastric and/or duodenal lesions (Gioè *et al.*, 2008). In patients with renal failure on haemodialysis, a positive correlation between serum 25-OH-vitamin D levels and *H. pylori* infection has been found.

A possible role of *H. pylori* has also been evaluated in some autoimmune diseases. Sterzl *et al.* (2008) assessed the relationship between *H. pylori* infection and autoimmune thyroiditis, suggesting a relation between *H. pylori* infection and the occurrence of anti-TPO autoantibodies and antigastric parietal cells autoantibodies, representing the thyro-gastric syndrome. Apan *et al.* (2007) found an association between seropositivity to *H. pylori* and CagA, and Behçet's disease. *H. pylori* eradication decreased the manifestations of Behçet's disease, such as oral and genital ulceration, arthritis/arthralgia and cutaneous findings. Other authors have not confirmed these results (Cakmak *et al.*, 2009).

Many studies have investigated *H. pylori* association with numerous extragastric diseases. The exact pathogenetic mechanisms are not clear yet and the causal role of *H. pylori* is not well established. Evidence that eradication of *H. pylori* leads to clinical improvement is missing. At present, like in gastroduodenal diseases, with which *H. pylori* has been associated, there is evidence indicating that *H. pylori* infection is one of key factors in the development of ITP and IDA (Figura *et al.*, 2010; Mitchell *et al.*, 2010). There is an increasing evidence of a possible role of *H. pylori* in different extragastric diseases, including IHD. Further well-designed prospective studies that control for the many factors are needed to verify those findings.

References

Abdel-Hady, H., Zaki, A., Badra, G., Lotfy, M., Selmi, C., Giorgini, A., El-Sayed, M., and Badr, R. (2007). *Helicobacter pylori* infection in hepatic encephalopathy: relationship to plasma endotoxins and blood ammonia. Hepatol. Res. *37*, 1026–1033.

Abdel-Hafez, H.Z., Mahran, A.M., Hofny, E.R., Attallah, D.A., Sayed, D.S., and Rashed, H.A. (2008). Is *Helicobacter pylori* infection associated with alopecia areata? J. Cosmet. Dermatol. *8*, 52–55.

Abdou, A.G., Elshayeb, E.I., Farag, A.G., and Elnaidany, N.F. (2009). *Helicobacter pylori* infection in patients with chronic urticaria: correlation with pathologic findings in gastric biopsies. Int. J. Dermatol. *48*, 464–469.

Abe, Y., Ohara, S., Koike, T., Sekine, H., Iijima, K., Kawamura, M., Imatani, A., Kato, K., and Shimosegawa, T. (2004). The prevalence of *Helicobacter pylori* infection and the status of gastric acid secretion in patients with Barrett's esophagus in Japan. Am. J. Gastroenterol. *99*, 1213–1221.

Achyut, B.R., Tripathi, P., Ghoshal, U.C., Moorchung, N., and Mittal, B. (2008). Interleukin-10 (-819C/T) and tumor necrosis factor-alpha (-308G/A) gene variants influence gastritis and lymphoid follicle development. Dig. Dis. Sci. *53*, 622–629.

Adam, P., Schmausser, B., Gobeler-Kolve, M., Muller-Hermelink, H.K., and Eck M. (2008). Gastric extranodal marginal zone B-cell lymphomas of MALT type exclusively express Toll-like receptor 4 in contrast to other lymphomas infiltrating the stomach. Ann. Oncol. *19*, 566–569.

Afsar, B., Ozdemir, F.N., Elsurer, R., and Sezer, S. (2007). *Helicobacter pylori* infection may increase renal resistive index. Med. Hypotheses. *69*, 956–957.

Aladag, I., Eyibilen, A., Guven, M., and Erkokmaz, U. (2008) Effects of *Helicobacter pylori* eradication on chronic nonspecific pharyngeal symptoms. J. Otolaryngol. Head Neck Surg. *37*, 623–627.

Ali, M., and Whitehead, M. (2008). Clearance of chronic psoriasis after eradication therapy for *Helicobacter pylori* infection. J. Eur. Acad. Dermatol. Venereol. *22*, 753–754.

Altintas, E., Ucbilek, E., Ulu, O., Sezgin, O., Uzer, C., Tataroglu, C., and Camdeviren, H. (2007). *Helicobacter pylori*-associated atrophic gastritis and carotid intima–media thickness: is there a link? Int. J. Clin. Pract. *61*, 810–814.

Amedei, A., Cappon, A., Codolo, G., Cabrelle, A., Polenghi, A., Benagiano, M., Tasca, E., Azzurri, A., D'Elios, M.M., Del Prete, G., *et al.* (2006). The neutrophil-activating protein of *Helicobacter pylori* promotes Th1 immune responses. J. Clin. Invest. *116*, 1092–1101.

Anderson, L.A., Murphy, S.J., Johnston, B.T., Watson, R.G.P., Ferguson, F.R., Bamford, K.B., Ghazy, A., McCarron, P., McGuigan, J., Reynolds, J.V., *et al.* (2008). Relationship between *Helicobacter pylori* infection and gastric atrophy and the stages of the oesophageal inflammation, metaplasia, adenocarcinoma sequence: results from the FINBAR case-control study. Gut *57*, 734–739.

Angrill, J., Sanchez, N., Agusti, C., Guilemany, J.M., Miquel, R., Gomez, J., and Torres, A. (2006). Does *Helicobacter pylori* have a pathogenic role in bronchiectasis? Respir. Med. *100*, 1202–1207.

Apan, T.Z., Gursel, R., and Dolgun, A. (2007). Increased seropositivity of *Helicobacter pylori* cytotoxin-associated gene-A in Behcet's disease. Clin. Rheumatol. *26*, 885–889.

Arias, E., Martinetto, H., Schultz, M., Ameriso, S., Rivera, S., Lossetti, O., and Sevlever, G. (2006). Seminested polymerase chain reaction (PCR) for detecting *Helicobacter pylori* DNA in carotid atheromas. Diagn. Mol. Pathol. *15*, 174–179.

Arisawa, T., Tahara, T, Shibata, T., Nagasaka, M., Nakamura, M., Kamiya, Y., Fujita, H., Nakamura, M., Yoshioka, D., Arima, Y., *et al.* (2007). Association between genetic polymorphisms in the cyclooxygenase-1 gene promoter and peptic ulcers in Japan. Int. J. Mol. Med. *20*, 373–378.

Arnold, D.M., Bernotas, A., Nazi, I., Stasi, R., Kuwana, M., Liu, Y., Kelton, J.G., and Crowther MA. (2009). Platelet count response to *H. pylori* treatment in patients with immune thrombocytopenic purpura with and without *H. pylori* infection: a systematic review. Haematologica *94*, 850–856.

Arslan, E., Atilgan, H., and Yavasoglu, I. (2009). The prevalence of *Helicobacter pylori* in obese subjects. Eur. J. Intern. Med. *20*, 695–697.

Asahi, A., Nishimoto, T., Okazaki, Y., Suzuki, H., Masaoka, T., Kawakami, Y., Ikeda, Y., and Kuwana, M. (2008) *Helicobacter pylori* eradication shifts monocyte Fcg receptor balance toward inhibitory FcγRIIB in immune thrombocytopenic purpura patients. J. Clin. Invest. *118*, 2939–2949.

Asaka, M., Sugiyama, T., Nobuta, A., Kato M., Takeda, H., and Graham, D.Y. (2001). Atrophic gastritis and intestinal metaplasia in Japan: results of a large multicenter study. Helicobacter *6*, 294–299.

Asensio-Sanchez, V.M., Rodriguez- Delgado, B., Garcia-Herrero, E., Cabo-Vaquera, V., and Garcia-Loygorri, C. (2008). Central serous chorioretinopathy as an extradigestive manifestation of *Helicobacter pylori* gastric infection. Arch. Soc. Esp. Oftalmol. *83*, 177–182.

Assadi, M., Saghari, M., Ebrahimi, A., Reza Pourbehi, M., Eftekhari, M., Nabipour, I., Abbaszadeh, M., Nazarahari, M., Nasiri, M., and Assadi, S. (2009). The relation between *Helicobacter pylori* infection and cardiac syndrome X: a preliminary study. Int. J. Cardiol. *134*, 24–25.

Attallah, A.M., Toson, E.A., Ibrahim, G.G., and Bakr, N.E. (2007). Detection of a *Helicobacter pylori* antigen in cerebrospinal fluid of patients with meningitis. J. Immunoassay Immunochem. *28*, 25–33.

Attia, E.A., Abdel Fattah, N.S., and Abdella, H.M. (2010). Upper gastrointestinal findings and detection of *Helicobacter pylori* in patients with oral lichen planus. Clin. Exp. Dermatol. *35*, 355–360.

Axon, A.T. (2004). Personal view: to treat or not to treat? *Helicobacter pylori* and gastro-oesophageal reflux disease – an alternative hypothesis. Aliment. Pharmacol. Ther. *19*, 253–261.

Ayada, K., Yokota, K., Hirai, K., Fujimoto, K., Kobayashi, K., Ogawa, H., Hatanaka, K., Hirohata, S., Yoshino, T., Shoenfeld, Y., *et al.* (2009a). Regulation of cellular immunity prevents *Helicobacter pylori*-induced atherosclerosis. Lupus *18*, 1154–1168.

Ayada, K., Yokota, K., Kobayashi, K., Shoenfeld, Y., Matsuura, E., and Oguma, K. (2009b). Chronic infections and atherosclerosis. Clin. Rev. Allergy Immunol. *37*, 44–48.

Badran, H.M., and Mahfouz, M.E. (2007). Cytotoxin-associated gene-A bearing strains of *Helicobacter pylori* and atrial fibrillation due to ischaemic origin: is there a link? Eur. J. Cardiovasc. Prev. Rehabil. *14*, 518–520.

Baggett, H.C., Parkinson, A.J., Muth, P.T., Gold, B.D., and Gessner, B.D. (2006) Endemic iron deficiency associated with *Helicobacter pylori* infection among school-aged children in Alaska. Pediatrics *117*, e396-e404.

Bai, Y., Wang, Z., Bai, X., Yu, Z., Cao, L., Zhang, W., and Ruan, C. (2009). Crossreaction of antibody against *Helicobacter pylori* urease B with platelet glycoprotein IIIa and its significance in the pathogenesis of immune thrombocytopenic purpura. Int. J. Hematol. *89*, 142–149.

Barkun, A.N., Bardou, M., Kuipers, E.J., Sung, J., Hunt, R.H., Martel, M., Sinclair, P., and International Consensus Upper Gastrointestinal Bleeding Conference Group. (2010). International consensus recommendations on the management of patients with nonvariceal upper gastrointestinal bleeding. Ann. Intern. Med. *152*, 101–113.

Bener, A., Micallef, R., Afifi, M., Derbala, M., Al-Mulla, H.M., and Usmani, M.A. (2007). Association between type 2 diabetes mellitus and *Helicobacter pylori* infection. Turk. J. Gastroenterol. *18*, 225–229.

Bernasconi, B., Karamitopoulou-Diamantis, E., Tornillo, L., Lugli, A., Di Vizio, D., Dirnhofer, S., Wengmann, S., Glatz-Krieger, K., Fend, F., Capella, C., *et al.* (2008). Chromosomal instability in gastric mucosa-associated lymphoid tissue lymphomas: a fluorescent *in situ* hybridization study using a tissue microarray approach. Hum. Pathol. *39*, 536–542.

Berrutti, M., Pellicano, R., Fagoonee, S., Astegiano, M., Smedile, A., Saracco, G., Repici, A., Leone, N., and Rizzetto, M. (2008). Potential relationship between *Helicobacter pylori* and ischaemic heart disease: any pathogenic model? Panminerva Med. *50*, 161–163.

Bhatt, D.L., Scheiman, J., Abraham, N.S., Antman, E.M., Chan, F.K., Furberg, C.D., Johnson, D.A., Mahaffey, K.W., Quigley, E.M., Harrington, R.A., *et al.* (2008a). ACCF/ACG /AHA 2008 expert consensus document on reducing the gastrointestinal risks of antiplatelet therapy and NSAID use: a report of the American College of Cardiology Foundation Task Force on Clinical Expert Consensus Documents. J. Am. Coll. Cardiol. *52*, 1502–1517.

Bhatt, D.L., Scheiman, J., Abraham, N.S., Antman, E.M., Chan, F.K., Furberg, C.D., Johnson, D.A., Mahaffey, K.W., Quigley, E.M., Harrington, R.A., *et al.* (2008b) ACCF/ACG/AHA 2008 expert consensus document on reducing the gastrointestinal risks of antiplatelet therapy and NSAID use. Am. J. Gastroenterol. *103*, 2890–2907.

de Boer, W., de Wit, N., Geldof, H., Hazelhoff, B., Bergmans, P., Smout, A., and Tytgat, G. (2006). Does *Helicobacter pylori* infection influence response rate or speed of symptom control in patients with gastroesophageal reflux disease treated with rabeprazole? Scand. J. Gastroenterol. *41*, 1147–1154.

Bornschein, J., Rokkas, T., Selgrad, M., and Malfertheiner, P. (2009). *Helicobacter pylori* and Clinical Aspects of Gastric Cancer, Helicobacter *14* (*Suppl 1*), 41–45.

Boyanova, L. (2007). Detection of *Helicobacter pylori* in symptomatic Bulgarian adults. Clin. Microbiol. Infect. *13*, 908–914.

Bunch, T.J., Day, J.D., Anderson, J.L, Horne, B.D., Muhlestein, J.B., Crandall, B.G., Weiss, J.P., Lappe, D.L., and Asirvatham, S.J. (2008). Frequency of *Helicobacter pylori* seropositivity and C-reactive protein increase in atrial fibrillation in patients undergoing coronary angiography. Am. J. Cardiol. *101*, 848–851.

Cakmak, S.K., Cakmak, A., Gul, U., Sulaimanov, M., Bingol, P., and Hazinedaroglu, M.S. (2009). Upper gastrointestinal abnormalities and *Helicobacter pylori* in Behcet's disease. Int. J. Dermatol. *48*, 1174–1176.

Cam, S., Ertem, D., Bahceciler, N., Akkoc, T., Barlan, I., and Pehlivanoglu, E. (2009). The interaction between *Helicobacter pylori* and atopy: does inverse association really exist? Helicobacter *14*, 1–8.

Canedo, P., Corso, G., Pereira, F., Lunet, N., Suriano, G., Figueiredo, C., Pedrazzani, C., Moreira, H., Barros, H., Carneiro, F., *et al.* (2009). The interferon gamma receptor 1(IFNGR1) 56C/T gene polymorphism is associated with increased risk of early gastric carcinoma. Gut *57*, 1504–1508.

Capelle, L.G., de Vries,A.C., Haringsma, J., Ter Borg, F., de Vries, R.A., Bruno, M.J., Bruno, M.J., van Dekken, H., Meijer, J., van Grieken, N.C., *et al.* (2010). The staging of gastritis with the OLGA system by using intestinal metaplasia as an accurate alternative for atrophic gastritis. Gastrointest. Endosc. *71*, 1150–1158.

Cardamone, M., Laex, G., Harari, M.D., Moss, W.P., and Oliver, M.R. (2008). Severe iron-deficiency anemia in adolescents: consider *Helicobacter pylori* infection. J. Paediatr. Child Health. *44*, 647–650.

Cardenas, V.M., Mulla, Z.D., Ortiz, M., and Graham, D.Y. (2006). Iron deficiency and *Helicobacter pylori* infection in the United States. Am. J. Epidemiol. *163*, 127–134.

Carter, D., Maor, Y., Bar-Meir, S., and Avidan, B. (2008). Prevalence and predictive signs for gastrointestinal lesions in premenopausal women with iron deficiency anemia. Dig. Dis. Sci. *53*, 3138–3144.

Chakravorty, M., Ghosh, A., Choudhury, A., Santra, A., Hembrum, J., and Roychoudhury, S. (2006). Interaction between IL1B gene promoter polymorphisms in determining susceptibility to *Helicobacter pylori* associated duodenal ulcer. Hum. Mutat. *27*, 411–419.

Chan, C.C, Shen, D., Mochizuki, M., Gonzales, J., Yuen, H., Guex-Crosier, Y., and Le Hoang, P. (2006). Detection of *H. pylori* and *Chlamydia pneumoniae* genes in primary orbital lymphoma. Trans. Am. Ophthalmol. Soc. *104*, 62–70.

Chan, F.K., Chung, S.C., Suen, B.Y., Lee, Y.T., Leung, W.K., Leung, V.K., Wu, J.C., Lau, J.Y., Hui, Y., Lai, M.S., *et al.* (2001). Preventing recurrent upper gastrointestinal bleeding in patients with Helicobader pylori infection who are taking low-dose aspirin or naproxen. New Engl. J. Med. *34, 4967*–973.

Chan, F.K., To, K.F., Wu, J.C., Yung, M.Y., Leung, W.K., Kwok, T., Hui, Y., Chan, H.L., Chan, C.S., Hui, E., *et al.* (2002) Eradication of *Helicobacter pylori* and risk of peptic ulcers in patients starting long-term treatment with non-steroidal anti-inflammatory drugs: a randomised trial. Lancet 359, 9–13.

Chang, C.C., Chen, S.H., Lien, G.S., Lou, H.Y., Hsieh, C.R., Fang, C.L., and Pan S. (2005). Eradication of *Helicobacter pylori* significantly reduced gastric damage in nonsteroidal anti-inflammatory drug-treated Mongolian gerbils. World J. Gastroenterol. *11,* 104–108.

Chang, M.C., Chang, Y.T., Wei, S.C., Kuo, C.H., Liang, P.C., and Wong, J.M. (2009). Autoimmune pancreatitis associated with high prevalence of gastric ulcer independent of *Helicobacter pylori* infection status. Pancreas *38*, 442–446.

Charlett, A., Dobbs, R.J., Dobbs, S.M., Weller, C., Ibrahim, M.A., Dew, T., Sherwood, R., Oxlade, N.L., Plant, J.M., Bowthorpe, J., *et al.* (2009). Blood profile holds clues to role of infection in a premonitory state for idiopathic Parkinsonism and of gastrointestinal infection i *n* established disease. Gut Pathog. *1,* 20.

Chen, Y., and Blaser, M.J. (2007). Inverse associations of *Helicobacter pylori* with asthma and allergy. Arch. Intern. Med. *167*, 821–827.

Chen, D.F., Hu, L., Yi, P., Liu, W.W., Fang, D.C., and Cao, H. (2008). *Helicobacter pylori* damages human gallbladder epithelial cells *in vitro*. World J. Gastroenterol. *4,* 6924–6928.

Cherian, S., Forbes, D.A., Cook, A.G., Sanfilippo, F.M., Kemna, E.H., Swinkels, D.W., and Burgner, D.P. (2008). An insight into the relationships between hepcidin, anemia, infections and inflammatory cytokines in pediatric refugees: a cross-sectional study. PLoS One 3, e4030.

Chey, W.D., and Wong, B.C. (2007). American College of Gastroenterology guideline on the management of *Helicobacter pylori* infection. Am. J. Gastroenterol. *102,* 1–18.

Cho, I., Blaser, M.J., François, F., Mathew, J.P., Ye, X.Y., Goldberg, J.D., and Bini, E.J. (2005). *Helicobacter pylori* and overweight status in the United States: data from the Third National Health and Nutrition Examination Survey. Am. J. Epidemiol. *162,* 579–584.

Choi, J.W. (2006). Serum-soluble transferrin receptor concentrations in *Helicobacter pylori*-associated iron-deficiency anemia. Ann. Hematol. *85,* 735–737.

Chuang, C-H., Sheu, B-S., Yang, H-B., Lee, S-C., Kao, A-W., Cheng, H-C., Chang, W-L., and Yao, W-J. (2009). Gender difference of circulating ghrelin and leptin concentrations in chronic *Helicobacter pylori* infection. Helicobacter *14,* 54–60.

Ciervo, A, Mancini, F., Sale, P., Russo, A., and Cassone, A. (2008). Real-time polymerase chain reaction and laser capture microdissection: an efficient combination tool for Chlamydophila pneumonate DNA quantification and localization of infection in atherosclerotic lesions. Int. J. Immunopathol. Pharmacol. *21,* 421–428.

Ciortescu, I., Sfarti, C., Stan, M., Graur, M., and Stanciu C. (2009). Prevalence of *Helicobacter pylori* infection in patients with diabetes mellitus. Rev. Med. Chir. Soc. Med. Nat. Iasi. *113,* 1048–1055.

Collodel, G., Moretti, E., Campagna, M.S., Capitani, S., Lenzi, C., and Figura, N. (2010). Infection by CagA-positive *Helicobacter pylori* strains alter the sperm quality of men with fertility disorders and increase the systemic levels of TNF-a. Dig. Dis. Sci. 55, 94–100.

Corley, D.A., Kubo, A., Levin, T.R., Block, G., Habel, L., Rumore, G., Quesenberry, C., Buffler, P., and Parsonnet, J. (2008a). *Helicobacter pylori* and gastroesophageal reflux disease: a case-control study. Helicobacter *13,* 352–360.

Corley, D.A., Kubo, A., Levin, T.R., Block, G., Habel, L., Zhao, W., Leighton, P., Rumore, G., Quesenberry, C., Buffler, P., and Parsonnet, J. (2008b). *Helicobacter pylori* infection and the risk of Barrett's oesophagus: a community-based study. Gut 57, 727–733.

Correia, M., Machado, M.C., and Ristimaki, A. (2009). Basic Aspects of Gastric Cancer. Helicobacter *14* (*Suppl 1*),36–40.

Crucius, J.B.A., Canzian, F., Capella, G., Peña, A.S., Pera, G., Sala, N., Agudo, A., Rico, F., Del Giudice, G., Palli, D., *et al.* (2008). Cytokine gene polymorphisms and the risk of adenocarcinoma of the stomach in the European prospective investigation into cancer and Nutrition (EPIC-EURGAST). Ann. Oncol. *19,* 1894–1902.

Cvorovic, L., Brajovic, D., Strbac, M., Milutinovic, Z., and Cvorovic, V. (2008). Detection of *Helicobacter pylori* in nasal polyps: preliminary report. J. Otolaryngol. Head Neck Surg. 37, 192–195.

De Luca, A., De Falco, M., Manente, L., Dattilo, D., Lucariello, A., Esposito, V., Gnarini, M., Citro, G., Baldi, A., Tufano, M.A., *et al.* (2008). *Helicobacter pylori* heat shock protein B (HspB) localizes *in vivo* in the gastric mucosa and MALT lymphoma. J. Cell. Physiol. *216,* 78–82.

Delaney, B., Harris, A., Innes, M., Oakes, R., Wilson, S., Roalfe, A., Bennett, C., and Forman, D. (2005). Eradication of *Helicobacter pylori* for non-ulcer dyspepsia. Cochrane Database Syst. Rev. *1,* CD002096.

Delaney, B.C., Qume, M., Moayyedi, P., Logan, R.F., Ford, A.C., Elliott, C., McNulty, C., Wilson, S., and Hobbs, F.D. (2008). *Helicobacter pylori* test and treat versus proton pump inhibitor in initial management of dyspepsia in primary care: multicentre randomised controlled trial (MRC-CUBE trial). BMJ. *336,* 651–654.

D'Elios, M.M., Appelmelk, B.J., Amedei, A., Bergman, M.P., and Del Prete, G. (2004). Gastric autoimmunity:

the role of *Helicobacter pylori* and molecular mimicry. Trends Mol. Med. *10*, 316–323.

Demir, M., Gokturk, H.S., Ozturk, N.A., Kulaksizoglu, M., Serin, E., and Yilmaz, U. (2008). *Helicobacter pylori* prevalence in diabetes mellitus patients with dyspeptic symptoms and its relationship to glycemic control and late complications. Dig. Dis. Sci. *53*, 2646–2649.

Diamantidis, M.D., Ioannidou-Papagiannaki, E., Kountouras, J., Mandala, E., Klonizakis, P., Frida-Michailidou, I., Tsapournas, G., Haralambidou-Vranitsa, S., and Klonizakis I. (2008). *Helicobacter pylori* infection might protect from the leukaemic transformation of myelodysplastic syndromes. Med. Hypotheses. *71*, 159–160.

Dixon, M.F., Genta, R.M., Yardley, J.H., and Correa P. (1996). Classification and grading of gastritis. The updated Sydney System. International Workshop on the Histopathology of Gastritis, Houston 1994. Am. J. Surg. Pathol. *20*, 1161–181.

Dowd, J.B., Zajacova, A., and Aiello, A. (2009). Early origins of health disparities: burden of infection, health, and socioeconomic status in US children. Soc. Sci. Med. *68*, 699–707.

Drossman, D.A. (2006). The functional gastrointestinal disorders and the Rome III process. Gastroenterology *130*, 1377–1390.

Du, MQ. (2007). MALT lymphoma: recent advances in aetiology and molecular genetics. J. Clin. Exp. Hematop. *47*, 31–42.

Duque, X., Moran. S., Mera, R., Medina, M., Martinez, H., Mendoza. M.E., Torres, J., and Correa, P. (2010). Effect of eradication of *Helicobacter pylori* and iron supplementation on the iron status of children with iron deficiency. Arch. Med. Res. *41*, 38–45.

Elitsur, Y., Lawrence, Z., and Triest, W.E. (2002). Distribution of *Helicobacter pylori* organisms in the stomachs of children with *H. pylori* infection. Hum. Pathol. 33, 1133–1135.

Elizalde, J.I., Mendez, A., Gomez, J., del Rivero, M., Gironella, M., Closa, D., Quintero, E., and Pique JM. (2003). Gastric mucosal blood flow changes in *Helicobacter pylori* infection and NSAID-induced gastric injury. Helicobacter *8*, 124–131.

El-Zimaity, H.M.T., Ota, H., Graham, D.Y., Akamatsu, T., and Katsuyama, T. (2002). Patterns of gastric atrophy in intestinal type gastric carcinoma. Cancer *94*, 1428–1436.

El-Zimaity, H.M.T. (2007). Recent advances in the histopathology of gastritis. Curr. Diagn. Pathol. *13*, 340–348.

Emilia G, Luppi M, Zucchini P, Morselli, M., Potenza, L., Forghieri, F., Volzone, F., Jovic, G., Leonardi, G., Donelli, A., *et al.* (2007). *Helicobacter pylori* infection and chronic immune thrombocytopenic purpura: long-term results of bacterium eradication and association with bacterium virulence profiles. Blood *110*, 3833–3841.

Eshraghian, A., Hashemi, S.A., Hamidian Jahromi, A., Eshraghian, H., Masoompour, S.M., Davarpanah, M.A., Eshraghian, K., and Taghavi, S.A. (2009). *Helicobacter pylori* infection as a risk factor for insulin resistance. Dig. Dis. Sci. *54*, 1966–1970.

Eskandarian, R., Malek, M., Mousavi, S.H., and Babaei, M. (2006). Association of *Helicobacter pylori* infection with cardiac syndrome X. Singapore Med. J. *47*, 704–706.

Evrengul, H., Tanriverdi, H., Kuru, O., Enli, Y., Yuksel, D., Kilic, A., Kaftan, A., Kirac, S., and Kilic, M. (2007). Elevated homocysteine levels in patients with slow coronary flow: relationship with *Helicobacter pylori* infection. Helicobacter *12*, 298–305.

Fagan, R.P., Dunaway, C.E., Bruden, D.L., Parkinson, A.J., and Gessner, D. (2009). Controlled, household-randomized, open-label trial of the effect of treatment of *Helicobacter pylori* infection on iron deficiency among children in rural Alaska: results at 40 months. J. Infect. Dis. *199*, 652–660.

Fall, K., Ye, W., and Nyren, O. (2006). Antibiotic treatment and risk of gastric cancer. Gut *55*, 793–796.

Fancy, T., Mathers, P.H., and Ramadan, H.H. (2009). Otitis media with effusion: a possible role for *Helicobacter pylori*? Otolaryngol. Head Neck Surg. *140*, 256–258.

Fass, R., and Gasiorowska, A. (2008). Refractory GERD: what is it? Curr. Gastroenterol. Rep. *10*, 252–257.

Fayed, S.B., Aref, M.I., Fathy, H.M., Abd El Dayem, S.M., Emara, N.A., Maklof, A., and Shafik, A. (2008). Prevalence of celiac disease, *Helicobacter pylori* and gastroesophageal reflux in patients with refractory iron deficiency anemia. J. Trop. Pediatr. *54*, 43–53.

Feeney, M.A., Murphy, F., Clegg, A.J., Trebble, T.M., Sharer, N.M., and Snook, J.A. (2002). A case-control study of childhood environmental risk factors for the development of inflammatory bowel disease. Eur. J. Gastroenterol. Hepatol. *14*, 529–534.

Fernandini-Paredes, G.G., Mezones-Holguin, E., Vargas-Gonzales, R., Pozo-Briceno, E., and Rodriguez-Morales, A.J. (2008). In patients with type 2 diabetes mellitus, are glycosylated hemoglobin levels higher for those with *Helicobacter pylori* infection than those without infection? Clin. Infect. Dis. *47*, 144–145.

Ferrara, M., Capozzi, L., and Russo, R. (2009). Effect of *Helicobacter pylori* eradication on platelet count in children with chronic idiopathic thrombocytopenic purpura. Hematology *14*, 292–296.

Ferreira, A.C., Isomoto, H., Moriyama, M., Fujioka, T., Machado, J.C., and Yamaoka, Y. (2008). Helicobacter and gastric malignancies. Helicobacter *13* (*Suppl 1*), 28–34.

Fetissov, S.O., Hamze Sinno, M., Coeffier, M., Bole-Feysot, C., Ducrotte P, Hokfelt, T., and Dechelotte, P. (2008). Autoantibodies against appetite-regulating peptide hormones and neuropeptides: putative modulation by gut microflora. Nutrition *24*, 348–359.

Fiddian-Green, R.G. (2007). *Helicobacter pylori* eradication and L-dopa absorption in patients with PD and motor fluctuations. Neurology *68*, 1085.

Figura, N., Palazzuoli, A., Faglia, S., Lenzi, C., Borrello, F., Palazzuoli, V., Nami, R., Dal Canto, N., De Regis, F., Vaira, D., *et al.* (2002). Infection by CagApositive *Helicobacter pylori* strains in patients with ischaemic heart disease: prevalence and association with exercise-induced electrocardiographic abnormalities. Dig. Dis. Sci. *47*, 831–836.

Figura, N., Franceschi, F., Santucci, A., Bernardini, G., Gasbarrini G., and Gasbarrini, A. (2010). Extragastric Manifestations of *Helicobacter pylori* Infection. Helicobacter *15 Suppl. 1*, 60–68.

Fischbach, W., Dragosics, B., Kolve-Goebeler, M.E., Ohmann, C., Greiner, A., Yang, Q., Böhm, S., Verreet, P., Horstmann, O., Busch, M., *et al.* (2000). Primary gastric B-cell lymphoma: results of a prospective multicenter study. The German-Austrian Gastrointestinal Lymphoma Study Group. Gastroenterology *119*, 1191–1202.

Fischbach, W., Goebeler-Kolve, M.E., Dragosics, B., Greiner, A., and Stolte, M. (2004). Long term outcome of patients with gastric marginal zone B cell lymphoma of mucosa associated lymphoid tissue (MALT) following exclusive *Helicobacter pylori* eradication therapy: experience from a large prospective series. Gut *53*, 34–37.

Fischbach, W., Goebeler, M.E., Ruskone-Fourmestraux, A., Wündisch, T., Neubauer, A., Raderer, M., Savio, A., and EGILS (European Gastro-Intestinal Lymphoma Study) Group. (2007) Most patients with minimal histological residuals of gastric MALT lymphoma after successful eradication of *Helicobacter pylori* can be managed safely by a watch and wait strategy: experience from a large international series. Gut *56*, 1685–7.

Ford, A., Delaney, B., Forman, D., and Moayyedi, P. (2004). Eradication therapy for peptic ulcer disease in *H. pylori* positive patients. Cochrane Database Syst. Rev. CD003840.

Franceschi, F., and Gasbarrini, A. (2007). *Helicobacter pylori* and extragastric diseases. Best Pract. Res. Clin. Gastroenterol. *21*, 325–334.

Franceschi, F., Sepulveda, A.R., Gasbarrini, A., Pola, P., Silveri, N.G., Gasbarrini, G., Graham, D.Y., and Genta, R.M. (2002). Cross-reactivity of antiCagA antibodies with vascular wall antigens: possible pathogenic link between *Helicobacter pylori* infection and atherosclerosis. Circulation *106*, 430–434.

Franceschi, F., Niccoli, G., Ferrante, G., Gasbarrini, A., Baldi, A., Candelli, M., Feroce, F., Saulnier, N., Conte, M., Roccarina, D., *et al.* (2009). CagA antigen of *Helicobacter pylori* and coronary instability: insight from a clinicopathological study and a meta-analysis of 4241 cases. Atherosclerosis *202*, 535–542.

Franchini, M., and Veneri, D. (2004). *Helicobacter pylori* infection and immune thrombocytopenic purpura: an update. Helicobacter *9*, 342–346.

Fuccio, L., Zagari, R.M., Minardi, M.E., and Bazzoli, F. (2007) Systematic review. *Helicobacter pylori* eradication for the prevention of gastric cancer. Aliment Pharmacol. Ther. *25*, 133–41.

Fuccio, L., Eusebi, L.H., Zagari, R.M., and Bazzoli, F. (2009). *Helicobacter pylori* eradication treatment reduces but does not abolish the risk of gastric cancer. Am. J. Gastroenterol. *104*, 3100–3102.

Fujimori, S., Kishida, T., Kobaiashi, T., Sekita, Y., Seo, T., Nagata, K., Tatsuguchi, A., Gudis, K., Yokoi, K., Tanaka, N., *et al.* (2005). *Helicobacter pylori* infection increases the risk of colorectal adenoma and adenocarcinoma, especially in women. J. Gastroenterol. *40*, 887–893.

Fukase, K., Kato, M., Kikuchi, S., Inoue, K., Uemura, N., Okamoto, S., Terao, S., Amagai, K., Hayashi, S., Asaka, M., and Japan Gast Study Group. (2008). Effect of eradication of *Helicobacter pylori* on incidence of metachronous gastric carcinoma after endoscopic resection of early gastric cancer: an open-label, randomised controlled trial. Lancet *372*, 392–397.

Fukuhara, N,. Nakamura, T., Nakagawa, M., Tagawa, H., Takeuchi, I., Yatabe, Y., Morishima, Y., Nakamura, S., and Seto, M. (2008). Chromosomal imbalances are associated with outcome of *Helicobacter pylori* eradication in t(11;18)(q21;q21) negative gastric mucosa–associated lymphoid tissue lymphomas. Genes Chromosomes Cancer *46*, 784–790.

Fullerton, D., Britton, J.R., Lewis, S.A., Pavord, I.D., McKeever, T.M., and Fogarty, A.W. (2009). *Helicobacter pylori* and lung function, asthma, atopy and allergic disease. A population-based cross-sectional study in adults. Int. J. Epidemiol. *38*, 419–426.

Furuta, T., and Delchier, J.-Ch. (2009). *Helicobacter pylori* and non-malignant diseases. Helicobacter *14 (Suppl 1)*, 29–35.

Gao, L., Week, M.N. and Nieters, A., and Brenner, H. (2009a). Association between a pro-inflammatory genetic profile and the risk of chronic atrophic gastritis among older adults from Germany. Eur. J. Cancer *45*, 428–434.

Gao, L.B., Rao, L., Wang, Y.Y., Liang, W.B., Li, C., Xue, H., Zhou, B., Sun, H., Li, Y., Lv, M.L., *et al.* (2009b). The association of interleukin-16 polymorphisms with IL16 serum levels and risk of colorectal and gastric cancer. Carcinogenesis *30*, 295–299.

Gao, X.Y., Kuang, H.Y., Liu, X.M., Duan, P., Yang, Y., and Ma, Z.B. (2009c). Circulating ghrelin/obestatin ratio in subjects with *Helicobacter pylori* infection. Nutrition *25*, 506–511.

Garcia-Gonzalez, M.A., Savelkoul, P.H., Benito, R., Santolaria, S., Crusius, J.B., Pena, A.S., and Lanas, A. (2005). No allelic variant associations of the IL-1 and TNF gene polymorphisms in the susceptibility to duodenal ulcer disease. Int. J. Immunogenet. *32*, 299–306.

Gencer, M., Ceylan, E., Yildiz Zeyrek, F., and Aksoy, N. (2007). *Helicobacter pylori* seroprevalence in patients with chronic obstructive pulmonary disease and its relation to pulmonary function tests. Respiration *74*, 170–175.

Gessner, B.D., Baggett, H.C., Muth, P.T., Dunaway, E., Gold, B.D., Feng, Z., and Parkinson, A.J. (2006). A controlled, household-randomized, open-label trial of the effect that treatment of *Helicobacter pylori* infection has on iron deficiency in children in rural Alaska. J. Infect. Dis. *193*, 537–546.

Gillen, D., Wirz, A.A., and McColl, K.E. (2004). *Helicobacter pylori* eraclication releases prolonged increased acid secretion following omeprazole treatment. Gastroenterology *126*, 980–988.

Gioè, F.P., Cudia, B., Romano, G., Cocchiara, G., Li Vecchi, V., Gioe, M.A., Cali, C., Lo Coco, L., Li Vecchi, M., and Romano, M. (2008). Role and clinical importance of *Helicobacter pylori* infection in hemodialysis patients. G. Chir. *29*, 81–84.

Giral, A., Ozdogan, O., Celikel, C.A., Tozun, N., Ulusoy, N.B., and Kalayci C. (2004). Effect of *Helicobacter pylori* eradication on antithrombotic dose aspirininduced gastroduodenal mucosal injury. J. Gastroenterol. Hepatol. *19*, 773–777.

Gisbert, J.P., and Pajares, J.M. (2005). Systematic review and meta-analysis: is 1-week proton pump inhibitorbased triple therapy sufficient to heal peptic ulcer? Aliment. Pharmacol. Ther. *21*, 795–804.

Gisbert, J.P., Calvet, X., Feu, F., Bory, F., Cosme, A., Almela, P., Santolaria, S., Aznarez, R., Castro, M., Fernandez, N., *et al.* (2007). Eradication of *Helicobacter pylori* for the prevention of peptic ulcer re-bleeding. Helicobacter *12*, 279–286.

Goo, M.J., Ki, M.R., Lee, H.R., Hong, I.H., Park, J.K., Yang, H.J., Yuan, D.W., Hwang, O.K., Do, S.H., Yoo, S.E., *et al.* (2008). Primary biliary cirrhosis, similar to that in human beings, in a male C57BL? 6 mouse infected with *Helicobacter pylori*. Eur. J. Gastroenterol. Hepatol. *20*, 1045–1048.

Goodman, K.J., Joyce, S.L., and Ismond, K.P. (2006). Extragastric diseases associated with *Helicobacter pylori* infection. Curr. Gastroenterol. Rep. *8*, 458–464.

Graham, D.Y., Opekun, A.R., Yamaoka, Y., Osato, M.S., and El-Zimaity, H.M. (2003). Early events in proton pump inhibitor-associated exacerbation of corpus gastritis. Aliment. Pharmacol. Ther. *17*, 193–200.

Grande, M., Cadeddu, F., Villa, M., Attinà, G.M., Muzi, M.G., Nigro, C., Rulli, F., and Farinon, A.M. (2008) *Helicobacter pylori* and gastroesophageal reflux disease. World J. Surg. Oncol. *6*, 74.

Grbesa, I., Marinkovic, M., Ivkic, M., Kruslin, B., Novak-Kujundzic, R., Pegan, B., Bogdanovic, O., Bedekovic, V., and Gall-Troselj, K. (2008). Loss of imprinting of IGF2 and H19, loss of heterozygosity of IGF2R and CTCF, and *Helicobacter pylori* infection in laryngeal squamous cell carcinoma. J. Mol. Med. *86*, 1057–1066.

Grünberger, B., Wöhrer, S., Streubel, B., Formanek, M., Petkov, V., Puespoek, A., Haefner, M., Hejna, M., Jaeger, U., Chott, A., *et al.* (2006). Antibiotic treatment is not effective in patients infected with *H. pylori* suffering from extragastric MALT lymphoma. J. Clin. Oncol. *24*, 1370–1375.

Gulhan, M., Ozyilmaz, E., Tarhan, G., Demirag, F., Capan, N., Erturk, A., Canbakan, S., Ayaslioglu, E., Gulhan, E., and Ahmed, K. (2007). *Helicobacter pylori* in bronchiectasis: a polymerase chain reaction assay in bronchoalveolar lavage fluid and bronchiectatic lung tissue. Arch. Med. Res. *38*, 317–321.

Gunji, T., Matsuhashi, N., Sato, H., Fujibayashi, K., Okumura, M., Sasabe, N., and Urabe, A. (2009). *Helicobacter pylori* infection significantly increases insulin resistance in the asymptomatic Japanese population. Helicobacter *14*, 144–150.

Gunji, T., Matsuhashi, N., Sato, H., Fujibayashi, K., Okumura, M., Sasabe, N., and Urabe, A. (2008). *Helicobacter pylori* infection is significantly associated with metabolic syndrome in the Japanese population. Am. J. Gastroenterol. *103*, 3005–3010.

Gwee, K.A., Teng, L., Wong, R.K., Ho, K.Y., Sutedja, D.S., and Yeoh, K.G. (2009). The response of Asian patients with functional dyspepsia to eradication of *Helicobacter pylori* infection. Eur. J. Gastroenterol. Hepatol. *21*, 417–424.

Hamed, S.A., Amine, N.F., Galal, G.M., Helal, S.R., Tag El-Din, L.M., Shawky, O.A., Ahmed, E.A., and Abdel Rahman, M.S. (2008). Vascular risks and complications in diabetes mellitus: the role of *Helicobacter pylori* infection. J. Stroke Cerebrovasc. Dis. *17*, 86–94.

Harvey, R.F., Lane, J.A., Nair, P., Egger, M., Harvey, I., Donovan, J., and Murray, L. (2010). Clinical trial: the prolonged beneficial effect of *Helicobacter pylori* eradication on dyspepsia consultations – the Bristol Helicobacter Project. Aliment. Pharmacol. Ther. *32*, 394–400.

Hellmig, S., Troch, K., Ott, S.J., Schwarz, T., and Folsch U.R. (2008). Role of *Helicobacter pylori* infection in the treatment and outcome of chronic urticaria. Helicobacter *13*, 341–345.

Hermansson, M., Ekedahl, A., Ranstam, J., and Zilling, T. (2009). Decreasing incidence of peptic ulcer complications after the introduction of the proton pump inhibitors, a study of the Swedish population from 1974–2002. BMC Gastroenterol. *9*, 25.

Honda, C., Adachi, K., Arima, N., Tanaka, S., Yagi, J., Morita, T., Tanimura, T., Furuta, K., and Kinoshita, Y. (2008). *Helicobacter pylori* infection does not accellerate the age-related progression of arteriosclerosis: a 4-year follow-up study. J. Gastroenterol. Hepatol. *23*, 373–378.

Hong, S.S., Jung, H.Y., Choi, K.D., Song, H.J., Lee, G.H., Oh, T.H, Jo, J.Y., Kim, K.J., Byeon, J.S., Myung, S.J., *et al.* (2006). A prospective analysis of low-grade gastric MALT lymphoma after *H. pylori* eradication. Helicobacter *11*, 569–573.

Houghton, J.M., Stoicov, C., Nomura, S., Rogers, A.B., Carlson, J., Li, H., Cai, X., Fox, J.G., Goldenring, J.R., and Wang, T.C. (2004). Gastric cancer originating from bone marrow derived cells. Science *306*, 1568–1571.

Huang, J.Q., Sridhar, S., and Hunt, R.H. (2002). Role of *Helicobacter pylori* infection and nonsteroidal anti-inflammatory drugs in peptic-ulcer disease: a meta-analysis. Lancet. *359*, 14–22.

Humar, B., Fukuzawa, R., Blair, V., Dunbier, A., More, H., Charlton, A., Yang, H.K., Kim, W.H., Reeve, A.E., Martin, I., *et al.*(2007). Destabilized adhesion in the gastric proliferative zone and Src kinase activation mark the development of early diffuse gastric cancer. Cancer Res. *67*, 2480–2489.

Humar, B., Blair, V., Charlton, A., More, H., Martin, I., and Guildford, P. (2009). E-cadherin deficiency initiates gastric signet-ring cell carcinoma in mice and man. Cancer Res. *69*, 2050–2056.

Hung, L.C., Ching, J.Y., Sung, J.J., To, K.F., Hui, A.J., Wong, V.W., Leong, R.W., Chan, H.L., Wu, J.C., Leung, W.K., *et al.* (2005). Long-term outcome of *Helicobacter pylori*-negative idiopathic bleeding ulcers: a prospective cohort study. Gastroenterology *128*, 1845–1850.

Inagaki, H. (2007). Mucosa-associated lymphoid tissue lymphoma: molecular pathogenesis and clinicopathological significance. Pathol. Int. *57*, 474–484.

Iriz, E., Crak, M.Y., Engin, E.D., Zor, M.H., Erer, D., Ozdogan, M.E., Turet, S., and Yener, A. (2008). Detection of *Helicobacter pylori* DNA in aortic and left internal mammary artery biopsies. Tex. Heart Inst. J. *35*, 130–135.

Isaeva, G.Sh., Abuzarova, E.R., Valeeva, Iu.V., Pozdeev, O.K., and Muraveva, E.V. (2009). *Helicobacter pylori* in patients with disorders of hepatobiliary system. Zh. Mikrobiol. Epidemiol. Immunobiol. *2*, 96–101.

Isaka T, Nakamura T, Tajika M, Kawai H, Imaoka H, Okamoto Y, Aoki M, Inoue H, Takahashi K, Mizuno, N., *et al.* (2007). API2-MALT1 chimeric transcript-positive gastroduodenal MALT lymphoma with subsequent development of adenocarcinoma as a collision tumour over a clinical course of 7 years. Histopathology *51*, 119–123.

Ishikawa, S., Ohta, T., and Hatakeyama, M. (2009). Stability of *Helicobacter pylori* CagA oncoprotein in human gastric epithelial cells. FEBS Lett. *21*, 583.

Ito, K., Yamaoka, Y., Ota, H., El-Zimaity, H., and Graham, D.Y. (2008) Adherence, internalization, and persistence of *Helicobacter pylori* in hepatocytes. Dig. Dis. Sci. *53*, 2541–2549.

Izzotti, A., Sacca, S.C., Bagnis, A., and Recupero, S.M. (2009). Glaucoma and *Helicobacter pylori* infection: correlations and controversies. Br. J. Ophthalmol. *93*, 1420–1427.

Jackson, S.C., Beck, P., Buret, A.G., O'Connor, P.M., Meddings, J., Pineo, G., and Poon, M-C. (2008). Long-term platelet responses to *Helicobacter pylori* eradication in Canadian patients with immune thrombocytopenic purpura. Int. J. Hematol. *88*, 212–218.

Jakobs, R., and Riemann, J.F. (2004). CagA-positive *Helicobacter pylori* strains and gastro-oesophageal reflux disease: still puzzling? Eur. J. Gastroenterol. Hepatol. *16*, 635–637.

Jarbol, D.E., Bech, M., Kragstrup, J., Havelund, T., and Schaffalitzky de Muckadell, O.B. (2006). Economic evaluation of empirical antisecretory therapy versus *Helicobacter pylori* test for management of dyspepsia: a randomized trial in primary care. Int. J. Technol. Assess Health Care. *22*, 362–371.

Jha, H.C., Prasad, J., and Mittal A. (2008). High immunoglobulin A seropositivity for combined Chlamydia pneumoniae, *Helicobacter pylori* infection, and high-sensitivity C-reactive protein in coronary artery disease patients in India can serve as atherosclerotic marker. Heart Vessels *23*, 390–396.

Jonaitis, L.V., Kiudelis, G., and Kupcinskas, L. (2004). Characteristics of patients with erosive and nonerosive GERD in high-*Helicobacter-pylori* prevalence region. Dis. Esophagus *17*, 223–227.

Kamangar, F., Dores, G.M., and Anderson, W.F. (2006). Patterns of cancer incidence, mortality, and prevalence across five continents: defining priorities to reduce cancer disparities in different geographic regions of the world. J. Clin. Oncol. *24*, 2137–2150.

Kamangar, F., Qiao, Y.L., Blaser, M.J., Sun, X.D., Katki, H., Fan, J.H., Perez-Perez, G.I., Abnet, C.C., Zhao, P., Mark, S.D., *et al.* (2007). *Helicobacter pylori* and oesophageal and gastric cancers in a prospective study in China. Br. J. Cancer *96*, 172–176.

Kamburov, V., Boyanova, L., Manova, S., Tadjer, S., and Krastev, Z. (2007). Diagnostic tests for evaluation of *Helicobacter pylori* infection in peptic ulcer bleeding. Bulgarian Hepatogastroenterol. *IX*, 7–12.

Kanbay, M., Kasapoglu, B., and Akcay, A. (2007). An occult risk factor for proteinuria: *Helicobacter pylori* infection. Med. Hypotheses. *69*, 709–710.

Kang, J.M., Kim, N., Lee, D.H., Park, J.H., Lee, M.K., Kim, J.S., Jung, H.C., and Song, I.S. (2009). The effects of genetic polymorphisms of IL-6, IL-8 and IL-10 on *Helicobacter pylori*-induced gastroduodenal diseases in Korea. J. Clin. Gastroenterol. *43*, 420–428.

Kaplan, M., Yavuz, S.S., Cinar, B., Koksal, V., Kut, M.S., Yapici, F., Gercekoglu, H., and Demirtas, M.M. (2006). Detection of *Chlamydia pneumoniae* and *Helicobacter pylori* in atherosclerotic plaques of carotid artery by polymerase chain reaction. Int. J. Infect. Dis. *10*, 116–123.

Kaptan, Z.K., Emir, H., Uzunkulaoglu, H., Yucel, M., Karakoc, E., Koca, G., Tuzuner, A., Samim, E., and Korkmaz, M. (2009). Determination of *Helicobacter pylori* in patients with chronic nonspecific pharyngitis. Laryngoscope *119*, 1479–1483.

Kawashima, J., Ohno, S., Sakurada, T., Takabayashi, H., Kudo, M., Ro, S., Kato, S., and Yakabi, K. (2009). Circulating acylated ghrelin level decreases in accordance with the extent of atrophic gastritis. J. Gastroenterol. *44*, 1046–1054.

Kaya, T., Ozan, H., and Ozakin, C. (2009). The presence of *Helicobacter pylori* in cervical preinvasive lesions. Clin. Exp. Obstet. Gynecol. *36*, 113–115.

Kaye, P.V., Garsed, K., Ragunath, K., Jawhari, A., Pick, B., and Atherton, J.C. (2008) The clinical utility and diagnostic yield of routine gastric biopsies in the investigation of iron deficiency anemia: a case-control study. Am. J. Gastroenterol. *103*, 2883–2889.

Kiltz, U., Zochling, J., Schmidt, W.E., and Braun, J. (2008). Use of NSAIDs and infection with *Helicobacter pylori* – what does the rheumatologist need to know? Rheumatology (Oxford) *47*, 342–347.

Kim, J.S., Chung, S.J., Choi, Y.S., Cheon, J.H., Kim, C.W., Kim, S.G., Jung, H.C., and Song, I.S. (2007). *H. pylori* eradication for low-grade gastric mucosa-associated lymphoid tissue lymphoma is more successful in inducing remission in distal compared with proximal disease. Br. J. Cancer. *7*, 1324–1328.

Kim, N., Cho, S.I., Yim, J.Y., Kim, J.M., Lee, D.H., Park, J.H., Kim, J.S., Jung, H.C., and Song I.S. (2006). The effects of genetic polymorphisms of IL-1 and TNF-A on *Helicobacter pylori*-induced gastroduodenal diseases in Korea. Helicobacter *11*, 105–112.

Kim, N., Lee, S.W., Cho, S.I., Park, C.G.,Yang, C.H., Kim, H.S., Rew, J.S., Moon, J.S., Kim, S., Park, S.H., *et al.* (2008a). The prevalence of and risk factors for erosive oesophagitis and non-erosive reflux disease: a nationwide multicentre prospective study in Korea. Aliment. Pharmacol. Ther. *27*, 173–185.

Kim, N., Park, Y.S., Cho, S.I., Lee, H.S., Choe, G., Kim, I.W., Won, Y.D., Park, J.H., Kim, J.S., Jung, H.C., *et al.* (2008b). Prevalence and risk factors of atrophic gastritis and intestinal metaplasia in a Korean population without significant gastroduodenal disease. Helicobacter *13*, 245–255.

Kim, Y.S., Park, S.W., Kim, M.H., Jang, E.J., Park, J.S., Park, S.J., Baik, H.W., Chung, G., and Hahm, K.B. (2008c). Novel single nucleotide polymorphism of the VEGF gene as a risk predictor for gastroduodenal ulcers. J. Gastroenterol. Hepatol. *23, Suppl. 2*, S131-S139.

Kinoshita, Y. (2007). Lifestyle-related diseases and *Helicobacter pylori* infection. Intern. Med. *46*, 105–106.

Kivistö, R., Linros, J., Rossi, M., Rautelin, H., and Hänninen, M.L. (2010). Characterization of multiple *Helicobacter bizzozeronii* isolates from a Finnish patient with severe dyspeptic symptoms and chronic active gastritis. Helicobacter *15*, 58–66.

Kodama, M., Murakami, K., Okimoto, T., and Fujioka, T. (2007). Immune response to CagA protein is associated with improved platelet count after *Helicobacter pylori* eradication in patients with idiopathic thrombocytopenic purpura. Helicobacter *12*, 36–42.

Koike, T., Ohara, S., Inomata, Y., Abe, Y., Iijima, K., and Shimosegawa, T. (2007). The prevalence of *Helicobacter pylori* infection and the status of gastric acid secretion in patients with gastroesophageal adenocarcinoma in Japan. Inflammopharmacology *15*, 61–64.

Komuro, A., Yashiro, M., Iwata, C., Morishita, Y., Johansson, E., Matsumoto, Y., Watanabe, A., Aburatani, H., Miyoshi, H., Kiyono, K., *et al.* (2009). Diffuse-type gastric carcinoma: progression, angiogenesis, and transforming growth factor b signaling. J. Natl. Cancer Inst. *101*, 592–604.

Konno, M. (2009). Iron-deficiency anaemia in children. Nippon Rinsho *67*, 2345–2351.

Kountouras, J., Zavos, C., and Chatzopoulos, D. (2005). A concept on the role of *Helicobacter pylori* infection in autoimmune pancreatitis. J. Cell. Mol. Med. *9*, 196–207.

Kountouras, J., Zavos, C., Chatzopoulos, D., and Katsinelos, P. (2008). New aspects of *Helicobacter pylori* infection involvement in gastric oncogenesis. J. Surg. Res. *146*, 149–158.

Kountouras, J., Boziki, M., Gavalas, E., Zavos, C., Grigoriadis, N., Deretzi, G., Tzilves, D., Katsinelos, P., Tsolaki, M., Chatzopoulos, D., *et al.* (2009). Eradication of *Helicobacter pylori* may be beneficial in the management of Alzheimer's disease. J. Neurol. *256*, 758–767.

Krasinskas, A.M., Yao, Y., Randhawa, P., Dore, M.P., and Sepulveda, A.R. (2007). *Helicobacter pylori* may play a contributory role in the pathogenesis of primary sclerosing cholangitis. Dig. Dis. Sci. *52*, 2265–2270.

Krause, I., Anaya, J.M., Fraser, A., Barzilai, O., Ram, M., Abad, V., Arango, A., Garcia, J., and Shoenfeld, Y. (2009). Anti-infectious antibodies and autoimmune-associated autoantibodies in patients with h type I diabetes mellitus and their close family members. Ann. N. Y. Acad. Sci. *1173*, 633–639.

Kucukazman, M., Yavuz, B., Sacikara, M., Asilturk, Z., Ata, N., Ertugrul, D.T., Yalcin, A.A., Yenigun, E.C., Kizilca, G., Okten, H., *et al.* (2009). The relationship between updated Sydney system score and LDL cholesterol levels in patients infected with *Helicobacter pylori*. Dig. Dis. Sci. *54*, 604–607.

Kudo, M., Gutierrez, O., El-Zimaity, H.M., Cardona, H., Nurgalieva, Z.Z., Wu, J. and Graham, D.Y. (2005). CagA in Barrett's oesophagus in Colombia, a country with a high prevalence of gastric cancer. J. Clin. Pathol. *58*, 259–262.

Kuipers, E.J., Nelis, G.F., Klinkenberg-Knol, E.C., Snel, P., Goldfain, D., Kolkman, J.J., Festen, H.P.M., Dent, J., Zeitoun, P., Havu, N., *et al.* (2004). Cure of *Helicobacter pylori* infection in patients with reflux oesophagitis treated with long term omeprazole reverses gastritis without exacerbation of reflux disease: results of a randomised controlled trial. Gut *53*, 12–20.

Kuo, S.H,. Chen, L.T., Wu, M.S., Lin, C.W., Yeh, K.H., Kuo, K.T., Yeh, P.Y., Tzeng, Y.S., Wang, H.P., Hsu, P.N., *et al.* (2008). Long-term follow-up of gastrectomized patients with mucosa-associated lymphoid tissue lymphoma: need for a revisit of surgical treatment. Ann. Surg. *247*, 265–269.

Kupcinskas, L., and Malfertheiner, P. (2005). *Helicobacter pylori* and non-malignant diseases. Helicobacter *10 (Suppl 1)*, 26–33.

Kurotsuchi, S., Ando, H., Iwase, A., Ishida, Y., Hamajima, N. and Kikkawa, F. (2008). The plausibility of *Helicobacter pylori*-related infertility in Japan. Fertil. Steril. *90*, 866–868.

Kusano, K., Tokunaga, O., Ando, T. and Inokuchi, A. (2007). *Helicobacter pylori* in the palatine tonsils of patients with IgA nephropathy compared with those of patients with recurrent pharyngotonsillitis. Hum. Pathol. *38*, 1788–1797.

Kwon, J.H., Chung, I.S., Son, H.S., Park, J.M., Cho, Y.K., Lee, I.S., Kim, S.W., and Choi, M.G. (2008). The relationship of gastrin, pepsinogen and *Helicobacter pylori* in erosive reflux esophagitis. Korean J. Gastroenterol. *51*, 159–166.

Labenz, J., Blum, A.L., Bolten, W.W., Dragosics, B., Rösch, W., Stolte, M., and Koelz, H.R. (2002). Primary prevention of diclofenac associated ulcers and dyspepsia by omeprazole or triple therapy in *Helicobacter pylori* positive patients: a randomised, double blind, placebo controlled, clinical trial. Gut *51*, 329–335.

Labenz, J., Jaspersen, D., Kulig, M., Leodolter, A., Lind, T., Meyer-Sabellek, W., Stolte, M., Vieth, M., Willich, S., and Malfertheiner, P. (2004). Risk factors for erosive esophagitis: a multivariate analysis based on the

ProGERD study initiative. Am. J. Gastroenterol. *99*, 1652–1656.

Lahner, E., and Annibale, B. (2009). Pernicious anemia: new insights from a gastroenterological point of view. World J. Gastroenterol. *15*, 5121–5128.

Lai, K.C., Lam, S.K., Chu, K.M., Wong, B.C., Hui, W.M., Hu, W.H., Lau, G.K., Wong, W.M., Yuen, M.F., Chan, A.O., *et al.* (2002). Lansoprazole for the prevention of recurrences of ulcer complications from long-term low-dose aspirin use. N. Engl. J. Med. *346*, 2033–2038.

Lai, K.C., Lau, C.S., Ip, W.Y., Wong, B.C., Hui, W.M., Hu, W.H., Wong, R.W., and Lam, S.K. (2003). Effect of treatment of *Helicobacter pylori* on the prevention of gastroduodenal ulcers in patients receiving long-term SAIDs: a double-blind, placebo-controlled trial. Aliment. Pharmacal. Ther. *17*, 799–805.

Lee, A. (2007). Early influences and childhood development. Does *Helicobacter* play a role? Helicobacter *12* (*Suppl 2*), 69–74.

Lee, W.Y., Yoon, W.T., Shin, H.Y., Jeon, S.H., and Rhee, P.L. (2008). *Helicobacter pylori* infection and motor fluctuations in patients with Parkinson's disease. Mov. Disord. *23*, 1696–700.

Lee, J.H., Choe, Y.H., and Choi, Y.O. (2009). The expression of iron-repressible outer membrane proteins in *Helicobacter pylori* and its association with iron deficiency anemia. Helicobacter *14*, 36–39.

van Leerdam, M.E. (2008). Epidemiology of acute upper gastrointestinal bleeding. Best. Pract. Res. Clin. Gastroenterol. *22*, 209–224.

Leung, W.K., To, K.F., Chan, F.K., Lee, T.L., Chung, S.C., and Sung J.J. (2000). Interaction of *Helicobacter pylori* eradication and non-steroidal anti-inflammatory drugs on gastric epithelial apoptosis and proliferation: implications on ulcerogenesis. Aliment. Pharmacal. Ther. *14*, 879–885.

Li, G.Q., Xia, H.H., Chen, M.H., Gu, Q., Wang, J.D., Peng, J.Z., Chan, A.O., Cho, C.H., So, H.L., Lam, S.K., *et al.* (2008). Effects of aspirin on the development of *Helicobacter pylori*-induced gastric inflammation and heterotopic proliferative glands in Mongolian gerbils. Helicobacter *1*, 20–29.

Lindkvist, B., Johansen, D., Borgstrom, A., and Manjer, J. (2008). A prospective study of *Helicobacter pylori* in relation to the risk for pancreatic cancer. BMC Cancer *8*, 321.

Liu, L., Liu, Y., Tong, W., Ye, H., Zhang, X., Cao, W., and Zhang, Y. (2007). Pathogen burden in essential hypertension. Circ. J. *71*, 1761–1764.

Longo-Mbenza, B., Nkondi Nsenga, J., and Vangu Ngoma, D. (2007). Prevention of the metabolic syndrome insulin resistance and the atherosclerotic diseases in Africans infected by *Helicobacter pylori* infection and treated by antibiotics. Int. J. Cardiol. *121*, 229–238.

Lu, C.C., Sheu, B.S., Chen, T.W., Yang, H.B., Hung, K.H., Kao, A.W., Chuang, C.H., and Wu, J.J. (2005). Host TNF-alpha-1031 and -863 promoter single nucleotide polymorphisms determine the risk of benign ulceration after *H. pylori* infection. Am. J. Gastroenterol. *100*, 1274–1282.

Lunetta, M., Fazio, G., Avena, V., Corrado, E., Sutera, L., Arnone, E., Novo, G., and Novo, S. (2009). *Helicobacter pylori* and atrial fibrillation: absence of correlations. J. Cardiovasc. Med. *10*, 4–5.

Luther, J., Dave, M., Higgins, P.D., and Kao, J.Y. (2010). Association between *Helicobacter pylori* infection and inflammatory bowel disease: a meta-analysis and systematic review of the literature. Inflamm. Bowel Dis. *16*, 1077–1084.

Lutsey, P.L., Pankow, J.S., Bertoni, A.G., Szklo, M., and Folsom, A.R. (2009). Serological evidence of infections and Type 2 diabetes: the Multi-ethnic study of atherosclerosis. Diabet. Med. *26*, 149–152.

Malfertheiner, P., Mégraud, F., O'Morain, C., Hungin, A.P., Jones, R., Axon, A., Graham, D.Y., Tytgat, G., and European *Helicobacter pylori* Study Group (EHPSG) (2002). Current concepts in the management of *Helicobacter pylori* infection–the Maastricht 2-2000 Consensus Report. Aliment. Pharmacol. Ther. *16*, 167–180.

Malfertheiner, P., Mégraud, F., O'Morain, C., Bazzoli, F., El-Omar, E., Graham, D., Hunt, R., Rokkas, T., Vakil, N., Kuipers, E.J., and The European Helicobacter Study Group (EHSG). (2007). Current concepts in the management of *Helicobacter pylori* infection: the Maastricht III Consensus Report. Gut *56*, 772–778.

Malfertheiner, P., Chan, F.K., and McColl, K.E. (2009). Peptic ulcer disease. Lancet *374*, 1449–1461.

Marinchev, L., Terziev, I., Ivanova, R., and Vladimirov, B. (2007). Rheumatic diseases and endoscopic changes in the upper gastrointestinal mucosa. Prac. Med. *38*, 12–20.

Mateva L., Tankova, Tz., Petrova, J., Aleksiev, A., Spassova, Z., and Stoinova, V. (2001). Characteristic of dyspepsia in diabetic patients. Comparison with dyspeptic patients without diabetes. Gut *49* (*Suppl III*), A1689.

McColl, K.L.M. (2007). *Helicobacter pylori* and esophageal cancer – not always protective. Gut *56*, 457–459.

McDonald, S., Greaves, L., Gutierrez-Gonzalez, L., Rodriguez-Justo, M., Deheragoda, M., Leedham, S.J., Taylor, R.W., Lee, C.Y., Preston, S.L., Lovell, M., *et al.* (2008). Mechanisms of field cancerization in the human stomach: the expansion and spread of mutated gastric stem cells. Gastroenterology *134*, 500–510.

McNamara, D., and El-Omar E. (2008). *Helicobacter pylori* infection and the pathogenesis of gastric cancer: a paradigm for host–bacterial interactions. Dig. Liver Dis. *40*, 504–509.

Miki, K., Fujishiro, M., Kodashima, S., and Yahagi, N. (2009). Long-term results of gastric cancer screening using the serum pepsinogen test method among an asymptomatic middle-aged Japanese population. Dig. Endosc. *21*, 78–81.

Misiuk-Hojfo, M., Michalowska, M., and Turno-Krecicka, A. (2009). *Helicobacter pylori* – a risk factor for the development of the central serous chorioretinopathy. Klin. Oczna. *111*, 30–32.

Mitchell, H.M., Kaakoush, N.O., and Sutton, P. (2010). Extragastric manifestations of *Helicobacter pylori* infection. In *Helicobacter pylori* in the 21st Century,

Sutton, P., and Mitchell, H.M., eds. (Cab International: Wallingford, UK), pp. 69–93.

Mizuno, S., Morita, Y., Inui, T., Asakawa, A., Ueno, N., Ando, T., Kato, H., Uchida, M., Yoshikawa, T., and Inui A. (2005). *Helicobacter pylori* infection is associated with colon adenomatous polyps detected by high-resolution colonoscopy. Int. J. Cancer *117*, 1058–1059.

Mizuno, S., Kobayashi, M., Tomita, S., Miki, I., Masuda, A., Onoyama, M., Habu, Y., Inokuchi, H., and Watanabe, Y. (2009). Validation of the pepsinogen test method for gastric cancer screening using a follow-up study. Gastric Cancer *12*, 158–163.

Mizuno, S., Miki, I., Ishida, T., Yoshida, M., Onoyama, M., Azuma, T., Habu, Y., Inokuchi, H., Ozasa, K., Miki, K., *et al.* (2010). Prescreening of a high-risk group for gastric cancer by serologically determined *Helicobacter pylori* Infection and atrophic gastritis. Dig. Dis. Sci. *55*, 3132–3137.

Moayyedi, P., Soo, S., Deeks, J., Forman, D., Mason, J., Innes, M., and Delaney, B. (2000). Systematic review and economic evaluation of *Helicobacter pylori* eradication treatment for non-ulcer dyspepsia. BMJ *321*, 659–664.

Moayyedi, P., Soo, S., Deeks, J., Delaney, B., Harris, A., Innes, M., Oakes, R., Wilson, S., Roalfe, A., Bennett, C., *et al.* (2006). Eradication of *Helicobacter pylori* for non-ulcer dyspepsia. Cochrane Database Syst. 2, CD002096.

Morgner, A., Schmelz, R., Thiede, C., Stolte, M., and Miehlke S. (2007). Therapy of gastric mucosa associated lymphoid tissue lymphoma. World. J. Gastroenterol. *13*, 3554–3566.

Moss, S.F., and Malfertheiner, P. (2007). Helicobacter and gastric malignancies. Helicobacter *12* (*Suppl 1*), 23–30.

Moyaert, H., Franceschi, F., Roccarina, D., Ducatelle, R., Haesebrouck F., and Gasbarrini, A. (2008). Extragastric manifestations of *Helicobacter pylori* infection: other helicobacters. Helicobacter *13* (*Suppl 1*), 47–57.

Müller, H., Rappel, S., Wündisch, T., Bayerdörfferm E., and Stolte, M. (2001). Healing of active, non-atrophic autoimmune gastritis by *H. pylori* eradication. Digestion *64*, 30–39.

Muhsen, K., and Cohen, D. (2008). *Helicobacter pylori* infection and iron stores: a systematic review and meta-analysis. Helicobacter *13*, 323–340.

Mulayim, B., Celik, N.Y., and Yanik, F.F. (2008). *Helicobacter pylori* infection detected by 14C-urea breath test is associated with iron deficiency anemia in pregnant women. J. Obstet. Gynaecol. Res. *34*, 980–985.

Murakami, K., Sato, R., Okimoto, T., Watanabe, K., Nasu, M., Kodama, M., and Fujioka, T. (2006). Maintenance therapy with H2-receptor antagonist until assessment of *Helicobacter pylori* eradication can reduce recurrence of peptic ulcer after successful eradication of the organism: prospective randomized controlled trial. J. Gastroenterol. Hepatol. *21*, 1048–1053.

Nakamura, M., Murayama, S.Y., Serizawa, H., Sekiya, Y., Eguchi, M., Takahashi, S., Nishikawa, K., Takahashi, T., Matsumoto, T., Yamada, H., *et al.* (2007a). Candidatus *Helicobacter heilmanii* from a cynomolgus monkey induces gastric mucosa-associated lymphoid tissue lymphomas in C57BL/6 mice. Infect. Immun. *75*, 1214–1222.

Nakamura, S., Ye, H., Bacon, C.M., Goatly, A., Liu, H., Banham, A.H., Ventura, R., Matsumoto, T., Iida, M., Ohji, Y., *et al.* (2007b). Clinical impact of genetic aberrations in gastric MALT lymphoma: a comprehensive analysis using interphase fluorescence *in situ* hybridisation. Gut *56*, 358–363.

Nakamura, S., Ye, H., Bacon, C.M., Liu, H., Goatly, A., Matsumoto, T., Koga, H., Umeno, J., Ohji, Y., Yao, T., *et al.* (2007c). Gastric MALT lymphoma with t(14;18)(q32;q21) involving IGH and BCL2 genes that responded to *Helicobacter pylori* eradication. J. Clin. Pathol. *60*, 1171–1173.

Nakamura, T., Seto, M., Tajika, M., Kawai, H., Yokoi, T., Yatabe, Y., and Nakamura S. (2008). Clinical features and prognosis of gastric MALT lymphoma with special reference to responsiveness to *H. pylori* eradication and API2-MALT1 status. Am. J. Gastroenterol. *103*, 62–70.

Nervi, G., Liatopoulou, S., Cavallaro, L.G., Gnocchi, A., Dal-Bo, N., Rugge, M., Iori, V., Cavestro, G.M., Maino, M., Colla, G., *et al.* (2006). Does *Helicobacter pylori* infection eradication modify peptic ulcer prevalence? A 10 years' endoscopical survey. World J. Gastroenterol. *12*, 2398–2401.

Nikolopoulou, A., Tousoulis, D., Antoniades, C., Petroheilou, K., Vasiliadou, C., Papageorgiou, N., Koniari, K., Stefanadi, E., Latsios, G., Siasos, G., *et al.* (2008). Common community infections and the risk for coronary artery disease and acute myocardial infarction: evidence for chronic overexpression of tumor necrosis factor alpha and vascular cells adhesion molecule-1. Int. J. Cardiol. *130*, 246–250.

Nyberg, A, Skagius, E., Nilsson, I., Ljungh, A., and Henriksson, A.E. (2008). Abdominal aortic aneurysm and infection with CagA positive strains of *Helicobacter pylori*. Scand. J. Infect. Dis. *40*, 204–207.

Ohnishi, M., Fukui, M., Ishikawa, T., Ohnishi, N., Ishigami, N., Yoshioka, K., Hasegawa, G., Yoshikawa, T., and Nakamura, N. (2008a). *Helicobacter pylori* infection and arterial stiffness in patients with type 2 diabetes mellitus. Metabolism *57*, 1760–1764.

Ohnishi, N., Yuasa, H., Tanaka, S., Sawa, H., Miura, M., Matsui, A., Higashi, H., Musashi, M., Iwabuchi, K., Suzuki, M., *et al.* (2008b). Transgenic expression of *Helicobacter pylori* CagA induces gastrointestinal and hematopoietic neoplasms in mouse. Proc. Natl. Acad. Sci. U.S.A. *105*, 1003–1008.

Oridate, N., Takeda, H., Asaka, M., Nishizawa, N., Mesuda, Y., Mori, M., Furuta, Y., and Fukuda, S. (2008). Acid-suppression therapy offers varied laryngopharyngeal and esophageal symptom relief in laryngopharyngeal reflux patients. Dig. Dis. Sci. *53*, 2033–2038.

Ozcan, C., Polat, A., Otag, F., and Gorur, K. (2009). Does *Helicobacter pylori* play a role in etiology of nasal polyposis? Auris Nasus. Larynx *36*, 427–430.

Ozdogru, I., Kalay, N., Dogan, A., Inanc, M.T., Kaya, M.G., Topsakal, R., Gul, I., Kutukoglu, I., Kilic, H., and Eryol, N.K. (2007). The relationship between *Helicobacter pylori* IgG titre and coronary atherosclerosis. Acta Cardiol, *62*, 501–505.

Ozyurt, M., Gungor, A., Ergunay, K., Cekin, E., Erkul, E., and Haznedaroglu, T. (2009). Real-time PCR detection of *Helicobacter pylori* and virulence-associated cagA in nasal polyps and laryngeal disorders. Otolaryngol. Head. Neck Surg. *141*, 131–135.

Pacifico, L., Anania, C., Osborn, J.F., Ferrara, E., Schiavo, E., Bonamico, M., and Chiesa, C. (2008). Long-term effects of *Helicobacter pylori* eradication on circulating ghrelin and leptin concentrations and body composition in prepubertal children. Eur. J. Endocrinol. *158*, 323–332.

Pasceri, V., Patti, G., Cammarota, G., Pristipino, C., Richichi, G., and Di Sciascio, G. (2006). Virulent strains of *Helicobacter pylori* and vascular diseases: a meta-analysis. Am. Heart J. *151*, 1215–1222.

Peitz, U., Raps, S., Plein, K., Leodolter, A., Hotz Dagger, J., and Malfertheiner, P. (2004). Long-term course of reflux symptoms following *Helicobacter pylori* eradication. Dtsch. Med. Wochenschr. *129*, 671–675.

Pellicano, R., Menard, A., Rizzetto, M., and Megraud F. (2008) Helicobacter species and liver diseases: association or causation? Lancet Infect Dis 8, 254–260.

Pellicano, R., Franceschi, F., Saracco, G., Fagoonee, S., Roccarina, D. and Gasbarrini, A. (2009). Helicobacters and extragastric diseases, Helicobacter *14* (*Suppl 1*), 58–68.

Pietroiusti, A., Giuliano, M., Magrini, A., Bergamaschi, A., and Galante, A. (2006). Cytotoxin-associated gene A strains of *Helicobacter pylori* represent a risk factor for the development of microalbuminuria in type 2 diabetes. Diabetes Care *29*, 1399–1401.

Pilotto, A., Di Mario, F., Franceschi, M., Battaglia, G., Germanà, B., Marin, R., and Valerio, G. (2000). Pantoprazole versus one-week *Helicobacter pylori* eradication therapy for the prevention of acute NSAID-related gastroduodenal damage in elderly subjects. Aliment. Pharmacol. Ther. *14*, 1077–1082.

Pirouz, T., Zounubi, L., Keivani, H., Rakhshani, N., and Hormazdi, M. (2009). Detection of *Helicobacter pylori* in paraffin-embedded specimens from patients with chronic liver diseases, using the amplification method. Dig. Dis. Sci. *54*, 1456–1459.

Pugliese, A., Beltramo, T., Todros, T., Cardaropoli, S., and Ponzetto, A. (2008). Interleukin-18 and gestosis: correlation with *Helicobacter pylori* seropositivity. Cell Biochem. Funct. *26*, 817–819.

Queiroz, D.M., Guerra, J.B., Rocha, G.A., Rocha, A.M., Santos, A., De Oliveira, A.G., Cabral, M.M., Nogueira, A.M., and De Oliveira, C.A. (2004). IL1B and IL1RN polymorphic genes and *Helicobacter pylori* cagA strains decrease the risk of reflux esophagitis. Gastroenterology *127*, 73–79.

Raderer, M., Streubel, B., Wöhrer, S., Häfner, M., and Chott, A. (2006). Successful antibiotic treatment of *H. pylori* negative gastric mucosa associated lymphoid tissue lymphomas. Gut *55*, 616–618.

Raghunath, A.S., Hungin, A.P., Wooff, D., and Childs, S. (2004). Systematic review: the effect of *Helicobacter pylori* and its eradication on gastro-oesophageal reflux disease in patients with duodenal ulcers or reflux oesophagitis. Aliment. Pharmacol. Ther. *20*, 733–744.

Rajkumar, S.V., Kyle, R.A., Plevak, M.F., Murray, J.A., and Therneau, T.M. (2002). *Helicobacter pylori* infection and monoclonal gammopathy of undetermined significance. Br. J. Haematol. *119*, 706–708.

Ramus, J.R., Gatenby, P.A.C., Caygill, C.P.J., and Watson A. (2007). *Helicobacter pylori* infection and severity of reflux-induced esophageal disease in a cohort of patients with columnar-lined esophagus. Dig. Dis. Sci. *52*, 2821–2825.

Reibman, J., Marmor, M., Filner, J., Fernandez-Beros, M.E., Rogers, L., Perez-Perez, G.I., and Blaser, M.J. (2008). Asthma is inversely associated with *Helicobacter pylori* status in an urban population. PLoS One 3, e4060.

Ren, J.S., Kamangar, F., Qiao, Y.L., Taylor, P.R., Liang, H., Dawsey, S.M., Liu, B., Fan, J.H., and Abnet, C.C. (2009). Serum pepsinogens and risk of gastric and oesophageal cancers in the general population nutrition intervention trial cohort. Gut *58*, 636–642.

Reszka, E, Jegier, B., Wasowicz, W., Lelonek, M., Banach, M., and Jaszewski, R. (2008). Detection of infectious agents by polymerase chain reaction in human aortic wall. Cardiovasc. Pathol. *17*, 297–302.

Rezaii, J., Tavakoli, H., Esfandiari, K., Asheg, H., Hasibi, M., Ghanei, G., Khosh-Batn, M., and Rashidi, A. (2008). Association between *Helicobacter pylori* infection and laryngohypopharyngeal carcinoma: a case control study and review of the literature. Head Neck *30*, 1624–1627.

Robertson, D.J., Sandler, R.S., Ahnen, D.J., Greenberg, E.R., Mott, L.A., Cole, B.F., and Baron, J.A. (2009). Gastrin, *Helicobacter pylori*, and colorectal adenomas. Clin. Gastroenterol. Hepatol. *7*, 163–167.

Robinson, K., Kenefeck, R., Pidgeon, E.L., Shakib, S., Patel, S., Poison, R.J., Zaitoun, A.M., and Atherton, J.C. (2008). *Helicobacter pylori*-induced peptic ulcer disease is associated with inadequate regulatory T cell responses. Gut *57*, 375–385.

Rokkas, T., Simsek, I., and Ladas, S. (2007). *Helicobacter pylori* and non-malignant diseases. Helicobacter *12* (*Suppl 1*), 20–22.

Roper, J., Francois, F., Shue, P.L., Mourad, M.S., Pei, Z., Olivares de Perez, A.Z., Perez-Perez, G.I., Tseng, C.H., and Blaser, M.J.(2008) Leptin and ghrelin in relation to *Helicobacter pylori* status in adult males. J. Clin. Endocrinol. Metab. *93*, 2350–2357.

Rosenstock, S., Jørgensen, T., Bonnevie, O., and Andersen, L. (2003). Risk factors for peptic ulcer disease: a population based prospective cohort study comprising 2416 Danish adults. Gut *52*, 186–193.

Sakuma, H., Nakamura, T., Uemura, N., Chiba, T., Sugiyama, T., Asaka, M., Akamatsu, T., Ueda, R., Eimoto, T., Goto, H., *et al.* (2007). Immunoglobulin VH gene analysis in gastric MALT lymphomas. Mod. Pathol. *20*, 460–466.

Salomaa-Räsänen, A., Kosunen, T.U., Aromaa, A.R., Knekt, P., Sarna, S., and Rautelin, H. (2010). A 'screen-and-treat' approach for *Helicobacter pylori* infection: a population-based study in Vammala, Finland. Helicobacter *15*, 28–37.

Sandven, I., Abdelnoor, M., Nesheim, B.I., and Melby, K.K. (2009). *Helicobacter pylori* infection and hyperemesis gravidarum: a systematic review and meta-analysis of case-control studies. Acta Obstet. Gynecol. Scand. *88*, 1190–1200.

Santos, I.S., Boccio, J., Davidsson, L., Hernandez- Triana, M., Huanca-Sardinas, E., Janjetic, M., Moya-Camarena, S.Y., Paez-Valery, M.C., Ruiz- Alvarez, V., Valencia, M.E., *et al.* (2009). *Helicobacter pylori* is not associated with anaemia in Latin America: results from Argentina, Brazil, Bolivia, Cuba, Mexico and Venezuela. Public Health Nutr. *12*, 1862–1870.

Sarker, S.A., Mahmud, H., Davidsson, L., Alam, N.H., Ahmed, T., Alam, N., Salam, M.A., Beglinger, C., Gyr, N., and Fuchs, G.J. (2008) Causal relationship of *Helicobacter pylori* with iron-deficiency anemia or failure of iron supplementation in children. Gastroenterology *135*, 1534–1542.

Sawayama, Y., Hamada, M., Otaguro, S., Maeda, S., Ohnishi, H., Fujimoto, Y., Taira, Y., and Hayashi, J. (2008). Chronic *Helicobacter pylori* infection is associated with peripheral arterial disease. J. Infect. Chemother. *14*, 250–254.

Scandellari, R., Allemand, E., Vettore, S., Plebani, M., Randi, M.L., and Fabris, F. (2009). Platelet response to *Helicobacter pylori* eradication therapy in adult chronic idiopathic thrombocytopenic purpura seems to be related to the presence of anticytotoxin-associated gene A antibodies. Blood Coagul. Fibrinolysis. *20*, 108–113.

Schaeverbeke, T., Broutet, N., Zerbib, F., Combe, B., Bertin, P., Lamouliatte, H., Perié, F., Joubert-Collin, M., and Mégraud, F. (2005). Should we eradicate *Helicobacter pylori* before prescribing an NSAID? Result of a placebo-controlled study. Am. J. Gastroenterol. *100*, 2637–2643.

Selgrad, M., Bornschein, J, Rokkas, T., and Malfertheiner, P. (2010). Clinical aspects of gastric cancer and *Helicobacter pylori*-screening, prevention, and treatment. Helicobacter *15* (*Suppl 1*), 40–45.

Seo, J.K., Ko, J.S., and Choi, K.D. (2002). Serum ferritin and *Helicobacter pylori* infection in children: a seroepidemiologic study in Korea. J. Gastroenterol. Hepatol. *17*, 754–757.

Seo, D.B., Kwon, K.S., Park, H.S., Lee, D.H., Kim, H.G., Shin, Y.W., Kim, Y.S., and Kim, J.M. (2007). Metachronous gastric MALT lymphoma and early gastric cancer: a case report. Korean J. Gastroenterol. *49*, 245–250.

Shibata, T., Arisawa, T., Tahara, T., Ohkubo, M., Yoshioka, D., Maruyama, N., Fujita, H., Kamiya, Y., Nakamura, M., Nagasaka, M., *et al.* (2009a). Selenoprotein S (SEPS1) gene −105G>A promoter polymorphism influences the susceptibility to gastric cancer in the Japanese population. BMC Gastroenterol. *9*, 2.

Shibata, T., Arisawa, T., Tahara, T., Yoshioka, D., Maruyama, N., Fujita, H., Kamiya, Y., Nakamura, M., Nagasaka, M., Iwata, M., *et al.* (2009b). Protective role of genetic polymorphism of heat shock protein 70-2 for gastric cancer risk. Dig. Dis. Sci. *54*, 70–74.

Shimatani, T., Moriwaki, M., Xu, J., Tazuma, S., and Inoue, M. (2006). Acidsuppressive effects of rabeprazole: comparing 10 mg and 20 mg twice daily in Japanese *Helicobacter pylori* -negative and -positive CYP2C19 extensive metabolisers. Dig. Liver Dis. *38*, 802–808.

Shirin, H., Leja, M., and Niv, Y. (2008). *Helicobacter pylori* and non-malignant diseases. Helicobacter *13* (*Suppl 1*), 23–27.

Siddiqui, N.R. (2009). *H. pylori*-induced higher C-reactive protein in obese African Americans. Artery Res. *3*, 39–42.

Sipponen, P., Kekki, M., Haapakoski, J., Ihamaki, T., and Siurala, M. (2002a). Gastric cancer risk in chronic atrophic gastritis: statistical calculations of cross-sectional data. Int. J. Cancer *35*, 173–177.

Sipponen, P., Ranta, P., Helske, T., Kääriäinen, I., Mäki, T., Linnala, A., Suovaniemi, O., Alanko, A., and Härkönen M. (2002b). Serum levels of amidated gastrin-17 and pepsinogen I in atrophic gastritis. An observational case-control study. Scand. J. Gastroenterol. *37*, 785–791.

Smith, B.R., and Stabile, B.E. (2005). Emerging trends in peptic ulcer disease and damage control surgery in the *H. pylori* era. Am. Surg. *71*, 797–801.

So, W.Y., Tong, P.C., Ko, G.T., Ma, R.C., Ozaki, R., Kong, A.P., Yang, X., Ho, C.S., Lam, C.C., and Chan, J.C. (2009). Low plasma adiponectin level, white blood cell count and *Helicobacter pylori* titre independently predict abnormal pancreatic betacell function. Diabetes Res. Clin. Pract. *86*, 89–95.

Somi, M.H., Fattahi, E., Fouladi, R.F., Karimi, M., Bonyadi, R., and Baballou, Z. (2008). An inverse relation between CagA+ strains of *Helicobacter pylori* infection and risk of erosive GERD. Saudi Med. J. *29*, 393–396.

Song, M.J., Park, D.I., Hwang, S.J., Kim, E.R., Kim, Y.H., Jang, B.I., Lee, S.H., Ji, J.S., and Shin, S.J. (2009). The prevalence of *Helicobacter pylori* infection in Korean patients with inflammatory bowel disease, a multicenter study. Korean J. Gastroenterol. *53*, 341–347.

Sonnenberg, A. (2007a). Time trends of ulcer mortality in Europe. Gastroenterology *132*, 2320–2327.

Sonnenberg, A. (2007b). Time trends of ulcer mortality in non-European countries. Am. J. Gastroenterol. *102*, 1101–1107.

Soylu, A., Ozkara, S., Alis, H., Dolay, K., Kalayci, M., Yasar, N., and Kumbasar, A.B. (2008). Immunohistochemical testing for *Helicobacter pylori* existence in neoplasms of the colon. BMC Gastroenterol. *8*, 35.

Stasi, R., Sarpatwari, A., Segal, J.B., Osborn, J., Evangelista, M.L., Cooper, N., Provan, D., Newland, A., Amadori, S., and Bussel, J.B. (2009). Effects of eradication of *Helicobacter pylori* infection in patients with immune thrombocytopenic purpura: a systematic review. Blood *113*, 1231–1240.

Stec-Michalska, K., Malicki, S., Michalski, B., Peczek, L., Wisniewska-Jarosinska, M., and Nawrot, B. (2009) Gastric ghrelin in relation to gender, stomach topography and *Helicobacter pylori* in dyspeptic patients. World J. Gastroenterol. *15*, 5409–5417.

Sterzl, I., Hrda, P., Matucha, P., Cerovska, J., and Zamrazil, V. (2008). Anti*Helicobacter pylori*, antithyroid peroxidase, antithyroglobulin and antigastric parietal cells antibodies in Czech population. Physiol. Res. *57 (Suppl 1)*, 135–141.

Sugimoto, T., Furukawa, T., Maeda, T., Somura, M., Uzu, T., and Kashiwagi A. (2007). Marked reduction of proteinuria after eradication of gastric *Helicobacter pylori* infection in a patient with membranous nephropathy: coincidental or associated? Intern. Med. *46*, 1483–1484.

Sumida, Y., Kanemasa, K., Yamaoka, Y., Imamura, S., Katoh, N., Nakashima, T., Tachiba, S., Mitsuyoshi, H., Itoh, Y., and Okanoue, T. (2006). Influence of *Helicobacter pylori* infection on iron accumulation in hepatitis C. Liver Int. *26*, 827–833.

Sung, J.J., Kuipers, E.J., and El-Serag, H.B. (2009). Systematic review: the global incidence and prevalence of peptic ulcer disease. Aliment. Pharmacol. Ther. *29*, 938–946.

Suzuki, T., Matsushima, M., Shirakura, K., Koike, J., Masui, A., Takagi, A., Shirasugi, Y., Ogawa, Y., Shirai, T., and Mine, T. (2008). Association of inflammatory cytokine gene polymorphisms with platelet recovery in idiopathic thrombocytopenic purpura patients after the eradication of *Helicobacter pylori*. Digestion *77*, 73–78.

Szklo, M., Ding, J., Tsai, M.Y., Cushman, M., Polak, J.F., Lima, J., Barr, R.G., and Sharrett, A.R. (2009). Individual pathogens, pathogen burden and markers of subclinical atherosclerosis: the multi-ethnic study of atherosclerosis. J. Cardiovasc. Med. (Hagerstown) *10*, 747–751.

Tahara, T., Arisawa, T., Wang, F., Shibata, T., Nakamura, M., Sakata, M., Hirata, I., and Nakano H. (2008). Toll-like receptor 2 (TLR)-196 to 174del polymorphism in gastro-duodenal diseases in Japanese population. Dig. Dis. Sci. *53*, 919–924.

Tahara, T., Arisawa, T., Shibata, T., Nakamura, M., Yamashita, H., Yoshioka, D., Okubo, M., Maruyama, N., Kamano, T., Kamiya, Y., *et al.* (2009). Effect of RANTES promoter genotype on the severity of intestinal metaplasia in *Helicobacter pylori*-infected Japanese subjects. Dig. Dis. Sci. *54*, 1247–1252.

Take, S., Mizuno, M., Ishiki, K., Nagahara, Y., Yoshida, T., Yokota, K, Oguma, K., Okada, H., and Yamamoto, K. (2009). *Helicobacter pylori* eradication may induce de novo, but transient and mild, reflux esophagitis: Prospective endoscopic evaluation. J. Gastroenterol. Hepatol. *24*, 107–113.

Takenaka, R., Okada, H., Kato, J., Makidono, C., Hori, S., Kawahara, Y., Miyoshi, M., Yumoto, E., Imagawa, A., Toyokawa, T., *et al.* (2007). *Helicobacter pylori* eradication reduced the incidence of gastric cancer, especially of the intestinal type. Aliment. Pharmacol. Ther. *25*, 805–812.

Talley, N.J., Fock, K.M., and Moayyedi, P. (2008). Gastric Cancer Consensus conference recommends *Helicobacter pylori* screening and treatment in asymptomatic persons from high-risk populations to prevent gastric cancer. Am. J. Gastroenterol. *103*, 510–514.

Tamer, G.S., Tengiz, I., Ercan, E., Duman, C., Alioglu, E., and Turku, O. (2009). *Helicobacter pylori* seropositivity in patients with acute coronary syndromes. Dig. Dis. Sci. *54*, 1253–1256.

Terai, S., Iijima, K., Kato, K., Suzuki, T., Yoshida, M., Koike, T., Kitagawa, Y., Imatani, A., Sekine, H., Ohara, S., *et al.* (2008) Long-term outcomes of gastric mucosa-associated lymphoid tissue lymphomas after *Helicobacter pylori* eradication therapy. Tohoku J. Exp. Med. *214*, 79–87.

Tokunaga, Y., Shirahase, H., Hoppou, T., Kitaoka, A., Tokuka, A., and Ohsumi, K. (2000). Density of *Helicobacter pylori* infection evaluated semiquantitatively in gastric cancer. J. Clin. Gastroenterol. *31*, 217–221.

Toros, S.Z., Toros, A.B., Yuksel, O.D., Ozel, L., Akkaynak, C., and Naiboglu, B. (2009). Association of laryngopopharyngeal manifestations and gastroesophageal reflux. Eur. Arch. Otorhinolaryngol. *266*, 403–409.

Treepongkaruna, S., Sirachainan, N., Kanjanapongkul, S., Winaichatsak, A., Sirithorn, S., Sumritsopak, R., and Chuansumrit, A. (2009). Absence of platelet recovery following *Helicobacter pylori* eradication in childhood chronic idiopathic thrombocytopenic purpura: a multi-center randomized controlled trial. Pediatr. Blood Cancer *53*, 72–77.

Trejo-de la, O.A., Torres, J., Perez-Rodriguez, M., Camorlinga-Ponce, M., Luna, L.F., Abdo-Francis, J.M., Lazcano, E., and Maldonado-Bernal, C. (2008). TLR4 singlenucleotide polymorphisms alter mucosal cytokine and chemokine patterns in Mexican patients with *Helicobacter pylori*-associated gastroduodenal diseases. Clin. Immunol. *129*, 333–340.

Tripathi, S., Ghoshal, U., Ghoshal, U., Mittal, B., Krishnani, N., Chourasia, D., Agasrwal, A., and Singh, K. (2008). Gastric carcinogenesis: possible role of polymorphisms of GSTM1, GSTT1, and GSTP1 genes. Scand. J. Gastroenterol. *43*, 431–439.

Tu, S., Bhagat, G., Cui, G., Takaishi, S., Kurt-Jones, E.A., Rickman, B., Betz, K.S., Penz-Oesterreicher, M., Bjorkdahl, O., Fox, J.G., *et al.* (2008). Overexpression of interleukin-1beta induces gastric inflammation and cancer and mobilizes myeloid-derived suppressor cells in mice. Cancer Cell *14*, 408–419.

Tummala, S., Keates, S., and Kelly, C.P. (2004). Update on the immunologic basis of *Helicobacter pylori* gastritis. Curr. Opin. Gastroenterol. *20*, 592–597.

Tunca, A., Ardicoglu, Y., Kargili, A., and Adam, B. (2007). Migraine, *Helicobacter pylori*, and oxidative stress. Helicobacter *12*, 59–62.

Uemura, N., Okamoto, S., Yamamoto, S., Matsumura, N., Yamaguchi, S., Yamakido, M., Taniyama, K., Sasaki, N., and Schlemper, R.J. (2001). *Helicobacter pylori* infection and the development of gastric cancer. N. Engl. J. Med. *345*, 784–789.

Vaira, D., Menegatti, M., and Miglioli, M. (1997). What is the role of *Helicobacter pylori* in complicated ulcer disease? Gastroenterology *113*, S78–S84.

Vakil, N.B., Traxler, B.M., and Levine, D. (2004). Symptom response and healing of erosive esophagitis with proton-pump inhibitors in patients with *Helicobacter pylori* infection. Am. J. Gastroenterol. *99*, 1437–1441.

Vakil, N., Talley, N.J., Stolte, M., Sundin, M., Junghard, O., and Bolling-Sternevald, E. (2006). Patterns of gastritis and the effect of eradicating *Helicobacter pylori* on gastro-oesophageal reflux disease in Western patients with non-ulcer dyspepsia. Aliment. Pharmacol. Ther. *24*, 55–63.

Vayisoglu, Y., Ozcan, C., Polat, A., Delialioglu, N., and Gorur, K. (2008). Does *Helicobacter pylori* play a role in the development of chronic adenotonsillitis? Int. J. Pediatr. Otorhinolaryngol. *72*, 1497–1501.

Venerito, M., Treiber, G., Wex, T., Kuester, D., Roessner, A., Di Mario, F., and Malfertheiner, P. (2006). Effects of low-dose aspirin on gastric erosions, cyclooxygenase expression and mucosal prostaglandin-E2 do not depend on *Helicobacter pylori* infection. Aliment. Pharmacol. Ther. *23*, 1225–1233.

Venerito, M., and Malfertheiner, P. (2010). Interaction of *Helicobacter pylori* infection and nonsteroidal anti-inflammatory drugs in gastric and duodenal ulcers. Helicobacter *15*, 239–250.

Venizelos, I., Tamiolakis, D., Bolioti, S., and Nikolaidou, S. (2007). Primary MALT lymphomas of the stomach: a pathological study of 18 cases. Rev. Esp. Enferm. Dig. *99*, 270–274.

Vergara, M., Catalan, M., Gisbert, J.P., and Calvet, X. (2005). Meta-analysis: role of *Helicobacter pylori* eradication in the prevention of peptic ulcer in NSAID users. Aliment. Pharmacol. Ther. *21*, 1411–1418.

Vivekanandan, P., and Torbenson, M. (2008). Low frequency of *Helicobacter* DNA in benign and malignant liver tissues from Baltimore, United States. Hum. Pathol. *39*, 213–216.

Vladimirov, B., Asenova, B., Kaparov, Al., Chetirska, M., Marinov, A., Terziev, I., and Ivanova, R. (2007). Noninvasive diagnosis of atrophic gastritis with Bohit Gastropanel® – preliminary results. Hepatogastroenterology *9*, 39–46.

de Vries, A.C., Van Driel, H.F., Richardus, J.H., Ouwendijk, M., Van Vuuren, A.J., De Man, R.A., and Kuipers, E.J. (2008a). Migrant communities constitute a possible target population for primary prevention of *Helicobacter pylori*-related complications in low incidence countries. Scand. J. Gastroenterol. *43*, 403–409.

de Vries, A.C., van Grieken, N.C., Looman, C.W., Casparie, M.K., de Vries, E., Meijer, G.A., and Kuipers, E.J. (2008b). Gastric cancer risk in patients with premalignant gastric lesions: a nationwide cohort study in the Netherlands. Gastroenterology *134*, 945–952.

de Vries, A.C., Kuipers, E.J., and Rauws, E.A. (2009) *Helicobacter pylori* eradication and gastric cancer: when is the horse out of the barn? Am J Gastroenterol *104*, 1342–1345.

de Vries, A.C., and Kuipers, E.J. (2010). *Helicobacter pylori* infection and nonmalignant diseases. Helicobacter *15 (Suppl 1)*, 29–33.

Vu, C., and Ng, Y.Y. (2000). Prevalence of *Helicobacter pylori* in peptic ulcer disease in a Singapore hospital. Singapore Med. J. *41*, 478–481.

Wang, S.Z., Shi, Y.N., Zhao, J., and Wang, Z.D. (2009). Effects of *Helicobacter pylori* on blood glucose fluctuation in type 2 diabetic patients. Zhonghua Yi Xue Za Zhi *89*, 958–961.

Watari, J., Das, K.K., Amenta, P.S., Tanabe, H., Tanaka, A., Geng, X., Lin,J.J., Kohgo, Y., and Das, K.M. (2008). Effect of eradication of *Helicobacter pylori* on the histology and cellular phenotype of gastric intestinal metaplasia. Clin. Gastroenterol. Hepatol. *6*, 409–417.

Wei, J., O'Brien, D., Vilgelm, A., Piazuelo, M.B., Correa, P., Washington, M.K., El-Rifai, W., Peek, R.M., and Zaika, A. (2008). Interaction of *Helicobacter pylori* with gastric epithelial cells is mediated by the p53 protein family. Gastroenterology *134*, 1412–1423.

Weston, A.P., Sharma, P., Mathur, S., Banerjee, S., Jafri, A.K., Cherian, R., McGregor, D., Hassanein, R.S., and Hall, M. (2004). Risk stratification of Barrett's esophagus:updated prospective multivariate analysis. Am. J. Gastroenterol. *99*, 1657–1666.

Wong, B.C., Lam, S.K., Wong, W.M., Chen, J.S., Zheng, T.T., Feng, R.E., Lai, K.C., Hu, W.H., Yuen, S.T., Leung, S.Y., *et al.* (2004) *Helicobacter pylori* eradication to prevent gastric cancer in a high-risk region of China: a randomized controlled trial. JAMA *291*, 187–194.

Wu, C.Y., Wu, M.S., Chen, C.J., Li, M.C., Lin, J.T., and Chen GH. (2005). The interaction of *H. pylori* infection and NSAIDs in cyclooxygenase-2 mRNA expression in gastric antral, corpus mucosa, and gastric ulcer. J. Clin. Gastroenterol. *39*, 50–55.

Wu, J.C. (2008). Gastroesophageal reflux disease: an Asian perspective. J. Gastroenterol. Hepatol. *23*, 1785–1793.

Wu, C.Y., Kuo, K.N., Wu, M.S., Chen, Y.J., Wang, C.B., and Lin, J.T. (2009a). Early *Helicobacter pylori* eradication decreases risk of gastric cancer in patients with peptic ulcer disease. Gastroenterology *137*, 1641–1648.

Wu, C.Y., Wu, C.H., Wu, M.S., Wang, C.B., Cheng, J.S., Kuo, K.N., and Lin, J.T. (2009b). A nationwide population-based cohort study shows reduced hospitalization for peptic ulcer disease associated with *H. pylori* eradication and proton pump inhibitor use. Clin. Gastroenterol. Hepatol. *7*, 427–431.

Yadav, M.K., Rishi, J.P., and Nijawan, S. (2008). Chronic urticaria and *Helicobacter pylori*. Indian J. Med. Sci. *62*, 157–162.

Yaghoobi, M., Farrokhyar, F., Yuan, Y., and Hunt, R.H. (2010). Is there an increased risk of GERD after *Helicobacter pylori* eradication?: a meta-analysis. Am. J. Gastroenterol. *105*, 1007–1013.

Yahav, J., Samra, Z., Blau, H., Dinari, G., Chodick, G., and Shmuely, H. (2006). *Helicobacter pylori* and Clostridium difficile in cystic fibrosis patients. Dig. Dis. Sci. *51*, 2274–2279.

Yamaji Y, Watabe Ft, Yoshida H, Kawabe T, Wada R, Mitsushima T., and Omata, M. (2009). High-risk population for gastric cancer development based on serum pepsinogen status and lifestyle factors. Helicobacter *14*, 81–86.

Yamaoka, Y., Kato, M., and Asaka, M. (2008). Geographic differences in gastric cancer incidence can be explained by differences between *Helicobacter pylori* strains. Intern. Med. *47*, 1077–1083.

Yanaoka, K., Oka, M., Ohata, H., Yoshimura, N., Deguchi, H., Mukoubayashi, C., Enomoto, S., Inoue, I., Iguchi, M., Maekita, T., *et al.* (2009). Eradication of *Helicobacter pylori* prevents cancer development in subjects with mild gastric atrophy identified by serum pepsinogen levels. Int. J. Cancer *125*, 2697–2703.

Yang, C.S., Cao, S.Y., He, X.J., Wang, Y.X., and Zhang, Y.L. (2007). Study of correlation between *Helicobacter pylori* infection and hyperammonemia and hepatic encephalopathy in cirrhotic patients. Zhongguo Wei Zhong Bing Ji Jiu Yi Xue *19*, 422–424.

Yavasoglu, I, Kucuk, M., Cildag, B., Arslan, E., Gok, M., and Kafkas, S. (2009). A novel association between polycystic ovary syndrome and *Helicobacter pylori.* Am. J. Med. Sci. *338*, 174–177.

Ye, W., Held, M., Lagergren, J., Engstrand, L., Blot, W.J., McLaughlin, J.K., and Nyren, O. (2004). *H. pylori* infection and gastric atrophy: risk of adenocarcinoma and squamous-cell carcinoma of the esophagus and adenocarcinoma of the gastric cardia. J. Natl. Cancer Inst. *96*, 388–396.

Yeomans, N.D., Lanas, A.I., Talley, N.J., Daneshjoo, R., Eriksson, B., Appelman-Eszczuk, S., Långström, G., Naesdal, J., Serrano, P., Singh, M., *et al.* (2005). Prevalence and incidence of gastroduodenal ulcers during treatment with vascular protective doses of aspirin. Aliment. Pharmacol. Ther. *22*, 795–801.

Yiannopoulou, K.G., Efthymiou, A., Karydakis, K., Arhimandritis, A., Bovaretos, N., and Tzivras, M. (2007). *Helicobacter pylori* infection as an environmental risk factor for migraine without aura. J. Headache Pain *8*, 329–333.

Yokota, S., Konno, M., Mino, E., Sato, K., Takahashi, M., and Fujii, N. (2008). Enhanced Fe ion-uptake activity in *Helicobacter pylori* strains isolated from patients with iron-deficiency anemia. Clin. Infect. Dis. *46*, e31-e33.

Yoshikawa, H., Aida, K., Mori, A., Muto, S., and Fukuda, T. (2007). Involvement of *Helicobacter pylori* infection and impaired glucose metabolism in the increase of brachial-ankle pulse wave velocity. Helicobacter *12*, 559–566.

Zhang, S., Guo, Y., Ma, Y., and Teng, Y. (2008). Cytotoxin-associated gene-A-seropositive virulent strains of *Helicobacter pylori* and atherosclerotic diseases: a systematic review. Chin. Med. J. (Engl.) *121*, 946–951.

Zhang, Y., Jin, M., Liu, B., Ma, X., Kaiyan, Y., Li, Q., and Chen, K. (2008). Association between H-RAS T81C genetic polymorphism and gastrointestinal cancer risk: a population based case-control study in China. BMC Cancer *8*, 256–263.

Zhou, Y., Li, N., Zhuang, W., Liu, G.J., Wu, T.X., Yao, X., Du, L., Wei, M.L., and Wu, X.T. (2007). P53 codon 72 polymorphism and gastric cancer: a meta-analysis of the literature. Int. J. Cancer *121*, 1481–1486.

Zhuo, X.L., Wang, Y., Zhuo, W.L., and Zhang, X.Y. (2008). Possible association of *Helicobacter pylori* infection with laryngeal cancer risk: an evidence-based meta-analysis. Arch. Med. Res. 39, 625–628.

Zhuo, W.L., Zhu, B., Xiang, Z.L., Zhuo, X.L., Cai, L., and Chen, Z.T. (2009). Assessment of the relationship between *Helicobacter pylori* and lung cancer: a meta-analysis. Arch. Med. Res. *40*, 406–410.

Zycinska, K., Wardyn, K.A., Zycinski, Z., and Smolarczyk, R. (2008). Correlation between *Helicobacter pylori* infection and pulmonary Wegener's granulomatosis activity. J. Physiol. Pharmacol. *6*, 845–851.

Helicobacter pylori Resistance to Antibiotics

9

Lyudmila Boyanova

Abstract

H. pylori resistance to antibiotics emerges most often from point mutations but also from efflux mechanisms, natural transformation, altered membrane permeability and, probably, β-lactamase. The resistance, especially that to clarithromycin and quinolones, often causes treatment failures. For this reason, if national or regional primary resistance rates are ≥15–20% for clarithromycin and ≥40% for metronidazole, the agents should be avoided for primary therapy of the infection unless susceptibility testing of the strains is carried out. Clarithromycin resistance-associated A2143G (A2147G) mutation perhaps predicts eradication failures. Moreover, heteroresistance in *H. pylori* strains has been reported for metronidazole, clarithromycin, amoxicillin and quinolones. From <10% to >37% of the strains exhibit mixtures of genotypes. Importantly, both clarithromycin and quinolone resistance rates have grown sharply in many countries and multidrug resistance has been found in <5% in Europe and >14% in Brazil and South Korea. High primary resistance rates to clarithromycin (20% to >40%) and fluoroquinolones (20% to >33%) have been reported mostly in developed countries. Conversely, high primary resistance to metronidazole (≥76%), amoxicillin (6% to >30%) and tetracycline (≥15%) has been observed in some developing countries. Primary resistance and its evolution often depend on the country and national antibiotic consumption, patient characteristics such as age, sex, disease, prior antibiotic use and comorbidity, strain characteristics such as virulence as well as other factors. Post-treatment resistance rates have been usually much higher, often more than three times higher for clarithromycin and both clarithromycin and metronidazole and >1.5 times higher for metronidazole and quinolones, compared with those of the primary resistance. In brief, worrying evolution of antibiotic resistance in *H. pylori* and disturbing multidrug resistance increasingly hamper the success of the eradication of the infection. Knowledge on current *H. pylori* resistance patterns and evolution at global and local levels is highly important to show the efficacy or need for changes in treatment regimens and to improve the overall eradication success that also means the cure of the individual patient.

Introduction

The criterion for treatment success on *H. pylori* infection differs from those for most other infections, which require control only several days after the end of antibiotic use as in the case of urinary tract infections. Eradication of *H. pylori* infection has been defined as lack of the infection for over one month after the end of anti-*H. pylori* therapy. The eradication of the infection is curative, healing peptic ulcers and over half of mucosa-associated lymphoid tissue (MALT) lymphoma cases and preventing development of gastric cancer (Malfertheiner *et al.*, 2007). It is necessary to analyse the eradication not only by per-protocol (PP) analysis i.e. in patients who have completed the treatment course but also by 'intention to treat' (ITT) analysis (in all treated patients) to consider all possible disadvantages of the therapeutic regimen (see Chapter 10).

However, current eradication rates by triple and quadruple regimens have been low (≤70%) in some studies (see Chapter 10). After several treatment failures, the choice of rescue regimens (proton pump inhibitor-PPI, amoxicillin and levofloxacin or furazolidone, or rifabutin), should be based on the strain susceptibility (Malfertheiner *et al.*, 2007). Presently, acceptable eradication rates can be achieved by sequential therapy, four-drug, three-antibiotic concomitant therapy or bismuth-containing quadruple therapy (Graham and Shiotani, 2008). In some studies, adding probiotics to the regimens has improved the eradication and has diminished the side-effects of regimens (see Chapter 9). Graham *et al.* (2008) have stressed the need to improve *H. pylori* eradication to reach >95% rate. Briefly, many attempts have been made to improve the anti-*H. pylori* therapy.

H. pylori resistance to antibiotics is the leading cause of eradication failures. Clarithromycin resistance has been shown to reduce sharply the eradication rate, by 70% with PPI–clarithromycin–amoxicillin regimen and by 47% with PPI-clarithromycin-metronidazole regimen (Megraud and Lehours, 2007). Likewise, fluoroquinolone resistance in *H. pylori* has decreased the eradication success of quinolone-based regimens by 41.7–66.7% compared with that of susceptible strains (Perna *et al.*, 2007; Nishizawa *et al.*, 2009). Conversely, metronidazole resistance has diminished the eradication rates of triple regimens only slightly (by 25%; Megraud, 2004). Impact of amoxicillin resistance on *H. pylori* eradication is controversial. All amoxicillin-resistant strains have been eradicated in South Korea (Hwang *et al.*, 2010); however, in a Japanese study, triple amoxicillin-based therapy has eradicated only 25% of amoxicillin-resistant strains versus 84.6% of the susceptible strains (Watanabe *et al.*, 2005). In addition to the antibiotic resistance, poor patient compliance and host polymorphism of some genes also influence the treatment success. Patients with CYP2C19 RM and IL-1B-511 C/C genotype have exhibited low PPI bioavailability and high gastric acid release and hence weak acid inhibition by PPIs in the eradication regimens (Sugimoto *et al.*, 2009). Other factors associated with eradication failure have been high gastric acidity, high *H. pylori* density, *cagA* negative strains, existence of intracellular bacteria and patient immunosuppression (Mégraud, 2010).

Importantly, many susceptibility-guided regimens have shown better eradication rates than the empiric treatment (Lamouliatte *et al.*, 2003; Megraud, 2004; Kawai *et al.*, 2008). Lamouliatte *et al.* (2003) have reported eradication rates by empirical regimens such as PPI–amoxicillin–clarithromycin (34.5–47.4%) or PPI-amoxicillin-metronidazole (63.2%) to be inferior to those by susceptibility-based regimens (74.3%). In another study, the pre-treatment clarithromycin susceptibility testing has increased the eradication rates by 22.9% compared with empiric treatment (Kawai *et al.*, 2008). Raymond *et al.* (2010) have suggested extensive pre-treatment susceptibility testing of *H. pylori* to clarithromycin in countries with high macrolide resistance in *H. pylori* such as France.

H. pylori is intrinsically resistant to glycopeptides, cefsulodin, polymyxins, nalidixic acid, trimethoprim, sulfonamides and antifungal agents (Megraud and Lehours, 2007). Many of these drugs are used in *H. pylori* selective media. However, although *H. pylori* is *in vitro* susceptible to numerous antibiotics, only few are suitable for its eradication as a result of their minimal inhibitory concentrations (MICs) and pharmacokinetics, including local or systemic activity and inhibition by gastric pH. For instance, unlike clarithromycin MIC_{90}, that of erythromycin has been shown to rise much at pH 5.5 (Megraud and Lehours, 2007). In a chemostat (thereby with longer generation time of the bacteria than that *in vitro*), the bactericidal activity of anti-*H. pylori* drugs has been reduced (Megraud and Lehours, 2007). Moreover, the specimen source can affect the detection of resistance. About one-third of Italian children with mixed infections have harboured clarithromycin-resistant strains in the gastric fundus but not in the antrum (Caristo *et al.*, 2008). In South Korea, antibiotic resistance in only one of the several gastric sites has been found in 16.4% of the strains from 67 unsucessfully treated patients (Kim *et al.*, 2006a). Susceptibility testing methods are described in Chapter 3. In fact, both susceptibility of *H. pylori* to antibiotics and treatment of the associated infection are specific owing to the single natural niche of the bacteria and their close adaptation to the host.

Resistance mechanisms

H. pylori resistance often emerges from point mutations in the bacterial DNA by vertical (to daughter cells) transmission and is selected by unsuitable or frequent antibiotic use. Mutations, responsible for the resistance can occur in *rrl* gene, encoding the 23S rRNA (for macrolides), *rdxA* and *frxA* (for metronidazole, although with controversial importance), *gyrA* (for quinolones), *rpoB* (for rifampin), *pbp1* (for amoxicillin) and 16S RNA (for tetracycline) (Megraud and Lehours, 2007). Transmission of resistance markers between *H. pylori* strains is also possible by natural transformation in cases of mixed *H. pylori* infections. Mixed infections have been reported in 0–55% of patients, often in countries with high infection prevalence (Ghose *et al.*, 2005). Most *H. pylori* strains are naturally competent for transformation, which is assisted by proteins encoded by the *comB* locus (Lin *et al.*, 2009). The natural transformation includes integration of short (1300 bp) asymmetric DNA fragments showing gaps at 3–9 kb regions, so, recombinations can occur simultaneously within the same locus (Lin *et al.*, 2009). Strikingly, conjugation-like transfer of chromosomally encoded antibiotic (streptomycin) resistance marker from *H. pylori* to *Campylobacter jejuni* has been reported (Oyarzabal *et al.*, 2007). This raises the question about the role of inter-species transfer of the resistance markers between *H. pylori* and other *Helicobacter* species or other closely related species.

Importantly, heteroresistance in *H. pylori* strains has been reported for amoxicillin, clarithromycin, metronidazole and quinolones (Bogaerts *et al.*, 2006; Matteo *et al.*, 2008). Heteroresistance can be due to mixed infections or coexistence of susceptible and resistant variants in the same strain. In a study of Cambau *et al.* (2009), one third of French strains has shown mixed genotypes, e.g. wild type + clarithromycin-resistant *rrl* alleles (10%), several resistant *rrl* alleles (5%), wild type + levofloxacin-resistant *gyrA* alleles (12%) and several wild-type *gyrA* alleles (18.5%). Similarly, heteroresistance to clarithromycin has been found in 26.7–63.2% of untreated Italian patients (De Francesco *et al.*, 2007). The heteroresistance can be unrecognized by some molecular techniques such as RFLP (see Chapter 3) or by phenotypic susceptibility testing methods, performed with low bacterial inocula, (see below) and thus can compromise the success of the eradication of the infection.

In addition, efflux pumps can contribute to clarithromycin, tetracycline, metronidazole and multidrug resistance in *H. pylori* (van Amsterdam *et al.*, 2005; Li and Dannelly, 2006; Hirata *et al.*, 2010; Zhang *et al.*, 2010). Antibiotic MICs and accumulation have been assessed by agar dilution in presence/absence of carbonyl cyanide m-chlorophenylhydrazone (CCCP), an inhibitor of proton motive force, and fluorescence of accumulated drug by spectrofluorometry (Anoushiravani *et al.*, 2009; Zhang *et al.*, 2010). Some non-specific proteins and HefABC-resistance-nodulation-division efflux pump, including outer membrane efflux proteins encoded by *hefA* and *hefC* genes, have been linked to multidrug resistance in *H. pylori* (Kutschke and de Jonge, 2005; Liu *et al.*, 2008; Zhang *et al.*, 2010). Falsafi *et al.* (2009) have detected efflux of many antibiotics (β-lactams, metronidazole, tetracycline and ciprofloxacin) in five multidrug-resistant clinical strains. Importantly, *H. pylori* efflux pumps can be inhibited by PPIs, most strongly by rabeprazole, which has decreased four times MICs of metronidazole and amoxicillin in the multidrug-resistant strains (Zhang *et al.*, 2010). As *H. pylori* efflux pump systems, overexpressed by prolonged antibiotic treatment, can promote multidrug resistance, the length of treatment should be chosen with caution. Interestingly, internalization of resistant *H. pylori* strains has been better than that of the susceptible strains (Lai *et al.*, 2006). The internalization of the strains can protect them from the activity of the antibiotics. It is known that MFD protein, encoded by *mfd* gene, participates in both *H. pylori* DNA repair and antibiotic resistance (Lee *et al.*, 2009). Moreover, β-lactamase has been found in amoxicillin-resistant *H. pylori* strain (see below).

Macrolides

Macrolides, bacteriostatic agents, including clarithromycin, bind to ribosomal peptidyl transferase region of 23S rRNA, encoded by the *rrl* gene, and suppress bacterial protein synthesis (Megraud and Lehours, 2007). Clarithromycin has the advantage to be the most active single drug against *H. pylori* with good diffusion in gastric tissues and synergism with acid suppressive agents and the disadvantages of growing resistance in many countries and considerable cost (De Francesco *et al.*, 2007; Cambau *et al.*, 2009).

H. pylori resistance to clarithromycin most often arises from point mutations in the peptidyltransferase region of the 23S rRNA gene. Frequency of spontaneous mutation in *H. pylori* by subinhibitory macrolide concentrations *in vitro* has been high (2×10^{-5} to 3.5×10^{-8} per cell division; Cantón *et al.*, 2001). Macrolide-resistant strains usually exhibit point mutations at two nucleotide positions: adenine to guanine transition at either position 2142 (also reported as 2146) or 2143 (2147, including A2142G, and A2143G), or adenine to cytosine transversion at position 2142 (A2142C) in 23S rRNA domain V of the 23S rRNA gene (Megraud and Lehours, 2007; Cambau *et al.*, 2009). The above-mentioned three mutations have been linked to most (90%) of the clarithromycin-resistant strains in Western countries. The point mutations generate conformational changes in the ribosome and ensuing low macrolide binding (Megraud and Lehours, 2007; Cambau *et al.*, 2009). Two copies of 23S rRNA gene are present in *H. pylori* DNA and mutations in either one or both copies can promote clarithromycin resistance (Moder *et al.*, 2007). The same mutations often affect both copies of the gene (Soltermann *et al.*, 2005; Slinger *et al.*, 2009). Importantly, the resistance usually affects all macrolides. Efflux pumps also can contribute to *H. pylori* resistance to clarithromycin. In 15 clarithromycin-resistant strains from Japan, both efflux pump and 23S rRNA point mutations have been linked to the resistance (Hirata *et al.*, 2010).

The most common mutation in *H. pylori* for clarithromycin resistance has been A2143G (0–100%, often >60%), followed by A2142G (0–87%, often <24%) and A2142C (0–20%, often <5%; Alarcon *et al.*, 2000; Megraud, 2004; Raymond *et al.*, 2007; Table 9.1). The last mutation seems to be underrated since it has not been evaluated in many studies. A2142G mutation has been linked to 1–8 times (often 2–8 times) higher clarithromycin MICs compared with A2143G mutation in most studies (Alarcon *et al.*, 2000; Ribeiro *et al.*, 2003; Vega *et al.*, 2003; Furuta *et al.*, 2007), while no such difference has been found by other authors (Moder *et al.*, 2007; Raymond *et al.*, 2007). In Spain, the mutations have been associated with clarithromycin MICs of 4–8 µg/ml for A2143G, 16–64 µg/ml for A2142G and 8–64 µg/ml for A2142C mutation (Alarcon *et al.*, 2000).

However, despite the lower macrolide MICs by A2143G- compared with A2142G mutation, it is important to stress that mainly A2143G mutation is a risk factor for treatment failure. Francavilla *et al.* (2010) have obtained *H. pylori* eradication by 7-day clarithromycin-based triple therapy in 50% of Italian children with strains showing A2143G mutation versus 89% in those without the mutation; and only the sequential therapy has overcome the resistance in 80% of the patients. In South Korea, six strains detected with A2143G mutations have not been eradicated by triple clarithromycin-based therapy (Hwang *et al.*, 2010).

Importantly, phenotypically expressed A2143G point mutation has been shown to predict the lowest eradication rate. De Francesco *et al.* (2010) have found low (30.7%) eradication rate in association with phenotypic *H. pylori* resistance (by E test) owing to A2143G mutation versus >80% by phenotypic susceptibility in strains with genotypic resistance. Therefore, the detection of A2143G point mutation can be clinically important for the treatment of the infection and should be proven by both phenotypic and polymerase chain reaction (PCR)-based methods.

Both clarithromycin-resistant and wild type sequences have been found in <5–>35% of *H. pylori* strains (Table 9.1). It should be stressed that mixed populations, even with low proportion of mutants, have not been eradicated in a study of Soltermann *et al.* (2005). Subpopulations with different clarithromycin resistance mutations also may be found in the same gastric biopsy (Raymond *et al.*, 2007). A2143G mutation, mixed with other mutations, has been rare (2.1–15.0%) in France, Japan and Brazil but common in Taiwan (44.4%;

Table 9.1 Prevalence of mutations in clarithromycin-resistant *H. pylori* strains in some studies

Region/ country	No. of strains	Method	Mutations (%)					Reference
			A2143G (A2147G)	A2142G (A2146G)	A2142C (A2146C)	Other	None	
Europe								
Germany	23	PyroSq	39.1 (alone)	30.4 (26.1 alone, 4.3 TMs)	4.3 (single copy)	30.4 (T2182C: 26.1 alone,[1] 4.3 TMs)	NA	Moder *et al.* (2007)
France	24	Scorpion PCR	66.7 (62.5 alone, 4.2 R+S)	25.0 (alone)	8.3 (R+S)	NA	NA	Burucoa *et al.* (2008)
France (untreated)	104	Scorpion PCR	79.8 (61.5 alone, 18.2 R+S)	17.3 (14.4 alone, 2.9 R+S)	2.9 (1.9 alone, 0.9 R+S)	0	0	Raymond *et al.* (2010)
France (treated)	34	Scorpion PCR	88.2 (82.3 alone, 5.9 R+S)	5.9 (2.9 alone, 2.9 R+S)	5.9 (2.9 alone, 2.9 R+S)	0	0	Raymond *et al.* (2010)
France	48	Genotype HelicoDR	72.9 (52.1 alone, 10.4 R+S, 4.2 TMs+S, 6.2 TMs)	25 (12.5 alone, 4.2 R+S, 2.1 TMs+S, 6.2 TMs)	2.1 (TMs+S)	NA	8.3	Cambau *et al.* (2009)
Italy	38	RT-PCR, Sq	71	15.7	13.3	0	0	De Francesco *et al.* (2007)
Italy	12	Sq, RA	0	0	0	58.3 (T2717C[1])	41.7	Fontana *et al.* (2002)
Poland	80	RFLP, Sq, mmPCR	62.5	24	7.5	2.5 (T2182C[1]), 1.2 (C2131T)	2.5	Dzierzanowska-Fangrat *et al.* (2005)
Spain	25	RFLP and mmPCR	40	40	20	0	0	Alarcon *et al.* (2000)
Spain (children)	28	RFLP	82.1	7.1	NA	NA	10.7	Alarcon *et al.* (2003)
Spain (children)	36	RFLP	80.6	5.6	NA	NA	13.9	Vega *et al.* (2003)
Spain (adults)	30	RFLP	46.7	36.7	NA	NA	16.7	Vega *et al.* (2003)
Spain	42	Sq	85.3	8.8	NA	5.9 (T2182C[1])	11.9	Agudo *et al.* (2010)
Switzerland (total)	27	RFLP, Sq, LpH	51.8 (alone) and mixed mutations	14.8 (alone) and mixed mutations	11.1	0	0	Soltermann *et al.* (2005)
North America								
Canada	24	PyroSq	54.2 (double-copy), 12.5 (single-copy)	16.7 (double-copy)	NA	NA	16.7	Slinger *et al.* (2009)

Table 9.1 continued

Region/ country	No. of strains	Method	Mutations (%) A2143G (A2147G)	A2142G (A2146G)	A2142C (A2146C)	Other	None	Reference
South America								
Brazil	20	RFLP	75 (60 alone, 15 TMs)	30 (15 alone)	NA	NA	10	Prazeres Magalhães *et al.* (2002)
Brazil	52	RFLP, Sq	11.5	82.7	0	5.8 (T2245C[1])	0	Ribeiro *et al.* (2003)
Asia								
China	16	RFLP, RAPD	100	0	0	0	0	Zheng *et al.* (2009)
Japan	93	AS-PCR	96.7, (59.1 alone, 35.5 R+S, 2.1 TMs)	5.3, (1.1 alone, 2.1 R+S, 2.1 TMs)	NA	NA	NA	Furuta *et al.* (2007)
South Korea	23	RFLP, Sq	13	87	0	0	0	Lee *et al.* (2005)
South Korea (untreated)	21	Sq	71.4	0	0	28.6 (T2182C[1])	0	Kim *et al.* (2008)
South Korea (treated)	64	Sq	7.8 alone, 70.3 TMs, 14.1 triple mutations	0	0	7.8 (T2182C[1] alone, 70.3 TMs, 14.1 triple mutations)	0	Kim *et al.* (2008)
Taiwan	18	Sq	61.1, (11.1 alone, 5.6 R+S, 44.4 TMs)	0	0	33.3 (T2182C[1]), 5.6 (C2227T)	0	Hu *et al.* (2009)
Africa								
Tunisia	42	PCR	88.1	11.9	0	0	0	Ben Mansour *et al.* (2010)

[1]Not associated with clarithromycin resistance in some studies.

Sq, sequencing; RA, restriction analysis; RT-PCR, real-time polymerase chain reaction; RLFP, random fragment length polymorphism; mmPCR-3′, mismatched PCR; AS-PCR, allele-specific PCR; LpH, line probe hybridization; RAPD, random amplified polymorphic DNA; R+S, resistant and susceptible sequences; TMs, two different mutations; TMs+S, two different mutations and susceptible sequence; NA, not available or not applicable.

Prazeres Magalhaes *et al.*, 2002; Furuta *et al.*, 2007; Cambau *et al.*, 2009; Hu *et al.*, 2009). This possibly reflects the dissimilar mixed infection rates in different countries. In South Korea, double (often A2143G + T2182C) and triple mutations, e.g. A2143G + T2182C + T2190C, have been commonly found only in post-treatment isolates (Kim *et al.*, 2008). However, the T2182C mutation did not contribute to clarithromycin resistance in *H. pylori* in a study by Moder *et al.* (2007).

There are geographical differences in the prevalence and type of the mutations for macrolide resistance in *H. pylori.* Unlike most data, A2142G mutation has been common (>80%) in Brazil and South Korea (Ribeiro *et al.*, 2003; Lee *et al.*, 2005). C2227T mutation has been found in clarithromycin-resistant strains in eastern Taiwan (Hu *et al.*, 2009). Mutations at positions 2182, 2223, 2244 or 2288 have been linked to *H. pylori* resistance to clarithromycin in Poland, Bangladesh, China, South Korea, Taiwan and Brazil, and those at positions 2115, 2144 and 2711 have been reported in Sweden and Italy (Dzierzanowska-Fangrat *et al.*, 2005; Rimbara *et al.*, 2007b; Kim *et al.*, 2008; Hu *et al.*, 2009). In Chile, C2147G transversion as well as G1939A, T1942C and A2142G transition in the peptidyl transferase region of the 23S rRNA have been present in macrolide-resistant *H. pylori* strains (Garrido and Toledo, 2007). Fontana *et al.* (2002) found no mutations at position 2142 to 2143 in 12 Italian strains but detected T2717C associated with low-level (MIC, 1 µg/ml) clarithromycin resistance.

However, the significance of some 'strange' mutations for the clarithromycin resistance is controversial as A2115G, G2141A and T2717C mutations have not been reported later in other studies (Burucoa *et al.*, 2008). Moreover, T2182C mutation has been found also in clarithromycin susceptible strains, although Kim *et al.* (2008) have proven its association with the resistance by transformation. In Brazil, T2245C mutation has been found in 5.8% of clarithromycin-resistant strains but has not been transferred by natural transformation (Ribeiro *et al.*, 2003). Therefore, the unusual mutations for clarithromycin resistance should be interpreted with caution. In addition, in some studies, no mutations have been found in 0–41.7% (often 0 to ≤10%) of clarithromycin-resistant strains (Fontana *et al.*, 2002; Vega *et al.*, 2003; De Francesco *et al.*, 2007; Moder *et al.*, 2007; Cambau *et al.*, 2009; Slinger *et al.*, 2009). This strongly suggests the presence of other mutations or other resistance mechanisms.

Prevalence of clarithromycin resistance-associated mutations can also vary within patient groups, e.g. between children and adults and within hospital centres. In Spain, A2143G mutation has been more common in children (80.5%) than in adults (46.7%; Vega *et al.*, 2003). In South Korea, overall mutation rates have been 15.8–31.3% with a frequency of A2143G mutation alone, ranging from 8.5% to 25% among five hospitals studied (Lee *et al.*, 2008). The differences can reflect different clustering of *H. pylori* strains or different selection pressure on the resistance by previous macrolide use.

Nitroimidazoles

Nitroimidazoles, e.g. metronidazole and tinidazole, are inactive pro-drugs, which nitro group is to be reduced intracellularly at the fifth position of the imidazole ring (Jenks and Edwards, 2002; Kaakoush *et al.*, 2009; Tsugawa *et al.*, 2011). Nitroso- and hydroxylamine derivatives, generated by the metronidazole reduction, degrade DNA and break its strands. *H. pylori* susceptibility to metronidazole depends on *H. pylori* redox proteins such as oxygen (O_2)-insensitive NADPH nitroreductase (RdxA), encoded by *rdxA* gene, NAD(P)H-flavin oxidoreductase (FrxA), encoded by *frxA* gene, and maybe FqrB, an oxidoreductase for flavodoxin reduction in the presence of NADPH (Olekhnovich *et al.*, 2009). As oxygen inhibits the metronidazole-reductase activity by direct competition or effects on flavin or other proteins, the metronidazole reduction requires good anaerobic conditions inside the bacterial cell (Olekhnovich *et al.*, 2009). Therefore, the nitroimidazoles are active against anaerobes and some microaerophiles such as *H. pylori* but not against aerobes and facultative anaerobes.

Metronidazole and tinidazole are interchangeable for *H. pylori* treatment. Nitazoxanide, a thiazolide drug without cross-resistance to metronidazole, shows potential for *H. pylori* eradication and should be additionally studied (Graham and Fischbach, 2010).

Advantages of metronidazole are its bactericidal activity, stability at low pH, low cost and partial activity against resistant strains, but disadvantages are rapid emergence of resistance, high resistance rates in some countries and side-effects of the agent (Friedman *et al.*, 2009). *H. pylori* easily (10^{-5}–10^{-6} per cell division) develops mutations by subinhibitory metronidazole concentrations *in vitro* (Cantón *et al.*, 2001). Importantly, half of *H. pylori* strains can acquire metronidazole resistance after monotherapy with the agent, which can also cause resistance to other drugs, e.g. clarithromycin (Kwon *et al.*, 2001; Su *et al.*, 2006). In addition, frequent and prolonged treatment of *H. pylori*-positive patients by nitroaromatic drugs is not recommended because of possible mutagenic potential (Friedman *et al.*, 2009).

Diminished metronidazole reduction, i.e. by mutations in either *rdxA* or *frxA* gene or both, can render *H. pylori* strains metronidazole resistant (Kaakoush *et al.*, 2009). However, the resistant strains can retain a residual susceptibility to the drug owing to additional (O_2-sensitive) nitroreductases (Megraud, 2004; Megraud and Lehours, 2007). Nevertheless, the metronidazole resistance in *H. pylori* cannot be linked only to *rdxA* and *frxA* mutations. In four matched pairs of metronidazole susceptible and resistant mutants, the mutants have exhibited mutations in *rdxA* or *frxA* gene, or in both genes, or no mutations (Kaakoush *et al.*, 2009). In a resistant mutant, many oxireduction proteins have shown isoelectric point changes. Suppression of hydantoin-utilizing proteins, which structure is similar to that of imidazolidine, and ferredoxin oxidoreductase reduction and ferric uptake regulator (Fur) mutations have been linked also to the metronidazole resistance (Kaakoush *et al.*, 2009; Tsugawa *et al.*, 2011). Mutated Fur in some metronidazole-resistant strains has been thought to derepress superoxide dismutase (SodB) expression and to help *H. pylori* to avoid reactive oxygen species generated by the nitro group reduction (Tsugawa *et al.*, 2011).

Natural transformation is another means of emergence of *H. pylori* resistance to metronidazole and other antibiotics in cases of mixed infections. In addition, efflux pumps (TolC-like proteins encoded by Hp0605 and Hp0971) can promote metronidazole resistance in *H. pylori* (Kutschke and de Jonge, 2005; van Amsterdam *et al.*, 2005).

Metronidazole resistance decreases if *H. pylori* is pre-incubated anaerobically in the presence of the drug for 2–12 h (Megraud and Lehours, 2007; see explanations above). As *H. pylori* harbours several nitroreductase enzymes to reduce the agent, treatment with high metronidazole dose may overcome the resistance (Graham and Shiotani, 2008). However, the highest (>256 µg/ml) level of metronidazole resistance apparently hampers the eradication much more than those of <32 µg/ml (Kist, 2007).

Amoxicillin

Amoxicillin is a bactericidal drug that inhibits bacterial peptidoglycan synthesis, blocking penicillin-binding proteins (PBPs). Amoxicillin has advantages of good activity at low pH values and lack of resistance even after treatment, but has also disadvantages of possible side-effects such as allergic reactions. Moreover, the amoxicillin, being a broad spectrum antibiotic, also can affect the balance of the normal microflora (Megraud, 2004; Megraud and Lehours, 2007).

Rare *H. pylori* resistance to amoxicillin has been associated mostly with mutations in *pbp-1a* gene with substitution Asn562→Tyr, multiple substitutions such as Ser414→Arg in the transpeptidase region of PBP1 and mutations in PBP2 and PBP3 encoding genes (Gerrits *et al.*, 2006; Rimbara *et al.*, 2007a, 2008). Changes in PBP-1A can result in amoxicillin MIC of 2 µg/ml (Matteo *et al.*, 2008). Changes in both PBP1 and PBP3 have led to higher resistance level than those in PBP1 only or those in PBP1 + PBP2. In addition, combined amino acid replacements in PBP1 and porins HopB and HopC have been shown to reduce the β-lactam accumulation (Co and Schiller, 2006). High-level β-lactam resistance in multidrug-resistant *H. pylori* strains has been linked to combinations of altered PBP 1A and diminished membrane permeability (Kwon *et al.*, 2003). *H. pylori* tolerance to amoxicillin has been also described in association with lack of PBP-D and very high MICs of the agent (Megraud and Lehours, 2007).

Fascinatingly, TEM β-lactamase, encoded by $bla_{\text{TEM-1}}$ gene, has been reported recently in *H. pylori* clinical strain from Taiwan, showing high-level amoxicillin resistance (Tseng *et al.*, 2009). It appears to be useful to test amoxicillin-resistant *H. pylori* strains by nitrocefin test because of a possible risk for spreading of β-lactamase in the close future.

Tetracyclines

Tetracyclines are bacteriostatic agents, which block the bacterial protein synthesis at ribosome level, binding to the small (30S) ribosome subunit. Tetracyclines have advantages of low cost, rare *H. pylori* resistance and synergism with bismuth salts, and the disadvantage of numerous side-effects of treatment (Megraud, 2004; Megraud and Lehours, 2007). In Taiwan, tigecycline, a new parenteral glycylcycline with stability to most resistance mechanisms, has been highly active against *H. pylori* strains with MIC_{90} as low as 0.06 μg/ml (Yang *et al.*, 2010). As the tigecycline is parenteral-only agent, approved by FDA for treatment of complicated intra-abdominal and complicated skin/skin structure infections, its use in treatment of multidrug-resistant *H. pylori* infections is questionable.

Tetracycline resistance has been linked to mutations in 16S rDNA, corresponding to the helix 31 region of 16S rRNA (Lawson *et al.*, 2005). Mutations, e.g. nucleotide replacements, mainly in adenine–guanine–adenine triplets at positions 926 to 928 (often A926G as well as A926C and A928C mutations), have been found in tetracycline-resistant strains (Lawson *et al.*, 2005; Castro-Fernández and Vargas-Romero, 2009).

Multiple mutations in the 16S rRNA gene lead to higher level of tetracycline resistance in *H. pylori* than double and single mutations. Triple mutations, e.g. replacement of AGA926-A928C by thymine, thymine and cytosine (AGA926–928→TTC) in the 16S rRNA gene, can promote high-level (MICs, 64–256 μg/ml) tetracycline resistance in *H. pylori*. Lower level resistance to the agent has been found in strains with double (tetracycline MICs, 8–256 μg/ml) and single mutations (0.75–16 μg/ml; Lawson *et al.*, 2005; Nonaka *et al.*, 2005; Castro-Fernández and Vargas-Romero, 2009). For instance, in UK, more than half of 18 strains with reduced tetracycline susceptibility have shown mutations in 16S rRNA, including most often A926G mutation (tetracycline MICs 2–16 μg/ml) as well as double mutation (A926T/A928C) and single A926C and A928C mutations (Lawson *et al.*, 2005).

Membrane permeability inhibitors and TetA efflux protein also participate in the emergence of tetracycline resistance in *H. pylori* (Castro-Fernández and Vargas-Romero, 2009; Vakil, 2009). Associated tetracycline resistance-related gene (*HP*1165) has been detected (Li and Dannelly, 2006). Efflux mechanism, dependent on the proton motive force, has been linked to tetracycline resistance in clinical strains. The efflux mechanism has been shown to decrease 2–17 times the accumulation of the tetracycline in most tetracycline-resistant strains (Anoushiravani *et al.*, 2009).

Fluoroquinolones

DNA gyrase is known to generate negative supercoils into the bacterial DNA and to affect the DNA replication, recombination and transcription. DNA gyrase is a tetrameric enzyme of two A subunits, encoded by *gyrA* gene, and two B subunits, encoded by *gyrB* gene (Nishizawa *et al.*, 2009). Fluoroquinolones are able to inhibit the A subunit of the DNA gyrase, their single target in *H. pylori*, as these bacteria lack genes for topoisomerase IV, unlike many other species (Glocker *et al.*, 2007a; Nishizawa *et al.*, 2009).

Fluoroquinolones exhibit advantages of bactericidal activity and synergism with PPIs; however, the disadvantage of some members is the growing resistance due to increasing quinolone use and side-effects such as phototoxicity, arthropathy, effect on blood glucose etc. (Miyachi *et al.*, 2006; Nishizawa *et al.*, 2009.) Ciprofloxacin resistance mutations have been shown to occur at 10^{-6} to 10^{-8} per cell division in *H. pylori* (Bogaerts *et al.*, 2006; Glocker *et al.*, 2007a). Within the quinolones, levofloxacin is often used in third-line regimens for eradication of the infection (see above). *H. pylori* cross-resistance to ciprofloxacin (MIC, >1 μg/ml), levofloxacin and moxifloxacin has been observed. Ciprofloxacin has been used in some studies to detect *H. pylori* resistance to levofloxacin, as the resistance rates to levofloxacin have been similar to slightly lower if compared with those of ciprofloxacin (Kim *et al.*, 2006a; Wang *et al.*, 2010).

Importantly, the newer quinolone drugs, e.g. garenoxacin and sitafloxacin have been more active against *H. pylori* than levofloxacin and have been able to inhibit ciprofloxacin/levofloxacin-resistant strains (Hung *et al.*, 2009; Suzuki *et al.*, 2009a). MIC_{90} of sitafloxacin (0.25 μg/ml) has been 16 times lower than that of gatifloxacin and four times lower than that of garenoxacin in the study of Suzuki *et al.* (2009a). In Taiwan, nemonoxacin and gemifloxacin have been up to twice more active *in vitro* against *H. pylori* than ciprofloxacin, levofloxacin, and moxifloxacin (Yang *et al.*, 2010). Recently, new organometallic drugs, combining different quinolones with bismuth, have been developed and have shown low MICs (0.5–1 μg/ml) against *H. pylori* (Shaikh *et al.*, 2009). These agents should be evaluated in further studies.

H. pylori resistance to quinolones is owing to point mutations in the quinolone resistance-determining region of *gyrA* gene (QRDR), mostly at codons 87 (Asn to Lys) or 91 (Asp to Gly, Asp to Asn, or Asp to Tyr; Tankovic *et al.*, 2003; Cambau *et al.*, 2009). In UK, most (67%) ciprofloxacin-resistant strains have been shown to harbour *gyrA* mutations at codon 91, most often Asp91Gly and Asp91Asn mutations (Chisholm and Owen, 2009). In France, *H. pylori* resistance to fluoroquinolones has been linked to *gyrA* mutations in 80–100% of the resistant strains (Cambau *et al.*, 2009). Most (87%) fluoroquinolone-resistant strains in France have harboured mutations at codons 87 or 91. The strains have shown *gyrA* mutation at codons 87 (50%) or 91 (41.7%) or at both codons (6.7%) and two mutations at codon 91 (1.7%; Cambau *et al.*, 2009). In addition, *gyrA* mutation at codon 86 has been also reported in quinolone-resistant strains (Cattoir *et al.*, 2007). In Taiwan, a double mutation at Ser70Cys (newly described mutation) and Asn87Ile has promoted high level of levofloxacin resistance (Yang *et al.*, 2010). Moreover, heterogeneous resistance has been found in >20% of resistant Belgian strains (Bogaerts *et al.*, 2006). Importantly, in China, 2 of 29 quinolone-resistant (MICs >12 μg/ml) strains have exhibited no mutation in *gyrA* and *gyrB* genes (Wang *et al.*, 2010), suggesting other resistance mechanisms.

Other antibiotics

Rifampicin and rifabutin, the latter agent being more active against *H. pylori* than the former drug, inhibit the β-subunit of the DNA-dependent RNA polymerase, encoded by *rpoB* gene, and hamper the RNA function and the ensuing protein synthesis of the bacteria (Glocker *et al.*, 2007b; Megraud and Lehours, 2007). Point mutation in *rpoB* gene at codons 524, 525, 530, 540, 545 and 585 and (seldom) 149 can lead to the infrequent *H. pylori* resistance to both rifabutin and rifampicin (Glocker *et al.*, 2007b; Megraud and Lehours, 2007; Mégraud and Corti, 2009). In Germany, the usual mutation (in 40.9% of strains) for rifabutin resistance in *H. pylori* has affected the codon 530 (Glocker *et al.*, 2007b).

Like metronidazole, nitrofurans (furazolidone and nitrofurantoin) are nitroheterocyclic/nitroaromatic agents and pro-drugs (Kwon *et al.*, 2001). The nitrofuran resistance has been associated with two enzymes i.e. pyruvate: flavodoxin oxidoreductase (POR) and 2-oxoglutarate: acceptor oxidoreductase (OOR), which participate in generation of acetyl coenzyme A (acetyl-CoA) and succinyl-CoA, respectively (Su *et al.*, 2006). Namely, the mutations G353A, A356G, and C357T in *porD* gene as well as A041G, A122G, and C349A(G) in *oorD* gene have been responsible for the resistance (Su *et al.*, 2006). It is, however, important that because of potential genotoxic and carcinogenetic risk, furazolidone has been banned by FDA and European Medicinal Agency, which corresponds to FDA in the European Union (De Francesco *et al.*, 2009).

In most cases, the nitrofuran resistance in *H. pylori* has coexisted with that to metronidazole but not vice versa. Although resistance to only one of the two nitrofurans has been present in some Spanish strains, the nitrofurantoin-resistant strains have been metronidazole resistant (Alarcón *et al.*, 2005). Similarly, in South Korea, all seven furazolidone and nitrofurantoin-resistant *H. pylori* strains evaluated have been metronidazole resistant in the study of Kwon *et al.* (2001).

Resistance rates and risk factors for the resistance

Primary resistance

H. pylori resistance can be classified as primary (in patients not previously treated for *H. pylori* eradication), post-treatment (in those treated earlier for the infection regardless of the antibiotic combination) or secondary (resistance acquired during treatment, including an antibiotic, to which the strain had been susceptible before treatment) resistance (Mégraud *et al.*, 1999). The discussion of the results from studies which have reported resistance in consecutive (untreated and treated) patients is limited to show better the risk factors for the primary resistance.

First treatment of *H. pylori* infection is usually empirical; therefore, knowledge of local primary *H. pylori* resistance and its evolution is clinically valuable. If local resistance rates are ≥15–20% for clarithromycin and ≥40% for metronidazole, the agents are unsuitable for primary therapy unless the susceptibility of the strain is tested (Malfertheiner *et al.*, 2007).

Primary antibiotic resistance in *H. pylori* varies widely among patient groups according to the country or region, prior use of the agent for other infections, patient age, sex, disease and comorbidity, strain characteristics and other factors (McMahon *et al.*, 2003; Megraud, 2004; Koivisto *et al.*, 2004). Most of these risk factors mirror previous individual or national use of the same or related antibiotics for other infections.

Generally, higher national consumption of an antibiotic or related drug shows an association, although not universal and not applicable for all drugs, with higher primary resistance rate to the agent (see below). Prior use of antibiotics, e.g. clarithromycin for respiratory infections, quinolones for urinary and respiratory infections and metronidazole for anaerobic (gynaecological and dental) or parasitic infections, increases the individual risk for primary *H. pylori* resistance of the patient. In a study by McMahon *et al.* (2003), the clarithromycin resistance has been linked to previous use of any macrolide agent and the metronidazole resistance has been associated with earlier metronidazole use. Therefore, for patients with past antibiotic use, the risk for associated *H. pylori* resistance should be taken into account to choose optimal treatment regimen.

Transmission of resistant strain to a new host, e.g. from the parent to his/her child, may also occur. For instance, in Portugal, because of common use of the newer fluoroquinolones in adults over two decades, fluoroquinolone resistance has affected 4.5% of strains from children (Lopes *et al.*, 2005; Mégraud, 2010). Similarly, tetracycline resistance in Bulgarian children has been linked to high national consumption of the drug more than 10 years before the study (Boyanova, 2009).

Resistance rates to clarithromycin

In adults, the prevalence of primary *H. pylori* resistance to clarithromycin has varied from 1–3% in Northern Europe and The Netherlands to ≥20% in Italy, France, Chile, Japan, Thailand, South Korea and South Africa (Table 9.2; Laine *et al.*, 2003; Megraud, 2004). Worryingly, since 2004, *H. pylori* resistance to clarithromycin has reached 30% in the strains from Japanese untreated patients and 38% in those from consecutive Taiwanese patients (Horiki *et al.*, 2009; Yang *et al.*, 2010). Conversely, no clarithromycin resistance has been reported (mostly in developing countries) in studies from India, Kuwait, Ethiopia and New Zealand, and the resistance rate in Malaysia has been only 3% (Ahmed *et al.*, 2004; Asrat *et al.*, 2004; Datta *et al.*, 2005; John Albert *et al.*, 2006).

Similarly, in children, common (>20%) clarithromycin resistance has been detected mainly in developed countries in Europe and Asia (Israel and Japan; Fig. 9.1). Primary clarithromycin resistance rates in European children have been disturbingly high (17.5–49.2%, often ≥20%) since 2000 (Table 9.3). The highest rates (>30%) have been reported in Spain, Portugal, Austria and Japan (Lopes *et al.*, 2005; Agudo *et al.*, 2009; Kato and Fujimura, 2010; Vécsei *et al.*, 2010). Primary clarithromycin resistance rate surpassed 40% in Japanese children in 2003–2007 and 49% in Spanish children in 2002–2006 (Agudo *et al.*, 2009; Kato and Fujimura, 2010). The elevated primary *H. pylori* resistance rates to macrolides in many developed countries result from frequent use of newer macrolides, e.g. clarithromycin and azithromycin to treat upper respiratory infections, acute otitis media and community-acquired pneumonia (Megraud, 2004; Zevit *et al.*, 2010; Fig. 9.1). In contrast, owing to the price of newer macrolides, they have been used seldom in the developing countries and there, the resistance rates to macrolides have been lower compared with those in the Western countries.

Frequently reported risk factors for primary clarithromycin resistance have been years (recent) of study, country/region (developed), age (children), patient sex (female), race and origin (related to the country or population), disease (non-ulcer), comorbidity (diabetes) and hospital centre, most of which mirror previous use of macrolides (Table 9.4).

In Europe, gradient of the primary clarithromycin resistance has pointed out a direction from the north (often 2% in adults) to the south (up to 32.5% in adults and 49.2% in children; Koivisto *et al.*, 2004; Storskrubb *et al.*, 2006; Agudo *et al.*, 2009) owing to high antibiotic consumption and

Table 9.2 Primary antibiotic resistance of *H. pylori* in dyspeptic adults over 10 years

Countries	No. of patients	Years	Methods	Resistance rates (%)									References
				CLA	MET	AMO	TET	QUI	RIF	FUR	MET + CLA	MDR[1]	
Northern Europe													
Denmark (total, data given for 2001–2003)	81	98–04	MDDM, Et	3	27	0	0	NA[#]	NA	NA	0	NA	Petersen *et al.* (2006)
Finland	292	00–02	Et	2	38	NA	NA	NA	NA	NA	NA	NA	Koivisto *et al.* (2004)
Finland	505	00–08	Et	8	41	0	0.2 (low level resistance)	7 (LV)	NA	NA	3.2	0.4	Kostamo *et al.* (2011)
Sweden	333	98–01	ADM	1.5	16.2	0	0.3	NA	NA	NA	0.6	0	Storskrubb *et al.* (2006)
Southern Europe													
Italy	420	NA	BST, ADM	23.6	35.5 (TIN)	0.2	NA	NA	0	NA	9.5	NA	Toracchio *et al.* (2005)
Italy	178	04–05	RT-PCR, Sq	21.3	NA	NA	NA	NA	NA	NA	NA	NA	De Francesco *et al.* (2007)
Italy	255	04–06	Et	16.9	29.4	NA	NA	19.1 (LV)	NA	NA	3.5	0.8	Zullo *et al.* (2007)
Central/Eastern Europe													
Bulgaria	828	05–08	BST, Et	18.4	26.5	1	4.4	9 (CP)	NA	NA	8	4.2	Boyanova (2009)
Croatia	238	NA	Et	10.8	NA	NA	NA	5.9 (MX)	NA	NA	NA	NA	Bago *et al.* (2007)
Croatia	467	02–05	ADM	8.2	32.9	NA	NA	NA	NA	NA	NA	NA	Filipec Kanizaj *et al.* (2009)
Hungary	238	95–06	FISH	17.3	NA	NA	NA	NA	NA	NA	NA	NA	Buzás *et al.* (2007)
Poland	130	01–04	Et	15	42	0	0	NA	NA	NA	9	NA	Dzierzanowska-Fangrat *et al.* (2005)
Slovenia	97	07–09	Et	17.5	18.6	0	0	3.1 (CP)	NA	NA	4.1	NA	Jeverica *et al.* (2010)
Western Europe													
Belgium	436	02	MDDM	3	31	0	0	NA	NA	NA	NA	NA	Aguemon *et al.* (2005)
France (review)	NA	NA	NA	≥20	NA	0	0	≥15	0	NA	NA	NA	Mégraud (2008)
France	455	04–07	MDDM, Et	19.1	58.9	0	0	12 (CP)	0.2	NA	13.2	3 (triple)	Raymond *et al.* (2010)
France	128	04–05	ADM	NA	NA	NA	NA	17.2 (CP)	NA	NA	NA	NA	Cattoir *et al.* (2007)
Germany	196	03	NA (review)	<10	~20	NA	NA	~12	<5	NA	<5	NA	Kist and Glocker (2004)
Germany	126	01–05	Et	NA	NA	NA	NA	9.5 (CP)	NA	NA	NA	0	Glocker *et al.* (2007a)
Ireland	182	07–08	Et	9.3	29.1	NA	NA	NA	NA	NA	NA	NA	O'Connor *et al.* (2010)
Switzerland	93	02–03	RFLP, Sq, LpH	9	NA	NA	NA	NA	NA	NA	NA	NA	Soltermann *et al.* (2005)

Netherlands	1125	97–02	MDDM	1	14.4	NA	NA	NA	NA	NA	NA	NA	Janssen *et al*. (2006)
UK	363	01–03	MDDM, Et	7	24	0	0.3	NA	NA	NA	4	NA	Elviss *et al*. (2004)
UK	1310	00–05	MDDM, Et	8.3–12.7	28.6–36.3	0	0.5	NA	NA	NA	4.4–8.4	NA	Chisholm *et al*. (2007)
North America													
USA	106	00–01	ADM	12.2	33.9	NA	NA	NA	NA	NA	NA	NA	Laine *et al*. (2003)
USA (Alaska)	125	98–02	Et	NA	NA	NA	NA	8.8 (LV)	NA	NA	NA	NA	Carothers *et al*. (2007)
South America													
Brazil	203	NA	ADM	9.8	53	NA	NA	NA	NA	NA	7.4	NA	QPrazeres Magalhães *et al*. (2002)
Brazil	155	NA	ADM	16	54	39	9	NA	NA	13	20	14.2 (12.3 triple, 1.9 quadruple)	Godoy *et al*. (2003)
Chile[¶]	50	NA	ADM	20	45	NA	27	NA	NA	NA	10	4	Vallejos *et al*. (2007)
Chile	50		ADM, RFLP, Sq	20	NA	NA	NA	NA	NA	NA	NA	NA	Garrido and Toledo (2007)
Colombia	53	06–07	Et	NA	72	NA	NA	NA	NA	NA	NA	NA	Henao *et al*. (2009)
Paraguay	46	02–03	Et	2.2	33	2.2	NA	NA	NA	NA	2.2	NA	Fariña *et al*. (2007)
Asia													
Bangladesh[2]	120	99–01	ADM	10	77.5	6.6	15	NA	NA	NA	NA	NA	Nahar *et al*. (2004)
China	157	NA	Et	18.5	45.9	NA	NA	NA	NA	NA	NA	NA	Cheng *et al*. (2006)
China	196	NA	Et, ADM	13.8	37.8	NA	NA	NA	NA	NA	8.7	NA	Wong *et al*. (2003)
China	79	07–09	Et, Sq	NA	NA	NA	NA	26.6 (CP, LV)	NA	NA	NA	NA	Wang *et al*. (2010)
India	67	00–01	ADM	0	85	0	0, 7.5 (low level)	NA	NA	0	NA	NA	Datta *et al*. (2005)
Iran	80	06	MDDM	6.2	26.2	2.5	3.8	8.8 (CP)	NA	NA	3.8	3.8 (quadruple[1])	Khashei *et al*. (2008)
Israel	110	00–01	Et	8.2	38.2	0.9	0	NA	NA	NA	NA	NA	Samra *et al*. (2002)
Japan[2]	1257	04–05	ADM	27.7	2.1[3]	0.08[4]	NA	NA	NA	NA	NA	NA	Kobayashi *et al*. (2007)
Japan	507	01–04	Et, Sq	NA	NA	NA	NA	15 (LV)	NA	NA	NA	NA	Miyachi *et al*. (2006)
Japan	3521	96–08	NA	30 (since 2004)	NA	0.03	0.06 (MIN)	NA	NA	NA	NA	NA	Horiki *et al*. (2009)

Table 9.2 continued

Countries	No. of patients	Years	Methods	Resistance rates (%) CLA	MET	AMO	TET	QUI	RIF	FUR	MET + CLA	MDR[1]	References
Kuwait	96	03–05	Et, Sq	0	69.8	0	0	NA	NA	NA	0	0	John Albert *et al*. (2006)
Malaysia	120	05–07	Et, RFLP	3.3	NA	NA	NA	NA	NA	NA	NA	NA	Ahmad *et al*. (2009)
Malaysia	107	NA	RFLP	2.9	NA	NA	NA	NA	NA	NA	NA	NA	Ho *et al*. (2010)
Singapore	120	NA	ADM	NA	31.7	NA	NA	NA	NA	NA	NA	NA	Lui *et al*. (2003)
South Korea	65	04–05	Et, Sq	13.8	66.2	18.5	12.3	33.8 (CP), 21.5 (LV)	NA	NA	13.8	23.1 (16.9 triple[1], 6.2 quadruple)	Kim *et al*. (2006a)
South Korea	114	00–02	Et, RFLP, Sq	20.2	NA	0	NA	NA	NA	NA	NA	NA	Lee *et al*. (2005)
South Korea	144	01–05	mD, Et	16.7	34.7	11.8	NA	NA	NA	NA	NA	16.7	Bang *et al*. (2007)
Taiwan	180	NA	Et	10.6	26.7	0	NA	9.4 (LV)	NA	NA	NA	NA	Chang *et al*. (2009)
Taiwan	210	98–07	ADM, Et, PCR	9.5	27.6	1	0.5	5.7 (CP, LV)	NA	NA	NA	NA	Hung *et al*. (2009)
Taiwan	134	01–04	Et	6.7	25.4	0	NA	NA	NA	NA	NA	NA	Poon *et al*. (2009)
Thailand	69	02–03	Et	23.2	NA	NA	NA	NA	NA	NA	NA	NA	Mahachai *et al*. (2006)
Turkey	61	NA	Et	16.4	NA	NA	NA	NA	NA	NA	NA	NA	Tüzün *et al*. (2008)
Vietnam (peptic ulcer)	205	NA	Et	1	76	0	0	NA	NA	NA	0.5	0	Wheeldon *et al*. (2004)
Africa													
Ethiopia	50	NA	Et	0	76	6	0	NA	NA	NA	NA	NA	Asrat *et al*. (2004)
South Africa (total)	200	NA	MDDM, ADM	20	95.5	2.5	32.5	0 (CP)	NA	NA	NA	NA	Tanih *et al*. (2010)
Tunisia	225	05–07	Et, PCR	14.6	56.8	0	NA	NA	NA	NA	12.9	NA	Ben Mansour *et al*. (2010)
New Zealand													
Total	50	NA	Et	0	20	0	0	NA	NA	NA	0	0	Ahmed *et al*. (2004)

[1]To different classes of antibiotics. [2]Children and adults, mostly adults. [3]MICs >8 μg/ml. [4]MICs >0.5 μg/ml.

Et, E test; ADM, agar dilution method; BST, breakpoint susceptibility testing; MDDM modified disc-diffusion method; mD, microdilution; FISH, fluorescence *in situ* hybridization; RT-PCR, real-time polymerase chain reaction; RFLP, random fragment length polymorphism; Sq, sequencing; LpH, line probe hybridization; AMO, amoxicillin; MET, metronidazole; TIN, tinidazole; CLA, clarithromycin; TET, tetracycline; QUI, fluoroquinolones; CP, ciprofloxacin; LV, levofloxacin; MX, moxifloxacin; RIF, rifabutin; FUR, nitrofurans such as nitrofurantoin or furazolidone, MDR, multidrug; NA, not available.

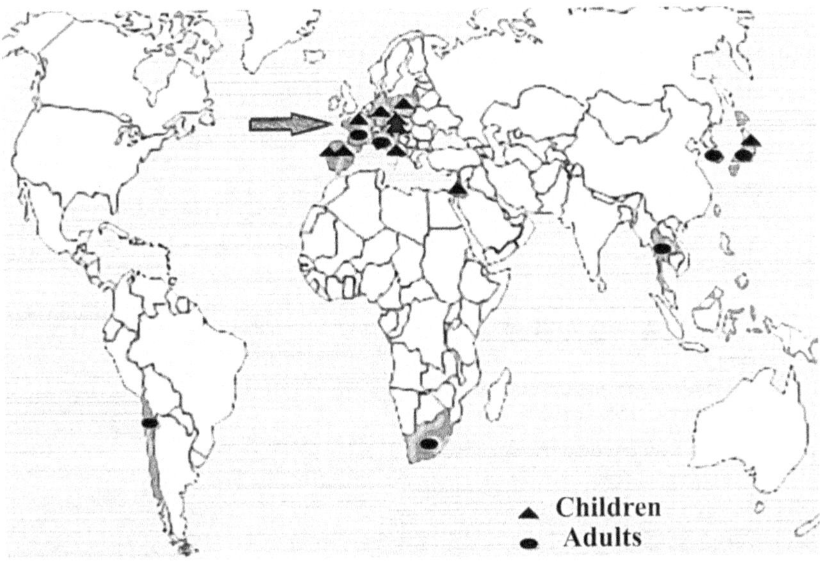

Figure 9.1 Countries reporting primary resistance rates of ≥20% for clarithromycin over the last 10 years.

less stringent antibiotic policy in some Southern European countries. Conversely, the low (4%) primary clarithromycin resistance in Egyptian children in 2002–2003 can reflect recent (18 months before the study) introduction of the drug in Egypt (Sherif *et al.*, 2004). No risk factors for the resistance have been reported in Turkey (Tüzün *et al.*, 2008). However, in European children, primary clarithromycin resistance rate has been higher in boys versus girls, younger (aged under 6 years) versus older children and in Southern Europe versus elsewhere (Koletzko *et al.*, 2006). The resistance has been more elevated in children as they are more often treated by macrolides for respiratory infections than the adults (Koletzko *et al.*, 2006).

It is obvious that national outpatient consumption of antibiotics can increase the selective pressure on the antibiotic resistance in the country. National outpatient consumption of macrolides, lincosamides and streptogramins (J01F) in Sweden, the Netherlands and Finland has been low [0.89–1.65 defined daily doses per 1000 inhabitants per day (DIDs) in 2007] and has been linked to low (1–2%) primary resistance rate to clarithromycin in *H. pylori* (Koivisto *et al.*, 2004; Janssen *et al.*, 2006; Storskrubb *et al.*, 2006; ESAC, 2007). Conversely, the higher (3.59–4.91 DIDs) consumption in Austria, Italy and France in 2007 has coincided with higher (≥20%) primary clarithromycin resistance rates (Toracchio *et al.*, 2005; ESAC, 2007; Mégraud, 2008; Vécsei *et al.*, 2010).

Previous use of macrolides even in the distant past also has been a risk factor for primary clarithromycin resistance in *H. pylori*. In the study of McMahon *et al.* (2003), about twice more (92%) patients harbouring clarithromycin-resistant strains had been treated by a macrolide drug in the 10 years before diagnosis compared with those with susceptible strains (57%). In addition to clarithromycin, previous use of erythromycin and, at less extent azithromycin, also can lead to development of *H. pylori* resistance to clarithromycin (see below), therefore the anamnesis of the patient should involve, if possible, the use of any macrolide agent in the past 10 years.

Patient disease also influences the resistance rates and the success of eradication of *H. pylori* infection. In France and according to a meta-analysis, less virulent (*cagA*-negative) strains have been linked to a 2-fold increased risk of eradication failure by a clarithromycin-based triple regimen compared with more virulent strains (Broutet *et al.*, 2001; Suzuki *et al.*, 2006). In accordance with these data, the *H. pylori* clarithromycin resistance rate has been higher in non-ulcer compared

Table 9.3 Primary antibiotic resistance of *H. pylori* in dyspeptic children over 10 years

Countries	No. of patients	Years	Method	Resistance rates (%)									References
				CLA	MET	AMO	TET	QUI	RIF	FUR	MET+ CLA	MDR[1]	
Europe	1037	99–02	Et, ADM	20	23	0.6	NA	NA	NA	NA	5.3	NA	Koletzko *et al*. (2006)
Southern Europe	184	99–02	Et, ADM	32.5	22.3	NA	NA	NA	NA	NA	NA	NA	Koletzko *et al*. (2006)
Italy	92	01–06	RT-PCR	29.3	NA	NA	NA	NA	NA	NA	NA	NA	Lionetti *et al*. (2009)
Portugal (total)	109	99–03	Et	39.4	16.5	0	0	4.5 (CP)	NA	NA	NA	NA	Lopes *et al*. (2005)
Spain	96	99–00	ADM	29.1	23.9	0	NA	NA	NA	NA	8.3	NA	Alarcon *et al*. (2003)
Spain	66	02–06	Et	49.2	32.8	0	0	3.6 (CP)	0	NA	15.4	NA	Agudo *et al*. (2009)
Central/ Eastern Europe	84	99–02	Et, ADM	17.5	23.8	NA	NA	NA	NA	NA	NA	NA	Koletzko *et al*. (2006)
Bulgaria	105	05–08	BST, Et	19	16.2	0	1.9	5.8 (CP)	NA	NA	6.7	1	Boyanova (2009)
Czech Republic	86	NA	ADM	21.2	NA	0	NA	NA	NA	NA	NA	NA	Sýkora *et al*. (2005)
Poland	139	01–04	Et	28	40	0	0	NA	NA	NA	16.5	NA	Dzierzanowska-Fangrat *et al*. (2005)
Western Europe	619	99–02	Et, ADM	18.2	24	NA	NA	NA	NA	NA	NA	NA	Koletzko *et al*. (2006)
Austria	153	02–08	Et	34	22.9	0	0.9	0 (LV)	0.9	NA	9.8	NA	Vécsei *et al*. (2010)
France	377	94–05	RFLP	22.8	36.7	0	NA	NA	NA	NA	7.9	NA	Kalach *et al*. (2007)
Germany	73	NA	FISH	22	NA	NA	NA	NA	NA	NA	NA	NA	Feydt-Schmidt *et al*. (2002)

Asia													
Iran	100	03–05	Et, MDDM	16	95	59	5	7 (CP)	NA	9	NA	NA	Rafeey *et al*. (2007)
Israel	36	05–07	Et	25	19	0	0	0 (LV)	NA	NA	0	0	Zevit *et al*. (2010)
Japan	61	99–07	mD	36.1	14.8	0	NA	NA	NA	NA	6.6	NA	Kato and Fujimura (2010)
Japan	55	NA	Et	21.8	9.1	0	NA	5.5 (LV)	NA	NA	NA	NA	Fujimura *et al*. (2004)
Africa													
Egypt	48	02–03	Et	4	100	2	NA	2 (CP)	NA	NA	4	NA	Sherif *et al*. (2004)
Tunisia	48	05–07	Et, PCR	18.8	25	0	NA	NA	NA	NA	16.7	NA	Ben Mansour *et al*. (2010)

[1]To different classes of antibiotics.

Et, E test; ADM, agar dilution method; BST, breakpoint susceptibility testing; MDDM modified disc-diffusion method; mD, microdilution; FISH, fluorescence *in situ* hybridization; RT-PCR, real-time polymerase chain reaction; RFLP, random fragment length polymorphism; Sq, sequencing; AMO, amoxicillin; MET, metronidazole; TIN, tinidazole; CLA, clarithromycin; TET, tetracycline; QUI, fluoroquinolones; CP, ciprofloxacin; LV, levofloxacin; MX, moxifloxacin; RIF, rifabutin; FUR, nitrofurans such as nitrofurantoin or furazolidone, MDR, multidrug; NA, not available.

Table 9.4 Some observed risk factors for antibiotic resistance in *H. pylori* in dyspeptic patients

Resistance to	Country/ region	No. of patients	Method	Risk factors for resistance	Reference
CLA	Austria[1]	153	Et	Patient origin (from Austria)	Vécsei *et al.* (2010)
	Bulgaria	266	BST	Hospital centre (possible association)	Boyanova *et al.* (2009)
	Bulgaria	1093	BST, Et	Year of study (2005–2008 vs. 1996–1999)	Boyanova and Mitov (2010)
	China	196	Et, ADM	Age (19–39 years vs. ≥40 years)	Wong *et al.* (2003)
	Europe[1]	991	Et, ADM	Sex (boys), age (<6 years), place of residence (South Europe)	Koletzko *et al.* (2006)
	Finland	342	Et	Prior antibiotic use (for respiratory/ dental infections)	Koivisto *et al.* (2004)
	Finland	505	Et	Women	Kostamo *et al.* (2011)
	France	2751	MDDM	Disease status (non-ulcer)	Broutet *et al.* (2003)
	Hungary	238	FISH	Age, sex (female), smoking	Buzás *et al.* (2007)
	Ireland	182	Et	Year of study (2007–2008 vs. 1997)	O'Connor *et al.* (2010)
	Italy	325	RT-PCR	Year of study (2004–2005 vs. 1989–1990)	De Francesco *et al.* (2007)
	Italy	255	Et	Disease status (non-ulcer)	Zullo *et al.* (2007)
	Israel[1]	36	Et	Previous use of macrolides (CLA or azithromycin)	Zevit *et al.* (2010)
	Japan	3521	NA	Year of study (2004–2008 vs. 1996), age (elderly), sex (female)	Horiki *et al.* (2009)
	Japan	3707	ADM	Year of study (2004–2005 vs. 2002–2003), sex (female)	Kobayashi *et al.* (2007)
	Poland	337	Et	Age (children), hospital centres (geographic regions)	Dzierzanowska-Fangrat *et al.* (2005)
	Portugal[1]	109	Et	Origin (European), previous *H. pylori* empiric therapy and amoxicillin MICs (≥0.016 μg/ml)	Lopes *et al.* (2005)
	Spain[1]	96	ADM	Age (younger children, 4–8 year-old, vs. older children)	Alarcon *et al.* (2003)
	Spain	118	Et, Sq	*vacA* s2/m2 genotype and CagA negative strains, birthplace, age	Agudo *et al.* (2010)
	Turkey	61	Et	None	Tüzün *et al.* (2008)
	Turkey	114	RT-PCR	Comorbidity (diabetes type 2)	Demir *et al.* (2009)
	UK	1310	Et, MDDM	Place of residence, sex (female), age (<45 years), year of study	Chisholm *et al.* (2007)
	USA (Alaska)	125	ADM	Previous use of macrolides (any macrolide) up to 10 years before diagnosis	McMahon *et al.* (2003)
MET	Bangladesh	120	ADM	Previous use of MET	Nahar *et al.* (2004)
	Bulgaria	266	BST	Birthplace, disease status (non-ulcer)	Boyanova *et al.* (2009)
	Europe[1]	1024	Et, ADM	Birthplace (outside Europe)	Koletzko *et al.* (2006)
	Finland	333	Et	Sex (female), prior antibiotic use (gynaecological infections)	Koivisto *et al.* (2004)
	Finland	505	Et	Women	Kostamo *et al.* (2011)
	India (ulcer)	259	Et	Place of residence	Thyagarajan *et al.* (2003)
	Ireland	182	Et	Sex (female)	O'Connor *et al.* (2010)
	Ireland	103	Et	Strain characteristics (*cagA*⁻)	Taneike *et al.* (2009)
	Italy	255	Et	Ethnicity (foreign)	Zullo *et al.* (2007)

Table 9.4 continued

Resistance to	Country/ region	No. of patients	Method	Risk factors for resistance	Reference
	Singapore	120	ADM	Disease status (non-ulcer)	Lui *et al*. (2003)
	South Korea	113	ADM	Hospital centre/place of residence	Kim *et al*. (2006b)
	Spain[1]	96	ADM	Age (older children, 14–18 years old vs. younger children)	Alarcon *et al*. (2003)
	Sweden	333	ADM	Sex (female)	Storskrubb *et al*. (2006)
	Netherlands	1125	MDDM	Ethnicity (foreign)	Janssen *et al*. (2006)
	UK	402	MDDM, Et	Birthplace	Elviss *et al*. (2005)
	UK	1310	Et, MDDM	Place of residence, sex (female), age (<45 years), year of study	Chisholm *et al*. (2007)
	USA (Alaska)	125	ADM	Prior use of MET up to 8 years before diagnosis	McMahon *et al*. (2003)
QUI	Bulgaria	93	MDDM, BST	Years of study (2007–2009 vs. 1990–1995)	Boyanova *et al*. (2010)
	Finland	505	Et	None	Kostamo *et al*. (2011)
	Italy	255	Et	Age (>45 years); resistance to MET or CLA	Zullo *et al*. (2007)
	France	455	Et, PCR	Years of study (2006–2007 vs. 2004–2005)	Raymond *et al*. (2010)
	South Korea	113	ADM	Hospital centre/place of residence	Kim *et al*. (2006b)
	Taiwan	227	Et	Years of study (after 2004 vs. before 2004)	Chang *et al*. (2009)
MET + CLA	Bulgaria	266	BST	None	Boyanova *et al*. (2009)
	Europe[1]	1037	Et, ADM	None	Koletzko *et al*. (2006)
	Germany	264	NA	Disease status (non-ulcer)	Kist and Glocker (2004)

[1]Children.

CLA, clarithromycin; MET metronidazole; QUI, quinolones; Et, E test; ADM, agar dilution method; BST, breakpoint susceptibility testing; MDDM modified disc-diffusion method; FISH, fluorescence *in situ* hybridization; RT-PCR, real-time polymerase chain reaction; NA, not available.

with ulcer patients in many countries (Broutet *et al.*, 2003; Wong *et al.*, 2005; Zullo *et al.*, 2007), although a striking inverse association has been reported in Argentina (Vega *et al.*, 2010). A likely explanation of the elevated clarithromycin resistance in non-ulcer compared with ulcer patients is the high prevalence of virulent (*cagA*$^+$) strains in ulcer patients, their more rapid growth and closer contact with gastric epithelium, which can increase the access of antibiotics to the strains (Broutet *et al.*, 2001; Megraud, 2004). Therefore, Broutet *et al.* (2001) have recommended prolonged treatment for patients with *CagA*-negative strains. Similar association has been reported in Bulgaria and Ireland for metronidazole (see below).

Differences in primary *H. pylori* resistance rates to clarithromycin have been found according to the patient ethnicity, geographical regions and even hospital centres (Dzierzanowska-Fangrat *et al.*, 2005; Koletzko *et al.*, 2006; Chisholm *et al.*, 2007; Boyanova *et al.*, 2009). In Bulgaria, the primary clarithromycin resistance rate has been associated marginally with hospital centres (Boyanova *et al.*, 2009). The resistance rate has been higher (23.4%) in a hospital centre, where many (29.7%) patients have been with complicated ulcer disease or operated stomach than in other centres (12.9%). The differences in the clarithromycin resistance rates in different hospital centres have been explained partially by variations in prescribing practice, empiric antibiotic use during past hospitalizations and complex patient characteristics (Elviss *et al.*, 2004; Boyanova *et al.*, 2009).

Smoking also has been linked to higher clarithromycin resistance in *H. pylori* and lower eradication rate compared with those in non-smokers (Broutet *et al.*, 2003; Buzás *et al.*, 2007). The effect of smoking could be due to changed acid secretion but needs to be further evaluated (Graham and Shiotani, 2008).

Comorbidity may be an additional risk factor for *H. pylori* resistance. Success of *H. pylori* eradication in diabetic patients has been inferior to that in other patients. In Turkey, *H. pylori* eradication rate has been lower and clarithromycin resistance has been higher (64.3%) in patients with type 2 diabetes than in control group (35.7%; Demir *et al.*, 2009). The low eradication success in diabetic patients could be due to changed gastric microvasculature, diminished antibiotic absorption, gastroparesis and frequent antibiotic use for other infections (Ojetti *et al.*, 2005). Difference in eradication rates has been found mainly between diabetic and non-diabetic adults (Ojetti *et al.*, 2010). Unlike diabetic children, the diabetic adults often suffer from frequent infections and delayed gastric emptying.

Resistance rates to metronidazole

Primary *H. pylori* resistance rates to metronidazole are low (<10%) in Japan, moderate (20–40%) in most European countries and the USA but usually high (>40%) in some European (Finland, France and Poland) and many developing countries, reaching 70–100% in Iranian and Egyptian children and adults from Ethiopia, South Africa, Bangladesh, India, Vietnam and Colombia (Tables 9.2 and 9.3, Fig. 9.2). The resistance rates can be associated with prior use of nitroimidazoles for parasitic diseases such as giardiasis and amoebiasis in some developing countries or for gynaecological diseases such as trichomoniasis and bacterial vaginosis in many countries (Nahar *et al.*, 2004; Koivisto *et al.*, 2004). These factors have promoted higher resistance rates to metronidazole in the developing countries and in women than in men (Koivisto *et al.*, 2004; Megraud, 2004; Yuen *et al.*, 2005; Tanih *et al.*, 2010). For the same reasons, metronidazole resistance rates also vary within a country. In India, metronidazole resistance has been related to geographical regions (Thyagarajan *et al.*, 2003). Nitroimidazole use for dental infections also can render *H. pylori* strains resistant to metronidazole.

Prior use of metronidazole, even in the distant past, can result in *H. pylori* resistance to the agent.

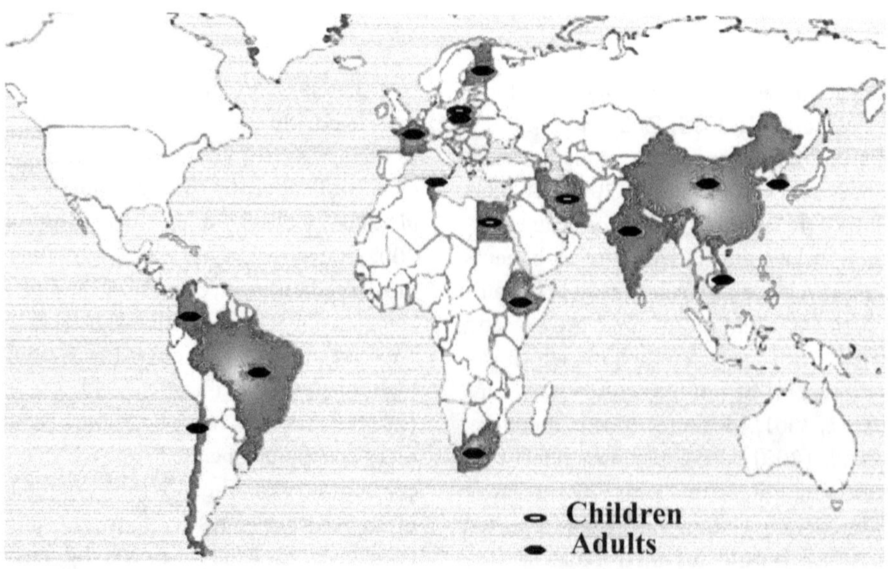

Figure 9.2 Countries reporting primary resistance rates of >40% for metronidazole over the last 10 years.

McMahon *et al.* (2003) reported metronidazole use 6–8 years before diagnosis in 21% of patients with metronidazole-resistant strains compared with only 3% of those harbouring susceptible strains. As for clarithromycin, data on prior metronidazole consumption over 10 years should be included, whenever possible, in the patient anamnesis to choose appropriate treatment.

Other risk factors for primary metronidazole resistance have been birthplace (in developing countries), age (older children or adults), disease status (non-ulcer) and strain (*cagA*-negative type; Table 9.4). For instance, a higher primary metronidazole resistance rate has been found in European children born outside Europe than in those born in Europe (Koletzko *et al.*, 2006). An association between metronidazole resistance and ethnical groups also has been observed in other (Elviss *et al.*, 2005; Yuen *et al.*, 2005), but not all, studies (Lopes *et al.*, 2005; Onder *et al.*, 2007). In Bulgaria, the primary metronidazole resistance rate has been found to be lower in patients born in villages (12.7%) than in those born in towns (27.6%; Boyanova *et al.*, 2009). Apparently, the above-mentioned differences reflect the different use of nitroimidazoles either in childhood, when the infection is acquired most often, or in adulthood, highlighting again the differences between developing and developed regions.

Like for the clarithromycin, the impact of gastric disease on primary resistance to metronidazole also should be considered. In Bulgaria, the primary metronidazole resistance rate has been lower (17%) in peptic ulcer than in non-ulcer patients (28.3%; Boyanova *et al.*, 2009). Lui *et al.* (2003) also have reported the non-ulcer patients to harbour three times more often (56.4%) metronidazole-resistant strains than the ulcer patients (19.8%). In Irish patients, Taneike *et al.* (2009) have associated higher overall metronidazole resistance rates with *cagA* negative strains compared with those with *cagA*$^+$ genotype. The higher metronidazole resistance rate in the non-ulcer patients compared with those with ulcers may be owing to the lower rate of virulent (*cagA+/vacA* s1) strains in the non-ulcer patients, such as in the case of clarithromycin resistance (Broutet *et al.*, 2003; Megraud, 2004). In general, lower damage of gastric mucosa and mucous layer in non-ulcer compared with ulcer patients has been suggested to reduce the access of antibiotics to the bacteria (Lui *et al.*, 2003).

Resistance rates to fluoroquinolones

Quinolones have been used in children and adolescents, but less often than in adults as the agents have the potential to damage cartilage or tendons (Zevit *et al.*, 2010). For this reason, primary fluoroquinolone resistance rates in *H. pylori* in children are lower than those in adults. Primary fluoroquinolone resistance rates in children have been only 0–7% (Rafeey *et al.*, 2007; Zevit *et al.*, 2010). In adults, resistance rates have been 0–3.1% in South Africa and Slovenia, higher (≥15%) in studies from Italy, France and Japan, and >20–33.8% in South Korea, China and one Italian study (Kim *et al.*, 2006a; Miyachi *et al.*, 2006; Cattoir *et al.*, 2007; Perna *et al.*, 2007; Zullo *et al.*, 2007; Mégraud, 2008; Tanih *et al.*, 2010; Wang *et al.*, 2010). Quinolone resistance has been a problem mainly in the developed countries.

Prior use of quinolone to treat urogenital or respiratory infections is a key risk factor for *H. pylori* resistance. Low (0–5.9%) fluoroquinolone resistance rates in Israeli children and in adults from Slovenia and Croatia can be linked to low (1.12–1.37 DIDs in 2007) outpatient quinolone (J01M) consumption, whereas higher (17–19%) resistance rates in adults from Italy and France can be associated with higher (2.19–3.53 DIDs in 2007) quinolone consumption (Samra *et al.*, 2002; Bago *et al.*, 2007; Cattoir *et al.*, 2007; ESAC, 2007; Zullo *et al.*, 2007; Jeverica *et al.*, 2010). Importantly, Carothers *et al.* (2007) reported a levofloxacin resistance rate of 2% in patients with no earlier use of fluoroquinolones but a 12–50-fold higher (≥25%) rate in 10 patients with more than one previous prescription of the drugs.

Other risk factors for *H. pylori* resistance to fluoroquinolones have been the years of study (recent years), implying an overall rise in resistance, age (>45 years), hospital centre and place of residence, and also the resistance to metronidazole or clarithromycin (Kim *et al.*, 2006b; Zullo *et al.*, 2007; Chang *et al.*, 2009). Importantly, levofloxacin resistance has been higher in clarithromycin or metronidazole-resistant strains that have shown high MICs of the agents, as compared with susceptible strains (Zullo *et al.*, 2007). In Taiwan, of nine levofloxacin-resistant strains, four have been metronidazole resistant, three have been clarithromycin resistant, and two have exhibited amoxicillin resistance (Yang *et al.*, 2010). This cross-resistance can be partially explained by simultaneous mutations, e.g. by the property of metronidazole to cause resistance in unrelated antibiotics (see above) and by efflux pumps but remains to be further evaluated.

Resistance rates to other antibiotics

In European and many other countries, primary *H. pylori* resistance to amoxicillin has been absent or negligible (0–<2%; Tables 9.2 and 9.3). However, in some developing countries (Ethiopia and Bangladesh) and in South Korea, the resistance rates have been higher (6–18.5%), reaching 39% in Brazilian adults and 59% in Iranian children (Godoy *et al.*, 2003; Asrat *et al.*, 2004; Nahar *et al.*, 2004; Kim *et al.*, 2006a; Rafeey *et al.*, 2007). Amoxicillin resistance can result from multiple prior treatments with the drug for other infections (Megraud, 2004).

Like the primary amoxicillin resistance, that to tetracycline has been low (0–9%, often <5%) in both children and adults except for that in South Korea (12.3%), Bangladesh (15%), Chile (27%) and South Africa (32.5% in consecutive patients; Nahar *et al.*, 2004; Kim *et al.*, 2006a; Vallejos *et al.*, 2007; Tanih *et al.*, 2010). Tetracycline use is discouraged for children aged below 8 years to prevent enamel discoloration but is employed in combined regimens in older children, adolescents and adults (Megraud, 2004; Zevit *et al.*, 2010). Thus, the primary tetracycline resistance rates have been very low (0–1.9%) in children, except for that (5%) in Iran (Rafeey *et al.*, 2007).

Local or national tetracycline consumption is a risk factor for the associated primary *H. pylori* resistance. Low level (MIC, 2 μg/ml) primary tetracycline resistance in 7.5% of strains from Kolkata (India), an endemic cholera region, has been explained by tetracycline use to treat the cholera (Datta *et al.*, 2005). In some countries with low (<1.1 DIDs) national outpatient use of tetracyclines in 2007 such as Israel, Spain and Slovenia, no primary tetracycline resistance in *H. pylori* has been reported (Agudo *et al.*, 2009; Jeverica *et al.*, 2010; Zevit *et al.*, 2010). However, this association is not universal. High (3.34 DIDs) tetracycline consumption in France in 2007 cannot be linked to the presence of tetracycline resistance in *H. pylori* (ESAC, 2007; Raymond *et al.*, 2010).

In Bulgaria, the primary tetracycline resistance rate in children (1.9%) and adults (4.4%) has been slightly higher than that (≤0.9%) in other European countries and can be explained by the high (>4.2 DIDs) national outpatient use of tetracyclines (J01A) in 1994–1999, although decreasing to 2.45 DIDs in 2007 (ESAC, 2007; Boyanova and Mitov, 2010; Vécsei *et al.*, 2010). This suggests that the decrease in *H. pylori* resistance rate to an antibiotic may take many years after the decline in the national consumption of the agent.

Primary *H. pylori* resistance to rifampicin/rifabutin has been uncommon (0-<5%) in most studies (Glocker *et al.*, 2007b; Chisholm and Owen, 2009; Vécsei *et al.*, 2010) as the agents are used seldom in current practice because of potential side-effects and rapid development of resistance in monotherapy. However, prior rifampicin use can increase the resistance rates or levels. Within 94 Japanese strains, high rifabutin MICs have been seen in all seven strains from patient previously treated by the agent versus 0% of those from a university hospital (Suzuki *et al.*, 2009b).

Furazolidone and nitrofurantoin-resistant *H. pylori* strains are also uncommon (Kwon *et al.*, 2001). In Bulgaria, the primary nitrofurantoin resistance has been only 1.8% in adults and 0% in children, as the drug has been used very seldom in the country (Boyanova *et al.*, 2008). Similarly, in Spain, resistance rates to both furazolidone (1.8%) and nitrofurantoin (0.6%) have been low (Alarcón *et al.*, 2005). Slightly higher furazolidone resistance rate has been found in Chinese adults (8.7%), Iranian children (9%) and Brazilian adults (13%; Godoy *et al.*, 2003; Su *et al.*, 2006; Rafeey *et al.*, 2007).

Selective pressure on the nitrofuran resistance rates in *H. pylori* by both furazolidone and metronidazole use has been observed. In China, *H. pylori* resistance to furazolidone has been detected in patients who had been treated by metronidazole beforehand (Su *et al.*, 2006). Moreover, furazolidone use as an antiseptic in fishponds in some

Chinese regions and the high fish consumption have increased *H. pylori* resistance rate to furazolidone. As a result, the furazolidone resistance in *H. pylori* in China has varied widely from 0–8.7% to 40% depending on the region (Su *et al.*, 2006).

Double and multi-drug resistance rates

Primary *H. pylori* resistance to both clarithromycin and metronidazole strongly compromises the success of many eradication regimens. The double resistance has been reported usually in <10% of strains in most countries but has reached 13.2–20% of the strains from Brazilian and French adults and Spanish, Tunisian and Polish children (Godoy *et al.*, 2003; Dzierzanowska-Fangrat *et al.*, 2005; Agudo *et al.*, 2009; Ben Mansour *et al.*, 2010; Boyanova and Mitov, 2010; Raymond *et al.*, 2010). In some studies, the double resistance to clarithromycin and metronidazole has been associated with female sex, older age, ethnicity and non-ulcer diseases (Kist and Glocker, 2004). Therefore, the risk factors for the primary resistance to both clarithromycin and metronidazole are among those for the single antibiotics.

Primary multidrug (triple and quadruple) resistance in *H. pylori* to antibiotics from different classes is still unusual and has been 0–1% in some countries but higher (3–4.2%) in Bulgaria, France, Chile and Iran (Table 9.2). Worrying primary multidrug resistance rates (14.2–23.1%) have been reported in Brazilian and South Korean adults (Godoy *et al.*, 2003; Kim *et al.*, 2006a; Bang *et al.*, 2007). High primary multidrug resistance may result from uncontrolled use of many antibiotics in some countries or from frequent individual use of many antibiotics for non-*pylori* infections.

Resistance in treated patients

Failure of first-line therapy sharply increases the antibiotic resistance rate in *H. pylori* and hampers the success of the next treatment regimens. As the term 'secondary resistance' is used often for all treated patients, mainly the post-treatment resistance is mentioned below.

Post-treatment resistance rates to antibiotics have been much higher, usually >3 times for clarithromycin and clarithromycin + metronidazole, and >1.5 times for metronidazole and quinolones, compared with those of the primary resistance (Samra *et al.*, 2002; Wong *et al.*, 2003; Dzierzanowska-Fangrat *et al.*, 2005; Soltermann *et al.*, 2005; Toracchio *et al.*, 2005; Agudo *et al.*, 2009; Boyanova, 2009; Raymond *et al.*, 2010). The post-treatment resistance rates to a given antibiotic are lower, of cause, than those of the secondary resistance. For example, in Bulgaria, the post-treatment resistance rate to clarithromycin has been 45.2% versus 57.1% for the secondary resistance rate (Boyanova, 2009). Post-treatment resistance to metronidazole or clarithromycin have been reported often in ≥57% of the strains (Table 9.5). For instance, in France, the post-treatment rates have been 3.6 times higher for clarithromycin, 4 times higher for clarithromycin + metronidazole and 1.7 times higher for quinolones compared with the primary resistance rates (Raymond *et al.*, 2010). Megraud (2004) has stated that clarithromycin use has promoted secondary resistance in *H. pylori* more often (in two out of three of the strains) than azithromycin use (23%), although the latter agent in combinations has provided poor eradication with 62% failures.

With increasing use of quinolones for *H. pylori* eradication, post-treatment resistance rates to these agents have increased most often to >17% according to the regimen (Table 9.5). In Japan, the overall post-treatment resistance rate to gatifloxacin in patients treated by first-, second- or third- line regimens has reached 47.9% (Nishizawa *et al.*, 2006).

Post-treatment amoxicillin resistance has been usually only <2.5%; however, Hsu *et al.* (2008) have reported high *H. pylori* resistance rates to the agent (in 17% of patients) and to levofloxacin (in 22%) after unsuccessful treatment with rabeprazole, bismuth subcitrate, amoxicillin and levofloxacin following two earlier unsuccessful treatments. Post-treatment levofloxacin resistance has been associated with that to metronidazole and clarithromycin (Branca *et al.*, 2004; Nishizawa *et al.*, 2006).

Multidrug resistance in treated patients is a topic of high concern. In Bulgarian adult patients, the multidrug resistance rate after treatment (15.4%) has been 3.7-fold higher than that

Table 9.5 Post-treatment and secondary *H. pylori* resistance in some studies

Countries	Patients	Years	No. of patients	Method	Resistance rates (%) CLA	MET	AMO	TET	QUI	RIF	FUR	MET + CLA	MDR[1]	References
Post treatment														
Europe	Children	99–02	196	Et, ADM	42	35	0.6	NA	NA	NA	NA	15.3	NA	Koletzko *et al*. (2006)
Austria	Children	02–08	19	Et	63.2	42.1	0	0	5.3	0		26.3	NA	Vécsei *et al*. (2010)
Bulgaria	Adults	05–08	124	BST, Et	45.2	50.8	2.4	10.6	14.5 (CP)	NA	NA	28.2	15.4	Boyanova (2009)
France	Adults	04–07	75	Et, PCR	68	74.6	0	0	20 (CP)	0	NA	53.3	10.6 (triple)	Raymond *et al*. (2010)
Germany	Adults, one therapy	03	24	NA (review)	~30	~45	NA	NA	~20	0	NA	~25	NA	Kist and Glocker (2004)
Germany	Adults, >1 treatments	03	22	NA (review)	~70	~55	NA	NA	NA	~5	NA	>55	NA	Kist and Glocker (2004)
Germany	Adults	01–05	805	Et	NA	NA	NA	NA	17.1	NA	NA	NA	11.8	Glocker *et al*. (2007a)
Germany	Total (mostly treated)	03–06	1585	ADM, MDDM	NA	NA	NA	NA	NA	1.4	NA	NA	NA	Glocker *et al*. (2007b)
Italy	Adults	NA	104	BST, ADM	76	64.4 (TIN)	1.9	NA	NA	1	NA	62.5	NA	Toracchio *et al*. (2005)
Italy	Adults	NA	67	Et	91	82.1	0	0	22.4 (LV)	NA	NA	77.6	NA	Branca *et al*. (2004)
Poland	Total	01–04	68	Et	54	68	0	0	NA	NA	NA	46	NA	Dzierzanowska-Fangrat *et al*. (2005)
Slovenia	Adults	07–09	372	Et	90.8	86.5	0	0	13.7 (CP)	NA	NA	NA	NA	Jeverica *et al*. (2010)
Spain	Children	02–06	35	Et	70.6	41.2	0	0	0	0	NA	26.5	NA	Agudo *et al*. (2009)
Switzerland	Adults	02–03	21	RFLP, Sq	90	NA	NA	NA	NA	NA	NA	NA	NA	Soltermann *et al*. (2005)
The Netherlands	Adults	97–02	58	MDDM	5.3	43	NA	NA	NA	NA	NA	NA	NA	Janssen *et al*. (2006)

Asia														
China	Adults	NA	24	Et, ADM	88	88	NA	NA	NA	NA	NA	75	NA	Wong *et al.* (2003)
Israel	Adults	NA	70	Et	65.7[2]	57.1[2]	NA	NA	18.6 (LV)	NA	NA	32.8[2]	12.8	Yahav *et al.* (2006)
Israel	Adults	00–01	28	Et	46.4	60.7	0	0	NA	NA	NA	28.6	NA	Samra *et al.* (2002)
Israel	Children	05–07	19	Et	42	52	0	0	0 (LV)	NA	NA	37	0	Zevit *et al.* (2010)
Japan	Adults	04–05	48	ADM	79.2	6.3	NA	NA	47.9 (GT)	NA	NA	4.2	4.2	Nishizawa *et al.* (2006)
South Korea	Adults	00–02	23	Et, RFLP, Sq	100	NA	0	NA	NA	NA	NA	NA	NA	Lee *et al.* (2005)
South Korea	Adults	04–05	67	Et, Sq	85.1	70.1	31.3	0	35.8 (CP), 32.8 (LV)	NA	NA	73.1	53.7 31.3 triple, 22.4 quadruple	Kim *et al.* (2006a)
South Korea	Adults	04–05	67	Et, Sq	85.1	NA	31.3	NA	NA	NA	NA	NA	NA	Kim *et al.* (2006a)
Taiwan	Adults	NA	47	Et	78.7	NA	0	NA	17 (LV)	NA	NA	NA	NA	Chang *et al.* (2009)
Turkey	Adults	NA	24	Et	27.2	NA	NA	NA	NA	NA	NA	NA	NA	Tüzün *et al.* (2008)
Vietnam	Adults (ulcer)	NA	63	Et	52	100	NA	NA	NA	NA	NA	NA	NA	Wheeldon *et al.* (2004)
Vietnam	Adults (ulcer)	NA	63	Et	52	100	NA	NA	NA	NA	NA	NA	NA	Wheeldon *et al.* (2004)
Secondary														
Bulgaria	Adults	05–08	16	BST, Et	57.1	88.8	0	NA	NA	NA	NA	NA	NA	Boyanova (2009)
Japan	Adults, dual therapy	NA	9	Et	88.9 (8/9)	NA	NA	NA	NA	NA	NA	NA	NA	Murakami *et al.* (2002)
Japan	Adults, triple therapy	NA	41	Et	51.2	NA	NA	NA	NA	NA	NA	NA	NA	Murakami *et al.* (2002)

[1]To different classes of antibiotics.

[2]Secondary resistance.

Et, E test; ADM, agar dilution method; BST, breakpoint susceptibility testing; MDDM modified disc-diffusion method; RFLP, random fragment length polymorphism; Sq, sequencing; AMO, amoxicillin; MET, metronidazole; TIN, tinidazole; CLA, clarithromycin; TET, tetracycline; QUI, fluoroquinolones; CP, ciprofloxacin; LV, levofloxacin; RIF, rifabutin; FUR, nitrofurans such as nitrofurantoin or furazolidone, MDR, multidrug; NA, not available; GT, gatifloxacin.

(4.2%) before treatment (Boyanova, 2009). In South Korea, very high post-treatment resistance rates have been reported to almost all antibiotics, including clarithromycin (85.1%), metronidazole (70.1%), amoxicillin (31.3%) and levofloxacin (32.8%), leading to triple resistance in >30% and quadruple resistance in >20% of strains (Kim *et al.*, 2006a). The recommendation of the Maastricht III Consensus Report (Malfertheiner *et al.*, 2007), stating that the rescue therapy after two treatment failures should be based on antimicrobial susceptibility testing, and the implementation of the antibiotic policy are important measures to counteract the risk for increase in multidrug resistance in *H. pylori*.

Evolution of the resistance

Over the last decade, along with the rising use of macrolides in many countries, especially developed ones, clarithromycin resistance rates have increased in Japan, South Korea, Ireland, Italy, Spain, Austria, Bulgaria and Hungary, whereas the rates have been stable in the UK and are falling in the Netherlands (Elviss *et al.*, 2004; Kim *et al.*, 2004; Janssen *et al.*, 2006; Buzás *et al.*, 2007; De Francesco *et al.*, 2007; Kobayashi *et al.*, 2007; Boyanova *et al.*, 2008; Boyanova and Mitov, 2010; O'Connor *et al.*, 2010; Vécsei *et al.*, 2010).

The 'jump' in primary clarithromycin resistance rates has been highest in South Korea (4.9- to 10-fold over 9 years), Italy (5-fold over 11 years in children) and Japan (1.5-fold over 4 years in adults and 1.3-fold over 8 years in children; Kim *et al.*, 2004; Bang *et al.*, 2007; De Francesco *et al.*, 2007; Kobayashi *et al.*, 2007; Lionetti *et al.* 2009; Kato and Fujimura, 2010). The resistance rate has increased less rapidly in Spain (1.7 times over 7 years in children), Austria (1.7-fold over 11 years in children) and Bulgaria (1.8-fold over 12 years; ESAC, 2007; Boyanova *et al.*, 2008; Agudo *et al.*, 2009; Vécsei *et al.*, 2010). Newer macrolides were introduced in Bulgaria in 1993 and the national consumption of macrolides increased from 0.8 DIDs in 2004 to 2.78 DIDs in 2007, which could explain the 3.1-fold increase in macrolide resistance rate from 1990–95 (6.2%) to 2007–09 (19.4%; Boyanova *et al.*, 2010). Conversely, in Finland, the primary resistance rates to clarithromycin were not significantly associated with macrolide consumption, although showing a tendency to decrease from 16% in 2003 to 4% in 2008 along with the overall decreasing consumption of the agents (from >2.5 DIDs in 2002–03 to about 1.8 DIDs in 2008; Kostamo *et al.*, 2011).

In Japan, growing macrolide use for pulmonary, ear, nose and throat infections was linked to increasing clarithromycin resistance rates in *H. pylori*, from 18.9% in 2002–2003 to 27.7% in 2004–2005 (Kobayashi *et al.*, 2007). In Italy, the primary resistance to clarithromycin increased more sharply (from 5.6% in 1995–2000 to 29.3% in 2001–2006) in children than in adults (twice from 1989–90 to 2004–05; De Francesco *et al.*, 2007; Lionetti *et al.* 2009). In France, worrying overall resistance (46% for clarithromycin and 25% for levofloxacin) was reported in consecutive patients in 2007–08 (Cambau *et al.*, 2009).

Importantly, the rise in antibiotic resistance can worsen the eradication success. In South Korea, the primary clarithromycin resistance rate has increased 10-fold (from 1.6% in 1996 to 16.7% in 2004–05) since 1996, leading to a decline in the eradication rate of the infection (Bang *et al.*, 2007). Conversely, the clarithromycin resistance rate decreased in the Netherlands from 1997 to 2002 and in Denmark, although only slightly (from 11% in 1998–2000 to 3% in 2001–03; Janssen *et al.*, 2006; Petersen *et al.*, 2006). This can be linked to the low antibiotic consumption in the Netherlands with sales of macrolides being four times lower than those in France, Spain, Italy and Greece. In Taiwan, the clarithromycin resistance rate was slightly lower in 2000–04 (6.7%) than in 1998–2000 (10.7%), which could be associated with the 1.4-fold increased use of the agent from 1999 to 2004 together with reduced (by 71%) erythromycin consumption from 2000 (0.45 DIDs) to 2003 (0.13 DIDs; Poon *et al.* 2009). Therefore, the cautious use of older macrolide agents, in addition to that of newer macrolides, can prevent the evolution of clarithromycin resistance in *H. pylori*.

Primary metronidazole resistance in *H. pylori* has been stable, mostly in the developed countries, e.g. in Austria, Bulgaria, Finland, the Netherlands, UK and Japan (Elviss *et al.*, 2004; Janssen *et al.*, 2006; Kobayashi *et al.*, 2007; Boyanova *et al.*,

2008; Vécsei *et al.*, 2010; Kostamo *et al.*, 2011). However, in Bulgarian untreated patients, the overall metronidazole MIC_{50} in 1990–1995 decreased about 4-fold by 2001–04 and about 8-fold by 2007–09 (Boyanova *et al.*, 2010). The changes in MICs of a given antibiotic are useful to foresee probable upcoming change in the resistance to the agent. The primary metronidazole resistance rate decreased from 43.3% in 1994–98 to 32% in 1999–2005 in French children (Kalach *et al.*, 2007). Similarly, the overall resistance rates to metronidazole dropped 2.5 times, from 27.7% in 1999 to 11.1% in 2003, in Portuguese children (Lopes *et al.*, 2005). However, in South Korea, metronidazole MICs have increased significantly and the resistance rates have increased slightly (1.2-fold over 16 years) owing to increasing metronidazole use and abuse for parasitic or genital infections (Kim *et al.*, 2004). Similarly, in Iran, the overall metronidazole resistance raised from 33% in 1997–2000 to 55.6% in 2005–08 (Siavoshi *et al.*, 2010). In Taiwan, the national consumption of metronidazole has been reduced since the introduction of antibiotic policy, resulting in decreasing overall metronidazole resistance rate in *H. pylori* (from 41.7% in 1998–2000 to 25.4% in 2001–2004) and associated increase in eradication by metronidazole-based regimens (Poon *et al.*, 2009). Likewise, in Japan, the low metronidazole resistance rates (4.9% in 2002–2003 and 3.3% in 2004–05) can be explained by the restriction of metronidazole use as an antiprotozoal drug (Kobayashi *et al.*, 2007).

Primary amoxicillin resistance has shown an increase in MICs in South Korea, stability or slight increase in Japan and a decrease in Bulgarian children (Kim *et al.*, 2004; Kobayashi *et al.*, 2007; Boyanova *et al.*, 2008; Boyanova and Mitov, 2010). However, in Bulgaria, the amoxicillin resistance was occasional, as it was found in children only in 1996–1999 and in adults only in 2007–2009 (Boyanova *et al.*, 2010). In Japan, however, the rate of *H. pylori* strains with amoxicillin MICs of 0.04–1 μg/ml increased from 1985 to 2003 (Watanabe *et al.*, 2005). No strain isolated in 1985–1996 was amoxicillin resistant, while the resistance rate since 1997 has varied between 0% and 8.8% (Watanabe *et al.*, 2005). Concern about future rise in amoxicillin resistance in *H. pylori* is the first detection of its β-lactamase (see above).

Primary *H. pylori* resistance to tetracycline has increased 2-fold in South Korea over 16 years but has been stable in Bulgaria (with MIC_{50} decreasing 3.8-fold since 1990–1995) owing to a 1.7-fold decrease in tetracycline consumption since 1994–1999 (Boyanova and Mitov, 2010; Boyanova *et al.*, 2010). In the UK, no increase in tetracycline resistance rate has been found (Lawson *et al.*, 2005). Conversely, in Iran, a sharp increase from 0% in 1997–2000 to 38.1% in 2005–2008 in overall tetracycline resistance was observed (Siavoshi *et al.*, 2010).

Because of the growing national consumption of fluoroquinolones, primary *H. pylori* resistance to the agents also has increased, especially in the developed countries such as France, Germany, South Korea and Taiwan (Boyanova and Mitov, 2010). In France, the primary *H. pylori* resistance to ciprofloxacin doubled from 7.3% in 2004–2005 to 14.1% in 2006–2007 (Raymond *et al.*, 2010). In Germany, overall ciprofloxacin resistance rates also doubled from 2003 (11.2%) to 2005 (22.1%; Glocker *et al.*, 2007a). In Finland, a tendency to increase has been shown for the primary resistance rates (Kostamo *et al.*, 2011). In Taiwan, *H. pylori* resistance rate to ciprofloxacin/levofloxacin jumped from 2.8% in 1998–2003 to 11.8% in 2004–2007 (Hung *et al.*, 2009).

Evaluating long-term trends in antibiotic resistance is useful to predict forthcoming changes in resistance patterns. In Bulgaria, the rise in ciprofloxacin resistance has been found only by extending the study period to 20 years. Primary quinolone resistance rate increased 7.7-fold from 1990–1995 (1.2%) to 2007–2009 (9.2%), probably because of the rising outpatient consumption of quinolones in Bulgaria from 0.034 DIDs in 1999 to 2.03 DIDs in 2007 (Boyanova *et al.*, 2010).

From 2002 to 2008, double resistance to clarithromycin and metronidazole showed no evolution in Austrian children (Vécsei *et al.*, 2010). Similarly, *H. pylori* prevalence of multidrug resistance showed no evolution in Bulgaria but exhibited a tendency to increase in South Korea from 1996 to 2004–2005 (Bang *et al.*, 2007; Boyanova, 2009) as in this country, the overall *H. pylori* resistance to many antibiotics such as clarithromycin, tetracycline, ciprofloxacin and levofloxacin increased from 2003–2005 to 2007–2009 (Hwang *et al.*, 2010).

In conclusion, *H. pylori* resistance to antibiotics often emerges from point mutations but also by efflux mechanisms, natural transformation, altered membrane permeability and, probably, β-lactamase. High primary resistance rates to clarithromycin and quinolones have been observed in *H. pylori*, mainly in the developed countries, and high rates of resistance to metronidazole, amoxicillin and tetracycline have been common in many developing countries. Primary *H. pylori* resistance depends on many patient or strain characteristics, most of which reflect the national and individual antibiotic consumption. Worryingly, both clarithromycin and quinolone resistance rates have grown sharply in many countries and multidrug resistance has been found in <5% to >14% of the strains. Post-treatment resistance rates to clarithromycin, metronidazole and quinolones have been often 1.5 to >3-fold higher than those of the primary resistance. Current *H. pylori* resistance to antibiotics does and will compromise more and more the treatment success. It is necessary to maintain frequently updated data on global and local resistance patterns of *H. pylori* to antibiotics, especially clarithromycin and fluoroquinolones, to choose, to improve or to change the eradication regimens aiming at higher treatment success. For this purpose, development and application of new eradication regimens and stringent implementation of the antibiotic policy are strongly required. Molecular and phenotypic susceptibility testing methods should be combined to detect the most important A2143G mutation, which could decrease sharply the treatment success of standard clarithromycin-based regimens. Susceptibility testing of *H. pylori* to antibiotics becomes increasingly important in the era of growing resistance and should be used more widely and more frequently, especially in the countries, exhibiting high resistance rates or multidrug resistance in *H. pylori*.

References

Agudo, S., Alarcón, T., Cibrelus, L., Urruzuno, P., Martínez, M., and López-Brea, M. (2009). High percentage of clarithromycin and metronidazole resistance in *Helicobacter pylori* clinical isolates obtained from Spanish children. Rev. Esp. Quimioter. *22*, 88–92.

Agudo, S., Pérez-Pérez, G., Alarcón, T., and López-Brea, M. (2010). High prevalence of clarithromycin resistance *Helicobacter pylori* strains and risk factors associated with resistance in Madrid, Spain. J. Clin. Microbiol. *48*, 3703–3707.

Aguemon, B., Struelens, M., Devière, J., Denis, O., Golstein, P., Salmon, I., and Nagy, N. (2005). Primary antibiotic resistance and effectiveness of *Helicobacter pylori* triple therapy in ulcero-inflammatory pathologies of the upper digestive tract. Acta Gastroenterol. Belg. *68*, 287–293.

Ahmad, N., Zakaria, W.R., Abdullah, S.A., and Mohamed, R. (2009). Characterization of clarithromycin resistance in Malaysian isolates of *Helicobacter pylori*. World J. Gastroenterol. *15*, 3161–3165.

Ahmed, D., Brooks, H., McConnell, M., and Barbezat, G. (2004). Antibiotic resistance in *Helicobacter pylori:* is it a problem in New Zealand? N.Z. Med. J. *117*, U1022.

Alarcon, T., Domingo, D., Prieto, N., and Lopez-Brea, M. (2000). PCR using 3′-mismatched primers to detect A2142C mutation in 23S rRNA conferring resistance to clarithromycin in *Helicobacter pylori* clinical isolates. J. Clin. Microbiol. *38*, 923–925.

Alarcon, T., Vega, A.E., Domingo, D., Martínez, M.J., and López-Brea, M. (2003). Clarithromycin resistance among *Helicobacter pylori* strains isolated from children: prevalence and study of mechanism of resistance by PCR-restriction fragment length polymorphism analysis. J. Clin. Microbiol. *41*, 486–499.

Alarcón, T., de la Obra, P., Domingo, D., García-Campos, J.A., Díaz-Regañón, J., and López-Brea, M. (2005). *In vitro* activity of furazolidone and nitrofurantoin in *Helicobacter pylori* clinical isolates and study of mutation rate. Rev. Esp. Quimioter. *18*, 313–318.

Anoushiravani, M., Falsafi, T., and Niknam, V. (2009). Proton motive force-dependent efflux of tetracycline in clinical isolates of *Helicobacter pylori*. J. Med. Microbiol. *58*, 1309–1313.

Asrat, D., Kassa, E., Mengistu, Y., Nilsson, I., and Wadström, T. (2004). Antimicrobial susceptibility pattern of *Helicobacter pylori* strains isolated from adult dyspeptic patients in Tikur Anbassa University Hospital, Addis Ababa, Ethiopia. Ethiop. Med. J. *42*, 79–85.

Bago, P., Vcev, A., Tomic, M., Rozankovic, M., Marusić, M., and Bago J. (2007). High eradication rate of *H. pylori* with moxifloxacin-based treatment, a randomized controlled trial. Wien. Klin. Wochenschr. *119*, 372–378.

Bang, S.Y., Han, D.S., Eun, C.S., Kim, J.E., Ahn, S.B., Sohn, J.H., Jeon, Y.C., and Kang, J.O. (2007). Changing patterns of antibiotic resistance of *Helicobacter pylori* in patients with peptic ulcer disease. Korean J. Gastroenterol. *50*, 356–362.

Ben Mansour, K., Burucoa, C., Zribi, M., Masmoudi, A., Karoui, S., Kallel, L., Chouaib, S., Matri, S., Fekih, M., Zarrouk, S., *et al.* (2010). Primary resistance to clarithromycin, metronidazole and amoxicillin of *Helicobacter pylori* isolated from Tunisian patients with peptic ulcers and gastritis: a prospective multicentre study. Ann. Clin. Microbiol. Antimicrob. *9*, 22.

Bogaerts, P., Berhin, C., Nizet, H., and Glupczynski, Y. (2006). Prevalence and mechanisms of resistance to fluoroquinolones in *Helicobacter pylori* strains from patients living in Belgium. Helicobacter *11*, 441–445.

Boyanova, L., Gergova, G., Nikolov, R., Davidkov, L., Kamburov, V., Jelev, C., and Mitov I. (2008). Prevalence and evolution of *Helicobacter pylori* resistance to 6 antibacterial agents over 12 years and correlation between susceptibility testing methods. Diagn. Microbiol. Infect. Dis. *60*, 409–415.

Boyanova, L. (2009). Prevalence of multidrug-resistant *Helicobacter pylori* in Bulgaria. J. Med. Microbiol. *58*, 930–935.

Boyanova, L., Ilieva, J., Gergova, G., Spassova, Z., Nikolov, R., Davidkov, L., Evstatiev, I., Kamburov, V., Katsarov, N., and Mitov I. (2009). Evaluation of clinical and socio-demographic risk factors for antibacterial resistance of *Helicobacter pylori* in Bulgaria. J. Med. Microbiol. *58*, 94–100.

Boyanova, L., and Mitov, I. (2010). Geographic map and evolution of primary *Helicobacter pylori* resistance to antibacterial agents. Expert. Rev. Anti Infect. Ther. *8*, 59–70.

Boyanova, L., Nikolov, R., Gergova, G., Evstatiev, I., Lazarova, E., Kamburov, V., Panteleeva, E., Spassova, Z., and Mitov, I. (2010). Two-decade trends in primary *Helicobacter pylori* resistance to antibiotics in Bulgaria. Diagn. Microbiol. Infect. Dis. *67*, 319–326.

Branca, G., Spanu, T., Cammarota, G., Schito, A. M., Gasparini, A., Gasparini, G. B., and Fadda, G. (2004). High levels of dual resistance to clarithromycin and metronidazole and *in vitro* activity of levofloxacin against *Helicobacter pylori* isolates from patients after failure of therapy. Int. J. Antimicrob. Agents *24*, 433–438.

Broutet, N., Marais, A., Lamouliatte, H., de Mascarel, A., Samoyeau, R., Salamon, R., and Mégraud, F. (2001). *cagA* status and eradication treatment outcome of anti*Helicobacter pylori* triple therapies in patients with nonulcer dyspepsia J. Clin. Microbiol. *39*,1319–1322.

Broutet, N., Tchamgoue, S., Pereira, E., Lamouliatte, H., Salamon, R., and Mégraud, F. (2003). Risk factors for failure of *Helicobacter pylori* therapy-results of an individual data analysis of 2751 patients. Aliment. Pharmacol. Ther. *17*, 99–109.

Burucoa, C., Garnier, M., Silvain, C., and Fauchère, J.L. (2008). Quadruplex real-time PCR assay using allele-specific scorpion primers for detection of mutations conferring clarithromycin resistance to *Helicobacter pylori*. J. Clin. Microbiol. *46*, 2320–2326.

Buzás, G.M., Lotz, G., and Kiss, A. (2007). The epidemiology of clarithromycin resistance of *Helicobacter pylori* infection in Hungary. Orv. Hetil. *148*, 1461–1467.

Cambau, E., Allerheiligen, V., Coulon, C., Corbel, C., Lascols, C., Deforges, L., Soussy, C.J., Delchier, J.C., and Megraud, F. (2009). Evaluation of a new test, genotype HelicoDR, for molecular detection of antibiotic resistance in *Helicobacter pylori*. J. Clin. Microbiol. *47*, 3600–3607.

Cantón, R., de Argila, C.M., de Rafael, L., and Baquero, F. (2001). Antimicrobial resistance in *Helicobacter pylori*. Rev. Med. Microbiol. *12*, 47–61.

Caristo, E., Parola, A., Rapa, A., Vivenza, D., Raselli, B., Dondi, E., Boldorini, R., and Oderda, G. (2008). Clarithromycin resistance of *Helicobacter pylori* strains isolated from children' gastric antrum and fundus as assessed by fluorescent in-situ hybridization and culture on four-sector agar plates. *Helicobacter 13*, 557–563.

Carothers, J.J., Bruce, M.G., Hennessy, T.W., Bensler, M., Morris, J.M., Reasonover, A.L., Hurlburt, D.A., Parkinson, A.J., Coleman, J.M., and McMahon, B.J. (2007). The relationship between previous fluoroquinolone use and levofloxacin resistance in *Helicobacter pylori* infection. Clin. Infect. Dis. *44*, e5-e8.

Castro-Fernández, M., and Vargas-Romero, J. (2009). Infection with *Helicobacter pylori*. Prevalence, research and impact of antibiotic resistance. Rev. Esp. Enferm. Dig. *101*, 743–756.

Cattoir, V., Nectoux, J., Lascols, C., Deforges, L., Delchier, J. C., Megraud, F., Soussy, C. J., and Cambau, E. (2007). Update on fluoroquinolone resistance in *Helicobacter pylori:* new mutations leading to resistance and first description of a *gyrA* polymorphism associated with hypersusceptibility. Int. J. Antimicrob. Agents. *29*, 389–396.

Chang, W.L., Sheu, B.S., Cheng, H.C., Yang, Y.J., Yang, H.B., and Wu, J.J. (2009). Resistance to metronidazole, clarithromycin and levofloxacin of *Helicobacter pylori* before and after clarithromycin-based therapy in Taiwan. J. Gastroenterol. Hepatol. *24*, 1230–1235.

Cheng, H., Hu, F.L., and Li, J. (2006). Influence of resistance of *Helicobacter pylori* to antibiotics on the *Helicobacter pylori* eradication regimens. Zhonghua Yi Xue Za Zhi *86*, 2679–2682.

Chisholm, S.A., Teare, E.L., Davies, K., and Owen, R.J. (2007). Surveillance of primary antibiotic resistance of *Helicobacter pylori* at centres in England and Wales over a six-year period (2000–2005). Eurosurveillance *12*, E3-E4.

Chisholm, S.A., and Owen, R.J. (2009). Frequency and molecular characteristics of ciprofloxacin- and rifampicin-resistant *Helicobacter pylori* from gastric infections in the UK. J. Med. Microbiol. *58*, 1322–1328.

Co, E.M., and Schiller, N.L. (2006). Resistance mechanisms in an *in vitro*-selected amoxicillin-resistant strain of *Helicobacter pylori*. Antimicrob. Agents Chemother. *50*, 4174–4176.

Datta, S., Chattopadhyay, S., Patra, R., De, R., Ramamurthy, T., Hembram, J., Chowdhury, A., Bhattacharya, S.K., Berg, D.E., Nair, G.B., *et al.* (2005). Most *Helicobacter pylori* strains of Kolkata in India are resistant to metronidazole but susceptible to other drugs commonly used for eradication and ulcer therapy. Aliment. Pharmacol. Ther. 22, 51–57.

De Francesco, V., Margiotta, M., Zullo, A., Hassan, C., Giorgio, F., Burattini, O., Stoppino, G., Cea, U., Pace, A., Zotti, M., *et al.* (2007). Prevalence of primary clarithromycin resistance in *Helicobacter pylori* strains over a 15 year period in Italy. J. Antimicrob. Chemother. *59*, 783–785.

De Francesco, V., Ierardi, E., Hassan, C., and Zullo, A. (2009). Furazolidone therapy for *Helicobacter pylori:* is it effective and safe? World. J. Gastroenterol. *15*, 1914–1915.

De Francesco, V., Zullo, A., Ierardi, E., Giorgio, F., Perna, F., Hassan, C., Morini, S., Panella, C., and Vaira, D. (2010). Phenotypic and genotypic *Helicobacter pylori* clarithromycin resistance and therapeutic outcome: benefits and limits. J. Antimicrob. Chemother. *65*, 327–332.

Demir, M., Gokturk, H.S., Ozturk, N.A., Arslan, H., Serin, E., and Yilmaz, U. (2009). Clarithromycin resistance and efficacy of clarithromycin-containing triple eradication therapy for *Helicobacter pylori* infection in type 2 diabetes mellitus patients. South Med. J. *102*, 1116–1120.

Dzierzanowska-Fangrat, K., Rozynek, E., Celińska-Cedro, D., Jarosz, M., Pawłowska, J., Szadkowski, A., Budzyńska, A., Nowak, J., Romańczuk, W., Prosiecki, R., *et al.* (2005). Antimicrobial resistance of *Helicobacter pylori* in Poland: a multicentre study. Int. J. Antimicrob. Agents. *26*, 230–234.

Elviss, N.C., Owen, R.J., Xerry, J., Walker, A.M., and Davies, K. (2004). *Helicobacter pylori* antibiotic resistance patterns and genotypes in adult dyspeptic patients from a regional population in North Wales. J. Antimicrob. Chemother. *54*, 435–440.

Elviss, N.C., Owen, R.J., Breathnach, A., Palmer, C., and Shetty, N. (2005). *Helicobacter pylori* antibiotic-resistance patterns and risk factors in adult dyspeptic patients from ethnically diverse populations in central and south London during 2000. J. Med. Microbiol. *54*, 567–574.

ESAC – European Surveillance of Antimicrobial Consumption. (2007). ESAC Yearbook 2007. http://www.esac.ua.ac.be/

Falsafi, T., Ehsani, A., and Niknam, V. (2009). The role of active efflux in antibiotic – resistance of clinical isolates of *Helicobacter pylori.* Indian J. Med. Microbiol. *27*, 335–340.

Fariña, N., Kasamatsu, E., Samudio, M., Morán, M., Sanabria, R., and Laspina, F. (2007). Antimicrobial susceptibility of *H. pylori* strains obtained from Paraguayan patients. Rev. Med. Chil. *135*, 1009–1014.

Feydt-Schmidt, A., Rüssmann, H., Lehn, N., Fischer, A., Antoni, I., Störk, D., and Koletzko, S. (2002). Fluorescence *in situ* hybridization vs. epsilometer test for detection of clarithromycin-susceptible and clarithromycin-resistant *Helicobacter pylori* strains in gastric biopsies from children. Aliment. Pharmacol. Ther. *16*, 2073–2079.

Filipec Kanizaj, T., Katicic, M., Skurla, B., Ticak, M., Plecko, V., and Kalenic, S. (2009). *Helicobacter pylori* eradication therapy success regarding different treatment period based on clarithromycin or metronidazole triple-therapy regimens. Helicobacter *14*, 29–35.

Fontana, C., Favaro, M., Minelli, S., Criscuolo, A.A., Pietroiusti, A., Galante, A., and Favalli, C. (2002). New site of modification of 23S rRNA associated with clarithromycin resistance of *Helicobacter pylori* clinical isolates. Antimicrob. Agents Chemother. *46*, 3765–3769.

Francavilla, R., Lionetti, E., Castellaneta, S., Margiotta, M., Piscitelli, D., Lorenzo, L., Cavallo, L., and Ierardi, E. (2010). Clarithromycin-resistant genotypes and eradication of *Helicobacter pylori.* J. Pediatr. *157*, 228–232.

Friedman, G.D., Jiang, S.F., Udaltsova, N., Quesenberry, C.P. Jr, Chan, J, and Habel, L,A. (2009). Epidemiologic evaluation of pharmaceuticals with limited evidence of carcinogenicity. Int. J. Cancer. *125*, 2173–2178.

Fujimura, S., Kato, S., Iinuma, K., and Watanabe, A. (2004). *In vitro* activity of fluoroquinolone and the *gyrA* gene mutation in *Helicobacter pylori* strains isolated from children. J. Med. Microbiol. *53*, 1019–1022.

Furuta, T., Soya, Y., Sugimoto, M., Shirai, N., Nakamura, A., Kodaira, C., Nishino, M., Okuda, M., Okimoto, T., Murakami, K., *et al.* (2007). Modified allele-specific primer-polymerase chain reaction method for analysis of susceptibility of *Helicobacter pylori* strains to clarithromycin. J. Gastroenterol. Hepatol. *22*, 1810–1815.

Garrido, L., and Toledo, H. (2007). Novel genotypes in *Helicobacter pylori* involving domain V of the 23S rRNA gene. Helicobacter *12*, 505–509.

Gerrits, M.M., Godoy, A.P., Kuipers, E.J., Ribeiro, M.L., Stoof, J., Mendonça, S., van Vliet, A.H., Pedrazzoli, J., Jr., and Kusters, J.G. (2006). Multiple mutations in or adjacent to the conserved penicillin-binding protein motifs of the penicillin-binding protein 1A confer amoxicillin resistance to *Helicobacter pylori.* Helicobacter *11*, 181–187.

Ghose, C., Perez-Perez, G.I., van Doorn, L.J., Domínguez-Bello, M.G., and Blaser, M.J. (2005). High frequency of gastric colonization with multiple *Helicobacter pylori* strains in Venezuelan subjects. J. Clin. Microbiol. *43*, 2635–2641.

Glocker, E., Stueger, H.P., and Kist, M. (2007a). Quinolone resistance in *Helicobacter pylori* isolates in Germany. Antimicrob. Agents Chemother. *51*, 346–349.

Glocker, E., Bogdan, C., and Kist, M. (2007b). Characterization of rifampicin-resistant clinical *Helicobacter pylori* isolates from Germany. J. Antimicrob. Chemother. *59*, 874–879.

Godoy, A.P., Ribeiro, M.L., Benvengo, Y.H., Vitiello, L., Miranda Mde, C., Mendonca, S., and Pedrazzoli, J.Jr. (2003). Analysis of antimicrobial susceptibility and virulence factors in *Helicobacter pylori* clinical isolates. BMC Gastroenterol. 3, 20.

Graham, D.Y., and Shiotani, A. (2008). New concepts of resistance in the treatment of *Helicobacter pylori* infections. Nat. Clin. Pract. Gastroenterol. Hepatol. *5*, 321–331.

Graham, D.Y., Lu, H., and Yamaoka, Y. (2008). Therapy for *Helicobacter pylori* infection can be improved: sequential therapy and beyond. Drugs *68*, 725–736.

Graham, D.Y., and Fischbach, L. (2010). *Helicobacter pylori* treatment in the era of increasing antibiotic resistance. Gut 59, 1143–1153.

Henao, S.C., Otero, W., Ángel, L.A., and Martínez, J.D. (2009). Primary resistance to metronidazole in *Helicobacter pylori* isolates in adult patients in Bogotá, Colombia. Rev. Col. Gastroenterol. *24*, 10–15.

Hirata, K., Suzuki, H., Nishizawa, T., Tsugawa, H., Muraoka, H., Saito, Y., Matsuzaki, J., and Hibi, T. (2010). Contribution of efflux pumps to clarithromycin resistance in *Helicobacter pylori*. J. Gastroenterol. Hepatol. 25 (*Suppl 1*), S75–S79.

Ho, S.L., Tan, E.L., Sam, C.K., and Goh, K.L. (2010). Clarithromycin resistance and point mutations in the 23S rRNA gene in *Helicobacter pylori* isolates from Malaysia. J. Dig. Dis. *11*, 101–105.

Horiki, N., Omata, F., Uemura, M., Suzuki, S., Ishii, N., Iizuka, Y., Fukuda, K., Fujita, Y., Katsurahara, M., Ito, T., *et al.* (2009). Annual change of primary resistance to clarithromycin among *Helicobacter pylori* isolates from 1996 through 2008 in Japan. Helicobacter *14*, 86–90.

Hsu, P.I., Wu, D.C., Chen, A., Peng, N.J., Tseng, H.H., Tsay, F.W., Lo, G.H., Lu, C.Y., Yu, F.J., and Lai, K.H. (2008). Quadruple rescue therapy for *Helicobacter pylori* infection after two treatment failures. Eur. J. Clin. Invest. *38*, 404–409.

Hu, C.-T., Chiou, P.-Y., Wu, C.-C., Tseng, Y.-H., Chang, Y.-J., and Lin, N.-T. (2009). Analysis of resistance to clarithromycin and virulence markers in *Helicobacter pylori* clinical isolates from Eastern Taiwan. Tzu Chi Med. J. *21*, 123–128.

Hung, K.-H., Sheu, B.-S., Chang, W.-L., Wu, H.-M., Liu, C.-C., and Wu, J.-J. (2009). Prevalence of primary fluoroquinolone resistance among clinical isolates of *Helicobacter pylori* at a University Hospital in Southern Taiwan. Helicobacter *14*, 61–65.

Hwang, T.J., Kim, N., Kim, H.B., Lee, B.H., Nam, R.H., Park, J.H., Lee, M.K., Park, Y.S., Lee, D.H., Jung, H.C., *et al.* (2010). Change in antibiotic resistance of *Helicobacter pylori* strains and the effect of A2143G point mutation of 23S rRNA on the eradication of *H. pylori* in a single center of Korea. J. Clin. Gastroenterol. *44*, 536–543.

Janssen, M.J., Hendrikse, L., de Boer, S.Y., Bosboom, R., de Boer, W.A., Laheij, R.J., and Jansen, J.B. (2006). *Helicobacter pylori* antibiotic resistance in a Dutch region, trends over time. Neth. J. Med. *64*, 191–195.

Jenks, P.J., and Edwards, D.I. (2002). Metronidazole resistance in *Helicobacter pylori*. Int. J. Antimicrob. Agents. *19*, 1–7.

Jeverica, S., Tepeš, B., Ihan, A., and Skvarč, M. (2010). Primary resistance of *Helicobacter pylori*. Zdrav. Vestn. *79*, 25–30.

John Albert, M., Al-Mekhaizeem, K., Neil, L., Dhar, R., Dhar, P.M., Al-Ali, M., Al-Abkal, H.M., and Haridas, S. (2006). High prevalence and level of resistance to metronidazole, but lack of resistance to other antimicrobials in *Helicobacter pylori,* isolated from a multiracial population in Kuwait. Aliment. Pharmacol. Ther. *24*, 1359–1366.

Kaakoush, N.O., Asencio, C., Mégraud, F., and Mendz, G.L. (2009). A redox basis for metronidazole resistance in *Helicobacter pylori*. Antimicrob. Agents Chemother. *53*, 1884–1891.

Kalach, N., Serhal, L., Asmar, E., Campeotto, F., Bergeret, M., Dehecq, E., Spyckerelle, C., Charkaluk, M.L., Decoster, A., Dupont, C., *et al.* (2007). *Helicobacter pylori* primary resistant strains over 11 years in French children. Diagn. Microbiol. Infect. Dis. *59*, 217–222.

Kato, S., and Fujimura, S. (2010). Primary antimicrobial resistance of *Helicobacter pylori* in children during the past 9 years. Pediatr. Int. *52*, 187–190.

Kawai, T., Yamagishi, T., Yagi, K., Kataoka, M., Kawakami, K., Sofuni, A., Itoi, T., Sakai, Y., Moriyasu, F., Osaka, Y., *et al.* (2008). Tailored eradication therapy based on fecal *Helicobacter pylori* clarithromycin sensitivities. J. Gastroenterol. Hepatol. 23 (Suppl 2*)*, S171–174.

Khashei, R., Shojaei, H., Adibi, P., Shavakhi, A., Aslani, M.M., and Naser, A.D. (2008). Genetic diversity and drug resistance of *Helicobacter pylori* strains in Isfahan, Iran. Iran. J. Basic Med. Sci. *11*, 174–182.

Kim, J.M., Kim, J.S., Jung, H.C., Kim, N., Kim, Y.J., and Song, I.S. (2004). Distribution of antibiotic MICs for *Helicobacter pylori* strains over a 16-year period in patients from Seoul, South Korea. Antimicrob. Agents Chemother. *48*, 4843–4847.

Kim, J.M., Kim, J.S., Kim, N., Kim, S.G., Jung, H.C., and Song, I.S. (2006a). Comparison of primary and secondary antimicrobial minimum inhibitory concentrations for *Helicobacter pylori* isolated from Korean patients. Int. J. Antimicrob. Agents. *28*, 6–13.

Kim, N., Kim, J.M., Kim, C.H., Park, Y.S., Lee, D.H., Kim, J.S., Jung, H.C., and Song, I.S. (2006b). Institutional difference of antibiotic resistance of *Helicobacter pylori* strains in Korea. J. Clin. Gastroenterol. *40*, 683–687.

Kim, J.M., Kim, J.S., Kim, N., Kim, Y.J., Kim, I.Y., Chee, Y.J., Lee, C.H., and Jung, H.C. (2008). Gene mutations of 23S rRNA associated with clarithromycin resistance in *Helicobacter pylori* strains isolated from Korean patients. J. Microbiol. Biotechnol. *18*, 1584–1589.

Kist, M., and Glocker, E. (2004). ResiNet–a nationwide German sentinel study for surveillance and analysis of antimicrobial resistance in *Helicobacter pylori*. Eurosurveillance 9, 44–45.

Kist, M. (2007). *Helicobacter pylori*: primary antimicrobial resistance and first-line treatment strategies. Eurosurveillance *12*, E1-E2.

Kobayashi, I., Murakami, K., Kato, M., Kato, S., Azuma, T., Takahashi, S., Uemura, N., Katsuyama, T., Fukuda, Y., Haruma, K., *et al.* (2007). Changing antimicrobial susceptibility epidemiology of *Helicobacter pylori* strains in Japan between 2002 and 2005. J. Clin. Microbiol. 45, 4006–4010.

Koivisto, T.T., Rautelin, H.I., Voutilainen, M.E., Niemelä, S.E., Heikkinen, M., Sipponen, P.I., and Färkkilä, M.A. (2004). Primary *Helicobacter pylori* resistance to metronidazole and clarithromycin in the Finnish population. Aliment. Pharmacol. Ther. *19*, 1009–1017.

Koletzko, S., Richy, F., Bontems, P., Crone, J., Kalach, N., Monteiro, M.L., Gottrand, F., Celinska-Cedro,

D., Roma-Giannikou, E., Oderda, G., *et al.* (2006). Prospective multicentre study on antibiotic resistance of *Helicobacter pylori* strains obtained from children living in Europe. Gut 55, 1711–1716.

Kostamo, P., Veijola, L., Oksanen, A., Sarna, S., and Rautelin, H. (2011). Recent trends in primary antimicrobial resistance of *Helicobacter pylori* in Finland. Int. J. Antimicrob. Agents 37, 22–25.

Kutschke, A., and de Jonge, B.L. (2005). Compound efflux in *Helicobacter pylori*. Antimicrob. Agents Chemother. *49*, 3009–3010.

Kwon, D.H., Lee, M., Kim, J.J, Kim, J.G., El-Zaatari, F.A., Osato, M.S., and Graham, D.Y. (2001). Furazolidone- and nitrofurantoin-resistant *Helicobacter pylori*: prevalence and role of genes involved in metronidazole resistance. Antimicrob. Agents Chemother. *45*, 306–308.

Kwon, D.H., Dore, M.P., Kim, J.J., Kato, M., Lee, M., Wu, J.Y., and Graham, D.Y. (2003). High-level beta-lactam resistance associated with acquired multidrug resistance in *Helicobacter pylori*. Antimicrob. Agents Chemother. *47*, 2169–2178.

Lai, C.H., Kuo, C.H., Chen, P.Y., Poon, S.K., Chang, C.S., and Wang, W.C. (2006). Association of antibiotic resistance and higher internalization activity in resistant *Helicobacter pylori* isolates. J. Antimicrob. Chemother. *57*, 466–471.

Laine, L., Hunt, R., El-Zimaity, H., Nguyen, B., Osato, M., and Spénard, J. (2003). Bismuth-based quadruple therapy using a single capsule of bismuth biskalcitrate, metronidazole, and tetracycline given with omeprazole versus omeprazole, amoxicillin, and clarithromycin for eradication of *Helicobacter pylori* in duodenal ulcer patients: a prospective, randomized, multicenter, North American trial. Am. J. Gastroenterol. *98*, 562–567.

Lamouliatte, H., Megraud, F., Delchier, J.C., Bretagne, J.F., Courillon-Mallet, A., De Korwin, J.D., Fauchere, J.L., Labigne, A., Flejou, J.F., and Barthelemy, P. (2003). Second-line treatment for failure to eradicate *Helicobacter pylori*: a randomized trial comparing four treatment strategies. Aliment. Pharmacol. Ther. *18*, 791–797.

Lawson, A.J., Elviss, N.C., and Owen, R.J. (2005). Real-time PCR detection and frequency of 16S rDNA mutations associated with resistance and reduced susceptibility to tetracycline in *Helicobacter pylori* from England and Wales. J. Antimicrob. Chemother. *56*, 282–286.

Lee, J.H., Shin, J.H., Roe, I.H., Sohn, S.G., Lee, J.H., Kang, G.H., Lee, H.K., Jeong, B.C., and Lee, S.H. (2005). Impact of clarithromycin resistance on eradication of *Helicobacter pylori* in infected adults. Antimicrob. Agents Chemother. *49*, 1600–1603.

Lee, H.K., Chae, H.S., Kang, J.O., Lee, M.K., Sung, H., Kim, M.N., Lee, J., Lee, M., and Shim, K.N. (2008). Multicenter Study for the frequency of 23S rRNA point mutations associated with clarithromycin resistance in *Helicobacter pylori* in Korea. Korean J. Clin. Microbiol. *11*, 84–89.

Lee, G.H., Jeong, J.Y., Chung, J.W., Nam, W.H., Lee, S.M., Pak, J.H., Choi, K.D., Song, H.J., Jung, H.Y., and Kim, J.H. (2009). The *Helicobacter pylori* Mfd protein is important for antibiotic resistance and DNA repair. Diagn. Microbiol. Infect. Dis. *65*, 454–456.

Li, Y., and Dannelly, H.K. (2006). Inactivation of the putative tetracycline resistance gene *Hp1165* in *Helicobacter pylori* led to loss of inducible tetracycline resistance. Arch. Microbiol. *185*, 255–262.

Lin, E.A., Zhang, X.-S., Levine, S.M., Gill, S.R., Falush, D., and Blaser, M.J. (2009). Natural transformation of *Helicobacter pylori* involves the integration of short DNA fragments interrupted by gaps of variable size. PLoS Pathog. *5*, e1000337.

Lionetti, E., Fico, S., Leone, G., Fontana,C., Castellaneta, S., Indrio, F., Cavallo, L., Ierardi, E., Francavilla, R. (2009). Prevalence of *Helicobacter pylori* primary clarithromycin resistance over 11 years in Italian children and its relationship with consumption of macrolides. Dig. Liv. Dis. *41* (*Suppl.* 3), S201.

Liu, Z.Q., Zheng, P.Y., and Yang, P.C. (2008). Efflux pump gene *hefA* of *Helicobacter pylori* plays an important role in multidrug resistance. World J. Gastroenterol. *14*, 5217–5222.

Lopes, A.I., Oleastro, M., Palha, A., Fernandes, A., and Monteiro, L. (2005). Antibiotic-resistant *Helicobacter pylori* strains in Portuguese children. Pediatr. Infect. Dis J. *24*, 404–409.

Lui, S.Y., Yeoh, K.G., and Ho, B. (2003). Metronidazole-resistant *Helicobacter pylori* is more prevalent in patients with nonulcer dyspepsia than in peptic ulcer patients in a multiethnic Asian population. J. Clin. Microbiol. *41*, 5011–5014.

Mahachai, V., Thong-Ngam, D., Noophun, P., Tumwasorn, S., and Kullavanijaya, P. (2006). Efficacy of clarithromycin-based triple therapy for treating *Helicobacter pylori* in Thai non-ulcer dyspeptic patients with clarithromycin-resistant strains. J. Med. Assoc. Thai. *89* (*Suppl* 3), S74–S78.

Malfertheiner, P., Megraud, F., O'Morain, C., Bazzoli, F., El-Omar, E., Graham, D., Hunt, R., Rokkas, T., Vakil, N., Kuipers, E.J., and the European Helicobacter Study Group (EHSG). (2007). Current concepts in the management of *Helicobacter pylori* infection: the Maastricht III Consensus Report. Gut *56*, 772–781.

Matteo, M.J., Granados, G., Olmos, M., Wonaga, A., and Catalano, M. (2008). *Helicobacter pylori* amoxicillin heteroresistance due to point mutations in PBP-1A in isogenic isolates. J. Antimicrob. Chemother. *61*, 474–477.

McMahon, B.J., Hennessy, T.W., Bensler, J.M., Bruden, D.L., Parkinson, A.J., Morris, J.M., Reasonover, A.L., Hurlburt, D.A., Bruce, M.G., Sacco, F., *et al.* (2003). The relationship among previous antimicrobial use, antimicrobial resistance, and treatment outcomes for *Helicobacter pylori* infections. Ann. Intern. Med. *139*, 463–469.

Mégraud, F., Lehn, N., Lind, T., Bayerdörffer, E., O'Morain, C., Spiller, R., Unge, P., van Zanten, S.V., Wrangstadh, M., and Burman, C.F. (1999). Antimicrobial susceptibility testing of *Helicobacter pylori* in a large multicenter trial, the MACH 2 Study. Antimicrob. Agents Chemother. *43*, 2747–2752.

Megraud, F. (2004). *H pylori* antibiotic resistance, prevalence, importance, and advances in testing. Gut *53*, 1374–1384.

Megraud, F., and Lehours, P. (2007). *Helicobacter pylori* detection and antimicrobial susceptibility testing. Clin. Microbiol. Rev. *20*, 280–322.

Mégraud, F. (2008). *Helicobacter pylori:* bacteriological aspects, diagnostic methods, and susceptibility to antibiotics. Presse Medicale 37, 507–512.

Mégraud, F., and Corti, R. (2009). Bacterial resistance of *Helicobacter pylori* in the world in 2009. Acta Gastroenterol. Latinoam. *39*, 282–290.

Mégraud, F. (2010). Antimicrobial resistance and approaches to treatment. In *Helicobacter pylori* in the 21st Century, Sutton, P., and Mitchell, H., eds. (Cab International: Wallingford, UK), pp. 45–68.

Miyachi, H., Miki, I., Aoyama, N., Shirasaka, D., Matsumoto, Y., Toyoda, M., Mitani, T., Morita, Y., Tamura, T., Kinoshita, S., *et al.* (2006). Primary levofloxacin resistance and gyrA/B mutations among *Helicobacter pylori* in Japan. Helicobacter *11*, 243–249.

Moder, K.A., Layer, F., König, W., and König, B. (2007). Rapid screening of clarithromycin resistance in *Helicobacter pylori* by pyrosequencing. J. Med. Microbiol. *56*, 1370–1376.

Murakami, K., Fujioka, T., Okimoto, T., Sato, R., Kodama, M., and Nasu, M. (2002). Drug combinations with amoxycillin reduce selection of clarithromycin resistance during *Helicobacter pylori* eradication therapy. Int. J. Antimicrob. Agents. *19*, 67–70.

Nahar, S., Mukhopadhyay, A.K., Khan, R., Ahmad, M.M., Datta, S., Chattopadhyay, S., Dhar, S.C., Sarker, S.A., Engstrand, L., Berg, D.E., *et al.* (2004). Antimicrobial susceptibility of *Helicobacter pylori* strains isolated in Bangladesh. J. Clin. Microbiol. *42*, 4856–4858.

Nishizawa, T., Suzuki, H., Kurabayashi, K., Masaoka, T., Muraoka, H., Mori, M., Iwasaki, E., Kobayashi, I., and Hibi T. (2006). Gatifloxacin resistance and mutations in *gyrA* after unsuccessful *Helicobacter pylori* eradication in Japan. Antimicrob. Agents Chemother. *50*, 1538–1540.

Nishizawa, T., Suzuki, H., and Hibi T. (2009). Quinolone-based third-line therapy for *Helicobacter pylori* eradication. J. Clin. Biochem. Nutr. *44*, 119–124.

Nonaka, L., Connell, S.R., and Taylor, D.E. (2005). 16S rRNA mutations that confer tetracycline resistance in *Helicobacter pylori* decrease drug binding in *Escherichia coli* ribosomes. J. Bacteriol. *187*, 3708–3712.

O'Connor, A., Taneike, I., Nami, A., Fitzgerald, N., Murphy, P., Ryan, B., O'Connor, H., Qasim, A., Breslin, N., and O'Moráin, C. (2010). *Helicobacter pylori* resistance to metronidazole and clarithromycin in Ireland. Eur. J. Gastroenterol. Hepatol. 22, 1123–1127.

Ojetti, V., Migneco, A., Silveri, N.G., Ghirlanda, G., Gasbarrini, G., and Gasbarrini, A. (2005). The role of *H. pylori* infection in diabetes. Curr. Diabetes Rev. *1*, 343–347.

Ojetti, V., Pellicano, R., Fagoonee, S., Migneco, A., Berrutti, M., and Gasbarrini, A. (2010). *Helicobacter pylori* infection and diabetes. Minerva Med. *101*, 115–119.

Oleastro, M., Ménard, A., Santos, A., Lamouliatte, H., Monteiro, L., Barthélémy, P., and Mégraud, F. (2003). Real-time PCR assay for rapid and accurate detection of point mutations conferring resistance to clarithromycin in *Helicobacter pylori*. J. Clin. Microbiol. *41*, 397–402.

Olekhnovich, I.N., Goodwin, A., and Hoffman, P.S. (2009). Characterization of the NAD(P)H oxidase and metronidazole reductase activities of the RdxA nitroreductase of *Helicobacter pylori*. FEBS J. *276*, 3354–3364.

Onder, G., Aydin, A., Akarca, U., Tekin, F., Ozutemiz, O., and Ilter, T. (2007). High *Helicobacter pylori* resistance rate to clarithromycin in Turkey. J. Clin. Gastroenterol. *41*, 747–750.

Oyarzabal, O.A., Rad, R., and Backert, S. (2007). Conjugative transfer of chromosomally encoded antibiotic resistance from *Helicobacter pylori* to *Campylobacter jejuni*. J. Clin. Microbiol. *45*, 402–408.

Perna, F., Zullo, A., Ricci, C., Hassan, C., Morini, S., and Vaira, D. (2007). Levofloxacin-based triple therapy for *Helicobacter pylori* re-treatment, role of bacterial resistance. Dig. Liver Dis. *39*,1001–1005.

Petersen, A.M., Gjøde, P., Vinge, O.D., Jensen, S., and Krogfelt, K.A. (2006). *Helicobacter pylori* antimicrobial resistance and risk factors in Denmark 1998–2004: no need for concern? Helicobacter *11*, 210–211.

Poon, S.K., Lai, C.H., Chang, C.S., Lin, W.Y., Chang, Y.C., Wang, H.J., Lin, P.H., Lin, H.J., and Wang, W.C. (2009). Prevalence of antimicrobial resistance in *Helicobacter pylori* isolates in Taiwan in relation to consumption of antimicrobial agents. Int. J. Antimicrob. Agents. *34*, 162–165.

Prazeres Magalhães, P., De Magalhães Queiroz, D.M., Campos Barbosa, D.V., Aguiar Rocha, G., Nogueira Mendes, E., Santos, A., Valle Corrêa, P.R., Camargos Rocha, A.M., Martins Teixeira, L., and Affonso de Oliveira, C. (2002). *Helicobacter pylori* primary resistance to metronidazole and clarithromycin in Brazil. Antimicrob. Agents Chemother. *46*, 2021–2023.

Rafeey, M., Ghotaslou, R., Nikvash, S., and Hafez, A.A. (2007). Primary resistance in *Helicobacter pylori* isolated in children from Iran. J. Infect. Chemother. *13*, 291–295.

Raymond, J., Burucoa, C., Pietrini, O., Bergeret, M., Decoster, A., Wann, A., Dupont, C., and Kalach, N. (2007). Clarithromycin resistance in *Helicobacter pylori* strains isolated from French children, prevalence of the different mutations and coexistence of clones harboring two different mutations in the same biopsy. Helicobacter *12*, 157–163.

Raymond, J., Lamarque, D., Kalach, N., Chaussade, S. and Burucoa, C. (2010). High level of antimicrobial resistance in French *Helicobacter pylori* isolates. Helicobacter *15*, 21–27.

Ribeiro, M.L., Vitiello, L., Miranda, M.C., Benvengo, Y.H., Godoy, A.P., Mendonca, S., and Pedrazzoli, J., Jr. (2003). Mutations in the 23S rRNA gene are associated with clarithromycin resistance in *Helicobacter pylori* isolates in Brazil. Ann. Clin. Microbiol. Antimicrob *21*, 2–11.

Rimbara, E., Noguchi, N., Kawai, T., and Sasatsu, M. (2007a). Correlation between substitutions in penicillin-binding protein 1 and amoxicillin resistance in *Helicobacter pylori*. Microbiol. Immunol. *51*, 939–944.

Rimbara, E., Noguchi, N., Kijima, H., Yamaguchi, T., Kawai, T., and Sasatsu, M. (2007b). Mutations in the 23S rRNA gene of clarithromycin-resistant *Helicobacter pylori* from Japan. Int. J. Antimicrob. Agents. *30*, 250–254.

Rimbara, E., Noguchi, N., Kawai, T., and Sasatsu, M. (2008). Mutations in penicillin-binding proteins 1, 2 and 3 are responsible for amoxicillin resistance in *Helicobacter pylori*. J. Antimicrob. Chemother. *61*, 995–998.

Samra, Z., Shmuely, H., Niv, Y., Dinari, G., Passaro, D.J., Geler, A., Gal, E., Fishman, M., Bachor, J., and Yahav, J. (2002). Resistance of *Helicobacter pylori* isolated in Israel to metronidazole, clarithromycin, tetracycline, amoxicillin and cefixime. J. Antimicrob. Chemother. *49*, 1023–1026.

Shaikh, A.R., Giridhar, R., Megraud, F., and Yadav, M.R. (2009). Metalloantibiotics: synthesis, characterization and antimicrobial evaluation of bismuth–fluoroquinolone complexes against *Helicobacter pylori*. Acta Pharm. *59*, 259–271.

Sherif, M., Mohran, Z., Fathy, H., Rockabrand, D.M., Rozmajzl, P.J., and Frenck, R.W. (2004). Universal high-level primary metronidazole resistance in *Helicobacter pylori* isolated from children in Egypt. J. Clin. Microbiol. *42*, 4832–4834.

Siavoshi, F., Saniee, P., Latifi-Navid, S., Massarrat, S., and Sheykholeslami, A. (2010). Increase in resistance rates of *H. pylori* isolates to metronidazole and tetracycline–comparison of three 3-year studies. Arch. Iran. Med. *13*, 177–187.

Slinger, R., Yan, L., Chan, F., Forward, K., Cooper-Lesins, G., Best, L., Haldane, D., and Veldhuyzen van Zanten, S. (2009). Pyrosequencing assay to rapidly detect clarithromycin resistance mutations in Canadian *Helicobacter pylori* isolates. Can. J. Gastroenterol. *23*, 609–612.

Soltermann, A., Perren, A., Schmid, S., Eigenmann, F., Güller, R., Weber, K.B., Meier, J., Eichenberger, P., and Komminoth. P. (2005). Assessment of *Helicobacter pylori* clarithromycin resistance mutations in archival gastric biopsy samples. Swiss Med. Wkly. *135*, 327–332.

Storskrubb, T., Aro, P., Ronkainen, J., Wreiber, K., Nyhlin, H., Bolling-Sternevald, E., Talley, N.J., Engstrand, L., and Agreus, L. (2006). Antimicrobial susceptibility of *Helicobacter pylori* strains in a random adult Swedish population. Helicobacter *11*, 224–230.

Su, Z., Xu, H., Zhang, C., Shao, S., Li, L., Wang, H., Wang, H., and Qiu, G. (2006). Mutations in *Helicobacter pylori porD* and *oorD* genes may contribute to furazolidone resistance. Croat. Med. J. *47*, 410–415.

Sugimoto, M., Furuta, T., and Yamaoka, Y. (2009). Clarithromycin resistance, tumor necrosis factor alpha gene polymorphism and mucosal inflammation affect *H. pylori* eradication success. J. Gastroenterol. Hepatol. *24*, 1725–1732.

Suzuki, T., Matsuo, K., Sawaki, A., Ito, H., Hirose, K., Wakai, K., Sato, S., Nakamura, T., Yamao, K., Ueda, R, *et al.* (2006). Systematic review and meta-analysis: importance of CagA status for successful eradication of *Helicobacter pylori* infection. Aliment. Pharmacol. Ther. *24*, 273–280.

Suzuki, H., Nishizawa, T., Muraoka, H., and Hibi, T. (2009a). Sitafloxacin and garenoxacin may overcome the antibiotic resistance of *Helicobacter pylori* with *gyrA* mutation. Antimicrob. Agents Chemother. *53*, 1720–1721.

Suzuki, S., Suzuki, H., Nishizawa, T., Kaneko, F., Ootani, S., Muraoka, H., Saito, Y., Kobayashi, I., and Hibi, T. (2009b). Past rifampicin dosing determines rifabutin resistance of *Helicobacter pylori*. Digestion *79*, 1–4.

Sýkora, J., Valecková, K., Amlerová, J., Siala, K., Dedek, P., Watkins, S., Varvarovská, J., Stozický, F., Pazdiora, P., and Schwarz, J. (2005). Effects of a specially designed fermented milk product containing probiotic *Lactobacillus casei* DN-11,4001 and the eradication of *H. pylori* in children: a prospective randomized double-blind study. J. Clin. Gastroenterol. *39*, 692–698.

Taneike, I., Nami, A., O'Connor, A., Fitzgerald, N., Murphy, P., Qasim, A., O'Connor, H., and O'Morain, C. (2009). Analysis of drug resistance and virulence-factor genotype of Irish *Helicobacter pylori* strains: is there any relationship between resistance to metronidazole and *cagA* status? Aliment. Pharmacol. Ther. *30*, 784–790.

Tanih, N.F., Okeleye, B.I., Naidoo, N., Clarke, A.M., Mkwetshana, N., Green, E., Ndip, L.M., and Ndip, R.N. (2010). Marked susceptibility of South African *Helicobacter pylori* strains to ciprofloxacin and amoxicillin: Clinical implications. S. Afr. Med. J. *100*, 49–52.

Tankovic, J., Lascols, C., Sculo, Q., Petit, J.C., and Soussy, C.J. (2003). Single and double mutations in *gyrA* but not in *gyrB* are associated with low- and high-level fluoroquinolone resistance in *Helicobacter pylori*. Antimicrob. Agents Chemother. *47*, 3942–3944.

Thyagarajan, S.P., Ray, P., Das, B.K., Ayyagari, A., Khan, A.A., Dharmalingam, S., Rao, U.A., Rajasambandam, P., Ramathilagam, B., Bhasin, D., *et al.* (2003). Geographical difference in antimicrobial resistance pattern of *Helicobacter pylori* clinical isolates from Indian patients, Multicentric study. J. Gastroenterol. Hepatol. *18*, 1373–1378.

Toracchio, S., Capodicasa, S., Soraja, D.B., Cellini, L., and Marzio, L. (2005). Rifabutin based triple therapy for eradication of *H. pylori* primary and secondary resistant to tinidazole and clarithromycin. Dig. Liver Dis. *37*, 33–38.

Tseng, Y.S., Wu, D.C., Chang, C.Y., Kuo, C.H., Yang, Y.C., Jan, C.M., Su, Y.C., Kuo, F.C., and Chang, L.L. (2009). Amoxicillin resistance with beta-lactamase production in *Helicobacter pylori*. Eur. J. Clin. Invest. *39*, 807–812.

Tsugawa, H., Suzuki, H., Satoh, K., Hirata, K., Matsuzaki, J., Saito, Y., Suematsu, M., and Hibi, T. (2011). Two amino acids mutation of ferric uptake regulator (Fur) determines *Helicobacter pylori* resistance to metronidazole. Antioxid. Redox Signal. *14*, 15–23.

Tüzün, Y., Bayan, K., Yilmaz, S., Dursun, M., and Ozekinci, T. (2008). The prevalence of primary and secondary *Helicobacter pylori* resistance to clarithromycin and probable contributing cofactors, data from southeastern Anatolia. Hepatogastroenterology *55*, 289–293.

Vakil, N. (2009). Antimicrobial resistance and eradication strategies for *Helicobacter pylori*. Rev. Gastroenterol. Disord. *9*, E78-E83.

Vallejos, C., Garrido, L., Cáceres, D., Madrid, A.M., Defilippi, C., Defilippi, C., and Toledo, H. (2007). Prevalence of metronidazole, clarithromycin and tetracycline resistance in *Helicobacter pylori* isolated from Chilean patients. Rev. Med. Chil. *135*, 287–293.

Van Amsterdam, K., Bart, A., and van der Ende, A. (2005). A *Helicobacter pylori* TolC efflux pump confers resistance to metronidazole. Antimicrob. Agents Chemother. *49*,1477–1482.

Vécsei, A., Kipet, A., Innerhofer, A., Graf, U., Binder, C., Gizci, H., Hammer, K., Bruckdorfer, A., Huber, W.D., Hirschl, A.M., *et al.* (2010). Time trends of *Helicobacter pylori* resistance to antibiotics in children living in Vienna, Austria. Helicobacter *15*, 214–220.

Vega, A.E., Alarcón, T., Domingo, D., Martínez, M.J., and López-Brea, M. (2003). Detection of resistance to clarithromycin in clinical isolates of *Helicobacter pylori* from children and adults. Rev. Esp. Quimioter. *16*, 53–57.

Vega, A.E., Cortiñas, T.I., Puig, O.N., and Silva, H.J. (2010). Molecular characterization and susceptibility testing of *Helicobacter pylori* strains isolated in western Argentina. Int. J. Infect. Dis. *14* (*Suppl* 3), e85–e92.

Wang, L.-H., Cheng, H., Hu, F.-L., and Li, J. (2010). Distribution of *gyrA* mutations in fluoroquinolone-resistant *Helicobacter pylori* strains. World J. Gastroenterol. *16*, 2272–2277.

Watanabe, K., Tanaka, A., Imase, K., Tokunaga, K., Sugano, H., Kai, A., Ishida, H., Itoh, T., and Takahashi, S. (2005). Amoxicillin resistance in *Helicobacter pylori:* studies from Tokyo, Japan from 1985 to 2003. Helicobacter *10*, 4–11.

Wheeldon, T.U., Granström, M., Hoang, T.T., Phuncarg, D.C., Nilsson, L.E., and Sörberg, M. (2004). The importance of the level of metronidazole resistance for the success of *Helicobacter pylori* eradication. Aliment. Pharmacol. Ther. *19*, 1315–1321.

Wong, W.M., Gu, Q., Wang, W.H., Fung, F.M., Berg, D.E., Lai, K.C., Xia, H.H., Hu, W.H., Chan, C.K., Chan, A.O., *et al.* (2003). Effects of primary metronidazole and clarithromycin resistance to *Helicobacter pylori* on omeprazole, metronidazole, and clarithromycin triple-therapy regimen in a region with high rates of metronidazole resistance. Clin. Infect. Dis. *37*, 882–889.

Wong, W.M., Xiao, S.D., Hu, P.J., Wang, W.H., Gu, Q., Huang, J.Q., Xia, H.H., Wu, S.M., Li, C.J., Chen, M.H., *et al.* (2005). Standard treatment for *Helicobacter pylori* infection is suboptimal in non-ulcer dyspepsia compared with duodenal ulcer in Chinese. Aliment. Pharmacol. Ther. *21*, 73–81.

Yahav, J., Shmuely, H., Niv, Y., Bechor, J., and Samra, Z. (2006). *In vitro* activity of levofloxacin against *Helicobacter pylori* isolates from patients after treatment failure. Diagn. Microbiol. Infect. Dis. *55*, 81–83.

Yang, J.C., Lee, P.I., and Hsueh, P.R. (2010). *In vitro* activity of nemonoxacin, tigecycline, and other antimicrobial agents against *Helicobacter pylori* isolates in Taiwan, 1998–2007. Eur. J. Clin. Microbiol. Infect. Dis. *29*, 1369–1375.

Yuen, B., Zbinden, R., Fried, M., Bauerfeind, P., and Bernardi, M. (2005). Cultural recovery and determination of antimicrobial susceptibility in *Helicobacter pylori* by using commercial transport and isolation media. Infection *33*, 77–81.

Zevit, N., Levy, I., Shmuely, H., Samra, Z., and Yahav, J. (2010). Antibiotic resistance of *Helicobacter pylori* in Israeli children. Scand. J. Gastroenterol. *45*, 550–555.

Zhang, Z., Liu, Z.Q., Zheng, P.Y., Tang, F.A., and Yang, P.C. (2010). Influence of efflux pump inhibitors on the multidrug resistance of *Helicobacter pylori*. World J. Gastroenterol. *16*, 1279–1284.

Zheng, X.L., Hu, F.L., Wang, W.H., and Xu, L. (2009). Genotypic homogeneity and susceptibility to clarithromycin for clarithromycin-resistant *Helicobacter pylori* strains. Zhonghua Yi Xue Za Zhi *89*, 1558–1562.

Zullo, A., Perna, F., Hassan, C., Ricci, C., Saracino, I., Morini, S., and Vaira, D. (2007). Primary antibiotic resistance in *Helicobacter pylori* strains isolated in northern and central Italy. Aliment. Pharmacol. Ther. *25*, 1429–1434.

Treatment of *Helicobacter pylori*-associated Diseases

10

Borislav Vladimirov

Abstract

Treatment of *Helicobacter pylori* infection remains a significant clinical problem despite extensive research on the topic over the last 25 years. For *H. pylori* eradication, combined regimens of non-antibiotic (bismuth compounds and/or proton pump inhibitors) and usually two antibiotics are used. The antibiotic choice may involve amoxicillin, clarithromycin, metronidazole, tetracycline, levofloxacin, rifabutin and furazolidone. A number of factors such as duration of treatment, choice of antibiotics, new drug combinations, improved patient compliance and novel agents may help to improve the eradication rates. In this chapter, data on triple, quadruple, sequential and rescue therapeutic regimens are discussed. However, *H. pylori* resistance to antibiotics, poor patient compliance and host genetic polymorphism can strongly reduce the success of eradication. Recent data have shown that the rates of eradication have decreased worldwide. In recent years, many new drugs and combinations have been applied with aims to raise the eradication rates to more acceptable levels but more evidence is needed to support the routine use of the new antibiotics and modified treatment regimens for eradication of *H. pylori*.

Indications for *H. pylori* eradication

Treatment of *Helicobacter pylori* infection remains a significant clinical problem despite extensive research over the last 25 years. A number of factors such as duration of treatment, choice of antibiotics, new drug combinations, improved patient compliance and novel agents may help to improve eradication rates.

Proton pump inhibitor (PPI)-based triple therapy has been used as first-line treatment of choice for many years. In 1994, the National Institutes of Health Consensus Development Conference on *H. pylori* in Peptic Ulcer Disease determined the Consensus panel for the treatment of *H. pylori* infection. According to these guidelines, ulcer patients with *H. pylori* infection require treatment with antimicrobial agents in addition to antisecretory drugs whether on first presentation with the illness or on recurrence. This consensus panel also concluded that the value of treating non-ulcer dyspepsia patients with *H. pylori* infection remains to be determined and the relationship between *H. pylori* infection and gastric cancers requires further exploration (NIH Consensus Conference, 1994). After that, in the first, second and third Maastricht Consensus Report, the spectrum of diseases associated with *H. pylori* that has been suitable for *H. pylori* eradication has increased (Malfertheiner *et al.*, 2002, 2007). This has been confirmed by the evidence from both clinical and cost-effectiveness studies. Treatment regimens have been recommended to be simple, well tolerated and easy to comply with, leading to high eradication rates and cost-effectiveness.

According to many data, the treatment of *H. pylori* infection has led to successful eradication results in ulcer remission in 97% of gastric ulcer cases and in 98% of duodenal ulcer patients, compared with healing rates of 61% and 65%, respectively, among those with persistent infection. Treatment is superior to no treatment of recurrence of gastric and duodenal ulcer. Recurrent ulcer bleeding has been reduced by 17% and 4% by ulcer healing treatment alone and ulcer healing treatment followed by maintenance therapy, respectively. *H. pylori* eradication has been associated with 64% remission of mucosa-associated lymphoid tissue (MALT) lymphoma. Beneficial effects of 'test and treat' strategy have been also shown in patients with dyspeptic symptoms, gastro-oesophageal reflux disease (GORD), non-steroidal anti-inflammatory drugs (NSAIDs) and aspirin users, iron-deficiency anaemia, and idiopathic thrombocytopenic purpura (ITP) as well as in those who have undergone resection of early gastric cancer, subjects at risk for gastric cancer or first-degree relatives of patients with gastric cancer (Malfertheiner *et al.*, 2002, 2007).

According to the Maastricht-III consensus statements, an eradication treatment regimen can be accepted as effective when it reaches an intention-to-treat eradication rate of >80% (Malfertheiner *et al.*, 2007). Recent data have shown that eradication rates have decreased worldwide (O'Connor *et al.*, 2010). In addition, in recent years, eradication rates have decreased below the recommended level of 80% for the most common regimens used in clinical practice, especially because of the increasing antimicrobial resistance as a result of the widespread use of antibiotics (Egan *et al*,. 2007; O'Connor *et al.*, 2009; Vakil, 2009).

Another important predictor for *H. pylori* eradication failure is the poor compliance with medication (O'Connor *et al.*, 2010). Some authors have found very high compliance rate, which has not been confirmed by other authors (Jafri *et al.*, 2008; Lee *et al.*, 2010; Molina-Infante *et al.*, 2010). Most data have shown that poorer rate of compliance has correlated with decreased level of eradication success (Graham *et al.*, 1992; O'Connor *et al.*, 2010). There has been limited evidence to suggest that smoking, alcohol consumption and diet could adversely affect the likelihood of successful eradication (Chey and Wong, 2007). On the other hand, there are inter-individual differences in response to *H. pylori* infection including also genetics.

Many new data about indications for treatment of *H. pylori*-associated diseases, antimicrobial resistance and treatment regimens for *H. pylori* eradication have been accumulated. *H. pylori*-specific antibiotics, probiotics and vaccines may become part of the treatment of *H. pylori* infection, but there are no practical recommendations at this stage.

Indications for *H. pylori* eradication, shown as a strong recommendation in Maastricht II-2000 guidelines, have been reconfirmed in the updated statements of the Maastricht III Consensus Report (Malfertheiner *et al.*, 2007). *H. pylori* eradication has been strongly recommended in:

- all infected patients with a diagnosis of peptic ulcer disease (active or not, duodenal and gastric ulcer or complicated one, level of evidence 1a, grade of recommendation A);
- patients with low-grade gastric MALT lymphoma (level of evidence 1c, grade of recommendation A);
- patients with atrophic gastritis (level of evidence 2a, grade of recommendation B);
- after gastric cancer resection (level of evidence 3b, grade of recommendation B);
- the first-degree relatives of patients with gastric cancer (level of evidence 3b, grade of recommendation B); and
- patients who so wish (after full consultation with their physician, level of evidence 5, grade of recommendation A).

In addition, the updated recommendations for *H. pylori* eradication formulated in the Maastricht III Consensus Report have been as follows:

- *H. pylori* eradication is an appropriate option for patients infected with *H. pylori* and investigated non-ulcer dyspepsia (level of evidence 1a, grade of recommendation A).
- *H. pylori* test and treat is an appropriate option for patients with uninvestigated dyspepsia (level of evidence 1a, grade of recommendation A).
- The effectiveness of *H. pylori* test and treat is low in populations with a low *H. pylori* prevalence; in this situation, the test and treat strategy or empirical acid suppression is an

appropriate option (level of evidence 2a, grade of recommendation B).

- *H. pylori* eradication does not cause GORD (level of evidence 1b, grade of recommendation A).
- *H. pylori* eradication does not affect the outcome of PPI treatment in patients with GORD in Western populations (level of evidence 1b, grade of recommendation A).
- In GORD, routine testing for *H. pylori* is not recommended (level of evidence 1b, grade of recommendation A).
- *H. pylori* testing should be considered for patients receiving long-term maintenance treatment with PPIs (level of evidence 2b, grade of recommendation B).
- There is a negative association between *H. pylori* prevalence and GORD in Asia, but the nature of this relationship is uncertain (level of evidence 2b, grade of recommendation B).
- In patients receiving long-term NSAIDs and those with peptic ulcer and/or ulcer bleeding, PPI maintenance treatment is better than *H. pylori* eradication in preventing ulcer recurrence and/or bleeding (level of evidence 1b, grade of recommendation A).
- *H. pylori* eradication is of value in chronic NSAID users but insufficient to prevent NSAID related ulcer disease completely (level of evidence 1b, grade of recommendation A).
- In naive users of NSAIDs, *H. pylori* eradication could prevent peptic ulcer and or bleeding (level of evidence 1b, grade of recommendation A).

Eradication of *H. pylori* infection has been also recommended in unexplained iron deficiency anaemia as well as idiopathic thrombocytopenic purpura. *H. pylori* has shown no proven role in other extra intestinal diseases.

There is still a controversy regarding *H. pylori* infection in children. According to the consensus statements of 1st Consensus Conference in September 1998 and Maastricht II Consensus Report, the main goals of *H. pylori* eradication in children are to relieve upper gastrointestinal symptoms and/or to treat peptic ulcer disease (Drumm *et al.*, 2000; Malfertheiner *et al.*, 2002; Bourke *et al.*, 2005). Therapy for the prevention of complications later in life in children with no or unspecific minor symptoms could be postponed to a later time when safer therapeutic options are available (Malfertheiner *et al.*, 2002). Standard triple therapy by a PPI plus two antibiotics for 7–14 days is the treatment of choice, but the higher clarithromycin resistance rate limits the efficacy. Maastricht III Consensus Report has stated that recurrent abdominal pain is not an indication for a test and treat strategy for *H. pylori* infection in children (Malfertheiner *et al.*, 2007). Children with upper gastrointestinal symptoms or iron-deficiency anaemia (IDA), refractory to iron supplementation as well as subjects with family history of peptic disease or gastric cancer can first been investigated to exclude other causes of the symptoms. Then, they should be tested for *H. pylori* infection and should be treated if positive.

According to the revised Japanese guidelines for *H. pylori* infection, eradication therapy is strongly recommended for the treatment of gastric and duodenal ulcers, MALT lymphoma, and hyperplastic gastric polyps, as well as for prevention of *H. pylori*-associated gastric cancer (Asaka *et al.*, 2010). Idiopathic thrombocytopenic purpura and IDA have been also recommended for treatment.

Eradication of *H. pylori* infection with classical regimens

First-line therapy

Numerous clinical trials have confirmed the efficacy of combined therapy of PPI and two antibiotics as clarithromycin and amoxicillin/or metronidazole. Based on the analysis of published randomized and control trials, the European, North American and Asia-Pacific guidelines have recommended PPI-based triple therapy as a standard first-line therapy of choice (Lind *et al.*, 1996; Chey and Wong, 2007; Malfertheiner *et al.*, 2007; Fock *et al.*, 2009; Vakil, 2009).

Again, the 1st Maastricht Consensus Report has strongly recommended that treatment of *H. pylori* infection should be started with PPI based triple therapy composed of PPI (standard dose twice daily) and two of the following antimicrobial agents: clarithromycin (250 mg twice daily), nitroimidazoles, e.g. metronidazole (400 mg twice daily) or tinidazole (500 mg twice daily) and amoxicillin (1000 mg twice daily) in various combinations for at least 7 days (European *Helicobacter pylori* Study Group, the Maastricht Consensus Report, 1997). In the case of treatment failure, a retreatment regimen should be selected after consideration of previous treatment or microbial sensitivities, or both. Additionally, quadruple therapy of omeprazole plus classic bismuth based triple therapy (tripotassium dicitrato bismuthate plus metronidazole and tetracycline) also has been recommended in case of failure of triple therapy. Updated recommendation of the 2nd Maastricht Consensus Conference has defined well the steps of treatment strategy (Malfertheiner *et al.*, 2002). First-line therapy, recommended in the original first Maastricht Consensus Report, has been modified because many data have shown a similar efficacy with regimens based on ranitidine bismuth citrate and with those based on a PPI (Pipkin *et al.*, 1998; Gisbert *et al.*, 1999). Therefore, the recommended first-line therapy should be started with triple therapy including a PPI or ranitidine bismuth citrate (standard dose twice daily) plus clarithromycin (500 mg twice daily), and amoxicillin (1000 mg twice daily) or metronidazole (500 mg twice daily) for a minimum of 7 days. Standard doses of PPIs are as follows: omeprazole, esomeprazole, and rabeprazole – 20 mg; lansoprazole – 30 mg; pantoprazole – 40 mg; and that of ranitidine bismuth citrate is 400 mg.

A new principle guiding treatment has been that of second-line therapy, which should be used in cases of treatment failure. Quadruple therapy with a proton pump inhibitor (standard dose twice daily), bismuth subsalicylate or subcitrate (120 mg four times daily), metronidazole (500 mg thrice daily) and tetracycline (500 mg four times daily) for a minimum of 7 days has been recommended. Where bismuth is not available, second-line therapy should be started with PPI-triple therapy. All these strong recommendations have been based on level 1 or level 2 evidence (Misiewicz *et al.*, 1997; Unge, 1998; Laheij *et al.*, 1999; Lind *et al.*, 1999). In addition, in uncomplicated duodenal ulcer patients, it has been stated that *H. pylori* eradication therapy does not need to be followed by further antisecretory treatment, based on level 1 evidence (Malfertheiner *et al.*, 2002). A large, randomized, controlled trial has compared treatment with placebo or 3 weeks of omeprazole monotherapy, after 1-week standard first-line triple therapy, in patients with active duodenal ulcer (Tulassay *et al.*, 2001). Ulcer healing and symptoms relief rates of over 90% were found in both treatment arms. The Maastricht III Consensus Conference (Malfertheiner *et al.*, 2007) has stated that:

- There is a small advantage in using a PPI–clarithromycin-metronidazole combination instead of PPI–clarithromycin-amoxicillin as the first choice treatment (level of evidence 1a, grade of recommendation A),
- PPI–clarithromycin–amoxicillin or metronidazole treatment remains the recommended first choice treatment in populations where the prevalence of clarithromycin resistance is less than 15–20%. In populations where the prevalence of metronidazole resistance is less than 40%, PPI–clarithromycin–metronidazole is preferable. Quadruple therapies are alternative first choice treatments. The same first choice *H. pylori* treatments are recommended worldwide, although different doses may be appropriate (level of evidence 1b, grade of recommendation A),
- Bismuth-based quadruple therapies remain the best second choice treatment, if available. If not, a PPI, amoxicillin or tetracycline and metronidazole are recommended.
- Rescue treatment should be based on antimicrobial susceptibility testing (level of evidence 2c, grade of recommendation B).

Accepted standard first-line triple therapy for *H. pylori* eradication includes PPI (standard dose twice daily), clarithromycin (500 mg twice daily) and amoxicillin (1000 mg twice daily) or metronidazole (400 mg or 500 mg twice daily). This therapy as a 14-day course has been more effective by 7–12% than a 7-day course, based on a meta-analysis (Calvet *et al.*, 2000, 2002; Ford and Moayyedi, 2003). However, other studies and meta-analyses have found no clinical benefit by longer courses of treatment (Fuccio *et al.*, 2007; Egan *et al.*, 2007). Some studies have analysed the cost effectiveness of the different courses (Calvet *et al.*, 2001). According to the Maastricht III Consensus Report and ACG guideline, longer duration of treatment has been associated with better eradication rates (Malfertheiner *et al.*, 2007; Chey and Wong, 2007). Published guidelines in the USA have recommended 2 weeks of triple therapy or 10–14 days of quadruple therapy (PPI, bismuth, metronidazole, and tetracycline; Chey and Wong, 2007). A 7-day treatment may be acceptable where local studies show that it is effective, because many studies with standard first-line triple therapy for seven days, mainly from European countries, have confirmed that this has been a valid duration for this treatment (Megraud, 2005). A number of recent studies have shown conflicting evidence with regard to the length of treatment course (Katicic *et al.*, 2006; Paoluzi *et al.*, 2006; Kim *et al.*, 2007; Zagari *et al.*, 2007). A Croatian study, comparing 7-, 10-, and 14-day treatment courses of PPI, amoxicillin, and clarithromycin (PAC) or metronidazole (PAM), has found that >80% eradication rates (ITT analysis) have been achieved only by a 14- and 10-day course of PAC and a 14-day course of PAM, but not by a 7-day treatment course (Paoluzi *et al.*, 2006). An Italian study also has shown a 14-day PAC therapy to reach a higher eradication rate than 7-day or 14-day regimes with metronidazole (70% vs. 52%) and the same therapy for 7 days (70% vs. 57%). At per-protocol (PP) analysis, a 14-day therapy with omeprazole, amoxicillin and clarithromycin has shown a significantly higher eradication rate than a 7-day therapy with amoxicillin and metronidazole (77% vs. 62%) but no difference with 1-week of the same regime (66%; Katicic *et al.*, 2006). This study has concluded that 2-week therapies, independently of antibiotic combination, lead to a significant increase in *H. pylori* eradication rate compared with 1-week therapies. The compliance and tolerability have been similar for 1-week and 2-week treatment groups. However, another Italian study, comparing 1- and 2-week regimes, has reported similar results in terms of efficacy, safety and patient compliance with ITT eradication rates of 79.9% and 81.7%, respectively (Zagari *et al.*, 2007). Similarly, a Korean study has failed to show a benefit by the longer duration of treatment (Kim *et al.*, 2007). Neither the 7-day nor the 14-day course of PAC has achieved acceptable eradication rates by ITT analysis. Egan *et al.* (2007) have suggested that these conflicting findings could reflect varying resistance rates within the populations studied and have confirmed the need for more local reference centres to determine the best treatment regimes. In these studies, it is also noticeable that the eradication rate for standard triple therapy has not been adequate and is has been often <80%. Increased duration of therapy may not overcome the increasing bacterial resistance.

Standard triple therapy is effective in regions where clarithromycin resistance is low. Increased resistance to antibiotics used in the standard PPI triple therapy is the base of the different selection of antibiotic combinations. The eradication rates have been higher by combination of PPI–clarithromycin–metronidazole (97%) than by the combination of PPI–clarithromycin-amoxicillin (88%) as well as in the patients with clarithromycin resistance alone (50% vs. 18%, respectively; Lind *et al.*, 1996; Megraud, 2004). The increasing rate of clarithromycin resistance is probably due to greater use of clarithromycin in the community for respiratory tract infections (Romano *et al.*, 2008; Boyanova, 2009). In addition, the combination of PPI–clarithromycin–metronidazole in cases of metronidazole resistance has led to decrease in eradication rate only by 25% (Megraud, 2004). These data suggest that a PPI–clarithromycin–metronidazole combination can provide higher eradication rates than PPI–clarithromycin–amoxicillin but this better effect is nullified when metronidazole resistance reaches 40% (Megraud, 2005).

Another therapeutic option for first-line treatment is a 10- or 14-day course of bismuth-based quadruple regimens. Quadruple therapy is designed to overcome clarithromycin resistance. Quadruple therapies comprising PPI, metronidazole, tetracycline, and bismuth are effective alternative first-line treatments and may be advocated in areas of high antibiotic resistance (Calvet *et al.*, 2002; Dore *et al.*, 2006). In an important study, a 7-day course of triple therapy with omeprazole, amoxicillin and clarithromycin

(OAC) was compared with a 10-day course of bismuth-based treatment containing metronidazole, tetracycline and omeprazole (OBMT). The eradication rates achieved with bismuth-based quadruple therapy were significantly higher than those in the PPI-based triple-therapy group, i.e. 93.3% by OBMT and 69.6% by OAC in the PP population and 79.8% and 55.4% in the ITT population, respectively (Malfertheiner *et al.*, 2010). According to O'Connor *et al.* (2010), although these results are encouraging, they fall short of the 80% eradication rate based on ITT that is desirable according to the Maastricht Consensus. They also suggested that a disadvantage of this study was associated with different treatment duration, because better treatment results are usually seen with longer treatment duration. In addition, the optimum duration of bismuth-based treatment, containing metronidazole, tetracycline and omeprazole has not been standardized yet. A recent meta-analysis has shown a PP eradication level of 92.9% and ITT eradication of 89.7% and has confirmed that quadruple therapy is an effective, safe and well tolerated as a first-line treatment (Fischbach and Evans, 2007; Essa *et al.*, 2009). Despite that quadruple therapy has led to increase in eradication, it could possibly increase the resistance rates to both clarithromycin and metronidazole. It has been well established that rabeprazole-based 7-day triple therapy has achieved high level of *H. pylori* eradication, equivalent to 10 days of standard PPI-based therapy (Nagahara *et al.*, 2000; Luth *et al.*, 2001). One recent meta-analysis, comparing clinical potency of various PPIs, has shown the highest relative potency of rabeprazole, based on the mean 24-h gastric pH (Kirchheiner *et al.*, 2009; Savarino *et al.*, 2009). In addition, the favourable pharmacokinetic and pharmodynamic profile of rabeprazole has been associated with a higher eradication rates for *H. pylori* in both normal and poor metabolizers (Horn, 2004). CYP2C19 polymorphism plays an important role in *H. pylori* eradication therapy (Kuo *et al.*, 2010). Rabeprazole-based quadruple therapy has shown better efficacy than esomeprazole-based regimes. The results of our open-label pilot study have confirmed the rationality of the rabeprazole-based eradication regimens (Vladimirov *et al.*, 2003). The short-term 4-day therapy in this study has led to equal *H. pylori* eradication rate as the 7-day triple therapy (80.6% and 83.9%, respectively) with rabeprazole 20 mg daily, amoxicillin + clavulanic acid 2000 mg daily and clarithomycin 1000 mg/daily, followed by rabeprazole 20 mg/daily for another 7 days in Bulgarian patients with peptic disease.

Second-line therapy

Bismuth quadruple therapy is the most widely used salvage regimen for persistent *H. pylori* infection as a second choice treatment if not previously used (Chey and Wong 2007). Bismuth-based quadruple therapy, composed of a PPI, bismuth, tetracycline and metronidazole has exhibited an efficacy of 76% in second-line therapy (Hojo *et al.*, 2001). This regimen, however, has failed to eradicate *H. pylori* in up to 30% of patients (Egan *et al.*, 2007; Fischbach and Evans, 2007). In a cohort study, including 4763 patients treated by bismuth based eradication therapy, no serious side-effects have been reported (Ford *et al.*, 2008). However, some authors have reported this therapy to be associated with a high incidence of side-effects (Gisbert and De la Morena, 2006).

In Taiwan, a study of quadruple therapies PPI, bismuth, metronidazole and tetracycline has reported an eradication rate of 77%, similarly to a quadruple therapy where clarithromycin has replaced bismuth (Wu *et al.*, 2006). Bismuth-based therapy, used as 10-day (four times daily) regimen, has been compared to a 14-day (twice daily) regimen in a mix of first-line and second-line patients (Dore *et al.*, 2002.) The response rate has increased to 95% in the second regimen and an improvement of compliance and tolerability has been observed. In quadruple therapy, rabeprazole-based regimens have provided better efficacy than esomeprazole-based regimens (Kuo *et al.*, 2010). CYP2C19 polymorphism can also play an important role in quadruple therapy.

Because bismuth is not available in many countries, PPI triple treatments are acceptable as second choice treatment. For example, PPI–tetracycline–metronidazole has led to a 91% eradication rate (Realdi *et al.*, 1999). The combination of PPI–amoxicillin–metronidazole used as a second choice treatment has resulted in successful eradication in 81–89% and 64%–59%

for metronidazole-susceptible and -resistant strains, respectively, with a global eradication rate of 64–69% (Lamouliatte *et al.*, 2003). According to Egan *et al.* (2007), clarithromycin should be avoided in second-line treatment in most areas unless resistance tests confirm the *H. pylori* strain to be susceptible. In Japan, the PPI–amoxicillin–metronidazole regimen used as second-line treatment has been more successful by a number of trials, showing eradication rates of over 90%. The combination of ranitidine bismuth citrate with two antibiotics has been 70 to 80% effective in the case of first-line treatment failure (Uygun *et al.*, 2008).

One week of proton-pump inhibitor-based triple therapy, including amoxicillin and metronidazole, has been recommended as second-line eradication therapy in Japan according to the new revised guidelines (Asaka *et al.*, 2010)

Third-line therapy

Because of the variety of clinical situations and antibiotics available in different countries, no specific recommendation has been given for third-line treatment except to perform susceptibility testing (Malfertheiner *et al.*, 2007). Patients with eradication failure after two consecutive treatments, including clarithromycin and metronidazole, are most likely to have a strain resistant to one or both antibiotics. The successful eradication has been obtained by antibiotics chosen according to the susceptibility testing results rather than by those chosen empirically (Toracchio *et al.*, 2000; Lamouliatte *et al.*, 2003; Romano *et al.*, 2003). Some authors have found this strategy to be cost-effective (Breuer and Graham, 1999). Empirical treatment regimens following two treatment failures depend on initial treatment administered.

Numerous clinical trials have been published since the last Maastricht conference. The data have shown that the success of *H. pylori* eradication has decreased to less than 80%, especially for standard first-line triple therapy of two antibiotics and a proton pump inhibitor, partly due to the increasing *H. pylori* resistance to clarithromycin (Laine *et al.*, 1998; Kearney and Brousal, 2000; Calvet *et al.*, 2002; Vaira *et al.*, 2007; Chey and Wong, 2007; Malfertheiner *et al.*, 2010; Paoluzi *et al.*, 2010). Recent publications have found a decrease in eradication rates to 70% in many areas and lower than 60% in some regions (Kadayifci *et al.*, 2006; Saad and Chey, 2006; O'Connor *et al.*, 2009). A Turkish study that has determined eradication rates over ten-year period has shown a marked decrease in eradication since 2000. Pooled eradication rates for each year from 1996 through 2005 have been 79.4%, 83.7%, 81.8%, 81.8%, 75.1%, 61.3%, 65.6%, 65.1%, 55.3%, and 61.1%, respectively. In this study, eradication rates have not been affected by the duration of treatment, choice of PPI or indication for treatment (Kadayifci *et al.*, 2006). In the era of increasing clarithromycin use, the effectiveness of standard triple-therapy regimen for *H. pylori* eradication needs reassessment. In 2010, O'Connor *et al.* (2010) have stated that the standard first-line triple therapy, especially that including clarithromycin, might have gone down to come to an end.

In recent years, most authors have compared standard triple therapy with second-line schemes but also with non-standard new treatment regimens, for example subsequent-line therapies, and have found higher eradication rate using the non-standard regimens than with standard triple therapy (Dore *et al.*, 2002; O'Connor *et al.*, 2010).

Eradication of *H. pylori* infection with new regimens

Many authors have introduced new antibiotics or modified treatment regimens for *H. pylori* eradication. One of the most acceptable schemes has been a sequential treatment for 10 days consisting of a 5-day course of a PPI (standard dose twice daily) and amoxicillin (1000 mg twice daily), followed by additional 5 days of a PPI (standard dose twice daily), clarithromycin (500 mg twice daily), and tinidazole (500 mg twice daily). Sequential therapy has been proposed as an alternative to standard triple therapy for the eradication of *H. pylori* with the primary aim of overcoming clarithromycin resistance (Vakil and Vaira, 2008; Tong *et al.*, 2009). In the first part of therapy, amoxicillin affects the bacterial cell wall and, in this way, prevents the formation of the channels, and blocks clarithromycin from binding to the bacteria.

Data for sequential therapy use have shown a higher eradication rate (>80%) compared with standard first-line therapy for 7 days (<80%), independently of bacterial (*cagA* status and bacterial load) and host factors (underlying disease and smoking; De Francesco *et al.*, 2004; Zullo *et al.*, 2005; Scaccianoce *et al.*, 2006; Vaira *et al.*, 2007; Jafri *et al.*, 2008; O'Connor *et al.*, 2009). A meta-analysis from 10 recent clinical trials, including 2747 patients, also confirmed the higher eradication rates (93.4%) of sequential therapy compared with standard triple therapy (76.9%) as well as superior compliance regardless of the higher complexity for the patients (Jafri *et al.*, 2008; O'Morain *et al.*, 2009). In cases with clarithromycin-resistant strains, the eradication rate with sequential therapy has increased to 82.2% versus 40.6% for the standard triple therapy. In contrast, a recent Korean study failed to show a significant difference in eradication rates between 1-week standard therapy and sequential therapy. Eradication rates by ITT analysis have been 77.9% and 71.6%, respectively (Choi *et al.*, 2008). In this study, there were no significant differences in adverse events and treatment compliance between the two groups. Nevertheless, sequential therapy is now a first-line therapy for *H. pylori* in some countries. For example, the sequential regimen has been recommended as equivalent to standard 7- tp 14-day triple therapies in first-line treatment in the updated Italian guidelines on *H. pylori* management, because Italian studies have suggested that sequential therapy leads to higher eradication rates (about 90%) when compared with standard triple therapy (De Francesco *et al.*, 2004; Zullo *et al.*, 2005; Scaccianoce *et al.*, 2006; Caselli *et al.*, 2007; Vaira *et al.*, 2007). The sequential therapy has appeared promising in some European trials, while the regimen has not been validated yet in other European countries outside Italy as well as in North America (Chey and Wong, 2007; O'Connor *et al.*, 2009). Nevertheless, further randomized controlled trials in different countries and regions are needed before a sequential regimen is recommended as first-line *H. pylori* eradication treatment.

Two other classes of antibiotics have been proposed in the treatment of *H. pylori* infection: a fluoroquinolone (levofloxacin) and antituberculous agents (rifamycin and rifabutin). These antibiotics have been evaluated in first-choice treatments with PPI and amoxicillin (1000 mg twice daily) with a good success rate, as well as in second and third rescue therapy in different populations for 10–14 days (Howden and Hunt, 1998; Gisbert *et al.*, 2006; Saad *et al.*, 2006; O'Connor *et al.*, 2010; Romano *et al.*, 2010). For example, the combination of PPI plus levofloxacin (500 mg daily) and rifabutin (150 mg twice daily) triple treatment has reached 91% eradication rate (Vakil, 2009).

Levofloxacin has been applied in numerous studies with eradication rates of >80% by ITT analysis (Calvet *et al.*, 2002; Antos *et al.*, 2006; Dore *et al.*, 2006; Nista *et al.*, 2006; Gisbert *et al.*, 2007a,b). In a comparative study in Italy, the eradication rate of levofloxacin-based triple therapy (clarithromycin, levofloxacin, and esomeprazole) as a first-line treatment was significantly higher (87% by ITT and 91% by PP analysis) than that by the standard therapies, including clarithromycin, amoxicillin and esomeprazole or clarithromycin, metronidazole and esomeprazole (<80% in either ITT or PP analysis; Nista *et al.*, 2006.). The incidence of side-effects has been similar by both standard and levofloxacin treatment. In a study of patients with known antimicrobial sensitivity in a German population, a 7-day levofloxacin treatment gave an eradication rate of 92.2% and 86.7% on a PP and ITT analysis, respectively (Antos *et al.*, 2006). On the base of data on antimicrobial resistance and efficacy from the USA, it seems reasonable to consider levofloxacin-based triple therapy in circumstances where bismuth or clarithromycin-based therapies are not an option (Chey and Wong, 2007). In another study, a 10-day courses with levofloxacin or clarithromycin in both standard triple and sequential regimes have been compared (Molina-Infante *et al.*, 2010). Levofloxacin-based sequential therapy has shown the best eradication rate of 82.5%. In other recent studies, levofloxacin-based therapy with bismuth and PPI used as first-line alternative treatment has achieved eradication rates of 89% in ITT analysis and 84% in a PP analysis (Gisbert *et al.*, 2007b).

According to the Maastricht III Consensus Report, rifabutin is an antibiotic which can select resistance among *Mycobacterium* spp., so it must be used carefully (Malfertheiner *et al.*, 2007). *H. pylori* resistance to rifabutin may occur but is rare. However, rifabutin is a very important drug in the treatment of patients with multidrug-resistant tuberculosis. Using rifabutin to treat *H. pylori* infection may be dangerous because of the risk of increased resistance in *Mycobacterium tuberculosis* strains. It has already been illustrated that rifabutin efficacy is somewhat reduced by past rifampicin treatment (O'Connor *et al.*, 2010). In addition, rifabutin is associated with serious myelotoxicity as well as with ocular adverse events (Bhagat *et al.*, 2001; Apseloff, 2003).

Rifaximin is a semisynthetic, rifamycin-based non-systemic antibiotic. Rifaximin-based triple therapy has been evaluated in one Italian study (Gasbarrini *et al.*, 2006). The combinations of clarithromycin, rifaximin, and esomeprazole or levofloxacin, rifaximin and esomeprazole have reached eradication rates of 58% and 42%, respectively, by ITT analysis. The authors have concluded that rifaximin is not effective as first-line treatment in Italy.

Second-line therapy

Because of a fall in 20–30% of first-line therapy, an effective second-line treatment is very important. A number of options are available for second-line treatment that may overcome bacterial resistance. Various combinations are used as second-choice treatment, for which limited data exist (Malfertheiner *et al.*, 2007). Triple therapy, quadruple therapy and, more recently, levofloxacin-based therapy have been studied as second-line treatment (Egan *et al.*, 2007). Other alternative bismuth-based regimens have been proposed. For example, a combination of ofloxacin, azithromycin, omeprazole and bismuth has shown a higher eradication rate of 77.3% compared with 64.5% for amoxicillin, clarithromycin, bismuth and omeprazole (Minakari *et al.*, 2010). Shehada *et al.* (2007) have stated that the sequential therapy has not been effective in patients with previous treatment failures although larger studies are required.

Levofloxacin-based therapies are an effective alternative to quadruple therapy in second-line choice in some European countries and have been used very frequently in the recent years (Egan *et al.*, 2008; Gisbert and Javares, 2002; Gisbert, 2009). One Spanish study has found that a 10-day course with levofloxacin-based second-line therapy (on 300 patients) has led to 81% (by PP analysis) and 77% (by ITT analysis) eradication rates with good tolerability and a low side-effect rate of 22% (Gisbert *et al.*, 2008). A meta-analysis and systematic reviews have confirmed that the combination of levofloxacin (500 mg twice daily) and amoxicillin (1000 mg twice daily) with a PPI (twice daily) has been effective as second-line treatment with a mean eradication rate of 80% and less side-effects than the traditional quadruple therapy (19% vs. 44%, respectively; Gisbert and De la Morena, 2006; Gisbert *et al.*, 2006; Egan *et al.*, 2007). Levels of eradication have been higher with a 10-day than a 7-day regime. There has been evidence that fluoroquinolones and levofloxacin in particular are safe. Most adverse events have been mild to moderate and transient, affecting mainly the gastrointestinal system. Tendonitis and hepatotoxicity have been also reported (Spahr *et al.*, 2001; Van der Linden *et al.*, 2002). Some authors have suggested that a levofloxacin-based second-line therapy (together with omeprazole and clarithromycin) has been a potential alternative in the presence of penicillin allergy (Gisbert *et al.*, 2010). A study, comparing levofloxacin, amoxicillin, and omeprazole versus ranitidine bismuth citrate, tetracycline and metronidazole after treatment failure, has found comparable eradication rates. The incidence of side-effects with these two regimes has been also comparable (Gisbert *et al.*, 2007a). However, the concern for an increase in fluoroquinolone resistance when levofloxacin is used for *Helicobacter* eradication may be one of the future problems. A rapidly increasing rate of quinolone resistance has been reported in several countries (O'Connor *et al.*, 2009). In cases of previous quinolone resistance, levofloxacin may have limited use for *H. pylori* eradication.

Rifabutin (150 mg twice daily) has been used for patients with eradication failures in combination with PPI and amoxicillin (1000 mg twice daily) for 10 or 14 days. In one study, this therapy has reached 95% eradication rates as second-line therapy and 68% eradication for third- or subsequent line therapy (Van der Poorten and Katelaris, 2007). These data show that the efficacy of this regime has decreased with the increasing number of previous therapy failures.

Third-line treatment

According to the Maastricht III Consensus Report, third-line rescue treatment should be based on antimicrobial susceptibility testing (Malfertheiner *et al.*, 2007). It is very important to know that when clarithromycin resistance reaches 20% in some areas, the eradication of the infection is usually a failure (Malfertheiner *et al.*, 2007.). Most of the rescue regimens for *H. pylori* infection have been predominantly empirical. The most commonly used empiric rescue antibiotics are levofloxacin, rifabutin and furazolidone (Cianci *et al.*, 2006; Saad and Chey, 2008; Abbas *et al.*, 2009; Cheng and Hu, 2009; De Francesco *et al.*, 2009; Suzuki *et al.*, 2009).

Several studies have compared alternatives with bismuth-based quadruple salvage therapy such as levofloxacin or rifabutin-based therapy with rates of eradication ranging from 38% to 91%, as well as furazolidone, with rates of eradication ranging from 52% to 90% (Chey and Wong, 2007). Levofloxacin therapy as a third-line therapy has exhibited an eradication rate of 60% in patients with two failed treatments with standard triple therapy and bismuth-based quadruple therapy, respectively.

New data from patients with unsuccessful eradication by standard first-line or bismuth-based second-line therapy have shown that rifabutin-based third-line therapy has reached 79% eradication rate based on ITT analysis (Gisbert *et al.*, 2003; Gonzalez Carro *et al.*, 2007; Van der Poorten and Katelaris, 2007). In contrast, another large study has obtained an eradication rate of only 61% when rifabutin has been used in a third-line treatment (Qasim *et al.*, 2005). However, we should not forget the risk for increase in resistance in *Mycobacterium tuberculosis* strains. In patients with two previous treatment failures, levofloxacin has shown higher rate of eradication than rifabutin (Gisbert *et al.*, 2006).

A German study of over 100 patients with at least one previous treatment failure and *H. pylori* resistance to both metronidazole and clarithromycin has shown that triple therapy with esomeprazole (40 mg), moxifloxacin (400 mg), and rifabutin (300 mg), given once daily for 7 days has led to an ITT eradication rate of 77.7% (Miehlke *et al.*, 2008).

Furazolidone is a synthetic nitrofuran with antibacterial and antiprotozoal effects, which has shown some antibacterial activity in Gram-negative enteric organisms. Use of furazolidone alone or in combination with amoxicillin and PPI as a third line rescue treatment has shown 60% eradication rate in 10 patients, in whom previous eradication regimens, including also rifabutin, had failed (Qasim *et al.*, 2005). In contrast, Gisbert *et al.* (2010) have suggested that rifabutin-based rescue therapy could be used as empirical fourth-line strategy after multiple previous eradication failures with key antibiotics such as amoxicillin, clarithromycin, metronidazole, tetracycline, and levofloxacin. The eradication efficacy has increased to 83% by ITT when furazolidone has been applied with levofloxacin, but has decreased to 57% if used as a fourth-line therapy (Eisig *et al.*, 2009). A total of 65% eradication efficacy of furazolidone has been reported in one systematic review of furazolidone-based treatments for third and subsequent line eradication therapy (Buzas and Jozan, 2007).

In recent years, many new drugs and combinations have been applied with the aim of increasing eradication rates to more acceptable levels, but more evidence is needed to support the routine use of the new antibiotics and modified treatment regimens for eradication of *H. pylori*.

References

Abbas, Z., Yakoob, J., Abid, S., Jafri, W., Islam, M., Azam, Z., and Hilal, I. (2009). Furazolidone, co-amoxiclav, colloidal bismuth subcitrate, and esomeprazole for patients who failed to eradicate *Helicobacter pylori* with triple therapy. Dig. Dis. Sci. *54*, 1953–1957.

Antos, D., Schneider-Brachert, W., Bästlein, E., Hänel, C., Haferland, C., Buchner, M., Meier, E., Trump, F., Stolte, M., Lehn, N., *et al.* (2006). 7-day triple therapy of *Helicobacter pylori* infection with levofloxacin, amoxicillin, and high-dose esomeprazole in patients with known antimicrobial sensitivity. Helicobacter *11*, 39–45.

Apseloff, G. (2003). Severe neutropenia among healthy volunteers given rifabutin in clinical trials. Clin. Pharmacol. Ther. *74*, 591–592.

Asaka, M., Kato, M., Takahashi, S., Fukuda, Y., Sugiyama, T., Ota, H., Uemura, N., Murakami, K., Satoh, K., Sugano, K., and Japanese Society for Helicobacter Research. (2010). Guidelines for the management of *Helicobacter pylori* infection in Japan: 2009 revised edition. Helicobacter *15*, 1–20.

Bhagat, N., Read, R.W., Rao, N.A., Smith, R.E., and Chong, L.P. (2001). Rifabutin associated hypopyon uveitis in human immunodeficiency virus negative immunocompetent individuals. Ophthalmology *108*, 750–752.

Bourke, B., Ceponis, P., Chiba, N., Czinn, S., Ferraro, R., Fischbach, L., Gold, B., Hyunh, H., Jacobson, K., Jones, N.L., *et al.* (2005). Canadian Helicobacter Study Group Consensus Conference: Update on the approach to *Helicobacter pylori* infection in children and adolescents – an evidence-based evaluation. Can. J. Gastroenterol. *9*, 399–408.

Boyanova, L. (2009). Prevlance of multidrug-resistant *Helicobacter pylori* in Bulgaria. J. Med. Microbiol. *58*, 930–935.

Breuer, T., and Graham, D.Y. (1999). Costs of diagnosis and treatment of *Helicobacter pylori* infection: when does choosing the treatment regimen based on susceptibility testing become cost effective? Am. J. Gastroenterol. *94*, 725–729.

Buzas, G.M., and Jozan, J. (2007). Nitrofuran-based regimens for the eradication of *Helicobacter pylori* infection. J. Gastroenterol. Hepatol. *22*, 1571–1581.

Calvet, X., Garcıa, N., Lopez, T., Gisbert, J.P., Gene, E., and Roque, M. (2000). A meta-analysis of short versus long therapy with a proton pump inhibitor, clarithromycin and either metronidazole or amoxycillin for treating *Helicobacter pylori* infection. Aliment. Pharmacol. Ther. *14*, 603–609.

Calvet, X., Gene, E., Lopez, T., and Gisbert, J.P. (2001). What is the optimal length of proton pump inhibitor-based triple therapies for *H. pylori*? A cost-effectiveness analysis. Aliment. Pharmacol. Ther. *15*, 1067–1076.

Calvet, X., Ducons, J., Guardiola, J., Tito, L, Andreu, V., Bory, F., Guirao, R., and Group for Eradication Studies from Catalonia and Aragón (Gresca). (2002). One-week triple vs. quadruple therapy for *Helicobacter pylori* infection – a randomized trial. Aliment. Pharmacol. Ther. *16*, 1261–1267.

Caselli, M., Zullo, A., Maconi, G., Parente, F., Alvisi, V., Casetti, T., Sorrentino, D., Gasbarrini, G., and Working Group of the Cervia II Meeting. (2007). Cervia II Working Group Report 2006: guidelines on diagnosis and treatment of *Helicobacter pylori* infection in Italy. Dig. Liver Dis. *39*, 782–789.

Cheng, H., and Hu, F.L. (2009). Furazolidone, amoxicillin, bismuth and rabeprazole quadruple rescue therapy for the eradication of *Helicobacter pylori*. World. J. Gastroenterol. *15*, 860–864.

Chey, W.D., and Wong, B.C. (2007). Practice Parameters Committee of the American College ofGastroenterology American College of Gastroenterology guideline on the management of *Helicobacter pylori* infection. Am. J. Gastroenterol. *102*, 1808–1825.

Choi, W.H., Park, D.I., Oh, S.J., Baek, Y.H., Hong, C.H., Hong, E.J., Song, M.J., Park, S.K., Park, J.H., Kim, H.J., *et al.* (2008). Effectiveness of 10 day-sequential therapy for *Helicobacter pylori* eradication in Korea. Korean. J. Gastroenterol. *51*, 280–284.

Cianci, R., Montalto, M., Pandolfi, F., Gasbarrini, G.B., and Cammarota, G. (2006). Third-line rescue therapy for *Helicobacter pylori* infection. World J. Gastroenterol. *12*, 2313–2319.

Current European concepts in the management of *Helicobacter pylori* infection. The Maastricht Consensus Report. (1997). Gut *41*, 8–13.

De Francesco, V., Zullo, A., Hassan, C., Della Valle, N., Pietrini, L., Minenna, M.F., Winn, S., Monno, R., Stoppino, V., Morini, S., *et al.* (2004). The prolongation of triple therapy for *Helicobacter pylori* does not allow reaching therapeutic outcome of sequential scheme: a prospective, randomised study. Dig. Liver Dis. *36*, 322–326.

De Francesco, V., Ierardi, E., Hassan, C., and Zullo, A. (2009). Furazolidone therapy for *Helicobacter pylori*: is it effective and safe? World J. Gastroenterol. *15*, 1914–1915.

Dore, M.P., Graham, D.Y., Mele, R., Marras, L., Nieddu, S., Manca, A., and Realdi, G. (2002). Colloidal bismuth subcitrate-based twice-a-day quadruple therapy as primary or salvage therapy for *Helicobacter pylori* infection. Am. J. Gastroenterol. *97*, 857–860.

Dore, M.P., Maragkoudakis, E., Pironti, A., Tadeu, V., Tedde, R., Realdi, G., and Delitala, G. (2006). Twice-a-day quadruple therapy for eradication of *Helicobacter pylori* in the elderly. Helicobacter *11*, 52–55.

Drumm, B., Koletzko, S., and Oderda, G. (2000). *Helicobacter pylori* infection in children: a consensus statement. J. Pediatr. Gastroenterol. Nutr. *30*, 207–213.

Egan, B.J., Katicic, M., O'Connor, A., and Morain, C.A. (2007). Treatment of *Helicobacter pylori*. Helicobacter *12 (Suppl. 1)*, 31–37.

Eisig, J.N., Silva, F.M., Barbuti, R.C., Rodriguez, T.N., Malfertheiner, P., Moraes Filho, J.P., Moraes Filho, J.P., and Zaterka, S. (2009). Efficacy of a 7-day course of furazolidone, levofloxacin, and lansoprazole after failed *Helicobacter pylori* eradication. BMC Gastroenterol. *9*, 38.

Essa, A.S., Kramer, J.R., Graham, D.Y., and Treiber, G. (2009). Meta-analysis: four-drug, three-antibiotic, non-bismuth-containing "concomitant therapy" versus triple therapy for *Helicobacter pylori* eradication. Helicobacter *14*, 109–118.

Fischbach, L., and Evans, E.L. (2007). Meta-analysis: effect of antibiotic resistance on treatments for *H. pylori*. Aliment. Pharmacol. Ther. *26*, 343–357.

Fock, K.M., Katelaris, P., Sugano, K., Ang, T.L., Hunt, R., Talley, N.J., Lam, S.K, Xiao, S.D., Tan, H.J., Wu, C.Y., *et al.* (2009). Second Asia-Pacific Conference. Second Asia-Pacific Consensus Guidelines for *Helicobacter pylori* infection. J. Gastroenterol. Hepatol. *24*, 1587–1600.

Ford, A., and Moayyedi, P. (2003). How can the current strategies for *Helicobacter pylori* eradication therapy be improved? Can. J. Gastroenterol. *17 (Suppl B)*, 36–40B.

Ford, A.C., Malfertheiner, P., Giguere, M., Santana, J., Khan, M., and Moayyedi, P. (2008). Adverse events with bismuth salts for *Helicobacter pylori* eradication: systematic review and meta-analysis. World J. Gastroenterol. *14*, 7361–7370.

Fuccio, L., Minardi, M.E., Zagari, R.M., Grilli, D., Magrin, Ni., and Bazzoli, F. (2007). Meta-analysis: duration of first-line proton-pump inhibitor based triple therapy for *Helicobacter pylori* eradication. Ann. Int. Med. 2007 *147*, 553–562.

Gasbarrini, A., Lauritano, E.C., Nista, E.C., Candelli, M., Gabrielli, M., Santoro, M., Zocco, M.A., Cazzato, A., Finizio, R., Ojetti, V., *et al.* (2006). Rifaximin-based regimens for eradication of *Helicobacter pylori*: a pilot study. Dig. Dis. *24*, 195–200.

Gisbert, J.P., Pajares, J.M., and Valle, J. (1999). Ranitidine bismuth citrate therapy regimens for treatment of *Helicobacter pylori* infection: a review. Helicobacter *4*, 58–66.

Gisbert, J.P., and Javares, J.M. (2002). *Helicobacter pylori* "rescue" regimen when proton pump inhibitor based-triple therapies fail. Aliment. Pharmacol. Ther. *16*, 1047–1057.

Gisbert, J.P., Calvet, X., Bujanda, L., Marcos, S., Gisbert. J.L., and Pajares, J.M. (2003). 'Rescue' therapy with rifabutin after multiple *Helicobacter pylori* treatment failures. Helicobacter *8*, 90–94.

Gisbert, J.P., and De la Morena, F. (2006). Systematic review and meta-analysis: levofloxacin-based rescue regimens after *Helicobacter pylori* treatment failure. Aliment. Pharmacol. Ther. *23*, 35–44.

Gisbert, J.P., Gisbert, J.L., Marcos, S., Moreno-Otero, R., and Pajares, J.M. (2006). Third-line rescue therapy with levofloxacin is more effective than rifabutin rescue regimen after two *Helicobacter pylori* treatment failures. Aliment. Pharmacol. Ther. *24*, 1469–1474.

Gisbert, J.P., Gisbert, J.L., Marcos, S., Moreno-Otero, R., and Pajares, J.M. (2007a). Levofloxacin- vs. ranitidine bismuth citrate-containing therapy after *H. pylori* treatment failure. Helicobacter *12*, 68–73.

Gisbert, J.P., Fernandez Bermejo, M., Molina Infante, J., Perez Gallardo, B., Prieto Bermejo, A.B., Mateos Rodriguez, J.M., Robledo Andres, P., and Gonzalez Garcia G. (2007b). First-line triple therapy with levofloxacin for *Helicobacter pylori* eradication. Aliment. Pharmacol. Ther. *26*, 495–500.

Gisbert, J.P., Bermejo, F., Castro-Fernandez, M., Perez-Aisa, A., Fernandez-Bermejo, M., Tomas, A., Barrio, J., Bory, F., Almela, P., Sánchez-Pobre, P., *et al.* (2008). *H. pylori* Study Group of the Asociacion Espanola de Gastroenterologia. Second-line rescue therapy with levofloxacin after *H. pylori* treatment failure: a Spanish multicenter study of 300 patients. Am. J. Gastroenterol. *103*, 71–76.

Gisbert, J.P. (2009). Second-line tescue therapy of *Helicobacter pylori* infection: Second-line *H. pylori* 'rescue' therapy after failure of one eradication treatment. Ther. Adv. Gastroenterol. *2*, 331–356.

Gisbert, J.P., Perez-Aisa, A., Castro-Fernandez, M., Barrio, J., Rodrigo, L., Cosme, A., Cosme, A., Gisbert, J.L., Marcos, S., and Moreno-Otero, R. (2010). *Helicobacter pylori* first-line treatment and rescue option containing levofloxacin in patients allergic to penicillin. Dig. Liver Dis. *42*, 287–290.

Gonzalez Carro, P., Perez Roldan, F., De Pedro Esteban, A., Legaz Huidobro, M.L., Soto Fernández, S., Roncero Garcia Escribano, O., Esteban López-Jamar, J.M., Pedraza Martin, C. and Ruíz Carrillo, F. (2007). Efficacy of rifabutin-based triple therapy in *Helicobacter pylori* infected patients after two standard treatments. J. Gastroenterol. Hepatol. *22*, 60–63.

Graham, D.Y., Lew, G.M., Malaty, H.M., Evans, D.G., Evans, D.J. Jr., Klein, P.D., Alpert, L.C. and Genta, R.M. (1992). Factors influencing the eradication of *Helicobacter pylori* with triple therapy. Gastroenterology *102*, 493–496.

Hojo, M., Miwa, H., Nagahara, A. and Sato, N. (2001). Pooled analysis on the efficacy of the second-line treatment regimens for *Helicobacter pylori* infection. Scand. J. Gastroenterol. *36*, 690–700.

Horn, J. (2004). Review article: relationship between the metabolism and efficacy of proton pump inhibitors – focus on rabeprazole. Aliment. Pharmacol. Ther. *20 (Suppl 6)*, 11–19.

Howden, C., and Hunt, R.H. (1998). Guidelines for the management of *Helicobacter pylori*. Am. J. Gastroenterol. *93*, 2330–2338.

Jafri, N.S., Hornung, C.A., and Howden CW. (2008). Meta-analysis: sequential therapy appears superior to standard therapy for *Helicobacter pylori* infection in patients naive to treatment. Ann. Intern. Med. *148*, 923–931.

Kadayifci, A., Buyukhatipoglu, H., Cemil Savas, M., and Simsek I. (2006). Eradication of *Helicobacter pylori* with triple therapy: an epidemiologic analysis of trends in Turkey over 10 years. Clin. Ther. *28*, 1960–1966.

Katicic, M., Ticak, M., Prskalo, M., Naumovski-Mihalic, S., Filipec-Kanizaj, T., Skurla, B., Sabaric, B., Colic-Cvrlje, V., and Papa B. (2006). Eradication of *Helicobacter pylori* infection with two triple therapy regimes of 7, 10, and 14 days: four year experience. Helicobacter 2, A11.05, 386.

Kearney, D.J., and Brousal, A. (2000). Treatment of *Helicobacter pylori* infection in clinical practice in the United States: results from 224 patients. Dig. Dis. Sci. *45*, 265–271.

Kim, B.G., Lee, D.H., Ye, B.D., Lee, K.H., Kim, B.W., Kim, S.G., Kim, S.W., Kim, S.K., Kim, J.J., Kim, H.Y., *et al.* (2007). Comparison of 7-day and 14-day proton pump inhibitor-containing triple therapy for *Helicobacter pylori* eradication: neither treatment duration provides acceptable eradication rate in Korea. Helicobacter *12*, 31–35.

Kirchheiner, J., Glatt, S., Fuhr, U., Klotz, U., Meineke, I., Seufferlein, T., and Brockmöller J. (2009). Relative potency of proton-pump inhibitors-comparison of effects on intragastric pH. Eur. J. Clin. Pharmacol.*65*, 9–31.

Kuo, C.H., Wang, S.S., Hsu, W.H., Kuo, F.C., Weng, B.C., Li, C.J., Hsu, P.I., Chen, A., Hung, W.C., Yang, Y.C., *et al.* (2010). Rabeprazole can overcome the impact of CYP2C19 polymorphism on quadruple therapy. Helicobacter *15*, 265–272.

Laheij, R.J.F, van Rossum. L.G.M., Jansen. J.B., Straatman. H., and Verbeek., A.L. (1999). Evaluation of treatment regimens to cure *Helicobacter pylori* infection – a meta-analysis. Aliment. Pharmacol. Ther. *13*, 857–864.

Laine, L., Suchower, L., Frantz, J., Connors, A., and Neil, G. (1998). Twice-daily, 10-day triple therapy with omeprazole, amoxicillin, and clarithromycin for *Helicobacter pylori* eradication in duodenal ulcer disease: results of three multicenter, double-blind, United States trials. Am. J. Gastroenterol. *93*, 2106–2112.

Lamouliatte, H., Megraud, F., Delchier, J.C., Bretagne, J.F., Courillon-Mallet, A., De Korwin, J.D., Fauchère, J.L., Labigne, A., Fléjou, J.F., Barthelemy, P., and Multicentre Study Group. (2003). Second-line treatment for failure to eradicate *Helicobacter pylori*: a randomized trial comparing four treatment strategies. Aliment. Pharmacol. Ther. *18*, 791–797.

Lee, B.H., Kim, N., Hwang, T.J., Lee, S.H., Park, Y.S., Hwang, J.H., Kim, J.W., Jeong, S.H., Lee, D.H., Jung, H.C., and Song, I.S. (2010). Bismuth-containing quadruple therapy as second-line treatment for *Helicobacter pylori* infection: effect of treatment duration and antibiotic resistance on the eradication rate in Korea. Helicobacter *15*, 38–45.

Lind, T., Veldhuyzen van Zanten, S., Unge, P., Spiller, R., Bayerdörffer, E., O'Morain, C., Bardhan, K.D., Bradette, M., Chiba, N., Wrangstadh, M., *et al.* (1996). Eradication of *Helicobacter pylori* using one-week triple therapies combining omeprazole with two antimicrobials: the MACH I Study. Helicobacter *1*, 138–144.

Lind, T., Megraud, F., Unge, P., Bayerdörffer, E., O'Morain, C., Spiller, R., Veldhuyzen Van Zanten, S., Bardhan, K.D., Hellblom, M., Wrangstadh, M., *et al.* (1999). The MACH2 study: role of omeprazole in eradication of *Helicobacter pylori* with 1-week triple therapies. Gastroenterology *116*, 248–253.

van der Linden, P.D., Sturkenboom, M.C., Herings, R.M., Leufkens, H.G., and Stricker, B.H. (2002). Fluoroquinolones and risk of Achilles tendon disorders. B.M.J. *324*, 1306–1307.

Luth, S., Teyssen, S., Kolbel, C.B., and Singer, M.V. (2001). 4- day triple therapy with rabeprazole, amoxicillin and clarithromycin in the eradication of *Helicobacter pylori* in patients with peptic ulcer disease – a pilot study. Z. Gastroenterol. *39*, 279–281.

Malfertheiner, P., Mégraud, F., O'Morain, C., Hungin, A.P., Jones, R., Axon, A., Graham, D.Y., Tytgat, G., and European *Helicobacter pylori* Study Group (EHpSG). (2002). Current concepts in the management of *Helicobacter pylori* Infection – The Maastricht 2-2000 Consensus Report. Aliment. Pharmacol. Ther. *16*, 167–180.

Malfertheiner, P., Megraud, F., O'Morain, C., Bazzoli, F., El-Omar, E., Graham, D., Hunt, R., Rokkas, T., Vakil, N., Kuipers, E.J., and the European Helicobacter Study Group (EHSG). (2007). Current concepts in the management of *Helicobacter pylori* infection – The Maastricht III Consensus Report. Gut *56*, 772–778.

Malfertheiner, P., Megraud, F., Giguere, M., and Riviere, M. (2010). Quadruple therapy with bismuth subcitrate potassium, metronidazole, tetracycline, and omeprazole is superior to triple therapy with omeprazole, amoxicillin, and clarithromycin in the eradication of *Helicobacter pylori*. Gastroenterology *138 (Suppl 1)*, S33.

Megraud, F. (2004). *H. pylori* antibiotic resistance: prevalence, importance, and advances in testing. Gut *53*, 1374–1384.

Megraud, F. (2005). Update on therapeutic options for *Helicobacter pylori*-related diseases. Curr. Infect. Dis. Rep. 7, 115–120.

Miehlke, S., Schneider-Brachert, W., Kirsch, C., Morgner, A., Madisch, A., Kuhlisch, E., Haferland, C., Bästlein, E., Jebens, C., Zekorn, C., *et al.* (2008). One-week once-daily triple therapy with esomeprazole, moxifloxacin, and rifabutin for eradication of persistent *Helicobacter pylori* resistant to both metronidazole and clarithromycin. Helicobacter *13*, 69–74.

Minakari, M., Davarpanah Jazi, A.H., Shavakhi, A., Moghareabed, N., and Fatahi, F. (2010). A randomized controlled trial: efficacy and safety of azithromycin, ofloxacin, bismuth, and omeprazole compared with amoxicillin, clarithromycin, bismuth, and omeprazole as second-line therapy in patients with *Helicobacter pylori* infection. Helicobacter *15*, 154–159.

Misiewicz, J.J., Harris A.W., Bardhan, K.D., Levi, S., O'Morain, C., Cooper, B.T, Kerr, G.D., Dixon, M.F., Langworthy, H., and Piper, D. (1997). One week triple therapy for *Helicobacter pylori*: a multicenter comparative study. Gut *41*, 735–739.

Molina-Infante, J., Perez-Gallardo, B., Fernandez-Bermejo, M., Hernandez-Alonso, M., Vinagre, G., Dueñas, C., Mateos-Rodriguez, J.M., Gonzalez-Garcia, G., Abadia, E.G., and Gisbert JP. (2010) Clinical trial: clarithromycin vs. levofloxacin in first-line triple and sequential regimens for *Helicobacter pylori* eradication. Aliment. Pharmacol. Ther. *31*, 1077–1084.

NIH Consensus Conference. (1994). *Helicobacter pylori* in peptic ulcer disease. NIH consensus development panel on *Helicobacter pylori* in peptic ulcer disease. JAMA *272*, 65–69.

Nista, E.C., Candelli, M., Zocco, M.A., Cremonini, F., Ojetti, V., Finizio, R., Spada, C., Cammarota, G., Gasbarrini, G., and Gasbarrini A. (2006). Levofloxacin-based triple therapy in first-line treatment for *Helicobacter pylori* eradication. Am. J. Gastroenterol. *101*, 1985–1990.

O'Connor, A., Gisbert, J., and O'Morain, C. (2009). Treatment of *Helicobacter pylori* infection. Helicobacter *14* (Suppl. 1), 46–51.

O'Connor, A., Gisbert, J.P., McNamara, D., and O'Morain, C. (2010). Treatment of *Helicobacter pylori* infection. Helicobacter 15 (*Suppl 1*), 46–52.

O'Morain, C.A., and O'Connor, J.P. (2009). Is sequential therapy superior to standard triple therapy for the treatment of *Helicobacter pylori* infection? Nat. Clin. Pract. Gastroenterol. Hepatol. *6*, 8–9.

Paoluzi, P., Iacopini, F., Crispino, P., Nardi, F., Bella, A., Rivera, M., Rossi, P., Gurnari, M., Caracciolo, F., Zippi, M., and Pica R. (2006). 2-week triple therapy for *Helicobacter pylori* infection is better than 1-week in clinical practice: a large prospective single-center randomized study. Helicobacter *11*, 562–568.

Paoluzi, O.A., Visconti, E., Andrei, F., Tosti, C., Lionetti, R., Grasso, E., Ranaldi, R., Stroppa, I., and Pallone, F. (2010). Ten and eight-day sequential therapy in comparison to standard triple therapy for eradicating *Helicobacter pylori* infection: a randomized controlled study on efficacy and tolerability. J. Clin. Gastroenterol. *44*, 261–266.

Pipkin, G.A., Williamsson, R., and Wood JR. (1998). Review article: one week clarithromycin triple therapy regimens for eradication of *Helicobacter pylori*. Aliment. Pharmacol. Ther. *12*, 823–837.

Qasim, A., Sebastian, S., Thornton, O., Dobson, M., McLoughlin, R., Buckley, M., Buckley, M., O'Connor, H., and O'Morain C. (2005). Rifabutin- and furazolidone-based *Helicobacter pylori* eradication therapies after failure of standard first- and second-line eradication attempts in dyspepsia patients. Aliment. Pharmacol. Ther. *21*, 91–96.

Realdi, G., Dore, M.P., Piana, A., Atzei, A., Carta, M., Cugia, L., Manca, A., Are, B.M., Massarelli, G., Mura, I., *et al.* (1999). Pretreatment antibiotic resistance in *Helicobacter pylori* infection: results of three randomized controlled studies. Helicobacter *4*, 106–112.

Romano, M., Marmo, R., Cuomo, A., De Simone, T., Mucherino, C., Iovene, M.R, Montella, F., Tufano, M.A, Del Vecchio Blanco, C., and Nardone G. (2003). Pretreatment antimicrobial susceptibility testing is cost saving in the eradication of *Helicobacter pylori*. Clin. Gastroenterol. Hepatol. *1*, 273–278.

Romano, M., Iovene, M.R., Russo, M.I., Rocco, A., Salerno, R., Cozzolino, D., Pilloni, A.P., Tufano, M.A., Vaira, D., and Nardone, G. (2008). Failure of first-line eradication treatment significantly increases prevalence of antimicrobial-resistant *Helicobacter pylori* clinical isolates. J. Clin. Pathol. *61*, 1112–1115.

Romano, M., Cuomo, A., Gravina, A.G., Miranda, A., Iovene, M.R., Tiso, A., Sica, M., Rocco, A., Salerno, R., Marmo, R., *et al.* (2010). Empirical levofloxacin-containing versus clarithromycin-containing sequential therapy for *Helicobacter pylori* eradication: a randomised trial. Gut *59*, 1465–1470.

Saad, R., Schoenfeld, P., Hyungjin, M.K., and Chey, W.D. (2006). Levofloxacin-based triple therapy versus bismuth-based quadruple therapy for persistent *Helicobacter pylori* infection: a meta-analysis. Am. J. Gastroenterol. 3, 488–496.

Saad, R.J., and Chey, W.D. (2006). Treatment of *Helicobacter pylori* infection in 2006. Gastroenterol. Hepatol. Annu. Rev. *1*, 30–35.

Saad, R.J., and Chey, W.D. (2008). Persistent *Helicobacter pylori* infection after a course of antimicrobial therapy – what's next? Clin. Gastroenterol. Hepatol. *6*, 1086–1090.

Savarino, V., Di Mario, F., and Scarpignato, C. (2009). Proton pump inhibitors in GORD An overview of their pharmacology, efficacy and safety. Pharmacol, Res. *59*, 135–153.

Scaccianoce, G., Hassan, C., Panarese, A., Piglionica, D., Morini, S., and Zullo, A. (2006). *Helicobacter pylori* eradication with either 7-day or 10-day triple therapies, and with a 10-day sequential regimen. Can. J. Gastroenterol. *20*, 113–117.

Shehada, S., Srugo, I., and Shaoul, R. (2007). Failure of sequential therapy to eradicate *Helicobacter pylori* in previously treated subjects. Helicobacter *12*, 645–647.

Spahr, L., Rubbia-Brandt, L., Marinescu, O., Armenian, B., and Hadengue A. (2001). Acute fatal hepatitis related to levofloxacin. J. Hepatol. *35*, 308–309.

Suzuki, S., Suzuki, H., Nishizawa, T., Kaneko, F., Ootani, S., Muraoka, H., Saito, Y., Kobayashi, I., and Hibi, T. (2009). Past rifampicin dosing determines rifabutin resistance of *Helicobacter pylori*. Digestion *79*, 1–4.

Tong, J.L., Ran, Z.H., Shen, J., and Xiao, S.D. (2009). Sequential therapy vs. Standard triple therapies for *Helicobacter pylori* infection: a metaanalysis. J. Clin. Pharm. Ther. *34*, 41–53.

Toracchio, S., Cellini, L., Di Campli, E., Cappello, G., Malatesta, M.G., Ferri, A, Ciccaglione, A.F., Grossi, L., and Marzio, L. (2000). Role of antimicrobial susceptibility testing on efficacy of triple therapy in *Helicobacter pylori* eradication. Aliment. Pharmacol. Ther. *14*, 1639–1643.

Tulassay, Z., Kryszewski, A., Dite, P., Kleczkowski, D., Rudzinski, J., Bartuzi, Z., Hasselgren, G., Larkö, A., and Wrangstadh, M. (2001). One-week treatment with esomeprazole-based triple therapy eradicates *Helicobacter pylori* and heals patients with duodenal ulcer disease. Eur. J. Gastroenterol. Hepatol. *13*, 1457–1465.

Unge, P. (1998). Antimicrobial treatment of *H. pylori* infection – a pooled effcacy analysis of eradication therapies. Eur. J. Surg. *164* (Suppl. 582), 16–26.

Uygun, A., Ozel, A.M., Yildiz, O., Aslan, M., Yesilova, Z., Erdil, A, Bagci, S., and Gunhan, O. (2008). Comparison of three different second-line quadruple therapies including bismuth subcitrate in Turkish patients with non-ulcer dyspepsia who failed to eradicate *Helicobacter pylori* with a 14-day standard first-line therapy. J. Gastroenterol. Hepatol. *23*, 42–45.

Vaira, D., Zullo, A., Vakil, N., Gatta, L., Ricci, C., Perna, F., Hassan, C., Bernabucci, V., Tampieri, A., and Morini S. (2007). Sequential therapy versus standard triple-drug therapy for *Helicobacter pylori* eradication: a randomized trial. Ann. Intern. Med. *146*, 556–563.

Vakil, N., and Vaira, D. (2008). Sequential therapy for *Helicobacter pylori*: time to consider making the switch? JAMA *300*, 1346–1347.

Vakil, N. (2009). *H. pylori* treatment: new wine in old bottles? Am. J. Gastroenterol. *104*, 26–30.

Van der Poorten, D., and Katelaris, P.H. (2007). The effectiveness of rifabutin triple therapy for patients with difficult-to-eradicate *Helicobacter pylori* in clinical practice. Aliment. Pharmacol. Ther. *26*, 1537–1542.

Vladimirov, B., Terziev, I., and Ivanova, I. (2003). Rabeprazole, amoxiclave and clarithromycin – 4-day versus 7- day therapy in the eradication of *H. pylori*. In Eradication of *H. pylori*. 11th ICUR Meeting, Dubrovnik, 15–20 September 2003, pp. 279–284.

Wu, D.C., Hsu, P.I., Chen, A., Lai, K.H., Tsay, F.W., Wu, C.J., Lo, G.H., Wu, J.Y., Wu, I.C., Wang, W.M., *et al.* (2006). Randomized comparison of two rescue therapies for *Helicobacter pylori* infection. Eur. J. Clin. Invest. *36*, 803–809.

Zagari, R.M., Bianchi-Porro, G., Fiocca, R., Gasbarrini, G., Roda, E., and Bazzoli, F. (2007). Comparison of 1 and 2 weeks of omeprazole, amoxicillin and clarithromycin treatment for *Helicobacter pylori* eradication: the HYPER Study. Gut *56*, 475–479.

Zullo, A., Gatta, L., De Francesco, V., Hassan, C., Ricci, C., Bernabucci, V., Cavina, M., Ierardi, E., Morini, S., and Vaira, D. (2005). High rate of *Helicobacter pylori* eradication with sequential therapy in elderly patients with peptic ulcer: a prospective controlled study. Aliment. Pharmacol. Ther. *21*, 1419–1424.

11 Non-antibiotic Agents in the Treatment of *Helicobacter pylori* Infection

Lyudmila Boyanova

Abstract

As standard regimens for eradication of *H. pylori* infection often fail owing to antibiotic resistance or low patient compliance, there is an increasing need for new drugs such as non-antibiotic agents (NAAs) or optimized treatment regimens. Anti-*H. pylori* activity has been detected *in vitro* by many NAAs such as lactobacilli, *Saccharomyces boulardii,* lactoferrin, green tea, garlic, propolis, broccoli, resveratrol, plant oils etc. Many NAAs have been active against both antibiotic susceptible and resistant *H. pylori* strains. So far, cell line, animal or clinical trials have shown *H. pylori* eradication, although often low, using NAAs alone, e.g. black caraway, garlic, green tea, turmeric, broccoli and mastic gum. Combinations of standard regimens with NAAs such as lactobacilli, *S. boulardii,* Korea red ginseng, *N*-acetylcysteine, vitamin C, etc. have improved the eradication success often by >10% and most of them have reduced the side-effects of the regimens. Advantages of some NAAs are their antiadhesive, antiurease, anti-inflammatory and antitumour effects and bactericidal activity on *H. pylori.* Moreover, antiadhesive NAAs can prevent *H. pylori* colonization or reinfection. In addition, aspirin, monoclonal antibodies, redox protein inhibitors and synthetic antimicrobial peptidomimetics have been evaluated as applicants for therapeutic intervention. Synergy between NAAs is a very attractive topic, which should be evaluated, however, with caution. To choose an agent, it is necessary to consider the source, safety or toxicity, dose-depending deleterious effects and allergenicity of some NAAs as well as their quality control and evidence-based trials. At present, we await further studies, especially clinical trials, to suggest the choice, dosage, start time and duration of administration aiming at the routine use of the NAAs. In conclusion, NAAs probably carry greater than the expected potential for prophylaxis or adjuvant therapy of *H. pylori* infection and can help to control the rising bacterial resistance to antibiotics.

Introduction

H. pylori eradication may fail because of antibiotic resistance in the strain, low patient compliance and other factors. As *H. pylori* resistance has compromised the success of first-line eradication regimens in up to 50% of cases (see Chapter 9), there is an increasing need for safe and effective new regimens or compounds, e.g. non-antibiotic agents (NAAs). Activities of numerous NAAs have been evaluated. Topics of interest have been their antiadhesive properties, bacteriostatic and bactericidal activities against *H. pylori,* especially those against antibiotic-resistant strains, anti-inflammatory and anticancer effects, improvement of *H. pylori* eradication and decrease of side-effects of the standard regimens. However, it should be remembered that the *in vitro* activity against *H. pylori* does not always indicate *in vivo* effect of NAAs (Lesbros-Pantoflickova *et al.*, 2007; Wittschier *et al.*, 2007). Moreover, different methods used and various sources of the same agent can result in discrepancy between results (Zaidi *et al.*, 2009a). Studies on animal models and clinical trials are valuable to show the benefice of NAAs, which could be used for prophylaxis and treatment of the infection.

Lactobacilli

Probiotics such as lactobacilli, bifidobacteria and *Saccharomyces boulardii* are live non-pathogenic microorganisms that, when ingested in adequate amounts, exert health-promoting activities for the host. Probiotics have been used for management of many intestinal diseases (Servin, 2004). The lactobacilli, as a whole, are safe to the human organism except for the low risk of bacteraemia in severely immunocompromised patients or infants (Gupta and Garg, 2009). *Lactobacillus* spp. can persist in the stomach longer than most other bacteria because of their acid resistance and can show many anti-*H. pylori* activities. Among them are suppression of *H. pylori* urease activity, adhesion to gastric epithelial cells, secretion of short chain fatty acids, e.g. propionic, acetic, formic and, especially, lactic acid and also autolysins, bacteriocins, bacteriocins-like or other active substances such as hydrogen peroxide, proteinases, exopolysaccharides and cell wall components, induction of immunomodulation [inhibition of *H. pylori*-induced interleukin 8 (IL-8) release], etc. (de Vuyst *et al.*, 2004; Lesbros-Pantoflickova *et al.*, 2007; Zhou *et al.*, 2008). For instance, anti-*H. pylori* activity has been linked to autolysins of *Lactobacillus acidophilus*, heat-stable substance of *Lactobacillus johnsonii* La1 and *L. acidophilus* LB, heat-labile substance of *Lactobacillus casei* strain Shirota and bacteriocins of *Lactococcus lactis* BH5 and *Lactobacillus delbrueckii* subsp. *bulgaricus* (LDB) BB18 (Lorca *et al.*, 2001; Lesbros-Pantoflickova *et al.*, 2007; Simova *et al.*, 2009).

Viable cells and low pH are important for the activity of the lactobacilli. Viable *L. casei* strain Shirota and *Lactobacillus salivarius* cells have been required to inhibit *H. pylori in vitro* (Sgouras *et al.*, 2004; Ryan *et al.*, 2008). *L. acidophilus* and *Lactobacillus paraplantarum* KNUC25 activity has been associated mainly with lactic acid (Lorca *et al.*, 2001; Ki *et al.*, 2010). Three of six LDB strains evaluated also have been more active at low than at neutralized pH (Boyanova *et al.*, 2009). The most active strains have inhibited 86.7% of *H. pylori* strains at low pH and 66.7% at neutralized pH (Boyanova *et al.*, 2009; Fig. 11.1). Although *H. pylori* survives in low gastric pH, inhibition of the bacterial urease activity by the lactobacilli can increase the bacterial susceptibility to lactic acid (Servin, 2004; Sgouras *et al.*, 2004). Moreover, increasing the permeability of *H. pylori* outer membrane, the lactic acid can enhance the activity of the antibiotics (Servin, 2004).

It is noteworthy that the *in vitro* and *in vivo* activities of the probiotics may differ and the activity of the lactobacilli is extremely strain specific (Lesbros-Pantoflickova *et al.*, 2007; Lionetti *et al.*, 2010). Boyanova *et al.* (2009) have reported that all six LDB strains studied have inhibited a number of *H. pylori* strains, including those

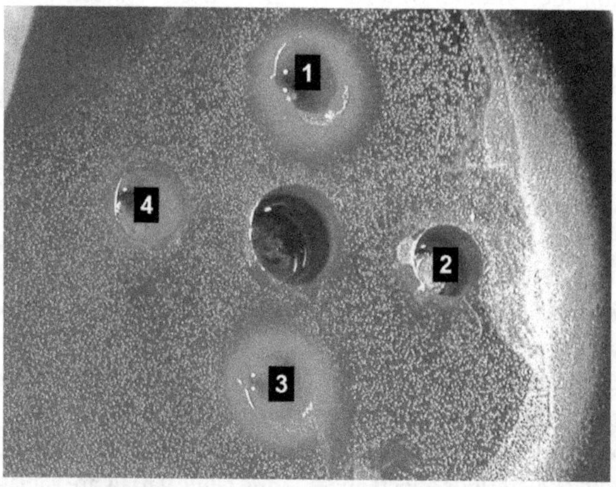

Figure 11.1 *H. pylori* growth inhibition by *Lactobacillus delbrueckii* subsp. *bulgaricus* strains in milk at pH 4 (wells 1 and 3) and lack of inhibition by another *L. delbrueckii* subsp. *bulgaricus* strain (well 2) as well as by sterile 10% skimmed cow's milk as the control (well 4).

resistant to antibiotics, however, the LDB activity has been highly strain-dependent. *Lactobacillus amylovorus, L. casei* Shirota and *L. salivarius* also have exhibited strain-dependent anti-*H. pylori* activity (de Vuyst *et al.*, 2004; Ryan *et al.*, 2008). Moreover, minimal inhibitory concentrations (MICs) of bacteriocins have ranged widely according to the *H. pylori* and lactic acid bacterial strains used with highest activity by lacticins A164 (produced by *Lactococcus lactis* subsp. *lactis* A164) and BH5 (produced by *L. lactis* BH5) compared with pediocin PO2 and leucocin K (Kim *et al.*, 2003). The MICs of the lacticins have oscillated widely from 0.097–0.39 µg/ml to 12.5–25 µg/ml according to the *H. pylori* strain tested. Therefore, the activity of the probiotics depends not only on the probiotic strain/product but also on its target strain, *H. pylori*.

It is worth mentioning that the lactobacilli can exert bactericidal activity on *H. pylori* (Table 11.1). *L. acidophilus* CRL 639 has been bactericidal in mixed cultures with *H. pylori* within 48 h, because of a proteinaceous substance released after cell lysis (Lorca *et al.*, 2001). Peptide extracts from *L. johnsonii* La1 have reduced the number of viable *H. pylori* ATCC 43504 colony-forming units (CFUs) by >2 log within 4 h (de Vuyst *et al.*, 2010). The extracts from *L. johnsonii* La1 and *L. amylovorus* DCE 471 have killed all viable *H. pylori* ATCC 43504 cells at 6 and 4 h, respectively, of incubation. In addition, *L. amylovorus* DCE 471 has been able to kill five of six clinical *H. pylori* strains within 4 h of incubation (de Vuyst *et al.*, 2010).

Importantly, the lactobacilli can reduce *H. pylori* density in the gastric mucosa and can prevent colonization or reinfection, inhibiting *H. pylori* adhesion to the gastric cells (Lionetti *et al.*, 2010). *L. paraplantarum* KNUC25, isolated from kimchi, a Korean food, has inhibited the adherence of *cagA*$^+$ strains to AGS gastric cell line (Ki *et al.*, 2010). Ingestion of *L. gasseri, Lactobacillus reuteri* ATCC 55730 and *L. casei* has reduced *H. pylori* density in animals or humans (Sakamoto *et al.*, 2001; Sgouras *et al.*, 2004; Francavilla *et al.*, 2008). In the meta-analysis of Lesbros-Pantoflickova *et al.* (2007), the probiotic use has improved gastritis and has decreased gastric *H. pylori* density in seven of nine studies. Four weeks' administration of *L. reuteri* decreased UBT in Japanese patients by about 70% (Imase *et al.*, 2007). Epidemiological studies of yoghurt consumption have confirmed the beneficial effect of probiotic consumption. In Mexico, lower *H. pylori* seropositivity has been reported in yoghurt consumers than in non-consumers (Ornelas *et al.*, 2007). In Poland, regular probiotic consumption has been reported to result in a 2-fold reduced risk for *H. pylori* reinfection (Jarosz *et al.*, 2009).

In some studies, adding lactobacilli to anti-*H. pylori* standard regimens has been shown to improve *H. pylori* eradication, to decrease side-effects of therapy and to inhibit antibiotic-resistant strains. Unlike Lionetti *et al.* (2010), other authors have reported that probiotics increase the eradication rates of standard regimens by 10% (5–15%) and reduce the side-effects of therapy by 23% compared with the same regimens used alone, although most often no eradication has been obtained using probiotics only (Lesbros-Pantoflickova *et al.*, 2007; Sachdeva and Nagpal, 2009). In a meta-analysis, Zou *et al.* (2009a) reported eradication in 82.3% of patients treated with lactobacilli and standard triple therapy compared with 77% in those not treated with lactobacilli using intention-to-treat (ITT) analysis. An omeprazole, amoxicillin, clarithromycin and *L. casei* DN-114001 regimen increased the eradication rate by 27.1% (ITT analysis) in Czech children compared with the regimen without probiotics. The probiotic-supplemented regimen resulted in eradication of five of the six clarithromycin-resistant strains (Sýkora *et al.*, 2005). Conversely, Scaccianoce *et al.* (2008) obtained merely 71% eradication rate using a 14-day triple therapy and probiotic mixture. So far, *L. casei* and *L. johnsonii* La1 appear to be the most active strains (Lesbros-Pantoflickova *et al.*, 2007).

In case of co-administration of eradication regimen with lactobacilli, possible inhibition of the probiotic strain by the antibiotics should be considered. For this reason, it is important to administer a high number of lactobacilli. Increased inocula of lactobacilli can reduce the antibiotic activity, increasing antibiotic MICs against all *Lactobacillus* spp. (Egervärn *et al.*, 2007). Although variable doses of lactobacilli have been used in some studies, the recommended dose is 5×10^9

Table 11.1 Examples of some non-antibiotic agents with activity against *H. pylori*

Agents	Activities on *H. pylori*						Reference
	Inhibitory (MIC, μg/ml)	Bactericidal (MBC, μg/ml)	Other[1]	Eradication (%) by the agent	Increase (%) in eradication of standard therapy	Decrease (%) in side-effects of standard therapy	
Lactobacillus strains and lactic acid bacteria	+ (0.097–25 lacticins)	+ (*L. amylovorus* DCE 471 and *L. johnsonii* La1 extracts at 4–6 h)[2]	AA, AU, AI, AT	0–3 (*L. johnsonii* La1)[2]	+ (by 5.3–15)[2]	+ (by ≤23)[2]	Kim *et al*. (2003) Lesbros-Pantoflickova *et al*. (2007), Gotteland *et al*. (2008), Zhou *et al*. (2008), de Vuyst *et al*. (2010)
Saccharomyces boulardii	NA	NA	AA, AI	NA	+ (by 11–12)	+	Cindoruk *et al*. (2007), Hurduc *et al*. (2009), McFarland (2010)
Bovine lactoferrin	+ (>500)	NA	AA	NA	+ (by 6–16.3)	+ (by 7.3)	de Bortoli *et al*. (2007), Tursi *et al*. (2007), Zou *et al*. (2009b)
Garlic extract	+ (4–6 allicin)	+ (6 allicin)[2]	AT	12 (garlic + PPI)	NA	NA	O'Gara *et al*. (2000), Ali *et al*. (2005), Fani *et al*. (2007)
Black caraway (*Nigella sativa*)	+	(62.5 at 60 min)	AI, AT	66.7 of 21 patients (with PPI)	NA	NA	Zaidi *et al*. (2009a), Salem *et al*. (2010)
Propolis	+ (MIC_{50}, ≤100)	NA	AU, AI, AT	NA (transitory inhibition in 17)	NA	NA	Boyanova *et al*. (2005), Coelho *et al*. (2007)
Green tea (extract, EGCG)	+ (MIC_{50}, 8–100 EGCG)	+ (50->100 EGCG at 48 h)[2]	AA, AU, AI, AT	≤60 (Catechins + sialic acid in AGS cells and mice)	NA	NA	Yanagawa *et al*. (2003), Song and Seong (2007), Yang *et al*. (2008), Shoae Hassani *et al*. (2009), Stoicov *et al*. (2009)
Turmeric (curcumin)	+ (5–50)	+ (25–62 at 60 min)	AA, AI	5.9–12	NA	+	O'Mahony *et al*. (2005), Di Mario *et al*. (2007), De *et al*. (2009), Zaidi *et al*. (2009a), Koosirirat *et al*. (2010)
Ginger (ME and AE)	+ (6.25–50 ME; 300 AE)	+ (≤100,000 at 30 min)	AI	NA	NA	NA	Mahady *et al*. (2003b) O'Mahony *et al*. (2005) Nanjundaiah *et al*. (2009)
Vitamin C	NA	NA	AI, AT	NA	+ (by 29)	None	Zojaji *et al*. (2009), Kohler *et al*. (2010)

Broccoli (sulforaphane)	+ (0.06–8; MIC_{90}, ≤4)	+ (10 at 24 h)	AU, AI, AT	8/11 human gastric grafts	NA	NA	Fahey *et al*. (2002), Haristoy *et al*. (2003), Yanaka *et al*. (2009)
Olive oil (ligstroside aglycon)	+	+ (1.3)	AT	NA	NA	NA	Romero *et al*. (2007)
Mastic gum fractions	+	+ (140–200 at 24 h)	AI	4/13 cases[2]	NA	NA	Paraschos *et al*. (2007), Kottakis *et al*. (2009), Dabos *et al*. (2010)
Grape extracts (resveratrol)	+ (6.5–12.5)	+ (6.5 to >12.5)	AI, AT	NA	NA	NA	Brown *et al*. (2009), Martini *et al*. (2009a)
Berry extracts	+	134–270 (blackberry leaves at 48 h)	AA	0 (cranberry juice), 2.9 (combined with *L. johnsonii* La1)	NA	NA	Zhang *et al*. (2005), Gotteland *et al*. (2008), Martini *et al*. (2009b)
Mallotus philippensis	+	+ (15.6–31.2 at 60 min)	NA	NA	NA	NA	Zaidi *et al*. (2009a)
Korea red ginseng (panaxdiol)	+ (50)	NA	AA, AI, AT	NA	+ (by 12.3)	NA	Kim *et al*. (2010)
Microalgal extract (*Spirulina*) (*PS*)	NA	NA	AA	Decrease (90) of *H. pylori* cells in mice	NA	NA	Loke *et al*. (2007)
DHNA	+ (1.6–3.2)	NA	NA	In germ-free mice	NA	NA	Nagata *et al*. (2010)
Liquorice roots	+ (<50–100 for GA)	+ (50–100 for GA at 4–24 h)	AA	NA	NA	NA	Krausse *et al*. (2004), Wittschier *et al*. (2007)
Aspirin	+ (MIC_{50}, 256)	+ (at 48 h)	AT	NA	NA	NA	Wang *et al*. (2003)
Eugenol	+ (2)	+ (4 at 6 h)	AI	NA	NA	NA	Ali *et al*. (2005)
Cinnamon (cinnamaldehyde)	+ (2)	+ (4 at 9 h)	AI	NA	NA	NA	Ali *et al*. (2005)
Acyl-lysyl oligomer $C_{12}K$-$2\beta_{12}$	+ (6.5–26 μM)	14.5–90 μM	NA	NA	NA	NA	Makobongo *et al*. (2009)
N-acetylcysteine	+ (≥5000)	NA	AA	NA	+ (by 45, pre-treatment)	NA	Huynh *et al*. (2004), Cammarota *et al*. (2010)

[1]Present.

[2]Absent in some studies.

MIC, minimal inhibitory concentration; MBC, minimal bactericidal concentration; AA, antiadhesive; AU, antiurease; AI, anti-inflammatory; AT, antitumour; NA, not available; EGCG, epigallocatechin-gallate; GA, glycyrrhetinic acid; DHNA, 1,4-dihydroxy-2-naphthoic acid; PPI, proton pump inhibitors; ME, methanol extract; AE, aqueous extract.

CFU/day for no fewer than 5 days (Gupta and Garg, 2009). In addition, the lactobacilli should be taken at least 2–4 h after the antibiotic dose. In co-administration with antibiotics, antimicrobial resistance in probiotic strains can help with their survival but, on the other hand, may be transferable to other species (Masco *et al.*, 2006).

Use of probiotic combinations is a 'two-edged sword'. Sheu *et al.* (2002) and Ziemniak (2006) obtained better results using regimens combined with AB-yoghurt with equal amounts of *L. acidophilus* La5, *Bifidobacterium lactis* Bb12, LDB and *Streptococcus thermophilus* or with a combination of *L. acidophilus* and *Lactobacillus rhamnosus* compared with a standard regimen alone. Sheu *et al.* (2006) reported an increase (by 14%) in eradication rate using a quadruple regimen after 4 weeks' pre-treatment with AB-yoghurt compared with quadruple therapy alone. However, probiotic combinations should be carefully selected as they can show pro-inflammatory or anti-inflammatory effects. Evaluating *L. rhamnosus* GG, *L. rhamnosus* Lc705, *Propionibacterium freudenreichii* subsp. *shermanii* Js and *Bifidobacterium breve* Bb99, Myllyluoma *et al.* (2008) have reported that all probiotics have inhibited *H. pylori* adhesion to human (Caco-2) colon cancer cell line and most of them have inhibited *H. pylori*-induced cell damage. Importantly, although *L. rhamnosus* GG and Lc705, and *P. freudenreichii* subsp. *shermanii* Js inhibited IL-8 secretion from the infected cells, *B. breve* Bb99 and the probiotic combinations increased IL-8 release, and *L. rhamnosus* GG enhanced the secretion of prostaglandin E_2 (PGE_2). Some probiotic combinations have increased the release of pro-inflammatory cytokines in the *H. pylori*-infected gastric mucosa, e.g. the levels of IL-8, prostaglandin E_2 and leukotriene B_4. In the combinations, the pro-inflammatory effect of a single probiotic has dominated over the anti-inflammatory effect of the other probiotics (Myllyluoma *et al.*, 2008). Therefore, predominant activity of the probiotic combinations should be evaluated *in vitro* with care before their recommendation.

Many lactobacilli such as *L. casei*, *L. acidophilus*, *L. rhamnosus*, LDB and *L. plantarum* possess additional advantages such as anti-inflammatory and antimutagenic properties (Perdigón *et al.*, 2002; Caldini *et al.*, 2005). The lactobacilli have been shown to reduce the gastric inflammation by production of conjugated linoleic acids via dissociation of IκB kinase (IKK)-γ and heat shock protein 90 (Hsp90) complex (Kim *et al.*, 2008). The activity has been explained by increased expression of suppressors of cytokine signalling (SOCS)-2 and SOCS-3 because of phosphorylation of signal transducers and activation of transcription or owing to inhibition of Janus kinase 2 phosphorylation, which signals negatively SOCS-2/SOCS-3 (Lee *et al.*, 2010). *H. pylori*-related expression of inflammatory mediators, e.g. tumour necrosis factor-α (TNF-α) and IL-8 has been reduced by *L. plantarum*, *L. rhamnosus* and *L. acidophilus* (Lee *et al.*, 2010). Immunomodulation has been found also by *L. casei* L26 in a mouse model (Zhang *et al.*, 2008). Similarly, LDB has been able to suppress the IL-8 release, activated by *H. pylori* lipopolysaccharide interaction with toll-like receptor 4 (TLR4; Zhou *et al.*, 2008).

Briefly, many *Lactobacillus* strains could be used to prevent *H. pylori* reinfection, to reduce the bacterial colonization, gastric inflammation and *H. pylori*-induced injuries in infected subjects, to inhibit antibiotic-resistant strains, to enhance eradication rates and to reduce the side-effects of the standard therapy. However, the choice of the best *Lactobacillus* strains for routine use and the optimal start time and duration of their administration should be determined in extensive clinical trials.

Other probiotics

Saccharomyces boulardii, bifidobacteria and propionibacteria are also probiotics with potential benefit for *H. pylori* infection. Collado *et al.* (2005) have reported one-fourth of the 24 faecal *Bifidobacterium* strains evaluated to inhibit both antibiotic susceptible and-resistant *H. pylori* strains. The activity has been associated with antimicrobial peptides produced by the bifidobacteria. Pre-incubation with *Bifidobacterium* strain BF-1 has suppressed *H. pylori*-induced IL-8 release in the gastric mucosa and has inhibited the nuclear factor-κB (NF-κB) signalling pathways (Shirasawa *et al.*, 2010). However, it has been reported that some dead probiotics such as bifidobacteria are able to induce an increase in TNF-α production (Adams, 2010). Therefore, both live and dead probiotics are generally active but their activities may be different (Adams, 2010). Culture filtrates of propionibacteria also can inhibit *H. pylori* growth. Quinones such as 1,4-dihydroxy-2-naphthoic acid (DHNA), produced by propionibacteria, appear as promising agents. DHNA has inhibited *H. pylori*, including clarithromycin-resistant strains, at very low (1.6–3.2 μg/ml) MICs, without affecting the normal intestinal flora (Nagata *et al.*, 2010). DHNA has been shown to alter *H. pylori* cellular respiration. Nagata *et al.* (2010) have reported that oral DHNA administration could eradicate *H. pylori* in germ-free mice.

Non-pathogenic probiotic yeast *S. boulardii* has been used for management of *Clostridium difficile*-associated disease (McFarland, 2010). *S. boulardii* can inhibit bacterial toxins, can protect cellular physiology, can reduce microbial attachment and can act as an immunomodulator (McFarland, 2010). However, the yeast should be used with caution in hospitalized patients as *S. boulardii* fungaemia following its use as probiotic has been reported in critically ill patients, in those at parenteral nutrition and in patients with central venous catheters (Venugopalan *et al.*, 2010).

S. boulardii, together with a 14-day triple therapy for *H. pylori* eradication, has reduced the rates of side-effects of treatment but has only slightly increased the eradication rate (71% vs. 59.7% in the control; Cindoruk *et al.*, 2007). Similarly, in children, adding *S. boulardii* to a standard regimen slightly improved (by 12%) eradication success (Hurduc *et al.*, 2009). In a systematic review and meta-analysis, the main advantage of *S. boulardii* was to reduce the side-effects of triple therapy, as shown in most studies (McFarland, 2010).

Lactoferrin

Lactoferrin is a natural glycoprotein with iron binding, antiadhesive, immunomodulating and antimicrobial properties, which can damage membranes of Gram-negative bacteria. The lactoferrin is present in saliva, blood, gall fluid, milk and gastric juice and mucosa. Bovine lactoferrin has been shown to inhibit *H. pylori* at a concentration of >500 μg/ml (Di Mario *et al.*, 2006; Tursi *et al.*, 2007). However, according to Huynh *et al.* (2009), only partial *H. pylori* growth inhibition has been reached by the agent. Adding both bovine lactoferrin and probiotics to a triple therapy has increased, using ITT analysis, *H. pylori* eradication from 72.3% to 88.6% and has reduced the side-effects of therapy in the study of de Bortoli *et al.* (2007). In a meta-analysis, *H. pylori* eradication was higher (86.6%) in patients with lactoferrin supplemented regimens than in those without lactoferrin (74.4%) by ITT analysis, and the rate of side-effects has been lower (9%) in the group with lactoferrin than that in the group without lactoferrin (16.3%; Zou *et al.*, 2009b). However, Zullo *et al.* (2007) have reported suboptimal eradication by rabeprazole + clarithromycin + tinidazole + bovine lactoferrin 200 mg, b.i.d. [72.2% with ITT and 76.5% with per protocol (PP) analysis]. Similarly, in another study, supplementation of quadruple therapy (ranitidine bismuth citrate + esomeprazole + amoxicillin + tinidazole) by bovine lactoferrin (200 mg b.i.d.) has reduced the side-effects of therapy but has increased the eradication rate only slightly (by 6%; Tursi *et al.*, 2007). Moreover, in a mouse model, both *H. pylori* growth and gastric inflammation have been even enhanced by lactoferrin treatment according to Huynh *et al.* (2009). In brief, as overall benefits of lactoferrin for *H. pylori* infection are somewhat controversial, further studies on this question are needed.

Propolis

Propolis (bee glue) contains resins (flavonoids and phenolic acids), wax, essential oils, pollen and other compounds (Viuda-Martos *et al.*, 2008). Antimicrobial activity of European propolis has been associated with flavonoids (pinocembrin and galangin) and esters of phenolic acids, and its antitumour activity has been due to esters of the caffeic acid (Viuda-Martos *et al.*, 2008). The propolis damages bacterial cytoplasmic membrane and inhibits bacterial motility and enzymatic activity (Bankova *et al.*, 2005). Anti-*H. pylori* effect of Brazilian propolis has been linked to lambdane-type diterpenes and prenylated phenolic compounds (Banskota *et al.*, 2001). Moreover, the bee glue can be bactericidal at high concentrations. Interestingly, the antimicrobial activities of propolis from different geographic regions have been similar (Bankova, 2005). The agent is relatively non-toxic, although allergic reactions have been reported. Synergism between propolis and antibiotics is especially important. Synergistic or additive effect of clarithromycin and either propolis or ginger (*Zingiber officinale*) extract on *H. pylori* has been detected (Nostro *et al.*, 2006).

In vitro activity of Bulgarian propolis on *H. pylori* has been observed (Boyanova *et al.*, 2005; Fig. 11.2). Growth of all *H. pylori* strains have been inhibited by 90 µl of 30% ethanolic extract of Bulgarian propolis per well in agar-well diffusion method. By agar dilution method, 100 and 300 µg/ml propolis have inhibited 57.1% and 76.2% of strains, respectively (Boyanova *et al.*, 2005). Bulgarian propolis has shown a dose-dependent activity against most *H. pylori* strains tested. Unfortunately, in a study of Coelho *et al.* (2007), the use of Brazilian propolis (20 drops of alcohol solution of propolis t.i.d. for a week) provided only transitory *H. pylori* inhibition in 3 of 18 patients. It is well known, however, that in monotherapy, even antibiotics lead to *H. pylori* eradication in only 10–50% of the cases. In addition, NAAs are better given for a long time (several months) to obtain optimal activity. Further studies with different propolis application, doses and frequency are requited. Given the synergism between propolis and antibiotics on *H. pylori in vitro* (Nostro *et al.*, 2006), propolis may be a supplement to *H. pylori* eradication regimens. Anti-inflammatory, antitumour, anaesthetic and tissue-regenerative properties (Bankova *et al.*, 2005; Nostro *et al.*, 2006), could be other advantages for propolis use in management of the infection, providing that further studies confirm the still expected clinical effect of this agent.

Broccoli

Broccoli (*Brassica oleracea*) compounds can reduce both *H. pylori* colonization and gastritis in infected mice and humans (Yanaka *et al.*, 2009). Sulforaphane in broccoli sprouts has shown bactericidal (minimal bactericidal concentration MBC, 10 µg/ml within 24 h) and antitumour activities (Fahey *et al.*, 2002; Haristoy *et al.*, 2003; Yanaka *et al.*, 2009). The sulforaphane has exhibited very low MIC_{90} (4 µg/ml) against antibiotic-resistant *H. pylori* strains (Fahey *et al.*, 2002). Large inhibitory zones (>5 cm) of a strain have been obtained by chloroform and hexane extract of fresh broccoli sprouts (Moon *et al.*, 2010). Of 18 sulforaphane compounds, two amines, six isothiocyanates and a nitrile have been most active (Moon *et al.*, 2010).

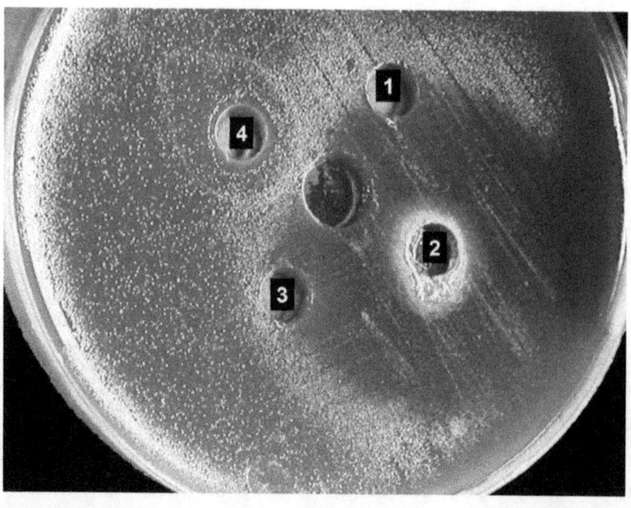

Figure 11.2 Strong inhibition of *H. pylori* growth by ethanol extract of propolis (60 µl/well) (well 2) and slight inhibition by the control, 96% ethanol (well 4). Wells 1 and 3 are empty.

Daily intake of sulforaphane-rich broccoli sprouts (70 g/day) for 2 months lessened both *H. pylori* colonization (detected by UBT and stool antigen test, SAT) and gastric inflammation (assessed by serum pepsinogens I and II), although the effects lasted less than 2 months after the end of treatment (Yanaka *et al.*, 2009). *H. pylori* was eradicated in 8/11 sulforaphane-treated human gastric xenografts in nude mice (Haristoy *et al.*, 2003). Results obtained so far with broccoli are promising and merit extensive evaluation.

Green tea

Hydrolysable tannins, mostly monomeric tannins such as TG-I, TG-II, geraniin and agrimoniin, of Asiatic plants have been *in vitro* active against *H. pylori* (MICs 3.1–50 µg/ml) owing to membrane-damaging activity and sequestering of metal ions needed for the bacterial growth (Funatogawa *et al.*, 2004; Vattem *et al.*, 2005). However, much attention has been drawn by green tea (*Camellia sinensis*) as a beverage, which use has been associated with long traditions in many countries and numerous suggested health benefits.

About 13–30% of the dry weight of green tea leaves is made up of the active substances, i.e. the polyphenolic catechins, including the most active compound, the epigallocatechin gallate (EGCG; Song and Seong, 2007). Tea catechins can intercalate into cell phopsholipid bilayers and some of them are able to penetrate the surface (Sirk *et al.*, 2008). Antienzymatic activity of the tea catechins is very important. Similarly to the quinolones, the tea catechins can suppress bacterial DNA gyrase; however they target gyrase B subunit (Gradisar *et al.*, 2007). Catechins also have been shown to reduce *H. pylori* urease activity and associated gastric cytotoxicity (Song and Seong, 2007). Inhibition of *H. pylori* urease activity has been linked to a hydroxyl group in the 5′-position of the B-ring in the catechin skeleton (Song and Seong, 2007). Non-fermented green tea extract (at 2500 µg/ml) and semi-fermented extract (at 3500 µg/ml) have been found to abolish the production of *H. pylori* urease subunits (Shoae Hassani *et al.*, 2009). Furthermore, the EGCG also has been able to block Toll-like receptor (TLR)-4 activation and to inhibit the extracellular signal response kinase 1/2 and NF-κB and therefore, to reduce the gastric inflammation as a result of *H. pylori* infection (Lee *et al.*, 2004). In addition, the green tea polyphenols have been reported to reduce the gastric cell apoptosis and carcinogenesis in a mouse model (Akai *et al.*, 2007). Naturally, a question can arise about the coexistence of common gastric cancer in East Asian countries and the usually high green tea consumption there (Song and Seong, 2007). So far, results on the inverse association between the green tea consumption and the gastric cancer incidence have been inconsistent (Boehm *et al.*, 2009). Anyway, high green tea consumption (usually more than ten cups per day) has been linked to low risk for gastric cancer or atrophic gastritis in 4 of 10 studies performed in Japan and China (Cabrera *et al.*, 2006).

Antiadhesive property of the green tea is highly important as *H. pylori* adhesion is an essential prerequisite for establishment of the infection. Green tea extract (maximum yield, 4% w/v), containing carbohydrates with 40% uronic acids and lack of catechins, has shown antiadhesive effect on *H. pylori* on host cell lines (IC_{50} for adhesion, 140–2300 µg/ml) with no unfavourable effects on bacteria of the normal intestinal flora such as *L. acidophilus* and *Bifidobacterium bifidum* (Lee *et al.*, 2009). Moreover, the green tea has been highly active against *H. pylori in vitro* (with MIC_{50} of 8–100 and MIC_{90} of 100 µg/ml for both EGCG and epicatechin gallate), but less active in a model of gerbils (Yanagawa *et al.*, 2003; Song and Seong, 2007). The difference could be due to the catechin instability, especially in low pH, low bioavailability or short gastric transit time. Therefore, better delivery of catechins and their combination with proton pump inhibitors are under evaluation (Song and Seong, 2007). Chinese tea (Lung Chen tea) catechins has been shown to inhibit *H. pylori* growth at 0.5% (w/w) in water, i.e. by only one cup of tea (Yee and Koo, 2000). Green tea, at a ratio of 3:1 to *H. pylori*, has shown a bactericidal effect on the bacteria (Stoicov *et al.*, 2009). Concentrations of 4000 µg/ml non-fermented and 5500 µg/ml semi-fermented green tea extract have killed *H. pylori* in the study of Shoae Hassani *et al.* (2009). The bactericidal activity of EGCG has been 50 and >100 µg/ml against two *H. pylori* strains after 48 h in the study of Yanagawa *et al.* (2003).

Metronidazole and clarithromycin-resistant *H. pylori* strains have been sensitive to EGCG and epicatechin gallate, and in combination with sub-MICs of amoxicillin, their anti *H. pylori* activity has increased. EGCG has shown an additive activity with metronidazole and clarithromycin and strongest increase with the amoxicillin (Yanagawa *et al.*, 2003), therefore, amoxicillin-based anti-*H. pylori* therapy can take advantage of the green tea supplementation.

Catechins have been favourable *in vivo* in animal models. In mice, green tea compounds have reduced both *H. pylori*-induced gastric damage and VacA toxin (Ruggiero *et al.*, 2007). Pre-treatment of pathogen-free BALB/c mice by catechins + sialic acid (72 h before *H. pylori* inoculation) has prevented fully the infection (Yang *et al.*, 2008). The sialic acid has been used to prevent additionally *H. pylori* adhesion to the host cells (Yang *et al.*, 2008). Even in infected mice, the green tea has decreased *H. pylori* number and the gastric inflammation, and post-treatment by catechins + sialic acid has eradicated the infection in most (up to 60%) animals (Yang *et al.*, 2008; Stoicov *et al.*, 2009). Similar positive results have been reported with *H. felis* by Stoicov *et al.* (2009). Mice, drinking green tea before *H. felis* infection, have shown no gastric inflammation and even the already infected mice, drinking the tea, have exhibited only mild inflammation (Stoicov *et al.*, 2009). Green tea can exhibit interaction with some drugs but often has only a few side-effects, mainly those associated with its caffeine content.

Therefore, the benefit of the green tea, one of the best non-antibiotic candidates for management of *H. pylori* infection, is expected to be proven in epidemiological and clinical studies in more countries and in large clinical trials. The evaluation of green tea to prevent *H. pylori* infection and reinfection and to enhance the eradication success of the standard regimens is strongly required.

Grape extracts

Grapes (*Vitis vinifera*) possess high levels of antioxidants and polyphenols such as resveratrol and exhibit anti-inflammatory and antimicrobial effects. The MBC of grape extracts against *H. pylori* has varied widely from 6.2 to 2000 μg/ml, resveratrol showing the highest (6.5–12.5 μg/ml) activity (Brown *et al.*, 2009; Martini *et al.*, 2009a). MICs of 100 μg/ml have been reported for proanthocyanes from grape seeds. In mice, red wine compounds inhibited the *H. pylori*-induced gastric damage and VacA exotoxin (Ruggiero *et al.*, 2007). Brown *et al.* (2009) show also the activity of muscadine *(Vitis rotundifolia)* grape skin extract (MICs, 256–512 μg/ml) and, at lower extend, that of muscadine seed (256–1024 μg/ml) against *H. pylori*. Resveratrol and ellagic acid have shown the highest activity against *H. pylori* (MICs, 6.25–50 μg/ml), whereas myricetin has been found to be inactive (Brown *et al.*, 2009). However, high overall phenolic amounts of the grape extracts do not always exhibit high anti-*H. pylori* efficacy (Brown *et al.*, 2009).

Importantly, the resveratrol and red wine extracts have been able to inhibit the virulent (*cagA*$^+$) *H. pylori* strains possibly because of the inhibition of Src kinases. However, it should be noted that the inhibition of *H. pylori* by red wine has been strain-specific (Daroch *et al.*, 2001). Although lacking antiadhesive properties, the resveratrol has strongly reduced the *H. pylori*-induced IL-8 release, reactive oxygen species production and morphological damage of the gastric cells (Zaidi *et al.*, 2009b). Ellagic acid has also exhibited anticancer effects (Mahady *et al.*, 2003a). However, it has been considered that at low concentrations (≤5 μM), resveratrol can exhibit cell-type specific procarcinogenic effect by activation of NF-κB signalling pathways as shown on MDA-MB-435s human cancer cells (Fukui *et al.*, 2010). Similarly, Calabrese *et al.* (2010) have stated that the resveratrol is a 'two-edged sword' as it can increase the tumour cell proliferation at low concentrations but can inhibit the proliferation at high concentrations. Conversely, in a large epidemiological study in Denmark, the relative risk ratio for gastric cancer (0.60 per glass of wine per day) was found to be lower in wine consumers than in non-consumers (Barstad *et al.*, 2005).

Apparently, resveratrol is active against *H. pylori* and can influence the clinical expression of the associated diseases, however, owing to some controversial considerations, concerning the pro- and antimutagenic activities of the agent, it should be additionally evaluated *in vitro* and in more epidemiological and clinical observations. Since resveratrol supplements are now commercially available, it is necessary to consider carefully the optimal dose, pharmacokinetics, interactions with other compounds and safety of the agent.

Plant oils

Many plant oil components belong to the terpenoid family. Those from cinnamon bark (*Cinnamomum zeylanicum*), lemongrass (*Cymbopagon citratus*), vervein (*Lippia citriodora*), white grapefruit (*Citrus paradisi*), clove leaf (*Eugenia caryophyllus*), manuka (*Leptosperrum scoparium*) and oregano (*Origanum vulgare*) have been active against *H. pylori in vitro* (Bergonzelli *et al.*, 2003). Terpenoids are not well water-soluble but penetrate the phospholipid bilayer of membranes. Terpenoid activity is due to damage of bacterial membrane lipids and proton motive force, and change in intracellular pH (Bergonzelli *et al.*, 2003). Inhibitory concentration (IC 2 μg/ml) of eugenol, a phenolic compound and main component of clove oil (*Eugenia caryophillis*), and cinnamaldehyde, an aromatic aldehyde and main component of bark extract of cinnamon (*Cinnamomum verum*) has been lower than those of garlic oil (≥8 μg/ml) and EGCG (≥50 μg/ml) in the study of Ali *et al.* (2005).

In vitro activity of virgin olive oil polyphenols against *H. pylori* has been reported with strongest bactericidal activity (1.3 μg/ml) obtained by the dialdehydic form of decarboxymethyl ligstroside aglycon (Romero *et al.*, 2007). Importantly, the bactericidal concentration of the olive oil has been lower than those of the tea, wine and plant extracts, furthermore, the agent has exhibited antitumour property. In addition, essential oil from fresh leaves of *Plinia cerrocampanensis* Barrie also has shown bactericidal activity against *H. pylori*, although with MICs and MBCs values of 62.5 μg/ml (Vila *et al.*, 2010).

Geraniol, an aliphatic terpene alcohol and main compound of *Rosa damascena* (rose) oil, also has been active against *H. pylori in vitro*. At 2 μg/ml, the rose oil and the geraniol have inhibited the growth of most (85.7% and 92%, respectively) strains (Boyanova and Neshev, 1999). Effect of Bulgarian rose oil preparations could be due to their aliphatic alcohols. Side-effects of geraniol are allergic reactions in some subjects. Rose oil also possesses anti-inflammatory, antiulcer, antitumour and spasmolytic properties (Boyanova and Neshev, 1999). However, its role as treatment adjuvant has not been evaluated.

Essential oil from *Apium nodiflorum* (Fool's watercress), a culinary herb, has exhibited an MIC of 12.5 μg/ml for *H. pylori* (Menghini *et al.*, 2010). The main components of the oil are limonene (27.7%), *p*-cymene (23.1%), myristicine (18.5%) and β-pinene (6.6%). By contrast, the oil extract of *Chamomilla recutita* (German chamomile) flowers is less active (MIC_{90} 125,000 μg/ml and MIC_{50}, 62500 μg/ml) against *H. pylori*; however, it has been shown to inhibit bacterial urease production (Shikov *et al.*, 2008). Essential oil of dried aerial parts of Iranian *Thymus caramanicus* (2.5% oil), containing carvacrol, *p*-cymene, thymol, γ-terpinene and borneol, has been found to induce large (>50 mm) zones of *H. pylori* inhibition by the disc diffusion method and to exhibit MICs of 14.5–58 μg/ml (Eftekhar *et al.*, 2009).

Berries

Cranberry (*Vaccinium macrocarpon*) juice is used to inhibit the adhesion of *Escherichia coli* to urinary epithelial cells. Cranberry extracts can act on the bacteria by acidification of bacterial membranes, damage of H^+-ATPase, quench of electrons from the electron transport chains, disruption of metabolism and weakening the membrane structure and stability (Vattem *et al.*, 2005). However, precaution in use of cranberry concentrate tablets is needed in patients at risk for calcium oxalate nephrolithiasis as the agent has been shown to increase urinary oxalate levels (Terris *et al.*, 2001).

Proanthocyanidins of cranberry and other berries can reduce *H. pylori* adhesion to human gastric mucosa (Lengsfeld *et al.*, 2004). The antiadhesive properties of black currant seed extracts on *H. pylori* have been due to acidic, high molecular weight galactans (Lengsfeld *et al.*, 2004). Chatterjee *et al.* (2004) have reported that extracts from many berries, e.g. raspberries, strawberries and cranberries, have inhibited *H. pylori* and have increased its clarithromycin susceptibility.

Isolated polyphenols and leaf extract of blackberry leaves (*Rubus ulmifolius*) also have been evaluated. MBCs of the extract against two *H. pylori* strains have been 1200–1500 μg/ml after 24 h of exposure and 134–270 μg/ml after 48 h exposure (Martini *et al.*, 2009b). The results suggest that not the occasional but the regular consumption of NAAs is desirable.

It is especially interesting that combination of several NAAs can increase their activity on *H. pylori*. Keenan *et al.* (2010) have used a broth dilution method with reading of the optical densities of the mixture or controls by spectrophotometry to assess the combined activity of NAAs on *H. pylori*. The authors have observed that black currant oil and, to less extent, Omega-3 oil and bilberries have increased the anti-*H. pylori* effects of broccoli sprouts and UMF20 manuka honey (Keenan *et al.*, 2010).

Conversely, combining cranberry and lactobacilli has not been advantageous. In a multicentric, randomized, double-blind study on 295 Chilean children; Gotteland *et al.* (2008) have compared the activity of cranberry juice (4%, v/v, 200 ml), *L. johnsonii* La (10^7 CFU/ml, 80 ml) and placebo in combinations for three weeks. Despite *H. pylori* temporary suppression, which has been ^{13}C UBT detected in 14.9% of cases by *L. johnsonii* La1 and in 16.9% by cranberry juice, the eradication rates have been minimal (only 3% for *L. johnsonii* La1, 0% for the cranberry juice and 2.9% for both agents). Another interesting clinical study has been performed by cranberry juice in humans. In the prospective, randomized, double-blind trial in China, Zhang *et al.* (2005) have found negative ^{13}C UBT results in 14.3% of 97 *H. pylori*-positive adults who had consumed 500 ml of cranberry juice daily for 90 days. Although a temporary suppression has been obtained by consumption of cranberry juice, this can be of some benefit for the regular consumer of the beverage.

Garlic

Activity of garlic (*Allium sativum*) merits much attention. Anti-*H. pylori* activity has been found by water garlic extracts, commercial garlic tablets, allyl sulfide compounds, oil-macerated garlic constituents and the most active garlic substance, allicin (diallyl thiosulfinate). MICs and MBCs on *H. pylori* have been very low (4–6 and 6 μg/ml) for allicin, median (8–32 μg/ml and 16–32 μg/ml, respectively) for garlic oil and higher (250–500 μg/ml) for garlic powder (O'Gara *et al.*, 2000).

However, the garlic possesses antithrombocytic (anticoagulant) properties and can prolong bleeding time; therefore, it should be used carefully in patients taking anticoagulants and should be avoided at least 10 days before surgery (Tattelman, 2005). In addition, most garlic components are volatile and crushing the garlic cloves may vaporize them (Fani *et al.*, 2007). Moreover, garlic, dill, black pepper, coriander and tea, boiled in water, have lacked bactericidal activity within 60 min (O'Mahony *et al.*, 2005). Allitridi of garlic oil, containing diallyl trisulfide and diallyl disulfide, carry a potential for cancer prevention. Allitridi have shown bacteriostatic activity on *H. pylori*, affecting its metabolism, including the main biosynthetic pathways. Even at subinhibitory concentrations, allitridi have suppressed CagA and VacA production (Liu *et al.*, 2010). Therefore, studies for prophylaxis and treatment of *H. pylori* infection have been carried out in humans. Administration of raw garlic (15 g) and omeprazole (20 mg, b.i.d.) to non-ulcer patients has resulted in *H. pylori* eradication in 12% (Fani *et al.*, 2007).

Long-term garlic consumption has been linked to lower anti-*H. pylori* antibody titre in infected subjects than in non-consumers (Lee *et al.*, 2008). Benefit of garlic in the management of *H. pylori* infection appears modest but important, taking into account its dual antibacterial and antitumour activity.

Mastic gum

Mastic gum (resin of *Pistacia lentiscus*) has been active against *H. pylori in vitro* and in animal models. Administration of Chios mastic extract (0.75 mg/day) without polymer to *H. pylori* SS1-infected mice over three months has resulted in 30-fold lower *H. pylori* colonization than in controls, although without improvement of gastritis (Paraschos *et al.*, 2007). The insoluble polymer (poly β-myrcene) has been removed to increase the bioavailability of the agent. Highest bactericidal activity has been obtained by acid fraction (MBC, 140 µg/ml) and isomasticadienolic acid (MBC 200 µg/ml) within 24 h (Paraschos *et al.*, 2007).

Bebb *et al.* (2003) failed to eradicate *H. pylori* in nine *H. pylori*-positive UK patients by mastic 1 g q.i.d. for 14 days. However, Dabos *et al.* (2010) achieved eradication of *H. pylori* infection in 4 of 13 Greek patients with a 14-day regimen with 350 mg t.i.d. of pure Chios mastic gum from *Pistacia lentiscus* var. *chia*. The variety *chia* grows only in the Greek island of Chios. Another favourable property of Chios mastic gum is the suppression of the neutrophil activation in the presence of Hp-NAP, possibly due to the binding of arabinogalactan proteins of the agent to two membrane proteins of neutrophils. This activity has been reported by Kottakis *et al.* (2009) by observation on five *H. pylori*-positive patients who have consumed Chios mastic gum for 2 months. As for other NAAs, more studies, including large groups of patients and prolonged administration of this agent may be useful.

Antiadhesive plant compounds

Major importance of antiadhesive plant compounds is to prevent *H. pylori* colonization or reinfection. However, the properties *in vitro* may differ from those *in vivo*. For instance, the antiadhesive activity of polysaccharides from immature okra fruits (*Abelmoschus esculentus*) against *Campylobacter jejuni in vitro* has not been proven *in vivo* in chicken (Wittschier *et al.* 2007). Discrepancy has been found also between the *in vitro* activities and the clinical effect of many NAAs such as some probiotics and berry extracts against *H. pylori*. Therefore, all benefits of NAAs should be confirmed by clinical trials.

In addition to green tea and berries, other plants also have shown antiadhesive properties against *H. pylori*. Polysaccharides from *Glycyrrhiza glabra* (liquorice), enriched with glucuronic acid, have reduced *H. pylori* adhesion *in vitro* by 60–70% (Wittschier *et al.*, 2007) and the major triterpene saponin of liquorice, glycyrrhizin, also has been shown to have anti-inflammatory properties. The MICs and MBCs of the most active compound of liquorice (glycyrrhetinic acid) have been 50–100 µg/ml at 4–24 h, while the liquorice extract has been active only at high concentrations (1600 µg/ml; Krausse *et al.*, 2004). The liquorice has been also active against clarithromycin-resistant *H. pylori* strains. Bactericidal liquorice concentrations of 1000–2000 µg/ml of polysaccharides can be obtained by liquorice root of 5–15 g daily (Wittschier *et al.*, 2009). Conversely, according to Weerasekera *et al.* (2008), liquorice, cinnamon, cumin and parsley have shown only incomplete inhibition of *H. pylori* growth. Sources and methods used in different studies can explain the variations in results. It should be noted that the toxic dose of the liquorice (>100 mg glycyrrhetinic acid a day) should be avoided (Krausse *et al.*, 2004).

Other antiadhesive plants are turmeric (*Curcuma longa*), borage (*Borago officinalis*), parsley (*Petroselinum crispum*), blackcurrant seeds (galactans), *Panax ginseng*, *Artemisia capillaries* (wormwood capillary, polysaccharide fractions) etc. (Lane *et al.*, 2010.) Curcumin, a polyphenol and yellow pigment of the Indian spice turmeric, acts as a non-competitive inhibitor of shikimate dehydrogenase (De *et al.*, 2009; see below). The curcumin has been able to inhibit IκB-α degradation, IKK-α and -β activity, NF-kB DNA-binding and activation and the resulting IL-8 release as well as the *H. pylori*-induced cell scattering in gastric epithelial cells (Foryst-Ludwig *et al.*, 2004; Zaidi *et al.*, 2009a). Similar effect has been reported by capsaicin, the key compound of hot red and chilli pepper, at 100 µmol/l, which is a non-toxic dose (Lee *et al.*, 2007). In mice, the curcumin has been able to reduce macromolecular leakage of the *H. pylori*-infected gastric cells and to repair the damaged tissues (De *et al.*, 2009; Sintara *et al.*, 2010). Curcumin MBC against *H. pylori* has been 25–50 µg/ml at 60 min (Zaidi *et al.*, 2009a).

In Italy, one-week therapy by curcumin (30 mg b.i.d.), bovine lactoferrin (100 mg b.i.d.), *N*-acetylcysteine (600 mg b.i.d.), and pantoprazole (20 mg b.i.d.) has resulted in only 12% of *H. pylori* eradication (Di Mario *et al.*, 2007). Again, in Thailand, the eradication rate in patients on one-week therapy with curcumin (turmeric tablets, 700 mg orally t.i.d.) has been only 5.9% versus 78.9% in the group with omeprazole, amoxicillin and metronidazole treatment course (Koosirirat *et al.*, 2010). However, at 2 μg/ml, the curcumin has shown antigenotoxic effect on hepatoma G2 (HepG2) cells, while at 8–16 μg/ml, the agent has been genotoxic (Cao *et al.*, 2007). The curcumin hepatotoxicity and genotoxicity at high doses, observed also in mice (Al Mofleh, 2010), should not be disregarded in future clinical trials.

In Sri Lanka, activity of culinary and medicinal plants, boiled in water like by cooking, has been reported (O'Mahony *et al.*, 2005). At 100,000 μg/ml, the turmeric and ginger have shown the quickest bactericidal activity within 15 min, followed by ginger, cumin (*Cuminum cyminum*), and chilli (*Capsicum anunum*), whereas the liquorice, oregano, black caraway (*Nigella sativa*) and borage have acted within 60 min (O'Mahony *et al.*, 2005). Slightly different results have been reported by Weerasekera *et al.* (2008) about the bactericidal action of the ginger (within 15 min) and chilli (within 60 min). Gingerol, a polyphenolic substance of the ginger root, has inhibited also $CagA^+$ strains (Al Mofleh, 2010). Besides the anti-*H. pylori* activity, partly linked to cinnamic acid, aqueous extract of ginger rhizome has shown ulcer-preventative effect by suppression of gastric H^+,K^+-ATPase and mucin protection in animals with swim/ethanol-stress-induced ulcers (Nanjundaiah *et al.*, 2009).

Antioxidant, antibacterial and anti-inflammatory effects have been attributed to Korea red ginseng (*Panax ginseng*). The plant has been able to inhibit *H. pylori* adherence to the gastric mucosa, using an acidic polysaccharide, and *H. pylori* LPS-induced TNF-α release by ginsenoside Rb1, and to reduce the oxidative stress-induced DNA mutation and *H. pylori*-induced IL-8 release in the gastric cells (Kim *et al.*, 2010). An important active fraction, containing panaxdiols, has shown anti-*H. pylori* MIC of 50 μg/ml (Kim *et al.*, 2010). Although 10 weeks' post-treatment by ginseng after a triple regimen has slightly improved the eradication (91.7% vs. 79.4% in patients without the agent), it has reduced DNA damage, apoptosis and gastric inflammation (Kim *et al.*, 2010). *Panax ginseng* ginsenosides, Rg3 and Rg5, have exhibited antitumour properties (Yun *et al.*, 2001). Against *H. pylori* infection, this plant encompasses multifarious favourable activities. It is also important that even an increase by only 10% in the eradication rate with standard therapy is beneficial and can affect a large number of patients.

Interestingly, polysaccharides of a commercially available microalga (*Spirulina*) have reduced *H. pylori* binding to gastric mucin at low pH (optimum, pH 2) and have reduced the bacterial density in mice by >90% (Loke *et al.*, 2007). *Pelargonium sidoides* (South African geranium) root extract, containing mostly polymeric proanthocyanidins, has shown variable antiadhesive effect on *H. pylori* at 1–10,000 μg/ml (Wittschier *et al.* 2007). *Antrodia camphorata* (jang-jy or niu-chang-chih) is a medicinal mushroom used in Taiwan. At 0.2 mM, methyl antcinate B, antcins K and A of fruiting bodies of *A. camphorata* have inhibited *H. pylori* growth with inhibitory zones of 13, 12 and 10 mm, respectively (Geethangili *et al.*, 2010). Some compounds have reduced *H. pylori* adhesion and invasion to AGS cells, and the NF-κB activation and IL-8 secretion in the cells (Geethangili *et al.*, 2010). Further studies should determine the possible application of these so far promising agents.

In addition to the numerous plant compounds, antiadhesive food agents are also those in milk, e.g. porcine milk (Le^b and sialylated Le^x glycans), buttermilk (glycopolypeptides), human milk (κ-casein), human and bovine milk (3′-sialyllactose) etc. (Lane *et al.*, 2010.)

Other non-antibiotic agents

In Pakistan, Zaidi *et al.* (2009a) have reported that 70% aqueous-ethanol extracts of *Curcuma amada* (rhizome), *Mallotus philippensis* (red Kamala, powder covering fruits), *Myristica fragrans* (nutmeg, seeds) and *Psoralea corylifolia* (babchi seeds) are bactericidal for *H. pylori* with MBCs of <62.5 μg/ml. Lowest MBCs (15.6–31.2 μg/ml) at 60 min have been observed by *Mallotus philippensis*. In addition, flavones of *Feijoa sellowiana* (pineapple, guava, native to South America) fruits have been reported to exhibit higher activity against *H. pylori* than metronidazole (Basile *et al.*, 2010).

Clinical study has been performed on black caraway (*Nigella sativa*), a spice and food preservative, growing in Middle East, Eastern

Europe and Asia. The plant possesses many active compounds, e.g. thymoquinone, thymohydroquinone, dithymoquinone, thymol, nigellicine and α-hedrin (Salem *et al.*, 2010). Importantly, in Saudi Arabia, Salem *et al.* (2010) have reached *H. pylori* eradication by only 2 g *N. sativa* + omeprazole daily for 4 weeks in 66.7% of 21 patients. Therefore, the anti-*H. pylori* activity of the black caraway alone and in combination with treatment regimens merits extensive evaluation on large patient groups. Intriguingly, the highest eradication rate, observed by 2 g of *N. sativa* has not been obtained by different doses (1 and 3 g daily) of the agent (Salem *et al.*, 2010), suggesting that the dosage of NAAs is of primary importance for the treatment success.

Antioxidants such as garcinol, a polyisoprenylated benzophenone from the fruit rind of *Garcinia* spp., and Protykin (*Polygonum cuspidatum* extract with 50% trans-resveratrol) alone have killed 91.4 and 87% *H. pylori* strains, respectively, and 90.8% of the strains in combination of both agents (Chatterjee *et al.*, 2005). In addition, anti-*H. pylori* effect of honey has been reported owing to its pH, hydrogen peroxide, osmotic effect and other substances (Manyi-Loh *et al.*, 2010). Activity of mountain honey from Cameroon, Capillano and Manuka honeys from New Zealand, South African honeys and Eco-honey from Kenya has been evaluated. Mountain honey and Manuka honey have been the most active agents against *H. pylori* (Ndip *et al.*, 2007). The best concentration of Mountain honey and South African honeys against *H. pylori* has been 75% v/v (Manyi-Loh *et al.*, 2010). Indeed, the antibacterial effect of the manuka honey has been associated with a unique manuka factor (UMF), showing activity corresponding to that of 20% phenol (w/v; Tonks *et al.*, 2007). However, in a study on wound healing, a 5.8-kDa factor of the manuka honey, interacting with TLR4, has been found to activate the production of the pro-inflammatory mediator, TNF-α (Tonks *et al.*, 2007). Both activities of the manuka honey should be considered with care.

Antioxidant vitamin C can increase *H. pylori* eradication success. Iranian study has reported favourable effect of adding 500 mg vitamin C/day to *H. pylori* eradication regimen of metronidazole, amoxicillin, bismuth and omeprazole (Zojaji *et al.*, 2009). *H. pylori* eradication rate has been higher (83% on PP analysis and 78% on ITT analysis) in the group with vitamin C than in the group without the vitamin (56.4% on PP analysis and. 48.8% on ITT analysis). Side-effects in both groups have been comparable. However, the possible risk for kidney stones and increased iron absorption should be considered with vitamin C supplementation (Filik, 2010). Epidemiological studies also suggest probable benefit of vitamin C for the management of *H. pylori* infection. Importantly, patients, consuming frequently fresh vegetables and fruits, have been at lower risk for *H. pylori* reinfection than the non-consumers, possibly owing to antioxidants such as vitamin C in the foods (Jarosz *et al.*, 2009). Therefore, the importance of antioxidants for the management of *H. pylori* infection is both prophylactic and therapeutic.

Active NAA against *H. pylori* has been *N*-acetylcysteine (NAC) as well. This drug is a mucolytic agent and thiol-containing antioxidant. NAC has inhibited *H. pylori* growth *in vitro*, has reduced both *H. pylori* production of extracellular polysaccharides and its gastric density in animal models, and has improved the eradication by omeprazole and amoxicillin even in smokers, for whom the eradication is usually lower than in non-smokers, when administered 1.2 g b.i.d. for 10 days (Huynh *et al.*, 2004). NAC pre-treatment prior a culture-guided antibiotic regimen has destroyed bacterial biofilm and has overcome *H. pylori* antibiotic resistance. Importantly, *H. pylori* has been eradicated in 65% of patients receiving NAC pre-treatment (PP and ITT analyses) versus only 20% of those treated without NAC (Cammarota *et al.*, 2010). In this study, only the patients with non-eradicated infection have exhibited biofilms on the gastric mucosa. This additional benefit of NAC i.e. the removal of *H. pylori* biofilms, can furthermore increase the success of the standard therapy.

According to the proverb, 'An apple a day keeps the doctor away', apples have shown both anti-*H. pylori* activity and improvement of the associated gastritis. Apple peel polyphenol-rich extract (APPE) has been reported to decrease *H. pylori*-induced vacuolation in HeLa cells (IC_{50}, 390 μg of gallic acid-GAE equivalents/ml) and to inhibit *H. pylori* attachment by 20–60% at ≥0.250 mg of GAE/ml. In mice, 150–300 mg/kg/day APPE have reduced *H. pylori* attachment. Orally administered apple peel polyphenols have decreased malondialdehyde levels and gastritis scores (Pastene *et al.*, 2010).

Few studies have been performed on the activity of St. John's Wort (*Hypericum perforatum*) on *H. pylori* infection. However, a butanol fraction of St. John's wort also has shown anti-*H. pylori* activity with MICs of 15.6–31.2 μg/ml (Reichling *et al.*, 2001). Inhibition has been linked to hyperforin, a phloroglucin derivative (Reichling *et al.*, 2001). Many other NAAs such as local plants have been able to inhibit *H. pylori* and should be further evaluated.

Some established drugs have been or should be re-evaluated in the management of *H. pylori* infection. Cytoprotective antiulcer agent, plaunotol, combined with clarithromycin, has exhibited either a synergistic effect against most (67%) of the clarithromycin-resistant *H. pylori* strains evaluated or additive effect against the remainder (Sasaki *et al.*, 2007). Given the increasing resistance rates to clarithromycin in *H. pylori* in many countries, the use of plaunotol–clarithromycin combinations can be more appreciated.

Aspirin can show bactericidal activity on *H. pylori* owing to cell lysis (Wang *et al.*, 2003). Aspirin at 400 μg/ml has reduced *H. pylori* CFUs by 2 log at 48 h, also suppressing the mutagenic effect of metronidazole. At 180 μg/ml, this agent has enhanced *H. pylori* susceptibility to many antibiotics. The activity has been linked to its acidity and changes of outer membrane proteins expression (Wang *et al.*, 2003; Zhang *et al.*, 2009). This also has resulted in increased endocellular concentration of the antibiotics (Zhang *et al.*, 2009). In a meta-analysis, the aspirin has been shown to protect *H. pylori* patients, especially Caucasians, from non-cardia gastric cancer, reducing the risk by about 38% compared with that in non-users (Yang *et al.*, 2010). However, this drug is known to exert ulcerogenic activity, which has been attempted to be reduced by new formulations of aspirin combined with natural phenolic acid antioxidants (Jiang *et al.*, 2010). The recommendations for *H. pylori* patients taking aspirin are given in Chapter 8. Activities of ecabet sodium (locally non-absorbable antiulcer agent) and rebamipide also have been reported (Di Mario *et al.*, 2006).

Different approach to control *H. pylori* infection has been reported by Horie *et al.* (2004), using yoghurt with antibodies against *H. pylori* urease from egg yolk of immunized hens. This yoghurt has reduced ^{13}C UBT DOB values of *H. pylori*-positive volunteers at 2 and 4 weeks. Monoclonal antibodies against *H. pylori* urease, acetylhydroxamic acid with chitosan coat for better delivery of the agent and colloidal bismuth subcitrate, which generates oxidative stress in *H. pylori* and has antienzymatic activity, all are established or promising agents with anti-*H. pylori* activity (Duckworth *et al.*, 2009).

Other inhibitors of *H. pylori* gene products also may be used for *H. pylori* therapy (Duckworth *et al.*, 2009). For this approach, the most appropriate targeted genes can be those present in all *H. pylori* strains, which are expressed *in vivo*. Using a laboratory chemical library, Han *et al.* (2006) have selected five inhibitors of shikimate dehydrogenase of *H. pylori*. Since shikimate pathway, needed for chorismate synthesis, is present in bacteria but absent in human body, the agents are attractive as applicants for *H. pylori* therapy. Specific redox protein (flavodoxin) inhibitors also have given good first results *in vitro* and in animal model as potential therapeutic agents against *H. pylori* infection (Cremades *et al.*, 2009). *H. pylori* flavodoxin is a redox protein, detected in bacteria but not in mammals, although being homologous to one domain of human P450 cytochrome reductase. Among 29 compounds that bind flavodoxin, Cremades *et al.* (2009) have selected three compounds to be most active against two *H. pylori* strains evaluated with MICs of 0.125–16 μg/ml and lack of toxicity in mice. Three of the compounds have been also rapidly bactericidal against *H. pylori* within only 2–4 h.

In addition, thienylthiazole and benzamide derivatives, and natural and synthetic pyloricidins, which exhibit strong inhibition on *H. pylori in vitro*, including the antibiotic-resistant strains, are under evaluation (Campo *et al.*, 2007). Pyloricidins have been isolated from *Bacillus* spp. and represent anti-*H. pylori* active peptide-like antibiotics with two unusual amino acids, one of which contributes to their activity (Nagano *et al.*, 2001).

Some cationic antimicrobial peptidomimetics have been highly active against many bacterial species. Although these agents can undergo proteolytic degradation, high *in vitro* stability in blood plasma has been reported for synthetic antimicrobial peptidomimetics (Haug *et al.*, 2008). Antibacterial effect of oligo-acyl-lysyl antimicrobial peptidomimetics against *H. pylori* has been detected, the most active peptide being C12K-2β_{12} (Makobongo *et al.*, 2009). This peptide has had MIC of 6.5–26 μM and MBC of 14.5–90 μM. These peptidomimetics have shown the advantage to be stable *in vitro* at a wide temperature range (4–95°C) and low pH (4.5), although with falling activity at pH 4.5 (Makobongo *et al.*, 2009).

In conclusion, some considerations are important for the studies on NAAs. *In vitro* studies on *H. pylori* should report both bacteriostatic and bactericidal activity of the NAAs, which should be confirmed later in animal models or better, in clinical trials. Importantly, to choose an agent, it is necessary to consider the source, safety or toxicity, dose-depending deleterious effects and allergenicity of some NAAs as well as their quality control and evidence-based trials. Now 'comes the turn' of the multifarious benefits of NAAs for the management of *H. pylori* infection. Numerous NAAs exhibit antiadhesive properties and thus can prevent *H. pylori* colonization or reinfection. These agents can be especially useful in the countries with high infection prevalence. Moreover, many NAAs reduce *H. pylori* colonization and gastric inflammation, acting on cytokine and chemokine release, e.g. on those associated with NF-κB signalling, reduce precancerous gastric changes and inhibit antibiotic-resistant strains (Lee *et al.*, 2008). These effects are very important to decrease the risk for development of malignancy in the infected patients. So far, cell line, animal or clinical trials have shown *H. pylori* eradication, although often modest, by NAAs alone, e.g. black caraway, garlic, green tea, turmeric, broccoli and mastic gum. Combinations of standard regimens with NAAs, e.g. lactobacilli, *S. boulardii*, Korea red ginseng as well as *N*-acetylcysteine and vitamin C have improved *H. pylori* eradication and most of the NAAs have reduced the side-effects of regimens. Synergy between NAAs is a very attractive topic, which, however, should be evaluated very carefully as in the case of the probiotic combinations. At present, we are in the hope of further studies, especially clinical trials, to suggest the choice, dosage, start time and duration of administration for the routine use of the NAAs. Finally, the NAAs possibly carry greater than the so far expected potential for prophylaxis or adjuvant therapy of the infection and can help to control the rising *H. pylori* resistance to antibiotics.

References

Adams, C.A. (2010). The probiotic paradox: live and dead cells are biological response modifiers. Nutr. Res. Rev. 23, 37–46.

Akai, Y., Nakajima, N., Ito, Y., Matsui, T., Iwasaki, A., and Arakawa, Y. (2007). Green tea polyphenols reduce gastric epithelial cell proliferation and apoptosis stimulated by *Helicobacter pylori* infection. J. Clin. Biochem. Nutr. *40*, 108–115.

Al Mofleh, I.A. (2010). Spices, herbal xenobiotics and the stomach: friends or foes? World J. Gastroenterol. *16*, 2710–2719.

Ali, S.M., Khan, A.A., Ahmed, I., Musaddiq, M., Ahmed, K.S., Polasa, H., Rao, L.V., Habibullah, C.M., Sechi, L.A., and Ahmed, N. (2005). Antimicrobial activities of eugenol and cinnamaldehyde against the human gastric pathogen *Helicobacter pylori*. Ann. Clin. Microbiol. Antimicrob. 4, 20.

Bankova, V. (2005). Recent trends and important developments in propolis research. eCAM. 2, 29–32.

Banskota, A.H., Tezuka, Y., Adnyana, I.K, Ishii, E., Midorikawa, K., Matsushige, K., and Kadota, S. (2001). Hepatoprotective and anti*Helicobacter pylori* activities of constituents from Brazilian propolis. Phytomedicine *8*, 16–23.

Barstad, B., Sørensen, T.I., Tjønneland, A., Johansen, D., Becker, U., Andersen, I.B., and Grønbaek, M. (2005). Intake of wine, beer and spirits and risk of gastric cancer. Eur. J. Cancer Prev. *14*, 239–243.

Basile, A., Conte, B., Rigano, D., Senatore, F., and Sorbo, S. (2010). Antibacterial and antifungal properties of acetonic extract of *Feijoa sellowiana* fruits and its effect on *Helicobacter pylori* growth. J. Med. Food. *13*, 189–195.

Bebb, J.R., Bailey-Flitter, N., Ala'Aldeen, D., and Atherton, J.C. (2003). Mastic gum has no effect on *Helicobacter pylori* load *in vivo*. J. Antimicrob. Chemother. 52, 522–523.

Bergonzelli, G.E., Donnicola, D., Porta, N., and Corthésy-Theulaz, I.E. (2003). Essential oils as components of a diet-based approach to management of *Helicobacter* infection. Antimicrob. Agents Chemother. 47, 3240–3246.

Boehm, K., Borrelli, F., Ernst, E., Habacher, G., Hung, S.K., Milazzo, S., and Horneber, M. (2009). Green tea (*Camellia sinensis*) for the prevention of cancer. Cochrane Database Syst. Rev. *8*, CD005004.

de Bortoli, N., Leonardi, G., Ciancia, E., Merlo, A., Bellini, M., Costa, F., Mumolo, M.G., Ricchiuti, A., Cristiani, F., Santi, S., *et al.* (2007). *Helicobacter pylori* eradication: a randomized prospective study of triple therapy versus triple therapy plus lactoferrin and probiotics. Am. J. Gastroenterol.*102*, 951–956.

Boyanova, L., and Neshev, G. (1999). Inhibitory effect of rose oil products on *Helicobacter pylori* growth *in vitro*: preliminary report. J. Med. Microbiol. *48*, 705–706.

Boyanova, L., Gergova, G., Nikolov, R., Derejian, S., Lazarova, E., Katsarov, N., Mitov, I., and Krastev, Z. (2005). Activity of Bulgarian propolis against 94 *Helicobacter pylori* strains *in vitro* by agar-well diffusion, agar dilution and disc diffusion methods. J. Med. Microbiol. *54*, 481–483.

Boyanova, L, Stephanova-Kondratenko, M., and Mitov, I. (2009). Anti*Helicobacter pylori* activity of *Lactobacillus delbrueckii* subsp. *bulgaricus* strains: preliminary report. Lett. Appl. Microbiol. *48*, 579–584.

Brown, J.C., Huang, G., Haley-Zitlin, V., and Jiang, X. (2009). Antibacterial effects of grape extracts on *Helicobacter pylori*. Appl. Environ. Microbiol. *75*, 848–852.

Cabrera, C., Artacho, R., and Giménez, R. (2006). Beneficial effects of green tea–a review. J. Am. Coll. Nutr. *25*, 79–99.

Calabrese, E.J., Mattson, M.P., and Calabrese, V. (2010). Resveratrol commonly displays hormesis: Occurrence and biomedical significance. Hum. Exp. Toxicol. *29*, 980–1015.

Caldini, G., Trotta, F., Villarini, M., Moretti, M., Pasquini, R., Scassellati-Sforzolini, G., and Cenci, G. (2005). Screening of potential lactobacilli antigenotoxicity by microbial and mammalian cell-based tests. Int. J. Food Microbiol. *102*, 37–47.

Cammarota, G., Branca, G., Ardito, F., Sanguinetti, M., Ianiro, G., Cianci, R., Torelli, R., Masala, G., Gasbarrini, A., Fadda, G., *et al.* (2010). Biofilm demolition and antibiotic treatment to eradicate resistant *Helicobacter pylori*: a clinical trial. Clin. Gastroenterol. Hepatol. *8*, 817–820.e3

Campo, S.M., Zullo, A., Hassan, C., and Morini, S. (2007). Antibiotic treatment strategies for *Helicobacter pylori* infection. Recent Pat. Anti Infect. Drug Discov. *2*, 11–17.

Cao, J., Jiang, L.P., Liu, Y., Yang, G., Yao, X.F., and Zhong, L.F. (2007). Curcumin-induced genotoxicity and antigenotoxicity in HepG2 cells. Toxicon *49*, 1219–1222.

Chatterjee, A., Yasmin, T., Bagchi, D., and Stohs, S. (2004). Inhibition of *Helicobacter pylori in vitro* by various berry extracts, with enhanced susceptibility to clarithromycin. Mol. Cell Biochem. *265*, 19–26.

Chatterjee, A., Bagchi, D., Yasmin, T., and Stohs, S.J. (2005). Antimicrobial effects of antioxidants with and without clarithromycin on *Helicobacter pylori*. Mol Cell Biochem. *270*, 125–130.

Cindoruk, M., Erkan, G., Karakan, T., Dursun, A., and Unal, S. (2007). Efficacy and safety of *Saccharomyces boulardii* in the 14-day triple anti*Helicobacter pylori* therapy: a prospective randomized placebo-controlled double-blind study. Helicobacter *12*, 309–316.

Coelho, L.G., Bastos, E.M., Resende, C.C., Paula e Silva, C.M., Sanches, B.S., de Castro, F.J., Moretzsohn, L.D., Vieira, W.L., and Trindade, O.R. (2007). Brazilian green propolis on *Helicobacter pylori* infection. a pilot clinical study. Helicobacter *12*, 572–574.

Collado, M.C., González, A., González, R., Hernández, M., Ferrús, M.A., and Sanz, Y. (2005). Antimicrobial peptides are among the antagonistic metabolites produced by *Bifidobacterium* against *Helicobacter pylori*. Int. J. Antimicrob. Agents. *25*, 385–391.

Cremades, N., Velázquez-Campoy, A., Martínez-Júlvez, M., Neira, J.L., Pérez-Dorado, I., Hermoso, J., Jiménez, P., Lanas, A., Hoffman, P.S., and Sancho, J. (2009). Discovery of specific flavodoxin inhibitors as potential therapeutic agents against *Helicobacter pylori* infection. ACS Chem. Biol. *4*, 928–938.

Dabos, K.J., Sfika, E., Vlatta, L.J., and Giannikopoulos, G. (2010). The effect of mastic gum on *Helicobacter pylori*: a randomized pilot study. Phytomedicine. *17*, 296–269.

Daroch, F., Hoeneisen, M., González, C.L., Kawaguchi, F., Salgado, F., Solar, H., and García, A. (2001). *In vitro* antibacterial activity of Chilean red wines against *Helicobacter pylori*. Microbios *104*, 79–85.

De, R., Kundu, P., Swarnakar, S., Ramamurthy, T., Chowdhury, A., Nair, G.B., and Mukhopadhyay, A.K. (2009). Antimicrobial activity of curcumin against *Helicobacter pylori* isolates from India and during infections in mice. Antimicrob. Agents Chemother. *53*, 1592–1597.

Di Mario, F., Cavallaro, L.G., and Scarpignato, C. (2006). 'Rescue' therapies for the management of *Helicobacter pylori* infection. Dig. Dis. *24*, 113–130.

Di Mario, F., Cavallaro, L.G., Nouvenne, A., Stefani, N., Cavestro, G.M., Iori, V., Maino, M., Comparato, G., Fanigliulo, L., Morana, E., *et al.* (2007). A curcumin-based 1-week triple therapy for eradication of *Helicobacter pylori* infection: something to learn from failure? Helicobacter *12*, 238–243.

Duckworth, M.J., Okoli, A.S., and Mendz, G.L. (2009). Novel *Helicobacter pylori* therapeutic targets: the unusual suspects. Expert Rev. Anti-infect. Ther. *7*, 835–867.

Eftekhar, F., Nariman, F., Yousefzadi, M., Hadiand, J., and Ebrahimi, S.N. (2009). Anti*Helicobacter pylori* activity and essential oil composition of *Thymus caramanicus* from Iran. Nat. Prod. Commun. *4*, 1139–1142.

Egervärn, M., Lindmark, H., Roos, S., Huys, G., and Lindgren, S. (2007). Effects of inoculum size and incubation time on broth microdilution susceptibility testing of lactic acid bacteria. Antimicrob. Agents Chemother. *51*, 394–396.

Fahey, J.W., Haristoy, X., Dolan, P.M., Kensler, T.W., Scholtus, I., Stephenson, K.K., Talalay, P., and Lozniewski, A. (2002). Sulforaphane inhibits extracellular, intracellular, and antibiotic-resistant strains of *Helicobacter pylori* and prevents benzo[a] pyrene-induced stomach tumors. Proc. Natl. Acad. Sci. U.S.A. *99*, 7610–7615.

Fani, A., Fani, I., Delavar, M., Fani, P., Eshrati, B., and Elahi, M. (2007). Combined garlic-omeprazole versus standard quadruple therapy for eradication of *Helicobacter pylori* infection. Indian J. Gastroenterol. *26*, 145–146.

Filik, L. (2010). Comment to 'The efficacy of *Helicobacter pylori* eradication regimen with and without vitamin C supplementation'. Dig. Liv. Dis. *42*, 596.

Foryst-Ludwig, A., Neumann, M., Schneider-Brachert, W., and Naumann, M., (2004). Curcumin blocks NF-B and the motogenic response in *Helicobacter pylori*-infected epithelial cells. Biochem. Biophys. Res. Commun. *316*, 1065–1072.

Francavilla, R., Lionetti, E., Castellaneta, S.P., Magistà, A.M., Maurogiovanni, G., Bucci, N., De Canio, A., Indrio, F., Cavallo, L., Ierardi, E., *et al.* (2008). Inhibition of *Helicobacter pylori* infection in humans by *Lactobacillus reuteri* ATCC 55730 and effect on eradication therapy: a pilot study. Helicobacter *13*, 127–134.

Fukui, M., Yamabe, N., Kang, K.S., and Zhu, B.T. (2010). Growth-stimulatory effect of resveratrol in human cancer cells. Mol. Carcinog. *49*, 750–759.

Funatogawa, K., Hayashi, S., Shimomura, H., Yoshida, T., Hatano, T., Ito, H., and Hirai, Y. (2004). Antibacterial activity of hydrolyzable tannins derived from medicinal plants against *Helicobacter pylori*. Microbiol. Immunol. *48*, 251–261.

Geethangili, M., Fang, S.H., Lai, C.H., Rao, Y.K., Lien, H.M., and Tzeng, Y.M. (2010). Inhibitory effect of *Antrodia camphorata* constituents on the *Helicobacter pylori*-associated gastric inflammation. Food Chem. *119*, 149–153.

Gotteland, M., Andrews, M., Toledo, M., Munoz, L., Caceres, P., Anziani, A., Wittig, E., Speisky, H., and Salazar, G. (2008). Modulation of *Helicobacter pylori* colonization with cranberry juice and *Lactobacillus johnsonii* La1 in children. Nutrition *24*, 421–426.

Gradisar, H., Pristovsek, P., Plaper, A., and Jerala, R. (2007). Green tea catechins inhibit bacterial DNA gyrase by interaction with its ATP binding site. J. Med. Chem. *50*, 264–271.

Gupta, V., and Garg, R. (2009). Probiotics. Indian J. Med. Microbiol. *27*, 202–209.

Han, C., Wang, L., Yu, K., Chen, L., Hu, L., Chen, K., Jiang, H., and Shen, X. (2006). Biochemical characterisation and inhibitor discovery of shikimate dehydrogenase from *H. pylori*. FEBS J. *273*, 4682–4692.

Haristoy, X., Angioi-Duprez, K., Duprez, A., and Lozniewski, A. (2003). Efficacy of sulforaphane in eradicating *Helicobacter pylori* in human gastric xenografts implanted in nude mice. Antimicrob. Agents Chemother. *47*, 3982–3984.

Haug, B.E., Stensen, W., Kalaaji, M., Rekdal, Ø., and Svendsen, J.S. (2008). Synthetic antimicrobial peptidomimetics with therapeutic potential. J. Med. Chem. *51*, 4306–4314.

Horie, K., Horie, N., Abdou, A.M., Yang, J.O., Yun, S.S., Chun, H.N., Park, C.K., Kim, M., and Hatta, H. (2004). Suppressive effect of functional drinking yoghurt containing specific egg yolk immunoglobulin on *Helicobacter pylori* in humans. J. Dairy Sci. *87*, 4073–4079.

Hurduc, V., Plesca, D., Dragomir, D., Sajin, M., and Vandenplas, Y. (2009). A randomized, open trial evaluating the effect of *Saccharomyces boulardii* on the eradication rate of *Helicobacter pylori* infection in children. Acta Paediatr. *98*, 127–131.

Huynh, H.Q., Couper, R.T., Tran, C.D., Moore, L., Kelso, R., and Butler, R.N. (2004). *N*-acetylcysteine, a novel treatment for *Helicobacter pylori* infection. Dig. Dis. Sci. *49*, 1853–1861.

Huynh, H.Q., Campbell, M.A., Couper, R.T., Tran, C.D., Lawrence, A., and Butler, R.N. (2009). Lactoferrin and desferrioxamine are ineffective in the treatment of *Helicobacter pylori* infection and may enhance *H. pylori* growth and gastric inflammation in mice. Lett. Appl. Microbiol. *48*, 517–522.

Imase, K., Tanaka, A., Tokunaga, K., Sugano, H., Ishida, H., and Takahashi, S. (2007). *Lactobacillus reuteri* tablets suppress *Helicobacter pylori* infection – a double-blind randomised placebo-controlled cross-over clinical study. Kansenshogaku Zasshi *81*, 387–393.

Jarosz, M., Rychlik, E., Siuba, M., Respondek, W., Ryzko-Skiba, M., Sajór, I., Gugała, S., Błazejczyk, T., and Ciok, J. (2009). Dietary and socioeconomic factors in relation to *Helicobacter pylori* re-infection. World J. Gastroenterol. *15*, 1119–1125.

Jiang, S., Ding, N., Zhang, W., Zhang, Y., Geng, M., and Li, Y. (2010). Novel conjugates of aspirin with phenolic acid as anti-inflammatory agents having significantly reduced gastrointestinal toxicity. Arch. Pharm. (Weinheim) *343*, 215–221.

Keenan, J.I., Salm, N., Hampton, M.B., and Wallace, A.J. (2010). Individual and combined effects of foods on *Helicobacter pylori* growth. Phytother. Res. *24*, 1229–1233.

Ki, M.R., Ghim, S.Y., Hong, I.H., Park, J.K., Hong, K.S., Ji, A.R., and Jeong, K.S. (2010). *In vitro* inhibition of *Helicobacter pylori* growth and of adherence of *cagA*-positive strains to gastric epithelial cells by *Lactobacillus paraplantarum* KNUC25 isolated from kimchi. J. Med. Food. *13*, 629–634.

Kim, T.S., Hur, J.W., Yu, M.A., Cheigh, C.I., Kim, K.N., Hwang, J.K., and Pyun, Y.R. (2003). Antagonism of *Helicobacter pylori* by bacteriocins of lactic acid bacteria. J. Food Prot. *66*, 3–12.

Kim, J.M., Kim, J.S., Kim, Y.J., Oh, Y.K., Kim, I.Y., Chee, Y.J., Han, J.S., and Jung, H.C. (2008). Conjugated linoleic acids produced by *Lactobacillus* dissociates IKK-gamma and Hsp90 complex in *Helicobacter pylori*-infected gastric epithelial cells. Lab. Invest. *88*, 541–552.

Kim, Y.J., Chung, J.W., Lee, S.J., Choi, K.S., Kim, J.H., and Hahm, K.B. (2010). Progression from chronic atrophic gastritis to gastric cancer; tangle, toggle, tackle with Korea red ginseng. J. Clin. Biochem. Nutr. *46*, 195–204.

Kohler, J.E., Blass, A.L., Liu, J., Tai, K., and Soybel, D.I. (2010). Antioxidant pre-treatment prevents omeprazole-induced toxicity in an *in vitro* model of infectious gastritis. Free Radic. Biol. Med. *49*, 786–791.

Koosirirat, C., Linpisarn, S., Changsom, D., Chawansuntati, K., and Wipasa, J. (2010). Investigation of the anti-inflammatory effect of *Curcuma longa* in *Helicobacter pylori*-infected patients. Int. Immunopharmacol. *10*, 815–818.

Kottakis, F., Kouzi-Koliakou, K., Pendas, S., Kountouras, J., and Choli-Papadopoulou, T. (2009). Effects of mastic gum *Pistacia lentiscus* var. Chia on innate cellular immune effectors. Eur. J. Gastroenterol. Hepatol. *21*, 143–149.

Krausse, R., Bielenberg, J., Blaschek, W., and Ullmann, U. (2004). *In vitro* anti*Helicobacter pylori* activity of Extractum liquiritiae, glycyrrhizin and its metabolites. J. Antimicrob. Chemother. *54*, 243–246.

Lane, J.A., Mehra, R.K., Carrington, S.D., and Hickey, R.M. (2010). The food glycome: a source of protection against pathogen colonization in the gastrointestinal tract. Int. J. Food Microbiol. *142*, 1–13.

Lee, K.M., Yeo, M., Choue, J.S., Jin, J.H., Park, S.J., Cheong, J.Y., Lee, K.J., Kim, J.H., and Hahm, K.B. (2004). Protective mechanism of epigallocatechin-3-gallate against *Helicobacter pylori*-induced gastric epithelial cytotoxicity via the blockage of TLR-4 signaling. Helicobacter *9*, 632–642.

Lee, I.O., Lee, K.H., Pyo, J.H., Kim, J.H., Choi, Y.J., and Lee, Y.C., (2007). Anti-inflammatory effect of capsaicin in *Helicobacter pylori*-infected gastric epithelial cells. Helicobacter *12*, 510–517.

Lee, J.H., Shim, J.S., Chung, M.S., Lim, S.T., and Kim, K.H. (2009). *In vitro* anti-adhesive activity of green tea extract against pathogen adhesion. Phytother. Res. *23*, 460–466.

Lee, J.S., Paek, N.S., Kwon, O.S., and Hahm, K.B. (2010). Anti-inflammatory actions of probiotics through activating suppressor of cytokine signaling (SOCS) expression and signaling in *Helicobacter pylori* infection: a novel mechanism. J. Gastroenterol. Hepatol. *25*, 194–202.

Lee, S.Y., Shin, Y.W., and Hahm, K.B. (2008). Phytoceuticals: mighty but ignored weapons against *Helicobacter pylori* infection. J. Dig. Dis. *9*, 129–139.

Lengsfeld, C., Deters, A., Faller, G., and Hensel, A. (2004). High molecular weight polysaccharides from black currant seeds inhibit adhesion of *Helicobacter pylori* to human gastric mucosa. Planta Med. *70*, 620–626.

Lesbros-Pantoflickova, D., Corthésy-Theulaz, I., and Blum, A.L. (2007). *Helicobacter pylori* and probiotics. J. Nutr. *137 (Suppl 2)*, 812S–8128S.

Lionetti, E., Indrio, F., Pavone, L., Borrelli, G., Cavallo, L., and Francavilla, R. (2010). Role of probiotics in pediatric patients with *Helicobacter pylori* infection: A comprehensive review of the literature. Helicobacter *15*, 79–87.

Liu, S., Sun, Y., Li, W., Yu, H., Li, X., Liu, Z., Zeng, J., Zhou, Y., Chen, C., and Jia, J. (2010). The antibacterial mode of action of allitridi for its potential use as a therapeutic agent against *Helicobacter pylori* infection. FEMS Microbiol. Lett. *303*, 183–189.

Loke, M.F., Lui, S.Y., Ng, B.L., Gong, M., and Ho, B. (2007). Antiadhesive property of microalgal polysaccharide extract on the binding of *Helicobacter pylori* to gastric mucin. FEMS Immunol. Med. Microbiol. *50*, 231–238.

Lorca, G.L., Wadstrom, T., Valdez, G.F., and Ljungh, A. (2001). *Lactobacillus acidophilus* autolysins inhibit *Helicobacter pylori in vitro*. Curr. Microbiol. *42*, 39–44.

Mahady, G. B., Pendland, S. L., and Chadwick, L. R. (2003a). Resveratrol and red wine extracts inhibit the growth of CagA+ strains of *Helicobacter pylori in vitro*. Am. J. Gastroenterol. *98*, 1440–1441.

Mahady, G.B., Pendland, S.L., Yun, G.S., Lu, Z.Z., and Stoia, A. (2003b). Ginger (*Zingiber officinale Roscoe*) and the gingerols inhibit the growth of Cag A+ strains of *Helicobacter pylori*. Anticancer Res. *23*, 3699–3702.

Makobongo, M.O., Kovachi, T., Gancz, H., Mor, A., and Merrell, D.S. (2009). *In vitro* antibacterial activity of acyl-lysyl oligomers against *Helicobacter pylori*. Antimicrob. Agents Chemother. *53*, 4231–4239.

Manyi-Loh, C.E., Clarke, A.M., Mkwetshana, N.F., and Ndip, R.N. (2010). Treatment of *Helicobacter pylori* infections: Mitigating factors and prospective natural remedies. Afr. J. Biotechnol. *9*, 2032–2042.

Masco, L., Van Hoorde, K., De Brandt, E., Swings, J., and Huys, G. (2006). Antimicrobial susceptibility of *Bifidobacterium* strains from humans, animals and probiotic products. J. Antimicrob. Chemother. *58*, 85–94.

Martini, S., D'Addario, C., Braconi, D., Bernardini, G., Salvini, L., Bonechi, C., Figura, N., Santucci, A., and Rossi, C. (2009a). Antibacterial activity of grape extracts on *cagA*-positive and -negative *Helicobacter pylori* clinical isolates. J. Chemother. *21*, 507–513.

Martini, S., D'Addario, C., Colacevich, A., Focardi, S., Borghini, F., Santucci, A., Figura, N., and Rossi, C. (2009b). Antimicrobial activity against *Helicobacter pylori* strains and antioxidant properties of blackberry leaves (*Rubus ulmifolius*) and isolated compounds. Int. J. Antimicrob. Agents. *34*, 50–59.

McFarland, L.V. (2010). Systematic review and meta-analysis of *Saccharomyces boulardii* in adult patients. World J. Gastroenterol. *16*, 2202–2222.

Menghini, L., Leporini, L., Tirillini, B., Epifano, F., and Genovese, S. (2010). Chemical composition and inhibitory activity against *Helicobacter pylori* of the essential oil of *Apium nodiflorum (Apiaceae)*. J. Med. Food. *13*, 228–230.

Moon, J.K., Kim, J.R., Ahn, Y.J., and Shibamoto, T. (2010). Analysis and anti*Helicobacter* activity of sulforaphane and related compounds present in broccoli (Brassica oleracea L.) sprouts. J. Agric. Food Chem. *58*, 6672–6677.

Myllyluoma, E., Ahonen, A.M., Korpela, R., Vapaatalo, H., and Kankuri, E. (2008). Effects of multispecies probiotic combination on *Helicobacter pylori* infection *in vitro*. Clin. Vaccine Immunol. *15*, 1472–1482.

Nagano, Y., Ikedo, K., Fujishima, A., Izawa, M., Tsubotani, S., Nishimura, O., and Fujino, M. (2001). Pyloricidins, novel anti*Helicobacter pylori* antibiotics produced by bacillus sp. II. Isolation and structure elucidation. J. Antibiot. (Tokyo) *54*, 934–947.

Nagata, K., Inatsu, S., Tanaka, M., Sato, H., Kouya, T., Taniguchi, M., and Fukuda, Y. (2010). The bifidogenic growth stimulator inhibits the growth and respiration of *Helicobacter pylori*. Helicobacter *15*, 422–429.

Nanjundaiah, S.M., Annaiah, H.N., and Dharmesh, S. (2009). Gastroprotective effect of ginger rhizome (*Zingiber officinale*) extract: role of gallic acid and cinnamic acid in H+, K+-ATPase/*H. pylori* inhibition and antioxidative mechanism. Evid. Based Complement. Alternat. Med. doi:10.1093/ecam/nep060

Ndip, R.N., Malange Takan, A.E., Echakachi, C.M., Malongue, A., Akoachere, J.F.T.K., Ndip, L.M., and Luma, H.N. (2007). In-vitro antimicrobial activity of selected honeys on clinical isolates of *Helicobacter pylori*. Afr. Health Sci. 7, 228–231.

Nostro, A., Cellini, L., Di Bartolomeo, S., Cannatelli, M.A., Di Campli, E., Procopio, F., Grande, R., Marzio, L., and Alonzo, V. (2006). Effects of combining extracts (from propolis or *Zingiber officinale*) with clarithromycin on *Helicobacter pylori*. Phytother. Res. *20*, 187–190.

O'Gara, E.A., Hill, D.J., and Maslin, D.J. (2000). Activities of garlic oil, garlic powder, and their diallyl constituents against *Helicobacter pylori*. Appl. Envir. Microbiol. *66*, 2269 – 2273.

O'Mahony, R., Al-Khtheeri, H., Weerasekera, D., Fernando, N., Vaira, D., Holton, J., and Basset, C. (2005). Bactericidal and antiadhesive properties of culinary and medicinal plants against *Helicobacter pylori*. World J. Gastroenterol. *11*, 7499–7507.

Ornelas, I.J., Galvan-Potrillo, M., and López-Carrillo, L. (2007). Protective effect of yoghurt consumption on *Helicobacter pylori* seropositivity in a Mexican population. Public Health Nutr. *10*, 1283–1287.

Pastene, E., Speisky, H., García, A., Moreno, J., Troncoso, M., and Figueroa, G. (2010). *In vitro* and *in vivo* effects of apple peel polyphenols against *Helicobacter pylori*. J. Agric. Food Chem. *58*, 7172–7179.

Paraschos, S., Magiatis, P., Mitakou, S., Petraki, K., Kalliaropoulos, A., Maragkoudakis, P., Mentis, A., Sgouras, D., and Skaltsounis, A.L. (2007). *In vitro* and *in vivo* activities of Chios mastic gum extracts and constituents against *Helicobacter pylori*. Antimicrob. Agents Chemother. *51*, 551–559.

Perdigón, G., Maldonado Galdeano, C., Valdez, J.C., and Medici, M. (2002). Interaction of lactic acid bacteria with the gut immune system. Eur. J. Clin. Nutr. *56 (Suppl 4)*, S21–S26.

Reichling, J., Weseler, A., and Saller, R. (2001). A current review of the antimicrobial activity of *Hypericum perforatum* L. Pharmacopsychiatry *34 (Suppl 1)*, S116–S118.

Romero, C., Medina, E., Vargas, J., Brenes, M., and De Castro, A. (2007). *In vitro* activity of olive oil polyphenols against *Helicobacter pylori*. J. Agric. Food Chem. *55*, 680–686.

Ruggiero, P., Rossi, G., Tombola, F., Pancotto, L., Lauretti, L., Del Giudice, G., and Zoratti, M. (2007). Red wine and green tea reduce *H pylori*- or VacA-induced gastritis in a mouse model. World J. Gastroenterol. *13*, 349–354.

Ryan, K.A., Daly, P., Li, Y., Hooton, C., and O'Toole, P.W. (2008) Strain-specific inhibition of *Helicobacter pylori* by *Lactobacillus salivarius* and other lactobacilli. J. Antimicrob. Chemother. *61*, 831–834.

Sachdeva, A., and Nagpal, J. (2009). Effect of fermented milk-based probiotic preparations on *Helicobacter pylori* eradication: a systematic review and meta-analysis of randomized-controlled trials. Eur. J. Gastroenterol. Hepatol. *21*, 45–53.

Sakamoto, I., Igarashi, M., Kimura, K., Takagi, A., Miwa, T., and Koga, Y. (2001). Suppressive effect of *Lactobacillus gasseri* OLL 2716 (LG21) on *Helicobacter pylori* infection in humans. J. Antimicrob. Chemother. *47*, 709–710.

Salem, E.M., Yar, T., Bamosa, A.O., Al-Quorain, A., Yasawy, M.I., Alsulaiman, R.M., and Randhawa, M.A. (2010). Comparative study of *Nigella Sativa* and triple therapy in eradication of *Helicobacter pylori* in patients with non-ulcer dyspepsia. Saudi J. Gastroenterol. *16*, 207–214.

Sasaki, M., Mizoshita, T., Mizushima, T., Inoue, H., Kamiya, T., Kataoka, H., Ogaswara, N., Wada, T., Kubota, E., Mori, Y., *et al.* (2007). Effect of plaunotol in combination with clarithromycin against clarithromycin-resistant *Helicobacter pylori in vitro* and *in vivo*. J. Antimicrob. Chemother. *60*, 1060–1063.

Scaccianoce, G., Zullo, A., Hassan, C., Gentili, F., Cristofari, F., Cardinale, V., Gigliotti, F., Piglionica, D., and Morini, S. (2008). Triple therapies plus different probiotics for *Helicobacter pylori* eradication. Eur. Rev. Med. Pharmacol. Sci. *12*, 251–256.

Servin, A.L. (2004). Antagonistic activities of lactobacilli and bifidobacteria against microbial pathogens. FEMS Microbiol. Rev. *28*, 405–440.

Shirasawa, Y., Shibahara-Sone, H., Iino, T., and Ishikawa, F. (2010). *Bifidobacterium bifidum* BF-1 suppresses *Helicobacter pylori*-induced genes in human epithelial cells. J. Dairy Sci. *93*, 4526–4534.

Sgouras, D., Maragkoudakis, P., Petraki, K., Martinez-Gonzalez, B., Eriotou, E., Michopoulos, S., Kalantzopoulos, G., Tsakalidou, E., and Mentis, A. (2004). *In vitro* and *in vivo* inhibition of *Helicobacter pylori* by *Lactobacillus casei* strain Shirota. Appl. Environ. Microbiol. *70*, 518–526.

Sheu, B., Wu, J., Lo, C., Wu, H., Chen, J., Lin, Y., and Lin, M. (2002). Impact of supplement with *Lactobacillus*- and *Bifidobacterium*-containing yoghurt on triple therapy for *Helicobacter pylori* eradication. Aliment. Pharmacol. Ther. *16*, 1669–1675.

Sheu, B.S., Cheng, H.C., Kao, A.W., Wang, S.T., Yang, Y.J., Yang, H.B., and Wu, J.J. (2006). Pretreatment with *Lactobacillus*- and *Bifidobacterium*-containing yoghurt can improve the efficacy of quadruple therapy in eradicating residual *Helicobacter pylori* infection after failed triple therapy. Am. J. Clin. Nutr. *83*, 864–869.

Shikov, A.N., Pozharitskaya, O.N., Makarov, V.G., and Kvetnaya, A.S. (2008). Antibacterial activity of *Chamomilla recutita* oil extract against *Helicobacter pylori*. Phytother. Res. *22*, 252–253.

Shoae Hassani, A.R., Ordouzadeh, N., Ghaemi, A., Amirmozafari, N., Hamdi, K., and Nazari, R. (2009). *In vitro* inhibition of *Helicobacter pylori* urease with non and semi fermented *Camellia sinensis*. Indian J. Med. Microbiol. *27*, 30–34.

Simova, E.D., Beshkova, D.B., and Dimitrov, Zh. P. (2009). Characterization and antimicrobial spectrum of bacteriocins produced by lactic acid bacteria isolated from traditional Bulgarian dairy products. J. Appl. Microbiol. *106*, 692–701.

Sintara, K., Thong-Ngam, D., Patumraj, S., Klaikeaw, N., and Chatsuwan, T. (2010). Curcumin suppresses gastric NF-kappaB activation and macromolecular leakage in *Helicobacter pylori*-infected rats. World J. Gastroenterol. *16*, 4039–4046.

Sirk, T.W., Brown, E.F., Sum, A.K., and Friedman, M. (2008). Molecular dynamics study on the biophysical interactions of seven green tea catechins with lipid bilayers of cell membranes. J. Agric. Food Chem. *56*, 7750–7758.

Song, J.M., and Seong, B.L. (2007). Tea catechins as a potential alternative anti-infectious agent. Expert Rev. Anti Infect. Ther. *5*, 497–506.

Stoicov, C., Saffari, R., and Houghton, J. (2009). Green tea inhibits *Helicobacter* growth *in vivo* and *in vitro*. Int. J. Antimicrob. Agents. *33*, 473–478.

Sýkora, J., Valecková, K., Amlerová, J., Siala, K., Dedek, P., Watkins, S., Varvarovská, J., Stozický, F., Pazdiora, P., and Schwarz, J. (2005). Effects of a specially designed fermented milk product containing probiotic *Lactobacillus casei* DN-11,4001 and the eradication of *H. pylori* in children: a prospective randomized double-blind study. J. Clin. Gastroenterol. *39*, 692–698.

Tattelman, E. (2005). Health effects of garlic. Am. Fam. Physician. 72, 103–106.

Terris, M.K., Issa, M.M., and Tacker, J.R. (2001). Dietary supplementation with cranberry concentrate tablets may increase the risk of nephrolithiasis. Urology 57, 26–29.

Tonks, A.J., Dudley, E., Porter, N.G., Parton, J., Brazier, J., Smith, E.L., and Tonks, A. (2007). A 5.8-kDa component of manuka honey stimulates immune cells via TLR4. J. Leukoc. Biol. *82*, 1147–1155.

Tursi, A., Elisei, W., Brandimarte, G., Giorgetti, G.M., Modeo, M.E., and Aiello, F. (2007). Effect of lactoferrin supplementation on the effectiveness and tolerability of a 7-day quadruple therapy after failure of a first attempt to cure *Helicobacter pylori* infection. Med. Sci. Monit. *13*, CR187–CR190.

Vattem, D.A., Lin, Y.-T., Ghaedian, R., and Shetty, K. (2005). Cranberry synergies for dietary management of *Helicobacter pylori* infections Process Biochem. *40*, 1583–1592.

Venugopalan, V., Shriner, K.A., and Wong-Beringer, A. (2010). Regulatory oversight and safety of probiotic use. Emerg. Infect. Dis. *16*, 1661–1665.

Vila, R., Santana, A.I., Pérez-Rosés, R., Valderrama, A., Castelli, M.V., Mendonca, S., Zacchino, S., Gupta, M.P., and Cañigueral, S. (2010). Composition and biological activity of the essential oil from leaves of *Plinia cerrocampanensis*, a new source of alpha-bisabolol. Bioresour. Technol. *101*, 2510–2514.

Viuda-Martos, M., Ruiz-Navajas, Y., Fernández-López, J., and Pérez-Alvarez, J.A. (2008). Functional properties of honey, propolis, and royal jelly. J. Food Sci. *73*, R117-R124.

de Vuyst, L., Makras, L., Avonts, L., Holo, H., Yi, Q., Servin, A., Fayol-Messaoudi, D., Berger, C., Zoumpopoulou, G., Tsakalidou, E., *et al.* (2004) Antimicrobial potential of probiotic or potentially probiotic lactic acid bacteria, the first results of the international European research project PROPATH of PROEUHEALTH cluster. Microb. Ecol. Health Dis. *16*, 125–130.

de Vuyst, L., Vincent, P., Makras, E., Leroy, F., and Pot, B. (2010). Peptide extracts from cultures of certain lactobacilli inhibit *Helicobacter pylori*. Probiotics Antimicro. Prot. *2*, 26–36.

Wang, W.H., Wong, W.M., Dailidiene, D., Berg, D.E., Gu, Q., Lai, K.C., Lam, S.K., and Wong, B.C. (2003). Aspirin inhibits the growth of *Helicobacter pylori* and enhances its susceptibility to antimicrobial agents. Gut *52*, 490–495.

Weerasekera, D., Fernando, N., Bogahawatta, L.B.A.E., Rajapakse-Mallikahewa, R., and D.J. Naulla, D.J. (2008). Bactericidal effect of selected spices, medicinal plants and tea on *Helicobacter pylori* strains from Sri Lanka. J. Natn. Sci. Foundation Sri Lanka *36*, 91–94.

Wittschier, N., Lengsfeld, C., Vorthems, S., Stratmann, U., Ernst, J.F., Verspohl, E.J., and Hensel, A. (2007). Large molecules as antiadhesive compounds against pathogens. J. Pharm. Pharmacol. *59*, 777–786.

Wittschier, N., Faller, G., and Hensel, A. (2009). Aqueous extracts and polysaccharides from liquorice roots (*Glycyrrhiza glabra* L.) inhibit adhesion of *Helicobacter pylori* to human gastric mucosa. J. Ethnopharmacol. *125*, 218–223.

Yanagawa, Y., Yamamoto, Y., Hara, Y., and Shimamura, T. (2003). A combination effect of epigallocatechin gallate, a major compound of green tea catechins, with antibiotics on *Helicobacter pylori* growth *in vitro*. Curr. Microbiol. *47*, 244–249.

Yanaka, A., Fahey, J.W., Fukumoto, A., Nakayama, M., Inoue, S., Zhang, S., Tauchi, M., Suzuki, H., Hyodo, I., and Yamamoto, M. (2009). Dietary sulforaphane-rich broccoli sprouts reduce colonization and attenuate gastritis in *Helicobacter pylori*-infected mice and humans. Cancer Prev. Res. (Phila). *2*, 353–360.

Yang, J.C., Shun, C.T., Chien, C.T., and Wang, T.H. (2008). Effective prevention and treatment of *Helicobacter pylori* infection using a combination of catechins and sialic acid in AGS cells and BALB/c mice. J. Nutr. *138*, 2084–2090.

Yang, P., Zhou, Y., Chen, B., Wan, H.W., Jia, G.Q., Bai, H.L., and Wu, X.T. (2010). Aspirin use and the risk of gastric cancer: a meta-analysis. Dig. Dis. Sci. *55*, 1533–1539.

Yee, Y.K., and Koo, M.W. (2000). Anti-*Helicobacter pylori* activity of Chinese tea: *in vitro* study. Aliment. Pharmacol. Ther.*14*, 635–638.

Yun, T.K., Lee, Y.S., Lee, Y.H., Kim, S.I., and Yun, H.Y.(2001). Anticarcinogenic effect of Panax ginseng

C.A. Meyer and identification of active compounds. J. Korean Med. Sci. *16* (*Suppl*), S6–S18.

Zaidi, S.F., Yamada, K., Kadowaki, M., Usmanghani, K., and Sugiyama, T. (2009a). Bactericidal activity of medicinal plants, employed for the treatment of gastrointestinal ailments, against *Helicobacter pylori*. J. Ethnopharmacol. *121*, 286–291.

Zaidi, S.F., Ahmed, K., Yamamoto, T., Kondo, T., Usmanghani, K., Kadowaki, M., and Sugiyama, T. (2009b). Effect of resveratrol on *Helicobacter pylori*-induced interleukin-8 secretion, reactive oxygen species generation and morphological changes in human gastric epithelial cells. Biol. Pharm. Bull. *32*, 1931–1935.

Zhang, L., Ma, J., Pan, K., Go, V.L.W., Chen, J., and You, W. (2005). Efficacy of cranberry juice on *Helicobacter pylori* infection: a double-blind, randomized placebo-controlled trial. Helicobacter *10*, 139–145.

Zhang, L., Su, P., Henriksson, A., O'Rourke, J., and Mitchell, H. (2008). Investigation of the immunomodulatory effects of *Lactobacillus casei* and *Bifidobacterium lactis* on *Helicobacter pylori* infection. Helicobacter *13*, 183–190.

Zhang, X.-P., Wang, W.-H., Tian, Y., Gao, W., and Li, J. (2009). Aspirin increases susceptibility of *Helicobacter pylori* to metronidazole by augmenting endocellular concentrations of antimicrobials. World J. Gastroenterol. *15*, 919–926.

Zhou, C., Ma, F.Z., Deng, X.J., Yuan, H., and Ma, H.S. (2008). Lactobacilli inhibit interleukin-8 production induced by *Helicobacter pylori* lipopolysaccharide-activated Toll-like receptor 4. World J. Gastroenterol. *14*, 5090–5095.

Ziemniak, W. (2006). Efficacy of *Helicobacter pylori* eradication taking into account its resistance to antibiotics. J. Physiol. Pharmacol. *57* (*Suppl* 3), 123–141.

Zojaji, H., Talaie, R., Mirsattari, D., Haghazali, M., Molaei, M., Mohsenian, N., Derakhshan, F., and Zali, M.R. (2009). The efficacy of *Helicobacter pylori* eradication regimen with and without vitamin C supplementation. Dig. Liver Dis. *41*, 644–647.

Zou, J., Dong, J., and Yu, X. (2009a). Meta-analysis: *Lactobacillus* containing quadruple therapy versus standard triple first-line therapy for *Helicobacter pylori* eradication. Helicobacter *14*, 97–107.

Zou, J., Dong, J., and Yu, X.F. (2009b). Meta-analysis: the effect of supplementation with lactoferrin on eradication rates and adverse events during *Helicobacter pylori* eradication therapy. Helicobacter *14*, 119–127.

Zullo, A., De Francesco, V., Scaccianoce, G., Manes, G., Efrati, C., Hassan, C., Maconi, G., Piglionica, D., Cannaviello, C., Panella, C., *et al.* (2007). *Helicobacter pylori* eradication with either quadruple regimen with lactoferrin or levofloxacin-based triple therapy: a multicentre study. Dig. Liver Dis. *39*, 806–810.

Abbreviations

ADM	agar dilution method
AI-2	autoinducer-2
alp genes	adherence-associated lipoproteins genes
AmiA	*N*-acetylmuramoyl-L-alanyl amidase
amiF gene	formamidase gene
ArsRS TCS	ArsRS two-component system
ASA	acetylsalicylic acid
AS-PCR	allele-specific polymerase chain reaction
ATF-2	activating transcription factor 2
babA gene	blood group antigen binding adhesin
BlU	bleeding ulcer
BST	breakpoint susceptibility testing
CAD	coronary artery disease
cagA gene	cytotoxin-associated gene A
cagPAI	cag pathogenicity island
CDT	cytolethal distending toxin
CFU	colony-forming unit
CHD	coronary heart disease
che genes	regulatory proteins for chemotaxis
CLO	*Campylobacter*-like organisms
CM motif	CagA multimerisation motif
CoAT	CoA transferase
COPD	chronic obstructive pulmonary disease
COX	cyclooxygenase
DC-SIGN	dendritic cell-specific intercellular adhesion molecule-3-grabbing non-integrin
DHNA	1,4-dihydroxy-2-naphthoic acid
DU	duodenal ulcer
dupA gene	duodenal ulcer-promoting gene
E test	epsilometer test
EGCG	epigallocatechin-gallate
EGF	epidermal growth factor
EIA	enzyme immune assay
ELISA	enzyme-linked immunosorbent assay
ENT	ear, nose, and throat
EPIYA motif	tyrosine residues within repeating Glu-Pro-Ile-Tyr-Ala
ERK	extracellular signal-regulated kinase
FAK	focal adhesion kinase
FISH	fluorescence *in situ* hybridization
FRET	fluorescence resonance energy transfer

frxA gene	gene encoding a NAD(P)H-flavin oxidoreductase
FTIR	Fourier transform infrared spectroscopy
Fur	ferric uptake regulator
FusA	translation elongation factor EF-G
futA and *futB* genes	α1,3-fucosyltransferase gene
GAE	gallic acid
GDS	patients with gastroduodenal symptoms
GEC	gastric epithelial cell
GORD	gastro-oesophageal reflux disease
GU	gastric ulcer
hBD-2	human beta-defensin 2
HP-NAP	neutrophil-activating protein
HSPs	heat-shock proteins
HtrA	high-temperature requirement A protein
HypA and HypB	hydrogenase accessory proteins
IBD	inflammatory bowel disease
IC_{50}	inhibitory concentration 50%
iceA gene	gene induced by contact with epithelium
IDA	iron-deficiency anaemia
IFN-γ	interferon γ
IHD	ischaemic heart disease
IKK	IκB kinase
IL	interleukin
iNOS	inducible nitric oxide synthase
IRMS	isotope ratio mass spectrometry
ITP	idiopathic thrombocytopenic purpura
ITT analysis	intention-to-treat analysis
KatA	catalase
KOR	α-ketoglutarate oxidoreductase
LARA	laser assisted ratio analyser
LDB	*Lactobacillus delbrueckii* subsp. *bulgaricus*
Le	Lewis blood group antigens
LPS	lipopolysaccharide
LT	heat-labile toxin
MALT lymphoma	mucosa-associated lymphoid tissue lymphoma
MAPK	mitogen-activated protein kinase
MBC	minimal bactericidal concentration
MAb	monoclonal antibody
MCA	monoclonal antibody-based tests
MDA-DNA	multiple displacement amplification of DNA
MDDM	modified disc-diffusion method
MHC	major histocompatibility complex
MIC	minimal inhibitory concentrations
MIP-2	macrophage inflammatory protein 2
MutY	antimutator DNA glycosylase
NAAs	non-antibiotic agents
NDIRS	non-dispersive isotope-selective infrared spectrometry
NFAT	nuclear factor of activated T cells
NF-κB	nuclear factor-κB

nikR gene	nickel-responsive transcription factor gene
nixA gene	nickel transporter gene
NO	nitric oxide
Nod1	nucleotide-binding oligomerization domain protein I
NPHS	non-*pylori Helicobacter* species
NSAIDs	non-steroidal anti-inflammatory drugs
OAC	omeprazole, amoxicillin and clarithromycin
OBMT	omeprazole, bismuth, metronidazole and tetracycline
oipA gene	inflammatory outer membrane gene
OMP	outer membrane protein
OR	odds ratio
PAb	polyclonal antibody-based
PAC	PPI, amoxicillin, and clarithromycin
PAM	PPI, amoxicillin, and metronidazole
PAMPs	pathogen-associated molecular patterns
PAR1	partitioning-defective 1
PBPs	penicillin-binding proteins
PCR	polymerase chain reaction
PCR-LiPA	PCR-line probe assay
Pgb proteins	plasminogen-binding proteins
PgdA	peptidoglycan deacetylase
PGE_2	prostaglandin E_2
PP analysis	per protocol analysis
PPI	proton pump inhibitor
PUD	peptic ulcer disease
PZ	plasticity zone
RAPD	randomly amplified polymorphic DNA
rdxA gene	gene encoding an oxygen-insensitive NADPH nitroreductase
RFLP	fragment length polymorphism method
RGD motif	Arg-Gly-Asp motif
R-M genes	restriction and modification genes
ROC	receiver operating characteristic
rocF gene	arginase gene
ROS	reactive oxygen species
RT-PCR	real-time PCR
RUT	rapid urease test
sabA gene	sialic acid binding adhesin gene
SAT	stool antigen test
SFKs	SRC family tyrosine kinases
SHP-2	Src homology 2-containing protein-tyrosine phosphatase 2
S-IgA	secretory IgA antibody
SNARE	soluble *N*-ethylmaleimide-sensitive factor attachment protein receptor
SodB	superoxide dismutase
T4SS	type IV secretory system
TGF-β1	transforming growth factor β1
Th1	type 1 T-helper cell
Th2	type 2 T-helper cell
Tipα	TNF-α-inducing protein
tlpA gene	methyl-accepting chemotaxis receptor protein gene

tlpC gene	methyl-accepting chemotaxis transducer gene
TLR	Toll-like receptor
TNF-α	tumour necrosis factor α
Treg	regulatory T cells
UBT	urea breath test
UHR	urea hydrolysis rate
UreA	urease A subunit
vacA gene	vacuolating cytotoxin gene
VAMP7	vesicle-associated membrane protein 7
V-ATPase	vacuolar-ATPase
VBNC bacteria	viable but non-culturable bacteria

Index

Note: page numbers in *italic* refer to tables or figures.

B

C

J

K

L

M

N

O

P

Q

R

S

T

U

Current Books of Interest

Microbial Biofilms: Current Research and Applications	2012
Bacterial Glycomics: Current Research, Technology and Applications	2012
Non-coding RNAs and Epigenetic Regulation of Gene Expression	2012
Brucella: Molecular Microbiology and Genomics	2012
Molecular Virology and Control of Flaviviruses	2012
Bacterial Pathogenesis: Molecular and Cellular Mechanisms	2012
Bunyaviridae: Molecular and Cellular Biology	2011
Emerging Trends in Antibacterial Discovery: Answering the Call to Arms	2011
Epigenetics: A Reference Manual	2011
Metagenomics: Current Innovations and Future Trends	2011
Nitrogen Cycling in Bacteria: Molecular Analysis	2011
The Biology of Paramyxoviruses	2011
Helicobacter pylori	2011
Microbial Bioremediation of Non-metals: Current Research	2011
Lactic Acid Bacteria and Bifidobacteria: Current Progress in Advanced Research	2011
Viruses and Interferon: Current Research	2011
Essentials of Veterinary Parasitology	2011
Hepatitis C: Antiviral Drug Discovery and Development	2011
Streptomyces: Molecular Biology and Biotechnology	2011
Alphaherpesviruses: Molecular Virology	2011
Recent Advances in Plant Virology	2011
Vaccine Design: Innovative Approaches and Novel Strategies	2011
Salmonella: From Genome to Function	2011
PCR Troubleshooting and Optimization: The Essential Guide	2011
Insect Virology	2010
Environmental Microbiology: Current Technology and Water Applications	2011
Sensory Mechanisms in Bacteria: Molecular Aspects of Signal Recognition	2010
Bifidobacteria: Genomics and Molecular Aspects	2010
Molecular Phylogeny of Microorganisms	2010
Nanotechnology in Water Treatment Applications	2010
Iron Uptake and Homeostasis in Microorganisms	2010
Caliciviruses: Molecular and Cellular Virology	2010
Epstein-Barr Virus: Latency and Transformation	2010
Anaerobic Parasitic Protozoa: Genomics and Molecular Biology	2010
Lentiviruses and Macrophages: Molecular and Cellular Interactions	2010
Microbial Population Genetics	2010
Borrelia: Molecular and Cellular Biology	2010
Influenza: Molecular Virology	2010
RNA Interference and Viruses: Current Innovations and Future Trends	2010
Retroviruses: Molecular Biology, Genomics and Pathogenesis	2010
Metagenomics: Theory, Methods and Applications	2010

www.ingramcontent.com/pod-product-compliance
Lightning Source LLC
LaVergne TN
LVHW062119200926
848484LV00007B/1714